The Biology of Disease

The Biology of Disease

Third Edition

Edited by
Paul G. Murray
Professor of Pathology and Head of the Department of Pathology
Royal College of Surgeons in Ireland - Medical University of Bahrain
Professor of Molecular Pathology
University of Limerick, Ireland, and Honorary
Professor, University of Birmingham
United Kingdom

Simon J. Dunmore
Honorary Clinical Senior Lecturer
School of Medicine, Medical Sciences and Nutrition
University of Aberdeen
Honorary Lecturer, Cardiovascular Sciences
School of Medicine, University of Edinburgh
NHS Scotland (Grampian)

Shantha Perera
Visiting Senior Lecturer
School of Life Sciences
University of Wolverhampton
Wolverhampton, UK

Library of Congress Cataloging-in-Publication Data
Names: Murray, Paul (Paul G.), editor.
Title: The biology of disease / edited by Paul G. Murray, Professor of Pathology and Head of the Department of Pathology, Royal College of Surgeons in Ireland - Medical University of Bahrain, Professor of Molecular Pathology, University of Limerick, Ireland, and Honorary Professor, University of Birmingham, United Kingdom, Simon J. Dunmore, Honorary Clinical Senior Lecturer School of Medicine, Medical Sciences and Nutrition, University of Aberdeen; Honorary Lecturer, Cardiovascular Sciences, School of Medicine, University of Edinburgh, NHS Scotland (Grampian), Shantha Perera, Visiting Senior Lecturer, School of Life Sciences, University of Wolverhampton, Wolverhampton, UK.
Description: Third edition. | Hoboken, NJ : Wiley, 2024. | Includes index.
Identifiers: LCCN 2024007511 (print) | LCCN 2024007512 (ebook) | ISBN 9781118354155 (paperback) | ISBN 9781119426394 (adobe pdf) | ISBN 9781119426356 (epub)
Subjects: LCSH: Pathology. | Diseases.
Classification: LCC RB111 .B48 2024 (print) | LCC RB111 (ebook) | DDC 616.07–dc23/eng/20240314
LC record available at https://lccn.loc.gov/2024007511
LC ebook record available at https://lccn.loc.gov/2024007512

Cover Design: Wiley
Cover Image: © Billion Photos/Shutterstock, Courtesy of Éanna Fennell and Ciara Leahy

Set in 9.5/12.5pt STIXTwoText by Straive, Pondicherry, India

SKY10080720_072924

Contents

List of Contributors

Numbers in [ ] denote authors of chapter or case study.

Ashraf Albishtawi
Medical Student,
Royal College of Surgeons in Ireland - Medical
University of Bahrain, Busaiteen 228,
Bahrain [9,CS19,CS40,CS50]

Fatima AlHashimi
Consultant Histopathologist,
King Hamad University Hospital, Bahrain,
and Senior Clinical Lecturer, Royal College of
Surgeons in Ireland - Medical University of
Bahrain [CS31,CS32]

Mohamed AlKhaja
Consultant Neurologist, Epileptologist and
Clinical Neurophysiologist,
King Hamad University Hospital,
Bahrain [CS15,CS25,CS56]

Jumana Turky K. Alrujaib
Medical Student,
Royal College of Surgeons in Ireland - Medical
University of Bahrain, Busaiteen 228,
Bahrain [CS15,CS25]

Rawaa Alsayegh
Clinical Educator in Medicine,
Royal College of Surgeons in Ireland - Medical
University of Bahrain, Busaiteen 228,
Bahrain [9,CS10,CS12,CS13,CS14,CS18,CS20,
CS22,CS33,CS36,CS40,CS54,CS55,CS57,CS58,
CS61,CS63,CS67,CS68]

Raja H. Alyusuf
Consultant Pathologist,
Department of Pathology, Salmaniya Medical
Complex, Kingdom of Bahrain and Royal
College of Surgeons in Ireland - Medical
University of Bahrain, Busaiteen 228,
Bahrain [CS22]

Stephen Atkin
Head, School of Postgraduate Studies
and Research,
Royal College of Surgeons in Ireland - Medical
University of Bahrain, Busaiteen 228,
Bahrain [13,CS70,CS71,CS72]

Shivani Bailey
Consultant in Paediatric Neuro-oncology,
Birmingham Children's Hospital and Senior
Clinical Research Fellow, Birmingham
Cancer Research UK Clinical Trials Unit,
University of Birmingham,
United Kingdom [CS54]

David Burns
Consultant Haematologist,
University Hospitals Plymouth NHS
Trust, Derriford Road, Crownhill,
Plymouth, Devon,
United Kingdom [CS33]

Alexandra E. Butler
Professor in Pathology,
Royal College of Surgeons in Ireland - Medical
University of Bahrain, Busaiteen 228,
Bahrain [13,CS70,CS71,CS72]

Hiu Kwan Carolyn Tang
Consultant Medical Oncologist,
Cambridge University NHS Foundation Trust,
Cambrdige, United Kingdom [CS51]

Dimitrios Chanouzas
Consultant Nephrologist,
Queen Elizabeth Hospital Birmingham,
Mindelsohn Way, Edgbaston, Birmingham,
United Kingdom [CS31,CS32]

Amen EL Cheikh Ammar
Consultant Pulmonologist and Senior
Lecturer, Royal College of Surgeons in
Ireland - Medical University of Bahrain
Busaiteen 228, Bahrain [CS2,CS27,CS69]

Ruth Clifford
Consultant Haematologist, University
Hospital Limerick, Clinical Professor, GEMS,
University of Limerick Hospitals Group,
University Hospital Limerick,
Ireland [CS16,CS52,CS55,CS57,CS58,
CS62,CS67]

Pippa G. Corrie
Consultant and Affiliated Associate Professor
in Medical Oncology,
Cambridge University Hospitals NHS
Foundation Trust and University of
Cambridge, United Kingdom [CS51,CS68]

Simon J. Dunmore
Honorary Clinical Senior Lecturer,
School of Medicine, Medical Sciences
and Nutrition,
University of Aberdeen; Honorary Lecturer,
Cardiovascular Sciences, School of Medicine,
University of Edinburgh; NHS Scotland
(Grampian) [12,13,CS70,CS71,CS72]

Kevin Dunne
Professor of Paediatrics,
Royal College of Surgeons in Ireland - Medical
University of Bahrain, Busaiteen 228,
Bahrain [CS43]

J. David M. Edgar
Consultant Immunologist, St. James's Hospital
and Trinity College, Dublin, Ireland
[6,CS20,CS23,CS30,CS31,CS33,CS39]

Éanna Fennell
Irish Research Council Postdoctoral Fellow,
Limerick Digital Cancer Research Centre,
School of Medicine, Bernal Institute and
Health Research Institute, University of
Limerick, Limerick,
Ireland [CS11,CS57]

Fidelma Fitzpatrick
Professor and Head of Department of
Clinical Microbiology, RCSI and Consultant
Microbiologist, Beaumont Hospital, Dublin,
Ireland [CS9,CS12,CS13]

Siobhan Glavey
Professor and Head of the Department of
Pathology, RCSI, Dublin, and Consultant
Haematologist,
Department of Haematology Beaumont
RCSI Cancer Centre, Dublin,
Ireland [CS16,CS62]

Alexander Glover
Clinical Research Fellow, Institute of
Immunology and Immunotherapy,
University of Birmingham
Birmingham,
United Kingdom [CS63]

Kate Glover
Principal Clinical Scientist,
Birmingham Women's and Children's NHS
Foundation Trust,
Mindelsohn Way, Birmingham,
United Kingdom [10,CS41,CS42,CS44,CS45,
CS46,CS49]

Lorraine Hartles-Spencer
Principal Clinical Scientist,
West Midlands Regional Genetics Laboratory,
Birmingham Women's Hospital,
Mindelsohn Way, Birmingham [10]

Jamal Hasan Hashem
Lecturer in Surgery, and Academic Director of
Artificial Intelligence,
Royal College of Surgeons in Ireland - Medical
University of Bahrain, Busaiteen 228,
Bahrain [CS10,CS17,CS28,CS38]

Eddie Jones
University Hospital Limerick,
St. Nessan's Road,
Dooradoyle, Limerick, Ireland [CS47,CS48]

Nadira D. Karunaweera
Chair and Senior Professor of Parasitology,
Department of Parasitology,
Faculty of Medicine,
University of Colombo,
Sri Lanka [4,CS5,CS7,CS8]

Catherine King
Renal Research Fellow,
Institute of Immunology and Immunotherapy,
University of Birmingham, Cancer Sciences
Building, Edgbaston,
Birmingham [CS31,CS32]

Nitya Kumar
Senior Lecturer in Public Health and
Epidemiology,
Royal College of Surgeons in Ireland - Medical
University of Bahrain, Bahrain [2]

Daniela Kurfürstová
Department of Clinical and Molecular
Pathology, Palacky University and University
Hospital Olomouc,
Czech Republic [CS66]

Ciara Leahy
Irish Research Council PhD Student,
Limerick Digital Cancer Research Centre,
School of Medicine, Bernal Institute and
Health Research Institute, University of
Limerick, Limerick, Ireland [CS63]

Mary Lynch Al Tarief
Consultant Cardiologist,
Mohammed Bin Khalifa Bin Salman
AlKhalifa Cardiac Center, Adjunct
Professor of Medicine, Royal College of
Surgeons in Ireland - Medical University
of Bahrain, Busaiteen 228,
Bahrain [CS35,CS36,CS37,CS40]

Laila Ayelet Mizrahi
Medical Student,
Royal College of Surgeons in Ireland - Medical
University of Bahrain, Busaiteen 228,
Bahrain [CS4,CS36,CS44]

Sara Mohamed
Clinical Educator in Medicine,
Royal College of Surgeons in Ireland - Medical
University of Bahrain, Busaiteen 228,
Bahrain [CS9,CS21,CS26,CS40,CS59]

Fatima Moh'D Muhsen Khader Atieh
Medical Student,
Royal College of Surgeons in Ireland - Medical
University of Bahrain, Busaiteen 228,
Bahrain [CS12,CS21,CS30,CS32,CS35,CS36,
CS39,CS59]

Paul G. Murray
Professor of Pathology and Head of the
Department of Pathology,
Royal College of Surgeons in Ireland - Medical
University of Bahrain,
Professor of Molecular Pathology,
University of Limerick, Ireland, and Honorary
Professor, University of Birmingham,
United Kingdom [1,3,4,5,6,7,8,9,10,11,14,CS1,
CS3,CS4,CS5,CS6,CS7,CS10,CS11,CS12,CS15,
CS16,CS17,CS18,CS19,CS21,CS24,CS25,CS26,
CS27,CS28,CS31,CS32,CS35,CS36,CS37,CS38,
CS39,CS43,CS44,CS50,CS52,CS54,CS55,CS56,
CS57,CS58,CS59,CS60,CS61,CS62,CS63,CS64,
CS65,CS68,CS69]

Shantha Perera
Visiting Senior Lecturer,
School of Life Sciences,
University of Wolverhampton,
United Kingdom [6,CS3]

Niamh Peters
Consultant Medical Oncologist,
University of Limerick Hospitals Group,
University Hospital Limerick,
St. Nessan's Road,
Dooradoyle, Limerick, Ireland [CS41,CS53,
CS59,CS60,CS61,CS66]

Matthew Pugh
Jean Shanks / Pathological Society
Intermediate Clinical Fellow, Associate
Clinical Professor, Honorary Consultant
Histopathologist,
University Hospitals Birimingham
Foundation Trust [CS3,CS11,CS55,CS57,
CS63,CS67,CS68]

Sara Bashar Qasrawi
Medical Student,
Royal College of Surgeons in Ireland - Medical
University of Bahrain, Busaiteen 228,
Bahrain [CS21]

Denis S. Quill
Associate Professor in Surgery, Department
of Surgery, Royal College of Surgeons in
Ireland - Medical University of Bahrain,
Busaiteen 228, Bahrain [CS50]

Sara Mohammed Ahmed Rady
Medical Student,
Royal College of Surgeons in Ireland - Medical
University of Bahrain, Busaiteen 228, Bahrain
[CS20,CS22,CS37,CS38]

Aisling Ross
Marie Skłodowska-Curie Actions
Research Fellow,
Bernal Institute and School of Medicine,
University of Limerick, Ireland, and Walter
and Eliza Hall Institute of Medical Research,
Parkville, Victoria, Australia [CS55]

Yusuf Abdulkarim Mohamed Shafeea Shakeeb
Medical Student,
Royal College of Surgeons in Ireland - Medical
University of Bahrain, Busaiteen 228,
Bahrain [CS6,CS53,CS70]

Omar Sharif
Consultant Gastroenterologist, American
Board in Gastroenterology and Hepatology,
Head of Internal Medicine, Head of Training,
Deputy Chief of Medical Staff
King Hamad University Hospital-Royal
Medical Services, Senior Clinical Lecturer,
Royal College of Surgeons in Ireland - Medical
University of Bahrain, Busaiteen 228,
Bahrain [CS19,CS24,CS26]

Patrick Stapleton
Consultant Microbiologist,
University Hospital Limerick,
St. Nessan's Road,
Dooradoyle, Limerick
[CS1,CS3,CS4,CS6]

Graham Taylor
Associate Professor in Viral and Cancer
Immunology, Institute of Immunology and
Immunotherapy,
University of Birmingham,
United Kingdom [5,6,CS11]

Zainab A. Toorani
Specialist Anatomical Pathologist,
Salmaniya Medical Complex,
Salmaniya Area, Bahrain
[CS15,CS24,CS25,CS26,CS27,CS37,
CS59,CS61]

Uwe Torsten
Professor of Obstetrics and Gynaecology,
Royal College of Surgeons in Ireland -
Medical University of Bahrain, Busaiteen 228,
Bahrain [CS11,CS44,CS45,CS51,CS61,
CS64,CS65]

Contributors to Previous Editions

We gratefully acknowledge the following authors who made contributions to previous editions, the text of which has been retained and/or modified in the current edition.

Ambinder RF
Ayres JG
Bareford D
Baumforth K
Bowman SJ
Broomfield A
Burnett D
Campbell CK
Carter R
Coleman IPL
Cramb R
Crocker CB
Crocker J
Davenport C
Digby JE
Dušek J
Hassan H
Hill FGH
Jewsbury JM
Johnson EM

Jones EL
Killington R
Kolar Z
Lawson SE
Lederman H
Maltby EL
Martin A
Maxton D
Macdonald F
Nye K
Palefsky J
Phillips JD
Rea C
Rylance P
Scott K
Singh B
Tarlow M
Todd I
Waters J
Young LS

Preface to the Third Edition

It is with great pleasure and enthusiasm that we introduce the third edition of *The Biology of Disease*. Since the publication of the second edition over 2 decades ago, the landscape of disease biology has undergone a profound transformation. In parallel, advances in medical science and clinical practice have reshaped our approach to diagnosis and treatment. This new edition reflects these seismic shifts, incorporating the latest insights and discoveries in the field. As such, most of the previous chapters and case studies have undergone extensive modification to ensure that they remain at the cutting edge of current knowledge. Moreover, we have added 34 new case studies, each carefully selected to illuminate the relevance of disease biology within modern day clinical practice.

Notable changes since the second edition reflect the unprecedented challenges that the global community now faces, including the emergence of new and drug resistant pathogens, an increasing recognition of the environmental factors, including climate change, that influence health, and the growing impact of diet and physical inactivity on the development of obesity and related disorders including cancer and type 2 diabetes. Our revamped content addresses these issues, providing readers with a holistic view of the complex interplay between biology, the environment and human health.

For the first time, we are pleased to introduce over 500 new multiple-choice questions, designed to test students' knowledge and understanding of the material, helping them reinforce their learning and prepare for assessments. Among these questions, you will find 'extra challenge' questions that encourage deeper critical thinking and problem-solving, making this edition an invaluable resource for medical students and learners across various health sciences disciplines. These MCQ are available as an online companion to this book.

We would like to express our heartfelt appreciation to all the contributors who have dedicated their expertise to make this edition possible, including those who laid the foundations of the first and second editions. Their collective wisdom and dedication have enriched this book immeasurably. We also extend our gratitude to the previous editors, Jonathan Phillips, Paul Kirk and John Crocker. We particularly acknowledge the contribution of our friend and colleague, Paul Nelson, who passed away soon after proposing this edition but whose vision inspired us to continue with this endeavour. We also thank the production team at Wiley for their expert support and guidance. We are also indebted to Professor Kevin Dunne who carefully reviewed the entire manuscript.

Finally, we acknowledge the unwavering support of our respective families and colleagues. Your encouragement and support have been instrumental in bringing this venture to fruition.

We hope that this third edition of *The Biology of Disease* will continue to serve as an essential companion for students and professionals alike, fostering a deeper understanding of the biology of disease and its clinical applications.

Paul G. Murray
Simon J. Dunmore
Shantha Perera
July 2024

Preface to the Second Edition

We were delighted to be given the opportunity to prepare a second edition of *The Biology of Disease*. Its appearance is timely in that, in the 5 years or so since the first edition, significant advances have been made in the basic sciences of cell biology and immunology, and in our understanding of the molecular mechanisms of disease. These developments are reflected in the relevant chapters. A new edition also offers the opportunity to review the aims of the book, though these still remain to produce a succinct volume with sufficient detail to enable a good understanding of the principles of disease biology.

In planning the second edition, we were keen to build upon the successful aspects of the first, particularly the use of case studies to amplify the points made in the chapters and to show how an understanding of the biology of disease translates into clinical practice. We have expanded the number of case studies to include a greater range of conditions and have reorganized some chapters to reflect this more clinically oriented approach. This is in line with developments in the medical curricula where students are exposed to clinical teaching at an early stage, and there is increased emphasis on clinically oriented problem solving. We anticipate that the revised book will also appeal to students on a range of other courses in the Health Sciences, including biomedical science, physiotherapy and other paramedical subjects, complementary therapies and nursing.

Finally, we would like to express our thanks to the Associate Editor team, to our colleagues in our respective Institutions, to the production team at Blackwell Science and not least to our families. All of these have supported us throughout, demonstrating good humour and patience at difficult times, and made significant contributions to the achievement of our goals.

Jonathan Phillips
Paul G. Murray
Paul Kirk
June 2000

Preface to the First Edition

The Biology of Disease aims to present the basic principles of disease processes in a form readily accessible to students trying to assimilate large volumes of information from a variety of sources. In conceiving this book, we recognized a need for a succinct volume that would give a broad yet sufficiently detailed account of the biology of disease. Acknowledging the expertise which lies in our medical and scientific colleagues (some working in education, some in research and others in clinical practice), we have drawn upon the experience and enthusiasm of a wide range of authoritative contributors. We feel that the book has benefited from this diversity of expertise, which has enabled us to cover many of the important topics in medicine today. Equally importantly, we were keen to adopt an accessible style of presentation and a common writing style, and we are grateful to all our contributors for their help in enabling us to achieve this aim.

From the outset, we felt it important to integrate the biological principles of disease processes with their clinical manifestations-the signs and symptoms which enable a diagnosis to be made. We have achieved this by ensuring that the principal chapters are clinically relevant and, where possible, that they bridge any gap between the biological and clinical features of disease. In addition, we have included clinical case studies in all the major sections of the book, each of which emphasizes the link between our current understanding of the basic science of the relevant condition and its clinical presentation. There is inevitable overlap and integration of topics in different sections of the book, so we have indicated cross-references where appropriate.

We anticipate that this book will be of use to a wide range of readers, particularly medical students approaching their clinical studies, students of clinical and biomedical sciences and students of other paramedical subjects. Sections of the book cover major areas of clinical science including epidemiology, immunology, infection, disorders of the blood, genetic diseases, oncology and mental health. Each section is complete in itself, enabling the book to be used selectively for the study of individual topics. We have tried to provide a succinct, yet comprehensive, overview of each topic and we hope that readers will be stimulated to seek out further, more detailed information. For this reason, each chapter concludes with key points and suggestions for further reading.

We are most grateful to all our contributors whose cooperation and expertise has been invaluable in achieving the aims of the book. We also thank our Associate Editor, John Crocker, for his help and guidance in the early stages of the editing process. Finally, we wish to acknowledge the support of our colleagues in the School of Health Sciences, of the editorial team at Blackwell Science and not least, that of our respective families, all of whom have helped us to nurture this venture to fruition.

Jonathan Phillips
Paul G. Murray

About the Companion Website

The Biology of Disease is accompanied by a companion website:

www.wiley.com/go/murray/biologyofdisease3e

The website includes:

- Multiple choice questions listed by section and divided into two levels of difficulty.
- The answers to these multiple choice questions, with an explanation.
- The figures as power point slides which may be used, for example, for teaching.

Part 1

Basic Principles of Disease and Epidemiology

1

The Nature of Disease

CHAPTER MENU

Introduction

This chapter considers how disease is defined, what types of factors cause disease and how diseases are classified. It considers why disease classification systems are important and how they are evolving in the era of personalised medicine. The chapter also serves as a foundation for later chapters which are focussed on the biological mechanisms underpinning disease development and progression.

Definition of Health and Disease

Health and disease can be difficult to define. Health is often described as the absence of disease and an individual may be in good health if there are no impediments to proper functioning or survival. The World Health Organization (WHO) has defined health as 'a state of complete physical, mental and social well-being and not merely the absence of disease or infirmity'. This latter definition is much broader and it is likely that most people would not be considered 'healthy' on the basis of these criteria.

Nevertheless, the WHO definition is useful since it acknowledges the importance of psychological and social well-being in the maintenance of health. Perhaps, a more realistic definition considers that health is a condition or quality of the human organism which expresses adequate functioning under given genetic and environmental conditions. This definition implies that an individual may be

The Biology of Disease, Third Edition. Edited by Paul G. Murray, Simon J. Dunmore, and Shantha Perera.
© 2024 John Wiley & Sons Ltd. Published 2024 by John Wiley & Sons Ltd.
Companion website: www.wiley.com/go/murray/biologyofdisease3e

considered healthy even if compromised in some way. An example here would be someone with Down syndrome who might well be considered healthy under the latter definition but not under the former.

Implicit in many definitions of health is the concept that efficient performance of bodily functions takes place in the face of a wide range of changing environmental conditions. Health in this context may be regarded as an expression of adaptability, and disease a failure thereof. Disease can also be defined as a pattern of responses to some form of insult or injury resulting in disturbed function and/or structural alteration.

Concepts of Normality

Individuals who are free from disease are often described as being 'normal'. It is important to recognise that normality does not always indicate health, but is merely an indication of the frequency of a given condition in a defined population. Some diseases occur with such frequency in the population that they might be considered to be 'normal', e.g. dental caries.

If we examine the distribution of an indicator of health, for example, the level of a particular analyte in the blood (e.g. haemoglobin), we can see it usually follows a *normal distribution* in a population (Fig. 1.1). Applying limits to the distribution curve produces two cut-off points. Below and above these points, haemoglobin levels may be considered abnormal. However, a value within the normal range may be considered pathological in a particular individual or under certain circumstances. Likewise, a small percentage of individuals falling within the abnormal zones will remain healthy and suffer no consequences as a result of their 'abnormally' high or low haemoglobin concentration. Furthermore, in adult females, the total blood haemoglobin concentration tends to be lower than in males, due in part to their monthly menstrual blood loss. Taking another example of blood cholesterol

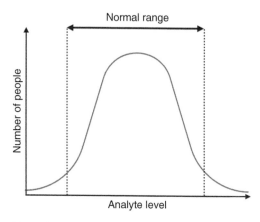

Figure 1.1   The normal distribution of blood levels of a typical analyte. Reference ranges are used to determine whether a patient's test results fall within the normal range, usually the central 95% of values derived from the results of tests done on a large number of healthy individuals. If blood levels fall outside the normal range, then this could indicate an underlying medical condition. Reference ranges can be influenced by a number of factors, including age and biological sex, and so reference ranges may be reported separately for different patient groups. Moreover, reference ranges may vary between different laboratories. For this reason, the reference ranges used in this book (Appendix 1) are provided only as a guide. *Source:* Created with BioRender.com.

concentration, the normal distribution in Western countries may not reflect levels that are ideal for the maintenance of health. For this reason, the term *reference range* instead of the *normal range* is preferred when defining the desired level of an analyte.

The rigid application of reference ranges can give rise to confusion, depending on how they are interpreted. For example, it is commonplace to show reference ranges for the *differential* white cell count (i.e. the different populations of white blood cells which make up the total leucocyte count) either as a percentage of the total or as the actual (or *absolute*) number in a given volume. When expressed as percentages, it is assumed that the total white cell count is normal. The use of absolute numbers is less ambiguous. For example, a normal lymphocyte count is in the range of $1.5–4.0 \times 10^9$/L or 20–45% when the total white cell count is normal ($4–11 \times 10^9$/L). If the total white cell count is raised to $15 \times 10^9/L$, a differential count showing lymphocytes as 15% of the total could be interpreted as indicating lymphopenia (reduction in lymphocyte count). In reality, the absolute lymphocyte count is normal (15% of $15 \times 10^9/L$ is $2.25 \times 10^9/L$), and the abnormally high cell population here could be neutrophils, for example. Typical reference ranges are given in Appendix 1 of this book, but should only be used as a guide, since reference ranges vary between different populations and even between different laboratories (for example, because of technical differences between different assays).

Promoting Health

Health promotion refers to the process of taking action to address the social, economic, environmental and behavioural factors that influence health and disease. The goal is to create an environment that encourages healthy behaviours and enables individuals to make informed decisions about their health. This may involve education and the provision of resources to help individuals adopt healthy lifestyles, promoting policies and programmes that support healthy behaviours and creating communities that are designed to support health and well-being. Health promotion activities may include initiatives such as providing access to healthy foods, promoting physical activity, providing resources for smoking cessation and supporting mental health and well-being. Health promotion may also involve efforts to address social determinants of health, such as poverty, lack of education and social isolation.

Onset of Disease

It can be difficult to be precise about the transition from health to disease because pathological changes with potential to cause disease are present in many apparently healthy people. For example, early *atheromatous* deposits are present in the arteries of a substantial proportion of symptom-free middle-aged adults, increasing their risk of cardiovascular disease. Most patients with coronary heart disease would date the onset of their illness from the first clinical manifestation (e.g. chest pain), rather than from the *hypertension* (increased blood pressure) which may have begun years before and which predisposed them to heart disease.

Causes of Disease

A great many agents and stimuli are implicated in the causation (*aetiology*) of disease. In the majority of cases, it is not possible to discover a single causative agent which always causes disease when present. For example, exposure to the microorganism *Mycobacterium tuberculosis* does not invariably result in tuberculosis; other factors (e.g. poor diet, reduced immunity, size of the infective dose) are important. However, tuberculosis cannot occur in the absence of the organism. Exposure to *M. tuberculosis* is therefore the *necessary* causal factor and the other factors are *subsidiary* causal factors.

Tuberculosis is an example of a disease for which the causal factors are well known. For other diseases, identification of the causal factors can represent a significant challenge. Often the initial search for a causal factor begins by examining the patterns of disease within human populations. This is *epidemiology* and is the subject of Chapter 2.

Types of Aetiological Factors

Aetiological factors may be broadly divided into *endogenous* factors (those which create a disturbance or imbalance from within) and *exogenous* factors (factors which threaten existence from the outside). Chromosome abnormalities giving rise to genetic disorders may be regarded as endogenous factors, whereas environmental insults are examples of exogenous aetiological agents. However, there is an overlap between these two groups. For example, some inherited chromosome abnormalities have been shown to be the result of parental exposure to mutagens in the environment, e.g. ionising radiation. Some genetic disorders then may ultimately be attributed to exposure to exogenous factors. Some disorders do not have a known cause and are often referred to as *idiopathic*. Diseases induced by medical intervention are referred to as *iatrogenic*.

Although modern humans, i.e. *Homo sapiens*, are a single species, they carry with them thousands of species of micro-organisms, including those that routinely inhabit the skin, lungs, saliva, oral mucosa, conjunctiva, biliary tract, gastrointestinal tract and other sites. Collectively, the bacteria, archaea, fungi, protists (single-celled organisms of the kingdom Protista, such as protozoa or simple algae) and viruses, living in or on humans, are known as the *microbiome*. Therefore, it is also important to consider how the microbiome, as an extension of the human organism, influences, and is itself influenced by, disease. The term *human metagenome* is sometimes used to refer to the collective genomes of these resident microorganisms.

Naming Diseases

With the recognition of new diseases such as COVID-19, it is timely to reflect on what basis diseases are named. Some diseases are given a name according to their symptoms, appearance, or other characteristics. For example, *psoriasis* which is a skin condition characterised by red, scaly patches on the skin (Fig. 1.2) comes from the Greek word 'psora', which means 'itch'. Other diseases are named after the person who first described them or identified their cause. For example, Alzheimer disease is named after Alois Alzheimer. Some diseases are named using

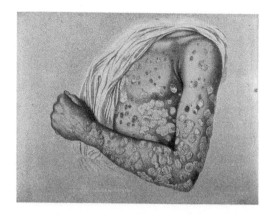

Figure 1.2 Psoriasis, so named after the Greek word 'psora', meaning 'itch'. This image shows diseased skin on the chest and arm of a man suffering from psoriasis. *Source:* Wellcome Collection/https://wellcomecollection.org/works/z2wkwnf6/images?id=ek5tq6qx / last accessed November 13, 2023.

an acronym that stands for the name of the condition. For example, AIDS is short for *acquired immunodeficiency syndrome*. Other diseases are named after the place where they were first identified or where they are particularly common. For example, Lyme disease is named after the town of Lyme in Connecticut, where it was first identified. Similarly, German measles (also known as rubella) gets its name because it was German physicians who first described this disease in the 1700s.

Classification of Disease

Diseases are classified either on the basis of the outward signs produced by disease (*manifestations*) or on the existence of a common aetiological agent. Most diseases are classified on manifestational criteria irrespective of the causative agents involved. Thus, a number of causal factors are implicated in the development of various types of carcinoma of the lung, including cigarette smoke, asbestos, coal smoke and other atmospheric pollutants, but patients with the manifestations of lung cancer are grouped together irrespective of the

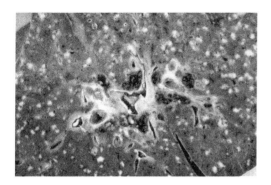

Figure 1.3 Miliary tuberculosis. This picture shows a close-up of a lung from an adult who was HIV-positive. The parenchyma is studded with 1–4 mm nodules of miliary tuberculosis granulomas. 'Miliary' refers to a form of tuberculosis in which there is widespread dissemination of lesions which have an appearance similar to millet seeds. *Source:* Wellcome Collection/https://wellcomecollection.org/works/wvtqyn8v/items / last accessed November 13, 2023.

involvement of one or more of these agents. Conversely, diseases caused by *M. tuberculosis* may produce different clinical manifestations in different individuals, yet all are classified as forms of tuberculosis (Fig. 1.3).

Classification of disease on the basis of shared aetiology may follow the identification of a new and important aetiological agent, particularly if this offers promise of major therapeutic or preventive advantage. For example, all forms of disease associated with infection with the SARS-CoV-2 virus are referred to as COVID-19. However, not all diseases are classified on the basis of a shared aetiology. For example, the identification of cigarette smoke as the most important cause of lung cancer did not promote a revision of the classification of lung cancer or of other diseases caused by smoking (e.g. emphysema and chronic bronchitis).

The classification of disease into discrete entities enables patients to be assigned to specific groups. The patients may then be treated in a similar fashion to other patients assigned previously to the same group, thus, at least in theory, improving the clinical outcome based on past experience. This is sometimes referred to as *stratified medicine*. This view may be extended on the basis that the development of disease, and its progression and response to treatment, should be regarded as unique to the individual. This has led to the concept of *personalised medicine*, in which an individual's profile, for example their genetics, is used to guide decisions with respect to prevention, diagnosis and treatment. The terms personalised and stratified medicines are often used interchangeably and can be considered under the broader heading of *precision medicine*.

Identifying Disease

We have seen how diseases are classified but have not yet considered how a particular set of features is first designated as a disease state. Doctors use *signs* (what the doctor sees or feels when carrying out a physical examination), *symptoms* (what the patient complains of) and a range of laboratory and clinical tests to determine whether a patient has a given disease. The taking of a thorough clinical history will determine whether there has been exposure to any potential aetiological agents. The existence of certain predisposing conditions may make the development of a disease more likely. Examples of these risk factors include certain genetic disorders, lifestyle, psychological and personality profile, age and environmental factors such as climate and pollution. Furthermore, the presence of one disease (e.g. type 2 diabetes; T2D) may predispose a patient to the development of a different disease (e.g. atherosclerosis).

One of the early steps in identifying a disease is to establish a range of diagnostic possibilities from which the eventual diagnosis will be selected. This is referred to as the *differential diagnosis*. The final diagnosis may sometimes be established shortly after clinical presentation, in other cases perhaps only after extensive use of laboratory and clinical tests, or occasionally may never be identified during the lifetime of the patient. It must be remembered that disease is a dynamic process and indicators of disease may vary as the disease progresses. Furthermore, in some patients, particularly those who are elderly, different diseases may co-exist (referred to as *co-morbidities*), thus confusing the diagnostic processes.

The impact of disease on an individual patient or population group may be measured as *morbidity* (i.e. its detrimental effects, such as pain or disability) or as *mortality* (death). *Prognosis* is the likely future for the patient in terms of length and quality of life. Prognosis depends on many factors, including the stage the disease has reached and the likely impact of therapy. Once a disease has been identified, treatment, aimed at relief of symptoms or, where feasible, cure, is usually initiated. The decision on how to manage the patient often involves input from a number of different healthcare professionals in the so-called *multidisciplinary team meetings*. Following treatment, patients may be cured of the disease or may enter *remission* (a symptom-free period) from which they may either *relapse* (when symptoms of the disease return) or be cured. Some diseases are not amenable to cure, but can be controlled, for example, by the administration of drugs. Some patients with advanced disease may receive *palliative* treatment only, with the aim of relieving their symptoms for the remainder of their life.

The Global Fight Against Disease

Significant progress has been made in reducing the incidence of, or even eradicating, certain diseases. One of the greatest success stories was the eradication of smallpox. This deadly virus had been a scourge on humanity for centuries, causing millions of deaths and leaving many more people permanently scarred or disabled (Fig. 1.4). Edward Jenner, widely regarded as the founder of immunology, was responsible for first showing that his cowpox 'vaccine' conferred specific immunity to smallpox (Fig. 1.5). Since then, a global effort centred upon widespread vaccination, as well as strict disease surveillance,

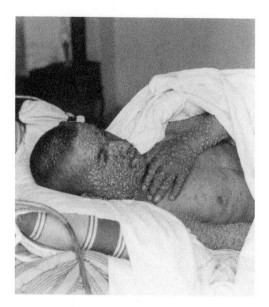

Figure 1.4 Smallpox, a once deadly disease. The heavily pockmarked face, arms and hands of a smallpox victim in Palestine, c. 1900–1925. By 1980, the disease was finally eradicated by the World Health Organization. *Source:* Everett Collection/Shutterstock.

led to the complete eradication of smallpox in 1980. Similarly, from the 1960s, the widespread use of poliovirus vaccines has prevented an estimated 30 million cases of paralysis. Efforts to eradicate poliovirus globally, initiated in 1988, have reduced the number of reported poliomyelitis cases from 35 000 in 1988 to less than 2100 by 2001. However, recent events such as the declaration of a state of emergency in New York following a reported case of poliomyelitis, the detection of polioviruses in wastewater samples in New York and London, and the resurgence of wild Poliovirus type 1 (WPV1) cases in Pakistan and imported cases in Malawi and Mozambique have highlighted that poliovirus remains a threat. Moreover, newly emergent infectious diseases, for example, Ebola, Zika and, most recently, coronavirus diseases, will continue to pose major challenges to global health in the years to come.

The rising incidence of non-communicable diseases perhaps presents an even a greater

Figure 1.5 Edward Jenner's cowpox 'vaccine' heralded the beginning of the science of immunology and of the use of vaccination to prevent infectious diseases. Edward Jenner vaccinating a boy. Oil painting by EE. Hillemacher, 1884. Hillemacher, Eugène-Ernest, 1818–1887. After many years of observations, Edward Jenner carried out his experiments on the effects of vaccines on human experimental subjects in 1796. On 14 May 1796, he inoculated cowpox from the arm of Sarah Nelmes into the arm of James Phipps, said to have been aged about 8 years. On 12 July 1796, he inoculated Phipps with smallpox, but the smallpox did not have any effect, demonstrating the protective effects of the cowpox. *Source:* Wellcome Collection/ https://wellcomecollection.org/works/tr7x4acf/items / last accessed November 13, 2023.

threat, with cancer, heart disease and diabetes becoming increasingly prevalent worldwide. Unlike infectious diseases, which can be tackled with vaccines and other targeted interventions, many non-infectious diseases arise as a consequence of complex interactions between genetics, lifestyle factors and environmental exposures. Thus, preventing and treating these diseases requires a multifaceted approach involving not only medical interventions but also social and economic policies to address underlying risk factors such as poor diet, sedentary lifestyles and environmental pollution. Moreover, many non-communicable diseases are chronic in nature requiring long-term management, which in turn places a significant burden on healthcare systems.

Mechanisms of Disease

Understanding the way in which disease begins and progresses is the major focus of research within the biomedical sciences. The assumption is made that an understanding of the biology of disease will contribute to an improvement in health care, in terms of both the prevention of disease and its treatment. In many cases, this has proved to be satisfyingly correct. For example, the recognition that type 1 diabetes (T1D) results from damage to insulin-secreting cells in the pancreas led to the development of insulin replacement therapy. However, an understanding of the biology of many important diseases (e.g. cancer) is only beginning to emerge and for the most part this has not yet contributed greatly to a decline in mortality rates.

Summary

The aim of this book is to instil an appreciation of the biological basis of disease. It is intended to provide undergraduate students of medicine, nursing and biomedical sciences, as well as other science and health professionals, with a fundamental understanding of the key biological mechanisms that result in disturbed homeostasis. The chapters deal with fundamental biological aspects of important groups of human diseases, while the clinical aspects of selected disease states and how these relate to the underlying disease biology are considered in the case studies. We hope that students will also take time to test what they have learned by attempting the multiple choice questions available online as part of this textbook.

Key Points
1) The World Health Organization (WHO) defines health as 'a state of complete physical, mental and social well-being and not merely the absence of disease or infirmity'. 2) Reference ranges refer to the range of values for a particular medical test or measurement considered to be within the normal range for a healthy individual. 3) Diseases can be caused by external factors such as infections, toxins or physical trauma, while others are caused by internal factors such as genetic alterations, immune dysfunction or metabolic imbalances. Many diseases are multifactorial. 4) Diseases can be named on the basis of the cause or the physical manifestations.	In the past, diseases were often named after the person who first described them or the place where they were first identified. 5) The classification of disease enables patients to be assigned to specific groups, allowing for improved clinical outcomes based on past experience and the potential for personalised or precision medicine based on individual profiles. 6) Doctors use signs, symptoms, clinical history and the results of laboratory tests to identify a disease and establish a differential diagnosis, a list of possibilities from which the eventual diagnosis can be made.

Further Reading

Kumar V., Abbas A.K. & Aster J.C. (2018) *Robbins Basic Pathology* (10th edn). Elsevier Saunders, Philadelphia, PA.

Kumar P. & Clark M. (2017) *Kumar and Clark's Clinical Medicine* (9th edn). Elsevier, Edinburgh.

Ralston S.H., Penman I.D., Strachan M.W.J. & Hobson R.P. (2018) *Davidson's Principles and Practice of Medicine* (23rd edn). Elsevier, Edinburgh.

2

Principles of Epidemiology

Introduction

Epidemiology is the study of the distribution and determinants of health and disease in populations. Epidemiologists use descriptive and analytical methods to determine the existence and extent of correlations between the distribution of disease and well-being, and agents which may affect this, such as the degree of exposure to environmental influences or the biological and sociological characteristics of individuals in populations. If the studies are precise enough, it may be possible to suggest causal links.

Epidemiology is fundamental in determining the aetiology of disease and, in turn, enhancing health. Its methods allow the assessment of disease burdens and trends, providing administrators and legislators with a rational basis for planning the management and development of health services at both a local and global scale. The subject is essential in the evaluation of regimens of treatment by clinical trials and in the efficient delivery of health care. This chapter introduces some of the basic concepts of epidemiology and outlines the more elementary methods it employs. These are further illustrated in the example at the end of this chapter.

Epidemiological Measures of Illness and Death

For the scientific study of disease and health in populations and comparisons between populations, anecdotal evidence and the simple counting up of a number of cases are

The Biology of Disease, Third Edition. Edited by Paul G. Murray, Simon J. Dunmore, and Shantha Perera.
© 2024 John Wiley & Sons Ltd. Published 2024 by John Wiley & Sons Ltd.
Companion website: www.wiley.com/go/murray/biologyofdisease3e

inadequate. Even to reach the stage of natural history requires the use of rates. To know that there are 100 cases of disease in population A and 1000 in population B enables no sensible conclusion as to whether A or B has a more serious problem. In fact, if the population of A is 1000 and that of B is 1 000 000, the problem is worse in A than in B, since the rate for A = 100/1000 and for B = 1000/1 000 000, or 0.1 compared to 0.001. For easy comparison, rates are usually multiplied by a factor. In this example, 1000 would be reasonable, so that the rates become (100/1000)×1000 for population A and (1000/1000 000)×1000 for population B. The rates show the real difference between the two populations. Any other factor could have been used, the choice is arbitrary and for convenience.

Morbidity Rates

These are measures of the illness in the population and are usually expressed as either *incidence rates* or *prevalence rates*.

Incidence

Conceptually, *incidence* answers the question 'how frequently are new cases of disease occurring?' Mathematically, the incidence of a disease is given by the number of new cases occurring over a specified time period divided by the population at risk for the disease during the same period.

The *incidence rate* per 10 000 would therefore be

$$\frac{\text{No. of new cases of a disease in a given time period}}{\text{Total population at risk over the same time period}} \times 10\,000$$

The population at risk consists of susceptible individuals in whom the disease is possible. Not all members of a population may be at risk and should not be included in the denominator. For instance, in calculating incidence rates for

ovarian cancer, it would be misleading to include males in the total population, since only females can be at risk of contracting this disease. The population at risk may also be decided based on other factors, for example, age, social class, ethnic group, occupation or geographical location.

Prevalence

Prevalence answers the question 'how many people have the disease?' It is computed by dividing the number of cases of disease existing in a population over a given time by the total mid-year population.

$$\text{Prevalence rate per 10 000} = \frac{\text{No. of cases existing during a given period}}{\text{Total mid-year population}} \times 10\,000$$

Prevalence may be expressed in two ways, *point prevalence* and *period prevalence*. As the term suggests, point prevalence is the number of cases of disease at a point in time. Period prevalence is the number of cases over an extended length of time. Fig. 2.1 shows the difference between these.

Impact of Duration of Disease on Incidence and Prevalence

Each new case of a disease joins an already existing number of cases and thus adds to the prevalence pool. The prevalence pool will remain at a given size if the number of new cases added is matched by old cases leaving the pool by recovering, being cured or dying. If a disease has low rates of recovery, cure or death, then people with the disease remain in the population for a longer time. Consequently, the prevalence of such diseases tends to be high.

Mortality Rates

Death rates may be useful indicators of the frequency of disease. If a disease has a very high case fatality or lasts only a short time

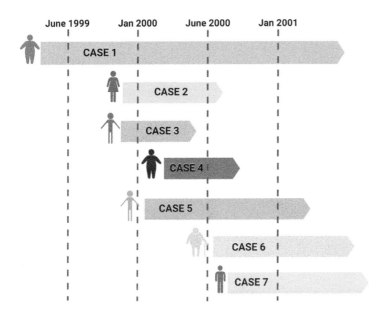

Figure 2.1 Duration of cases of a disease over a 2-year period illustrating the meaning of prevalence and incidence. Cases of disease recorded from 30 June 1999 to 30 June 2001 from a total population at risk of 100. Period prevalence from 1 January 2000 to 31 December 2000 was 7 (cases 1–7). Point prevalence on 30 June 2000 was 4 (cases 1, 2, 4 and 5). The incidence for the year 2000 was 4 (4, 5, 6 and 7). The incidence rate for the year 2000 was 4/100. *Source:* Created with BioRender.com.

from onset to death, the mortality rate is a good measure of the incidence of the disease. The most generalised form of mortality rate is that calculated from all causes of death over a given period (usually a year). This is represented as follows:

$$\frac{\text{Total no. of deaths in the population over a defined period}}{\text{No. of population at mid-period}}$$

The use of such crude rates can be misleading because they collapse a mass of demographic data into a single statistic. Table 2.1A gives the crude death rates for two resource-rich and two resource-poor countries in 1994. Contrary to expectation, the resource-rich countries have higher death rates than the resource-poor countries.

The explanation is that the demographic structures of the two pairs are very different. Table 2.1B shows that both Germany and the

United Kingdom had more than three times the proportion of people aged 65 years old or more than Brunei or Costa Rica. The chance of dying over any 1 year is much greater on average the older a person becomes, once past early childhood (Table 2.1C). Because the proportion of old people is higher in the two resource-rich countries, their death rates are inflated and do not reflect the better living conditions and state of health in those countries. It would not be unreasonable to compare crude death rates between Germany and the United Kingdom, because their demographic structures are similar, but it is meaningless to compare them to either Costa Rica or Brunei. For such a comparison, age-specific death rates would need to be calculated. Thus, the use of specific mortality (or morbidity) rates is often too clumsy in practice. A more convenient way of obtaining a fair comparison between countries is to *standardise* the crude rates using appropriate demographic factors which have a major bearing on health.

Table 2.1 Mortality rates.

(A)

Resource-rich countries death rate per 1000		Resource-poor countries death rate per 1000	
United Kingdom	11.4	Brunei	3.1
Germany	11.6	Costa Rica	3.7

(B)

	Population (%)			
Age group	Germany	United Kingdom	Brunei	Costa Rica
0–14	16.2	19.0	34.3	33.5
15–64	68.7	65.0	62.1	62.1
65+	15.1	15.5	3.6	4.4

(C)

Age group	Death rate per 1000
Under 1	6.1
1–4	0.3
5–14	0.1
15–24	0.5
25–34	0.7
35–44	1.4
45–54	3.4
55–64	9.8
65–74	27.7
75–84	68.1
85+	164.4

(A) Crude death rates for two resource-rich and two resource-poor countries. From WHO (1994). (B) Populations of two resource-rich and two resource-poor countries stratified by age for 1994. From WHO (1994) World Health Statistics Annual for 1994. WHO, Geneva. (C) Age-specific death rates for the UK 1996. From Office for National Statistics.

Standardisation of Rates

Standardisation is also referred to as *adjustment* and is used to compare mortality rates between two different populations. The two main factors used for the adjustment of rates are age and biological sex. Conceptually, the process may be thought of as removing the differences arising due to different age/sex distributions in the populations being compared. Standardisation may be calculated indirectly or directly. The *standardised mortality ratio* (*SMR*) is an indirect method calculated using the expression:

$$\frac{\text{Observed deaths in a population over a given time period}}{\text{Expected deaths in a standard population over the same period}} \times 100$$

Table 2.2 gives an example of such a calculation. *Standardised incidence ratios* may also be calculated by replacing mortality figures with incidence in the expression. Rates can be directly standardised for age or other features. In comparing two populations, the age-specific rates (or rates for other variables) found within them are applied to a third, reference (or standard), population. The recalculated rates are those which would have been found in each population had they had the same demographic structure as the reference population. The consequence is that the rates so adjusted for the two populations can then be directly (and fairly) compared with each other.

The reference population chosen is arbitrary. In practice, however, published reference populations are used such as the world standard population, or the European standard population (demographically, a relatively ageing population) found in World Health Organization (WHO) yearbooks.

Epidemiological Studies

Epidemiologists, like other scientists, employ study methods that range from descriptive through observational to experimental (or intervention) studies.

Descriptive Studies

To elucidate possible patterns of association between agents that may affect health positively or negatively, it is necessary to collect

Table 2.2 Calculation of standardised mortality ratio.

Age group	Death rate from lung cancer in England and Wales 1993 per million (SP)	Population for England and Wales 1995 (thousand) (PI)	Expected deaths (SP×PI)	Observed deaths
55–59	1.02	1321.6	1348.0	1288
60–64	1.98	1204.0	2383.9	2161
65–69	3.51	1107.2	3886.3	3533
70–74	4.86	970.3	4715.7	4519
75–79	6.15	622.2	3826.5	3602
80–84	7.11	409.6	2912.3	2099
85+	6.58	240.1	1579.9	1447
Total			20 652.6	18 649

SMR = observed/expected × 100 = (18 649/20 652.6) × 100 = 90.3%.
PI, population investigated; SP, standard population.
Shown are death rates from lung cancer in men in England and Wales (1995 compared to 1993) in the age range 55–85+. The SMR indicates an improvement in death rates for lung cancer in this age range over the 2-year period. From UK Office of National Statistics.

and collate data which relates such agents to different sectors of a population, e.g. age group, social class, periods of time when they are affected and the geographical areas in which the agent is distributed.

Often such studies rely on data which are routinely collected by government statistical services, health service authorities, case records and so on. Such data may show how rates of disease change among people in different occupations, parts of a city, or region, over time. This can lead to the discovery of important trends in disease and health and the identification and quantification of new and existing factors detrimental or advantageous to the health of a population. Figures 2.2–2.4 show examples of such data. The data may then be used to generate hypotheses on causation which may be tested by observational or experimental studies.

As can be seen in Fig. 2.2, over the period studied, the trends for the two forms of cancer are different; melanoma increased in incidence and was more common in women. Lung cancer incidence rates fell in men but were rising in women. Death rates for melanoma remained steady, indicating improved survival. For lung cancer, death rates paralleled the incidence

rates, that is, there was no improvement in survival. The lung cancer figures also show that more men were stopping smoking while more women were taking it up.

The histograms in Fig. 2.3 show the association of coronary heart disease mortality with social class in males in England and Wales. Premature deaths are more common in the lowest social grouping. In this group, mortality has hardly changed over the 23 years of records, whereas mortality in the highest social group has shown a considerable decrease. The distribution of SMRs shown in Fig. 2.4 indicates possible conditions adverse to health in the north and west of England and parts of London.

Descriptive studies are economical but can be no better than the raw data on which they are based. Records are not always adequately kept and there may be inaccuracy or lack of consistency in the diagnosis of disease between practitioners and between different countries. Morbidity figures record only those people presenting for medical advice; other cases, for various reasons, may not seek medical help and will not be recorded. Even records of mortality depend on the accuracy of the certification of cause of death. Changes in the definition of disease may also occur. WHO's International

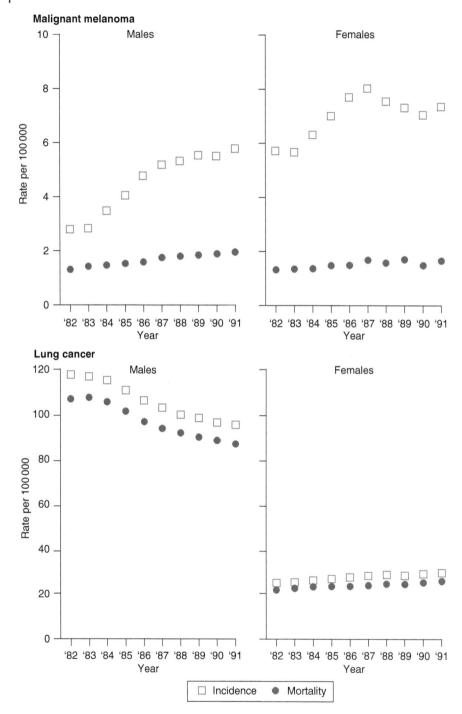

Figure 2.2 Standardisation: 3-year rolling directly standardised incidence and mortality rates per 100 000 (European population) for two forms of cancer over 12 years in the West Midlands. *Source:* West Midlands RHA Public Report, 1995.

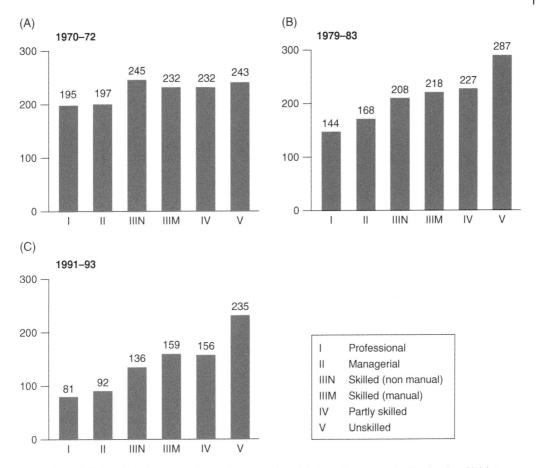

Figure 2.3 Relationship of coronary heart disease and social class. Figures are for England and Wales (European age-standardised rate per 100 000 population). *Source:* Drever and Whitehead (1997).

Classification of Diseases (ICD) is revised regularly, so that, for example, a particular cancer may move from one classification group to another in line with medical advances. As of writing, the ICD was in its 11th revision. The ICD-11, as it is therefore known, was accepted by WHO's World Health Assembly (WHA) in 2019, but officially came into effect only on 1 January 2022. Widespread adoption takes time; for example, ICD-10 was only adopted in the United States in 2015.

Observational Studies

Cross-Sectional Studies

Sometimes, available descriptive data may not be adequate to meet the purposes of an epidemiological study, and *cross-sectional studies* of a population may be necessary. These survey a sample of a population at a given point in time to look for associations between an exposure (e.g. smoking and a disease [e.g. asthma]). Such studies are economical to perform but are unable to establish cause directly since they determine the state of health and the presence or absence of the risk factor at the same time. They give a measure of the prevalence of a condition, and are able, if repeated, to show trends in behaviour, such as how smoking affects asthma risk. Progress towards the attainment of health targets is usefully monitored in this way and can provide helpful information to improve health promotion. While cross-sectional surveys give a

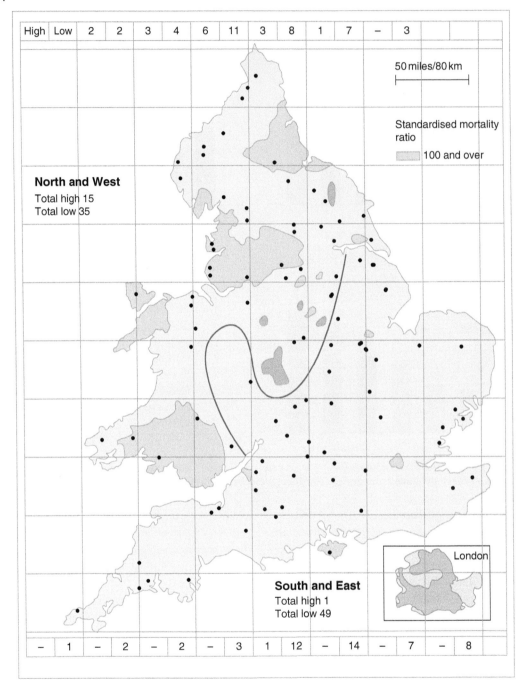

Figure 2.4 Relation of mortality to geographical location. England and Wales: mortality from all causes, males, 1959–1963. *Source:* Learmonth (1978).

snapshot of the health of a population and associated risk factors, to ascertain causal effects requires longer term *longitudinal studies*. If the survey is a snapshot, then longitudinal studies are akin to a movie comprising multiple snapshots. Longitudinal studies fall into two types: *case–control* and *cohort studies*.

Case-Control Studies

Case-control studies compare a sample of the population who already have the disease (*cases*) with a sample of the population who are currently unaffected by the condition (*controls*). The exposure of the cases and controls to the suspected agent is then determined either from records or by questionnaire and interview techniques.

Case–control studies are relatively inexpensive and can be completed relatively quickly. They are useful in indicating the aetiology of rare diseases since only small numbers of participants are needed as the cases already exist. Their main disadvantage is that since records and registers were not set up with a particular study in mind, they may be deficient in necessary detail or in the way data were collected or stored. Records may also be destroyed or misplaced. If interviews and questionnaires are used, people may have forgotten key details, they may not always be altogether truthful, particularly if embarrassing or intimate personal details are needed, or conversely, may make up details in an attempt, as they see it, to help the investigator. Case–control studies can usually only study one condition at a time (compared with cohort studies, see below). A further problem is that risk cannot be directly calculated, since the cases are an unknown sample of the whole population and incidence cannot therefore be determined.

To work out whether those with the disease show greater exposure to the suspected aetiological agent, numbers of cases and controls and their exposure are tabulated in a 2 × 2 table (Table 2.3). The proportion of all the cases in Table 2.3 who have been exposed to the agent is a/(a + b) and for the controls c/(c + d). These

Table 2.3 Form of data presentation in a case–control study.

	Exposed	Not exposed	Total
Cases	a	b	a + b
Controls	c	d	c + d
Total	a + c	b + d	

The letters represent the numbers recorded for each group.

ratios might be considered representing the absolute risk for the cases and the controls. This would follow if the whole population at risk was known and incidence could be calculated. In case–control studies which are retrospective, the groups a + b and c + d are not the total population exposed and not exposed, the cases and controls are at best a random sample of the total population.

Since absolute risk cannot be calculated, the ratio of risk between the cases and control (i.e. *relative risk*) cannot be directly calculated either. An estimate of relative risk can be made if the cases included in the study are a fair sample of all cases, the controls are an unbiased sample of the whole population and the disease is rare. This estimate is called the *odds ratio* and represents the ratio between the odds that a person with the condition has been exposed to the risk factor and the odds that a person without the condition has been exposed. Using the letters found in the 2 × 2 table, this can be written as:

$$\left(a / \left(a + b\right)\right) \div \left(c / \left(c + d\right)\right)$$

If the disease is rare, then 'a' will be very small compared to 'b', i.e. a + b ≈ b. Similarly, 'c' will be small compared to 'd' and the expression can be rewritten as (a/b) ÷ (c/d) or ad/bc. Fig. 2.5 illustrates this diagrammatically.

An odds ratio of 1 would indicate that the odds of having the condition are equal for those exposed and those not exposed, a ratio of 3 would indicate that the odds are three times greater for those exposed, and a figure less

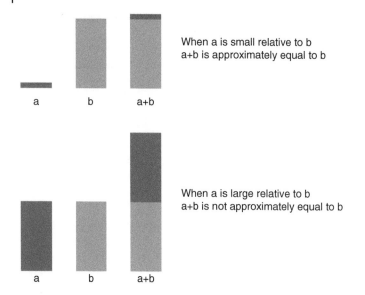

When a is small relative to b
a+b is approximately equal to b

When a is large relative to b
a+b is not approximately equal to b

Figure 2.5 Basis for deriving odds ratio equation. If the disease is rare, then a will be very small compared to b, i.e. a + b ≈ b.

than 1 that the odds are less for those exposed, in other words, the exposure is beneficial. An example of the use of such statistics, contributing to the case–control study of sporadic Creutzfeldt–Jakob disease in the United Kingdom, is given in Table 2.4. The odds ratio expressed in Table 2.4 indicates more than twice the risk for those with a family history of dementia. The result is difficult to interpret: *recall bias* may have influenced the outcome if relatives of cases (those with dementia) were more likely to remember other relatives with this condition than those of the control population.

Table 2.4 Exposure history of 216 cases of sporadic Creutzfeldt–Jakob disease and their controls.

	Family history of dementia	No family history of dementia	
Cases	35	181	
Controls	16	200	$N = 216$

Odds ratio = $(35 \times 200) \div (16 \times 181) = 2.4$

Cohort Studies

These begin with a defined population at risk, e.g. workers in a particular factory. This cohort of initially disease-free persons is then followed into the future and the effects of the exposure to the putative agent are recorded for the whole length of the study. Since the whole population is known, incidence rates can be worked out for those exposed and those not exposed. The advantage of cohort studies is that relative risk and attributable risk can be calculated (see below). They can also be used to study several diseases at the same time or to determine the rare causes of a condition. However, cohort studies are not suitable for investigating rare diseases because a very large population would have to be defined in the first place to ensure that sufficient cases occurred over the study period. They also have the disadvantage of often requiring a long time to complete to ensure that enough cases are recorded. Cohort studies are also less practicable than case–control studies for conditions with long incubation periods. For these reasons, they can be logistically difficult and expensive. Moreover, people may not complete the study, for example, they may leave the area, refuse to comply or die. If too many

Table 2.5 Form of data presentation in a cohort study.

	Exposed	Not exposed	Total
Condition present	p	q	p+q
Condition not present	r	s	r+s
Total	p+r	q+s	

people leave the study, it may be impossible to draw meaningful conclusions. There are other forms of study design, beyond the scope of this chapter, which combine the advantages of case–control and cohort studies. Table 2.5 illustrates how the results of a cohort study can be recorded.

In this type of study, the risk may be directly calculated. The *absolute risk* of contracting the condition for persons exposed is p/(p+r) and the *relative risk* is (p/(p+r)) ÷ (q/(q+s)). A value of 1 from this calculation would indicate no association between the exposure and the disease. Figures exceeding 1 indicate an association between the two; the higher the figure, the greater the association; if this is a disease, then the exposure increases the risk of developing the disease. A figure less than 1 indicates a beneficial effect of the exposure if the condition investigated is a disease. Briefly, relative risk is a measure of the strength of the association between exposure and the disease. It may be of little significance, e.g. a risk of 1 in a million is twice the risk relative to a risk of 1 in 2 million, but is still effectively negligible. To determine how much of the disease can be attributed to a particular agent requires a third measure of risk, the *attributable risk*. This is important in medical practice, particularly public health, because it indicates how much the incidence of the condition may be reduced if the risk factor were eliminated.

$$\text{Attributable risk} = \left(\begin{array}{c} \text{incidence of disease} \\ \text{for those exposed} \end{array} \right) - \left(\begin{array}{c} \text{incidence for those} \\ \text{not exposed} \end{array} \right)$$

Using the symbols in Table 2.5:

$$\text{Attributable risk} = \left(p / \left(p + r \right) \right) - \left(q / \left(q + s \right) \right)$$

For public health purposes, a further useful measure is the *attributable fraction in those exposed*. This can be calculated from the formula:

$$\text{Attributable fraction} = \frac{\left(\text{incidence in exposed} \right) - \left(\text{incidence in not exposed} \right)}{\text{incidence in exposed}}$$

Using the symbols in Table 2.5, this becomes:

$$\frac{\left(p / \left(p + r \right) \right) - \left(q / \left(q + s \right) \right)}{p / \left(p + r \right)}$$

The attributable fraction indicates the excess disease arising among those exposed which can be attributed to that exposure. As such, it indicates how much disease could be eliminated by removing the exposure.

Experimental Studies

In observational studies, the epidemiologist can test hypotheses only passively, being dependent on what data arise, or have arisen, from circumstance. In science, the strongest tests for hypotheses are those provided when the investigator is able to intervene actively by controlling variables. Randomized clinical trials are examples of such an experimental approach. All proposed new drug treatments are tested in this way and the method can be used to investigate other kinds of medical intervention and health service provision. In such trials, people are allocated randomly to control or intervention groups. A trial for determining the efficacy of a candidate drug for testicular cancer might be formulated as in Fig. 2.6.

To avoid *placebo effects*, it is essential that those in the trial do not know to which group (control or intervention) they have been assigned. To prevent subjective bias by the experimenters, it is also preferable that they

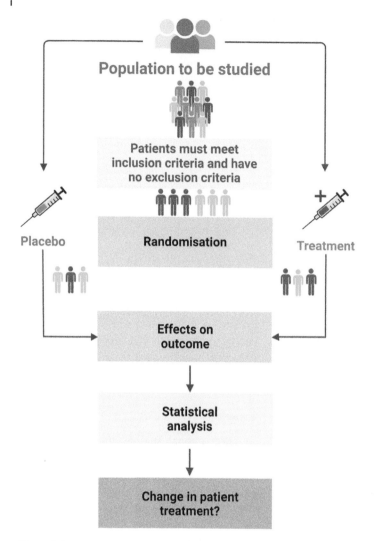

Figure 2.6   Outline of a trial for determining the efficacy of a candidate drug. Note: there must not be any difference obvious to the participants in the treatment of the two groups, e.g. tablets given to both groups should look the same in size, shape and colour so they are unaware of the group to which they have been allocated (*single blinding*). The clinical team should not know to which group participants have been allocated, nor be able to distinguish between the tablets (*double blinding*). *Source:* Created with BioRender.com

too are unaware of the allocation to groups. This process is called *blinding*. *Selection bias* occurs when those selected to participate are not representative of the population at risk. For example, if a study depends on volunteers from the population, bias will occur if those volunteering represented those who were more concerned about their health than the general population. Such people would be likely to be more careful in their dietary and drinking behaviour and more likely to take exercise and so would not necessarily be representative of the whole population. Bias will also occur if people dropping out of a study are not similar to those who remain in it. *Information bias* can arise in retrieving information from subjects or from records (an example is given in Table 2.4). Interactions between interviewers and subjects may affect response rates. Cases might be questioned more carefully by an interviewer than controls, being prompted by the interviewer to ransack their memories about past exposures more than those without the disease. Non-response to questionnaires can give rise to bias if the non-responders differ systematically from the responders.

Confounding

An agent may be correlated strongly with a disease, but may not be its actual cause, being merely independently correlated with the cause. Such agents are called *confounding variables*. For example, a study might show a strong correlation between a moderate daily intake of red wine and a significant reduction in the risk of cardiac disease. If, however, wine consumption is itself strongly correlated with a fish-based diet, and if there were independent evidence that this diet reduced heart disease, red wine consumption could be a confounder of the true situation (fish-based diet reduces heart disease).

Minimising Confounding Effects

Matching
In the example given above, the confounding effect of wine consumption could be removed by a study matching wine consumers with non-consumers for diet, age and biological sex. If too many matching variables are used, then the numbers that can be recruited to a study may be reduced below acceptable sample sizes for adequate statistical analysis. If matching is based on a variable which is actually associated with the disease, then real differences between groups will be eliminated.

Restriction
If a factor being investigated as a possible aetiological agent has an incidence that is greater in men than in women, then biological sex and the factor will be confounded. This can be avoided by restricting the study to one biological sex.

Stratification
To resolve the confounding between the effects of moderate intake of red wine and the effect of the fish-based diet on rates of heart disease, the data could be stratified as shown in Table 2.6.

Correlation or Causation?

Even if an epidemiological study has shown a strong association between an agent and a disease, it is not necessarily true that the agent is a *cause* of the condition. Statistical analysis may show that the correlation is significant, but this is not *proof* of actual cause.

The association may arise by chance, bias or confounding. If, in using evidence allowing rejection of a null hypothesis, probability levels are set too high, the hypothesis may be falsely rejected (*type I error*). If the level is set too low, the null hypothesis may be falsely retained (*type II error*).

In simple terms, A causes B if and only if

- A always precedes B in time.
- Changing A always changes B.
- The correlation between A and B is not due to both being independently correlated with a further agent, C.

In practice, it is impossible to be certain that there is no further underlying agent such as C, and it is impossible to perform the infinite number of observations of the relationship between A and B which are needed to give absolute proof of cause.

Even such guides as Koch's postulates for infectious disease (Chapter 4) may not give the cause of disease, since disease is often multifactorial in its pathogenesis. It is not possible to suffer from tuberculosis unless the bacillus is present (*necessary cause*), but the presence of the bacillus is not a *sufficient cause* for clinical manifestation of disease. The development of tuberculosis depends on other factors that include the characteristics of the infected person, such as genetic susceptibility or a compromised immune system, or over-crowded living conditions and malnutrition.

Table 2.6 Stratification of heart disease against red wine consumption and diet.

Amount of fish consumed per day	Heart disease rate/wine consumed per day			
	No wine consumed	250 mL	500 mL	1000 mL
(A)				
No fish	High	Low	Moderate	High
100 g	High	Low	Moderate	High
200 g	High	Low	Moderate	High
300 g	High	Low	Moderate	High
400 g	High	Low	Moderate	High
(B)				
No fish	High	High	High	High
100 g	Moderate	Moderate	Moderate	Moderate
200 g	Moderate	Moderate	Moderate	Moderate
300 g	Low	Low	Low	Low
400 g	Very low	Very low	Very low	Very low

(A) Expected type of distribution of rates if moderate wine consumption is the agent.
(B) Expected type of distribution of rates if fish consumption is the agent.

Although only imperfect knowledge is available in the real world, decisions still have to be made on regimens of treatment or the delivery of health care or the development of health promotion. Bradford Hill in 1965 devised nine criteria for making practical decisions on causality. These are guidelines and not absolute criteria.

Temporality
Cause must always precede effect in time. This is logically necessary, but not certainly provable.

Strength
What is the relative risk or odds ratio? The greater it is, the greater the likelihood of a causal association.

Consistency
If different studies, in different populations using different study designs, corroborate the conclusion, then cause becomes more likely. Inconsistency, however, does not eliminate

cause, since some effects may manifest only under certain circumstances.

Specificity
When a particular agent is associated with a disease, it may be held to be causal. However, the multifactorial aetiology of disease renders the finding of such specific agents unlikely since the agent may have more than one effect. Smoking, for example, is associated not only with lung cancer but also with other lung diseases, cardiovascular disease and other pathologies.

Dose–Response Relationship
Increasing exposure to the suspected agent will lead to an increase in disease if the agent is harmful or an increase in health if it is beneficial. Since confounding may produce the same response, cause is again not proven.

Plausibility
Does the postulated cause fit in with generally accepted scientific knowledge? This assumes

that current scientific knowledge contains enough information and is a full description of the possibilities which could exist.

Coherence

The association should not conflict with what is known of the natural history and biology of the condition. In effect, this is an opposite case to plausibility (above) and should be treated with the same caution.

Experiment

Experimentally changing exposures to the agent consistently changes the level of disease. Such correlation cannot logically provide proof of cause, but lack of it may cast doubt on a causal association.

Analogy

A very weak criterion. For example, simply because the use of a particular drug can lead to cancer does not necessarily imply that the use of a similar drug will do the same.

Apart from temporality, which is a logically necessary criterion, in general, these criteria cannot provide irrefutable evidence of cause.

The Bradford Hill criteria were one of the first to evaluate the strength of evidence supporting a causal relationship between two variables. However, in contemporary research, the field of *causal inference* has evolved significantly. While the Bradford Hill Criteria provided valuable qualitative guidelines, modern causal inference frameworks have become more quantitative and data driven. These frameworks, which involve advanced statistical modelling techniques, allow researchers not only to assess causality but also to quantify the strength of any causal effect. This shift towards quantitative approaches has enhanced the precision and rigor of causal assessments in various scientific disciplines, including epidemiology and biostatistics.

Key Points

1) Epidemiology is the study of the distribution of health and disease in populations.
2) The distribution is measured by the calculation of rates. Examples are given for the calculation of incidence, prevalence and mortality rates and how these rates may be indirectly or directly standardised.
3) Descriptive studies elucidate patterns of association between agents affecting health and populations.
4) Descriptive, observational and experimental studies can be used to test hypotheses derived from such data. The degree of risk associated with suspected aetiological agents can then be quantified. Statistically significant correlations between risk and disease may indicate a causal relation, but even after allowance is made for possible bias and confounding effects, a degree of uncertainty may remain in attributing actual cause to a particular agent.
5) Observational and experimental studies allow quantification of risk and may suggest causal relationships between exposure to possible aetiological agents and disease.
6) Strong associations are not invariably causal and may only be correlations due to errors such as bias or not eliminating confounding factors in the study design. Criteria which may in corroboration indicate causes are listed.
7) Measuring rates of disease and health in populations is helpful in rational decision-making on the distribution of health and medical services, as well as suggesting further ideas for investigating the aetiology of a disease.

Further Reading

Bonita R., Beaglehole R. & Kjellstrom T. (2006) *Basic Epidemiology* (2nd edn). World Health Organization.

Bradford-Hill A. (1965) The environment and disease: association or causation? *Proceedings of the Royal Society of Medicine* **58**: 295–300.

Friis R.H. (2017) *Epidemiology 101 (Essential Public Health)* (2nd edn). Jones & Bartlett Learning, Burlington, MA.

Merrill R.M. (2019) *Introduction to Epidemiology* (8th edn). Jones & Bartlett Learning, Burlington, MA.

Westreich D. (2019) *Epidemiology by Design: A Causal Approach to the Health Sciences*. Oxford University Press, New York, NY.

Example: Investigation of a Disease Outbreak; Epidemic Asthma Days in Barcelona

Throughout the 1980s, on a number of isolated days in Barcelona, unprecedented high numbers of hospital admissions for acute asthma occurred (Fig. 2.7). These were dramatic events and each day became known as an 'epidemic day'. The cause for the epidemic days was not clear at first, though there were a number of interesting clinical features about the attacks that gave some clues. Adults, for instance, were far more likely to be affected than children, and some attacks were very rapid in onset, recovery being equally rapid. A high proportion of attacks were so severe that mechanical ventilation in an intensive care unit was required and a higher-than-expected number of deaths occurred.

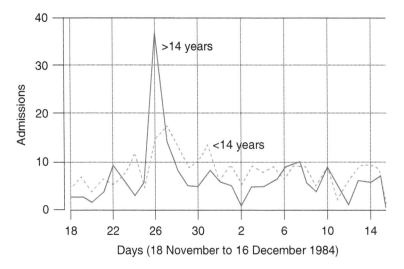

Figure 2.7   Daily number of adult and children asthma emergency room admissions from 18 November to 18 December 1984. From Anto *et al.* (1989).

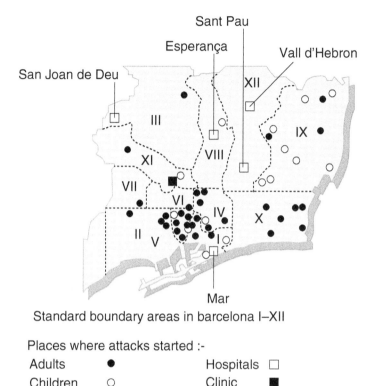

Standard boundary areas in barcelona I–XII

Places where attacks started :-

Adults	●	Hospitals	□
Children	○	Clinic	■

Figure 2.8   Places at which attacks of asthma started. From Anto *et al.* (1989).

Investigations

Investigators considered the possibility of an airborne cause for these epidemic days and were suspicious that emissions from the industrial area to the west of the city might be the cause. The first step in the investigation was a descriptive study mapping the geographical distribution of the patients at the time the attacks occurred (Fig. 2.8). The cases of asthma did not cluster around the industrial area as had been predicted, but around the docks. In the next stage, investigators considered what goods were being unloaded on epidemic and non-epidemic days. Initial analysis pointed to two types of goods, soya bean products and wheat, that were unloaded significantly more frequently on epidemic days compared to non-epidemic days. Analysis using a 2×2 table (Table 2.7) showed that no epidemic days occurred when soya products were *not* being

handled, though there were days when soya *was* being unloaded when epidemics *did not* occur. The relative risk was too high to be quantifiable.

In 1965, Bradford Hill identified nine criteria for an association to be regarded as causal. Thus, these epidemics were *temporally* related to soya being unloaded, the mechanism was *biologically plausible* and an *analogous* situation already existed, as castor bean dust was known to be a potent aeroallergen.

Having determined a strong epidemiological association, the next step was to find evidence of allergic sensitisation to soya in those admitted with acute asthma on epidemic days. Results from both skin-prick testing and the measurement of immunoglobulin E (IgE) levels (employing the allergenic protein in the husk of the bean as the antigen) showed that patients admitted with acute

Table 2.7 Unloading of soya bean on epidemic and non-epidemic days.

| | Soya | | |
	Handled	Not handled	Total
Epidemic days	13	0	13
Non-epidemic days	262	468	730
Totals	275	468	743

Figures are given as numbers of days.

Table 2.8 Effect of intervention on soya bean epidemic asthma (only days on which soya bean was unloaded are included).

	Total days	Days with high no. of asthma admissions	Epidemic days
Before cap installation	167	29	18
After cap installation	133	6	0
P value	<0.001	<0.001	<0.001

asthma on epidemic days were 30 times more likely to be sensitized to soya protein than those admitted on days when there was no epidemic.

Diagnosis and Intervention

On the basis of the epidemiological evidence and the clinical evidence of allergy, the cause of the epidemic days was thought to be soya bean. At the time of the outbreak, soya was being unloaded into two dockside silos, one without, and one with, a filter cap (the filter cap preventing the loss of soya dust into the atmosphere). When dust released from the uncapped silo was effectively eliminated by the addition of a filter cap (in epidemiological terms, an *intervention* to remove an *exposure*), no further epidemic days occurred (Table 2.8), providing ultimate *experimental* proof of a cause-and-effect relationship.

Discussion

These unusual epidemic days demonstrate the causal role of airborne allergens from a specific industrial source in causing asthma, in contrast to the more familiar domestic or general sources of allergen (e.g. cat dander in the home, grass pollen during the summer). The cause was identified by the practice of good epidemiology followed by appropriate immunological confirmation. The outcome of the identification of soya as the cause of the epidemic days was dramatic and epidemic days were eliminated.

Reference

Anto J.M., Sunyer J., Rodriguez-Roisin R., Suarez-Cervera M. & Vazquez L. (1989) Community outbreaks of asthma associated with inhalation of soyabean dust. *New England Journal of Medicine* **320**: 1097–1102.

Part 2

Cell Reproduction, Injury, and Death

3

Cell Reproduction, Injury, and Death

Introduction

Cellular reproduction, senescence and death are important physiological responses that are frequently perturbed in a variety of disease states. This chapter describes the processes of cellular reproduction, senescence and death. Cell death is considered with particular reference to apoptosis and to more recently recognised cell death mechanisms including necroptosis, ferroptosis and pyroptosis.

The Cell Cycle

The process of cellular reproduction is frequently expressed in terms of the *cell cycle* – a sequence of events that ultimately leads to cell division and the creation of two daughter cells from a single parental cell. The cell cycle is divided into four phases (Fig. 3.1). These are as follows:

1) *G1 (first gap)* phase is the period during which the cell is committed to cell cycle progression.
2) *S (synthetic)* phase during which the genetic material is replicated.
3) *G2 (second gap)* phase which separates S phase from M phase. This phase is important because it allows cells to repair errors that may have occurred during DNA replication before the cell divides.
4) *M (mitotic)* phase during which the cell constituents are segregated to each daughter cell.

A fifth phase, known as *G0*, is used to denote cells which are not dividing, a cell state known as *quiescence*.

The Biology of Disease, Third Edition. Edited by Paul G. Murray, Simon J. Dunmore, and Shantha Perera.
Companion website: www.wiley.com/go/murray/biologyofdisease3e

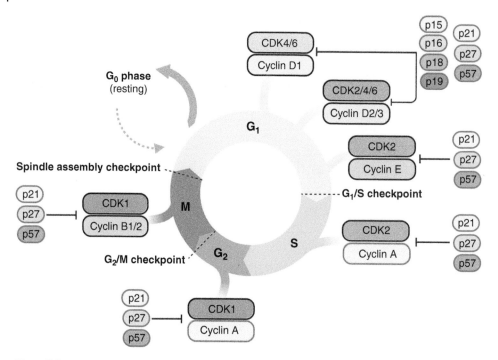

Figure 3.1 The cell cycle. Phases of the cell cycle are as follows: S (synthetic) phase during which the genetic material is replicated; M (mitotic) phase during which the cell constituents are segregated to each daughter cell. These are separated by two gap phases: G1 (first gap) which precedes the S phase and G2 (second gap) which separates S phase from M phase. G0 is used to denote cells, which have left the cell cycle. The concerted action of CDKs and cyclins drives the cell through the cell cycle. Different cyclin–CDK complexes are responsible for entry through different cell cycle phases. There are two main groups of CDK inhibitors: the p21 protein family (comprising p21, p27 and p57) and the INK4 family (p15, p16, p18 and p19). The CDK inhibitors act to suppress CDK-cyclin activation and inhibit progression through the cell cycle. *Source:* Created with BioRender.com.

The Stimulus for Cell Reproduction

Extracellular growth factors usually provide the external signal for a cell to divide. Thus, when growth factors are present in the extracellular environment, only cells expressing the corresponding receptors for the growth factor will respond and divide. This ensures that growth factor signals are only delivered to those cells required to divide. In addition, when the growth factor is no longer present, these cells will stop dividing. Binding of a growth factor to its receptor usually leads to the activation of the receptor. If this is a protein kinase, then activation of the receptor may mean that it is now able to phosphorylate other proteins (or itself), in turn activating them and

thereby transferring the growth signal. This second set of proteins, which is usually located within the cytoplasm, is often referred to as 'second messengers'. In turn, these activated second messengers activate other proteins and the sequence of activation continues until eventually a group of proteins in the nucleus, known as *transcription factors*, are activated. The group of transcription factors stimulated by these pathways then regulate the expression of genes required to execute the cell cycle.

Molecular Control of the Cell Cycle

The cell cycle is controlled by three main groups of regulatory proteins: *cyclin-dependent kinases* (CDKs), *cyclins* and *cyclin-dependent kinase inhibitors* (CDKIs). The CDKs consist of multiple members, including CDK1, CDK2, CDK4 and CDK6, among others, and the cyclin-dependent kinase-like genes (CDKL1–5). Cyclins, which partner with CDKs to regulate the cell cycle, are encoded by several genes and are grouped into various families, such as A, B, D and E. Lastly, the CDKIs act as negative regulators of the cell cycle by inhibiting the activity of cyclin–CDK complexes.

CDKs phosphorylate other proteins and thereby allow cells to progress through the cell cycle. As their name suggests, these kinases are dependent on the cyclins for their activity. Thus, a CDK can only be activated once its partner cyclin is present in sufficient concentration. The CDK is therefore deactivated once levels of its partner cyclin fall below a threshold concentration. Some CDKs are activated by more than one cyclin, so that CDK activity during the cell cycle may be maintained through the interaction of the CDK with different cyclins. The activation of CDK–cyclin complexes is further controlled by their phosphorylation with a complex comprising CDK7, cyclin H and CDK-activating kinase assembly factor, MAT1. This latter complex is often referred to as CDK-activating kinase (CAK).

During the early G1 phase and as a result of the growth factor stimulation described above, expression of the cyclin D genes is stimulated. There are three cyclin D-type proteins (cyclin D1–cyclin D3). Cyclin D1 is able to complex with both CDK4 and CDK6, whereas cyclin D2 and cyclin D3 are able to form a complex with CDK2, CDK4 and CDK6 (Fig. 3.1). The activated CDKs are then able to phosphorylate members of the retinoblastoma family of proteins, including the retinoblastoma (pRb) protein itself.

In the early part of the G1 phase, the pRb protein is present in a hypophosphorylated form and is able to bind to and inactivate members of the E2F family of transcription factors. The E2F transcription factors are responsible for stimulating DNA replication. Therefore, DNA synthesis will only take place once pRb has been fully phosphorylated and E2F

released. Activity of the CDKs 2/4/6-cyclin D complexes during early G1 begins the phosphorylation of pRb.

At the onset of the S phase, the concentrations of the cyclin D proteins decrease, whereas the concentration of cyclin E peaks, having already begun to rise during early G1. Cyclin E now binds to CDK2, effectively taking over the previous role of the cyclin D proteins in the activation of CDK2. The CDK2/cyclin E complex then completes the phosphorylation of pRb and DNA synthesis can now proceed. Thus, cyclin E is responsible for the initiation of S phase.

During the S phase, cyclin E is rapidly degraded and the activation of CDK2 is taken over by cyclin A. The cyclin A–CDK2 complex is necessary for the continuation of DNA replication during S phase. However, towards the end of the S phase, cyclin A preferentially activates CDK1. This marks the end of the S phase and the beginning of G2. Following G2, CDK1 activation by the cyclins B1 and B2 initiates and maintains mitosis. However, CDK1 activation at this stage is subject to several control steps, involving phosphorylation and dephosphorylation events. These control measures ensure that mitosis does not begin until DNA synthesis has occurred.

The CDK Inhibitors

The activation of CDKs by their partner cyclins must be tightly regulated in order to prevent perpetual cell reproduction. This is achieved by the CDKIs, which prevent progress through the cell cycle by blocking the activity of the CDKs. CDKIs usually act at specific stages of the cycle, known as 'checkpoints'. The two major checkpoints controlled by the CDKIs are at the transition from G1 to S phase and at the start of mitosis (G2 to M) (Fig. 3.1). There are two major families of CDKIs.

The p21 Family
The p21 family consist of p21 itself, p27 and p57. Some of these proteins are known by other names and often these alternative designations

are given as a superscript. Thus, p21 is also known as wild-type p53-activated fragment 1 (WAF1) or CDK2-interacting protein 1 (CIP1) and is often written as $p21^{WAF1/CIP1}$. Likewise, p27 and p57 are referred to as kinase inhibiting proteins 1 (KIP1) and 2 (KIP2), respectively. They are often written as $p27^{KIP1}$ and $p57^{KIP2}$. The p21 family are able to inhibit a wide range of CDKs. Following cellular trauma or DNA damage, p21 expression is induced by the p53 protein resulting in cell cycle arrest until the DNA damage can be repaired. Alternatively, if the damage is beyond repair, p53 can induce cell death by increasing the expression of pro-apoptotic proteins.

INK4 Family

In contrast to the p21-type proteins, members of the INK4 (Inhibitor of CDK4) family are more specific in their actions, preferentially inhibiting CDK4 and CDK6 during the G1 phase of the cell cycle. The INK4 family comprises $p15^{INK4A}$, $p16^{INK4B}$, $p18^{INK4C}$ and $p19^{INK4D}$. A related protein encoded from the same gene region as $p19^{INK4D}$ and referred to as $p19^{ARF}$ plays a role in p53-mediated cell cycle arrest.

A third major checkpoint within the cell cycle is the spindle assembly checkpoint (SAC), also known as the mitotic spindle checkpoint. When the SAC senses the lack of attachment at kinetochores, chromosome segregation is inhibited. Even a single unattached kinetochore is sufficient to induce the SAC. A normally functioning SAC is essential to maintain chromosome integrity.

Cell Senescence

Key experiments have demonstrated that fibroblasts will stop dividing after a finite number of cell doublings. The point at which cells no longer divide is known as *senescence*. It has been shown that most normal cells undergo senescence and that structures known as *telomeres* are key to this process.

Telomeres and Telomerase

Telomeres are repeated sequences of DNA located at the ends of chromosomes. Without telomeres, the chromosome ends would be recognised as DNA breaks and would be repaired. Thus, telomeres prevent chromosome fusion that would ultimately lead to massive genome instability.

Each time the cell divides the telomeres shorten. This is because of something called the *end-replication problem*. Essentially, when DNA is replicated, an RNA primer is used. This is a short sequence of RNA, which binds to the telomeres and initiates DNA replication. However, the telomere sequence that binds the RNA primer is not itself replicated. Thus, the telomeres progressively shorten. When the telomeres reach a critical length, the cell stops dividing and goes into senescence.

Some normal cell types are designed not to enter a senescent stage. These include stem cells, which possess an enzyme known as *telomerase* that can maintain telomere length. Some cancer cells express telomerase and this is one of the reasons why they can proliferate indefinitely.

Cell Injury and Adaptation

When the extracellular environment changes, cells can adapt in different ways:

- *Atrophy* is the decrease in the size or number of cells. This can be due to various reasons such as reduced functional demand, reduced blood supply, loss of innervation or inadequate nutrition. An example is the shrinking of unused muscles or the atrophy of the brain in certain neurodegenerative diseases, such as Alzheimer disease.
- *Hypertrophy* is an increase in the size of cells, leading to an enlargement of the organ or tissue and occurs in cells which can no longer divide. An example is the enlargement of muscle cells in the heart or skeletal muscle in response to increased workload.

- *Hyperplasia* is an increase in the number of cells, usually resulting in the enlargement of that organ or tissue. It is an adaptive response of cells that have the capability to divide. For instance, the endometrial lining undergoes hyperplasia during the menstrual cycle in response to hormones.
- *Metaplasia* is a change from one differentiated cell type to another. It often arises in response to chronic irritation or inflammation, allowing the tissue to better withstand the new environment. An example is the change from ciliated columnar epithelium to stratified squamous epithelium in the respiratory tract of a chronic smoker.

Failure to adapt can result in cell injury (Fig. 3.2). Some changes that follow cell injury are reversible if the injurious stimulus is removed. However, if the injury is too severe or prolonged, the cell will undergo irreversible changes, leading to cell death.

Reversible morphological changes include the following:

- Cell swelling, due to an influx of water into the cell, making it appear enlarged. It is an early sign of cell injury and is often seen in hypoxic conditions.
- Mitochondrial swelling and decreased mitochondrial function
- Detachment of ribosomes from the rough endoplasmic reticulum, which can affect protein synthesis.
- Vacuolation, in other words, the appearance of vacuoles in the cytoplasm.

Irreversible morphological changes include the following:

- *Severe mitochondrial damage:* While initial mitochondrial changes can be reversible, severe damage, especially to their membranes, may not be. Severe damage leads to the inability to produce ATP, which is vital for cell survival.

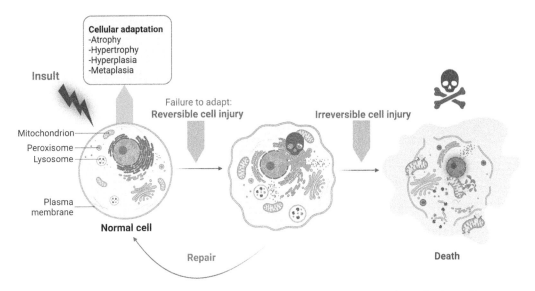

Figure 3.2 Failure of cell adaptation leading to cell injury. In response to a noxious stimulus, cells can adapt in various ways. Atrophy signifies a reduction in cell size or number, hypertrophy is an increase in cell size, hyperplasia is an increase in cell number and metaplasia is a change from one mature cell type to another. Cells that fail to adjust might suffer injury, leading to reversible changes that include cell swelling and mitochondrial dysfunction. Persistent or intense damage causes irreversible changes, including profound mitochondrial damage, nuclear alterations, lysosomal rupture and breaches in plasma membrane integrity, ultimately leading to cell death. *Source:* Created with BioRender.com.

- *Nuclear alterations:* Changes such as pyknosis (nuclear shrinkage with increased basophilia), karyorrhexis (nuclear fragmentation) and karyolysis (nuclear dissolution) indicate irreversible injury.
- *Lysosomal damage:* Lysosomes rupture and release their digestive enzymes into the cytoplasm, leading to autolysis.
- *Plasma membrane breach:* A severe breach in plasma membrane integrity leads to cell lysis.

The biochemical changes that occur within injured cells are well described in *ischaemia*, which is the deficiency of blood flow to a tissue or organ, often because of an obstruction in the blood vessels (Fig. 3.3). This lack of blood flow and, more crucially, the resulting deficiency of oxygen and nutrients, leads to reduced mitochondrial oxidative phosphorylation, which is the primary source of cellular ATP. In response, the cell shifts to anaerobic metabolism,

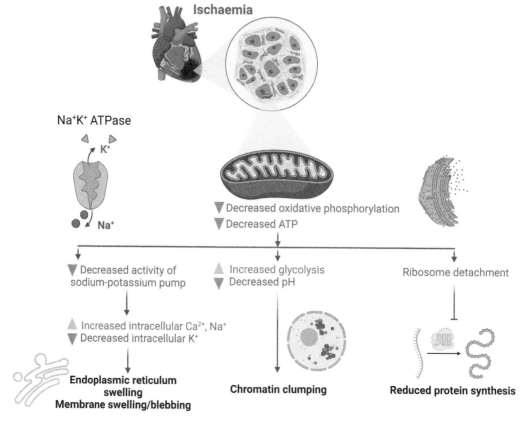

Figure 3.3 Ischaemic events instigate a cascade of cellular perturbations. Ischaemia results in a marked reduction in mitochondrial oxidative phosphorylation, the predominant pathway for adenosine triphosphate (ATP) synthesis. As a compensatory mechanism, cells transition to anaerobic glycolysis to maintain energy production. However, this metabolic shift is suboptimal, yielding a mere 2 ATP molecules per glucose molecule, in stark contrast to the 30–36 ATP molecules generated via oxidative phosphorylation. ATP deficit disrupts ribosomal adherence to the endoplasmic reticulum, impeding protein biosynthesis. Concurrently, the functionality of ATP-dependent ion transporters, notably the sodium-potassium ATPase pump, is compromised. This disruption causes ionic disequilibrium, characterised by an influx of sodium and water into the cell and a concomitant efflux of potassium, culminating in cytotoxic oedema. Moreover, ionic dysregulation elevates intracellular calcium concentrations, serving as a catalyst for the activation of destructive enzymes such as phospholipases, proteases and endonucleases. These enzymatic activities exacerbate the cellular damage. *Source:* Created with BioRender.com.

producing ATP through glycolysis. This process is less efficient and yields only 2 ATP molecules per glucose molecule, compared to the 30–36 ATP molecules produced by oxidative phosphorylation. Anaerobic metabolism results in the production of lactic acid, which decreases the intracellular pH and can disrupt cellular functions. ATP is necessary for many cellular processes, including the translation of proteins; as ATP levels fall, ribosomes detach from the endoplasmic reticulum. Decreased ATP levels also lead to the failure of ATP-dependent ion pumps, especially the sodium-potassium ATPase pump. This results in cellular swelling due to sodium and water influx and potassium efflux. This also increases intracellular calcium levels, which can activate enzymes such as phospholipases, proteases and endonucleases, which further damage the cell.

When oxygen becomes available again (as in reperfusion after ischaemia, known as *reperfusion injury*), the reintroduction of oxygen can lead to a burst of oxidative phosphorylation. This sudden increase can result in the formation of reactive oxygen species (ROS; Fig. 3.4). ROS are chemically reactive molecules containing oxygen, formed as natural by-products of the normal metabolism of oxygen. While they play crucial roles in cell signalling and homeostasis, an overproduction or inadequate removal of ROS can cause oxidative stress, leading to cellular damage. This oxidative damage is implicated in various diseases, including neurodegenerative disorders, cardiovascular diseases and cancers. ROS can be generated through a myriad of mechanisms, both within cells and externally. For example, enzymes like NADPH oxidases, which are associated with the plasma membrane, transport electrons across the membrane to react with oxygen, producing either superoxide or hydrogen peroxide. The cytochrome P450 enzymes, found in the endoplasmic reticulum, also contribute to ROS production during their metabolic activities. Peroxidases are another source of ROS. Exposure to radiation, especially ionising forms like X-rays and UV rays,

can induce ROS generation in exposed tissues. Environmental factors, including toxins and pollutants such as tobacco smoke, certain pesticides and some industrial solvents, also promote ROS formation.

Cell Death

Two major forms of cell death are recognised, namely, *apoptosis* and *necrosis*. Generally, necrosis is the type of cell death that occurs after irreversible cell injury that follows a noxious stimulus (described above). Apoptosis refers to *programmed* or *physiological* cell death. However, while the terms apoptosis and programmed cell death are often used synonymously, cell death can be programmed without undergoing the morphological characteristics of apoptosis. In addition, there is evidence that programmed cell death can lead to necrosis in some situations.

Necrosis

Necrosis is usually the result of severe injury to cells and can occur in response to a wide range of stimuli. Necrosis is characterised by membrane disruption, dissolution of organelles, including the nucleus, and lysosomal degeneration followed by activation of an inflammatory response (Fig. 3.5). Necrosis involving many cells can lead to the loss of tissue architecture.

Necrosis can occur in different morphological forms depending on the context. For example, *coagulative necrosis* is observed when tissue architecture is relatively well preserved and remains firm in texture. This type of necrosis is frequently seen in infarcted solid organs outside the central nervous system (CNS). In contrast, in *liquefactive necrosis*, there is dissolution of tissue leading to the production of viscous fluid. This form of necrosis is seen in the CNS and also in infections. *Caseous necrosis* is when the necrotic tissue has a 'cheese-like' appearance and is typically seen in tuberculosis

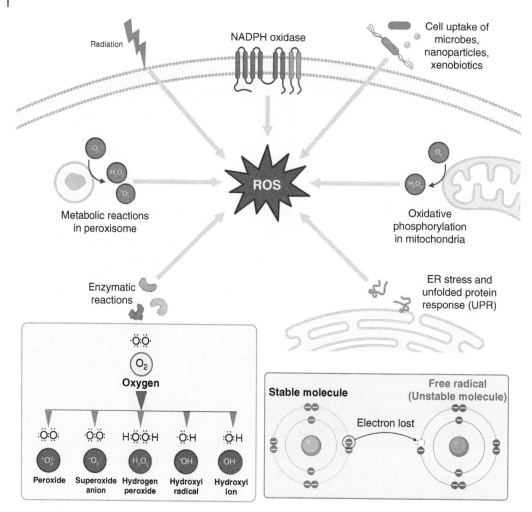

Figure 3.4 Reactive oxygen species. Reactive oxygen species (ROS) encompass a group of molecules formed by the partial reduction of oxygen. These molecules play a significant role in cellular functions, both as signalling molecules and as destructive agents. Examples of ROS are shown in the lower blue box. Many ROS are free radicals-atoms or molecules which possess unpaired electrons in their outer orbital (green box). However, not all ROS are free radicals, for example hydrogen peroxide (H_2O_2) is a ROS, but not a free radical. One of the primary sites for ROS generation is the mitochondria during oxidative phosphorylation. In this process, electrons are transferred through a series of protein complexes in the mitochondrial inner membrane. Premature transfer of these electrons to oxygen can cause formation of superoxide anions (O_2^-). These can further be converted to other ROS like H_2O_2 and hydroxyl radicals (•OH). Peroxisomes are organelles involved in lipid metabolism and the breakdown of long-chain fatty acids. During these metabolic processes, H_2O_2 is produced. Although peroxisomes contain catalase, an enzyme that breaks down H_2O_2, excessive production can overwhelm this system, leading to ROS accumulation. Another significant source of ROS is the enzyme NADPH oxidase, especially in immune cells. This enzyme is activated when cells, such as neutrophils, engulf microbes. The produced ROS aids in the destruction of the internalised pathogens. Various enzymatic reactions in the cell can also inadvertently produce ROS as by-products. For instance, the cytochrome P450 enzymes, involved in drug metabolism, can generate ROS. The endoplasmic reticulum (ER) is involved in protein folding and modification. When there is an accumulation of misfolded proteins, a condition termed ER stress occurs. This stress can increase ROS production, further exacerbating the cellular oxidative burden. *Source:* Created with BioRender.com.

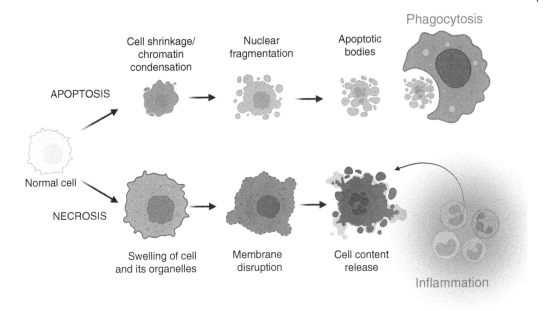

Figure 3.5 Differences between apoptosis and necrosis. Apoptosis and necrosis are two distinct forms of cell death. Apoptosis is a highly regulated and controlled form of cell death that is essential for maintaining tissue homeostasis and eliminating damaged or infected cells. Various internal and external stimuli, such as DNA damage, growth factor withdrawal or death receptor activation, initiate apoptotic signalling. The cell shrinks and nuclear material condenses and fragments, forming dense 'apoptotic bodies'. Phagocytic cells recognise and engulf apoptotic cells, which are then degraded and recycled within the phagocyte. This process prevents the release of potentially harmful cellular contents into the surrounding tissue and minimises inflammation. In contrast, necrosis is usually an uncontrolled form of cell death, typically resulting from severe cellular injury or stress. Initially, the cell membrane loses its integrity and becomes permeable, causing an influx of water and ions. This leads to cellular swelling, organelle dysfunction and mitochondrial damage. As the cell swells, the plasma membrane eventually ruptures, releasing cellular contents into the extracellular space which activate the inflammatory response. *Source:* Created with BioRender.com.

(Fig. 3.6). *Fat necrosis* can be seen in acute pancreatitis. In this form of necrosis, pancreatic lipases are released from damaged cells and generate fatty acids which complex with calcium to form soaps. These soaps appear as white 'chalky' deposits. *Fibrinoid (fibrin-like) necrosis* is seen, for example in immune-mediated vasculitis, and is due to the deposition of antigen–antibody complexes together with fibrin.

Apoptosis

In the early 1970s, it was proposed that cells died 'normally' by a form of cell 'suicide', although it took another 20 years and studies involving the nematode worm *Caenorhabditis elegans* to recognise the fundamental importance of apoptosis. Apoptosis occurs throughout life and plays a key role in tissue homeostasis. Apoptosis is also crucially important during development. For example, apoptosis is responsible for the loss of tadpole tails and for the removal of the interdigital webs in humans.

In contrast to necrosis, apoptosis does not normally induce an inflammatory response, thus avoiding further tissue damage. The morphological changes that accompany apoptosis are quite different from those seen in necrosis (Fig. 3.5). First, there is a loss of cell–cell adhesion and the cells become smaller and rounded, sometimes with lobulations, although some organelles remain intact. Cell shrinkage results from the loss of water and sodium ions. Within the nucleus, the chromatin condenses to form

(A)

(B)

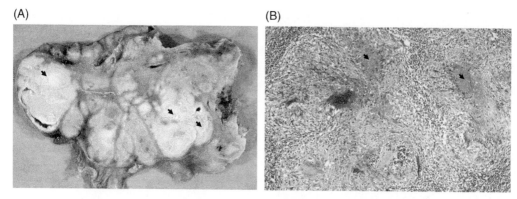

Figure 3.6 Example of caseous necrosis. (A) case of tuberculous lymphadenitis with a group of mesenteric lymph nodes massively expanded with caseous necrosis having a typical soft cheese-like appearance (arrows). The patient was HIV-positive. *Source:* Wellcome Collection / https://wellcomecollection.org/works/hesh7tz3/ items / last accessed November 13, 2023. (B) Haematoxylin and eosin (H&E) stained section showing necrosis seen as amorphous pink central areas (black arrows) surrounded by granulomatous tissue, including a Langhans giant cell (blue arrow). *Source:* Wellcome Collection / https://wellcomecollection.org/works/njunugy5 / last accessed November 13, 2023.

'half-moon' shaped structures and the nucleoli appear disorganised (Fig. 3.7). The nucleus itself may break up into multiple fragments. DNA fragmentation (into 180–200 base-pair pieces) caused by nuclear damage leads to the characteristic formation of apoptotic ladders when the fragments are resolved upon agarose gels and visualised under UV light. In the

tissue, the cell fragments appear as the so-called apoptotic bodies that are engulfed by phagocytes before the cell contents spill out. Apoptotic bodies may be seen within the cytoplasm of the consuming cells for up to 9 hours following phagocytosis. The rapid clearance and digestion of dead cells, even after large numbers of cells have died, may explain why apoptosis was overlooked for so long.

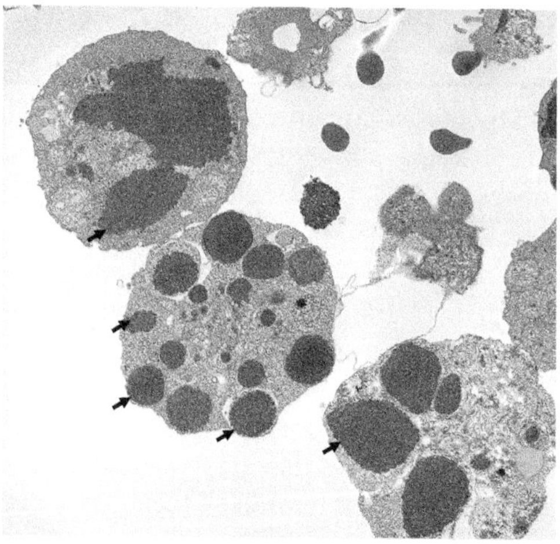

Figure 3.7 Electron micrograph of apoptotic cells. 'Transmission electron micrograph of cells undergoing apoptosis, showing typical apoptotic bodies (several arrowed)'. *Source:* Wellcome Collection / https://wellcomecollection.org/works/zbmz56rm / last accessed November 13, 2023.

Caspases

Studies involving *C. elegans* identified proteins encoded by three genes, two of which were required for apoptosis, namely *ced-3* and *ced-4*, and the third, *ced-9*, which prevented apoptosis. *Ced-3* encodes a protein, which is similar to the human interleukin-1β-converting enzyme (ICE), a proteolytic enzyme that cleaves the inactive precursor of IL-1β to an active cytokine. Many proteolytic enzymes were subsequently shown to be involved in apoptosis. Collectively, these proteolytic enzymes are known as *caspases* since they all possess cysteine in their active sites and cleave target proteins at specific aspartate residues. Later, ICE was re-designated caspase-1.

Caspases are produced as precursor enzymes, known as *proenzymes*, *zymogens* or

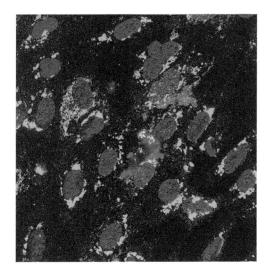

Figure 3.8 Caspase activity in a cell triggered to undergo apoptosis. A confocal micrograph showing caspase activity in a cell treated with 7-ketocholesterol to induce apoptosis. On receipt of signals that trigger apoptosis, there is a cascade of caspase activity in the cell, leading to the events that make up apoptosis. Nuclei are stained blue, cytochrome *c* green, and cleaved caspase-3 red. *Source:* Wellcome Collection / https://wellcomecollection.org/works/qfcq4pfu / last accessed November 13, 2023.

procaspases, and share amino acid and structural homology. They aggregate to form heterodimers and tetramers and can be activated autocatalytically (i.e. self-activated) or by other enzymes (usually other caspases) (Fig. 3.8). Caspases cause cell death by a number of mechanisms. They:

- Inactivate proteins, for example bcl-2, which protect cells from apoptosis.
- Directly disassemble cell structures, e.g. nuclear lamina.
- Reorganise and cleave the cytoskeleton.
- Prevent DNA replication and repair by cleaving many proteins important in cell maintenance and repair, such as poly (ADP ribose) polymerase (PARP) and DNA-dependent protein kinase (DNA-PK). At the same time, they activate a DNA fragmentation factor (DFF) which digests DNA.
- Induce signals which mark the cell for phagocytosis.

- Cut off contact with neighbouring cells.
- Disintegrate the cell into apoptotic bodies.

Owing to their ability to induce such changes, caspase activation must be strictly controlled. This is achieved in a variety of ways which include limiting the availability of substrate, formation of the caspases as inactive procaspases, the presence/absence of caspase inhibitors/cofactors, positive and negative feedback mechanisms, compartmentalisation of caspases and their cofactors and highly stringent substrate specificity.

Caspase activation may be subdivided into two classes: *initiator caspase activation* and *effector caspase activation*. A pro-apoptotic signal activates an initiator caspase that in turn activates an effector caspase, ultimately causing cell death. Different initiator caspases respond to different apoptotic signals, e.g. caspase 8 responds to the activation of death receptors (see later) and caspase 9 to the presence of cytotoxic agents. Initiator caspase activation relies on the binding of specific co-factors to the procaspase. For example, caspase 8 activation requires FADD (see later) whilst caspase 9 activation requires apoptotic protease activating factor-1 (APAF-1), cytochrome *c* and ATP (see later). An *apoptosome* is a large, multi-protein complex that forms during the process of apoptosis. The apoptosome plays a crucial role in the activation of the intrinsic (mitochondrial) apoptotic pathway. It is composed primarily of cytochrome *c* and APAF-1.

Control of Apoptosis

Apoptosis must be carefully controlled to ensure that cell death occurs only when and where necessary. Apoptosis is therefore regulated in a complex manner involving both extracellular and intracellular mechanisms.

Extracellular Controls

Death receptors provide the major mechanism for the extracellular control of apoptosis (Fig. 3.9). They detect extracellular death signals and activate the cell's apoptosis

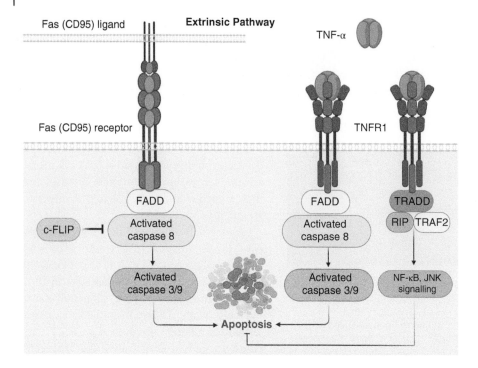

Figure 3.9 Extrinsic apoptosis. Left panel: Binding of the membrane-bound trimeric CD95L (FasL) (for example on cytotoxic T-cells) causes trimerisation of CD95 receptor (Fas) molecules and clustering of their cytoplasmic DDs on the target cell. FADD (Fas-associated death domain) binds and recruits/activates caspase 8 (initiator caspase). Caspase 8, also known as FADD-like IL-1β-converting enzyme (FLICE), then activates downstream effector caspases, for example, caspases 3 or 9, and the cell undergoes apoptotic death. The FLICE-inhibitory proteins (FLIPs) negatively regulate caspase activation. Right panel: TNFα binding to TNFR1 causes trimerisation of the receptors and aggregation of the cytoplasmic DDs. These can either recruit FADD which leads to apoptosis via the mechanism outlined for CD95 or recruit TRADD (tumour necrosis factor receptor-associated death domain). TRADD can recruit other signalling molecules, for example, TRAFs (tumour necrosis factor receptor-associated factors) or RIP (receptor-interacting protein), which activate pro-survival pathways (e.g. NF-κB and JNK). *Source:* Created with BioRender.com.

machinery. Death receptors belong to the tumour necrosis factor receptor (TNFR) super-family and contain a cytosolic death domain (DD) which engages the intracellular apoptotic machinery. Examples of death receptors are CD95 (also known as Fas or Apo-1), TNFR1 (also known as p55 or CD120a), death receptor 3 (DR3, Apo-3, TRAMP or LARD), DR4 and DR5 (Apo-2, TRAIL-R2, TRICK2 or KILLER). CD95 and TNFR1 are discussed below.

CD95 mediates cell death in a number of situations including;

- The peripheral deletion of mature T-cells following an immune response.

- Killing of virus-infected or cancer cells by cytotoxic T-lymphocytes (CTLs) or natural killer (NK) cells.
- Killing of inflammatory cells.

The ligand for CD95 (CD95L) exists as a trimer and binds to three CD95 receptor molecules resulting in clustering of the cytoplasmic DDs (Fig. 3.9). The FADD (Fas-associated death domain) adapter protein then binds via its own DD to the clustered DDs of the CD95 receptor molecules. FADD possesses a death effector domain (DED) which is able to recruit the pre-cursor form of caspase 8 (also known as FLICE or MACH). The resulting oligomerisation of caspase 8 leads to its activation by self-cleavage. Caspase 8 then activates downstream caspases such as caspase 9 resulting in apoptotic cell death. FLICE (caspase 8) inhibitory proteins (FLIPs) can inhibit caspase 8 activation. Human cells encode c-FLIP. The Kaposi sarcoma-associated herpesvirus (KSHV) encodes a viral FLIP (v-FLIP). Other cytoplasmic proteins, e.g. Daxx, can also bind to CD95. Daxx can induce the stress-activated c-JunNH2-terminal kinase (JNK) pathway that can lead to the potentiation of CD95-mediated apoptosis.

TNF-α is a cytokine produced by activated macrophages and T-cells in response to infection. It has many different effects that are dependent on the target cell type. In some cells, TNFα induces apoptosis. Thus, as with CD95, TNFα binding to TNFR1 results in trimerisation of the receptors and the aggregation of cytoplasmic DDs. FADD can bind in a manner analogous to CD95, leading to apoptosis. However, the DDs may also bind an adapter protein known as TRADD (tumour necrosis factor receptor-associated death domain), which recruits other signalling molecules, including the TRAFs (tumour necrosis factor receptor-associated factors) and RIP (receptor-interacting protein). This initiates very different signalling pathways leading to the activation of NF-κB and JNK/AP-1 and, in turn, the suppression of apoptosis. Thus, a single receptor may either induce or suppress apoptosis in different situations (Fig. 3.9).

Intracellular Controls

Intracellular regulators, and in particular the Bcl-2 family of proteins, play a major role in the control of apoptosis (Fig. 3.10). The Bcl-2 family has been conserved throughout evolution. All members possess at least one of four conserved Bcl-2 homology domains (BH1–BH4) and can be split into those which suppress apoptosis and those which promote apoptosis (Table 3.1). Pro-survival members all contain at least BH1 and BH2 domains, while pro-apoptotic members must contain the BH3 domain. Pro-apoptosis and anti-apoptosis

Bcl-2 family members can heterodimerise and block each other's function. The ratio of suppressors to promoters therefore determines the apoptotic susceptibility of the cell. Some pro-survival proteins bind to caspase activators and inhibit their function, e.g. Bcl-xL binds to APAF-1 preventing the activation of caspase 9. Pro-apoptotic proteins, e.g. Bik, may free APAF-1 from death inhibitors. Many Bcl-2 family members have a hydrophobic tail and can bind to or insert into the membranes of mitochondria, endoplasmic reticulum and the nucleus. Some members of the Bcl-2 family can form pores in membranes, in a similar way to that caused by bacterial toxins, and thus alter membrane permeability, membrane potential or even puncture the membrane completely. Bcl-2 itself prevents mitochondrial cytochrome c release and subsequent activation of the apoptosome and caspase 9 (Fig. 3.10). On the other hand, Bax promotes apoptosis by binding to the mitochondria and causing the release of cytochrome c. Additionally, Bax is induced by p53, providing a mechanism for the induction of apoptosis following cellular stress or DNA damage.

The Bcl-2 family are themselves regulated by cytokines and other death survival signals. The levels of particular proteins may be regulated by increased or decreased gene expression. Alternatively, regulation may occur via protein modification, for example, phosphorylation or protease cleavage. Bcl-2 can be activated by phosphorylation while the phosphorylation of Bad inhibits its ability to prevent the pro-survival role of Bcl-xL.

Stimulation of the RAS pathway leads to the activation of phosphaditylinositol-3 kinase which in turn activates Akt. Akt is responsible for the phosphorylation of Bad. Bcl-2 cleavage by caspases inactivates the inhibitory function of Bcl-2, while Bid cleavage by caspases leads to the release of cytochrome c from mitochondria. Both of these lead to increased apoptosis. Another family of intracellular apoptosis regulators, the inhibitor of apoptosis proteins (IAPs), directly inhibit caspases (primarily

Intrinsic Pathway

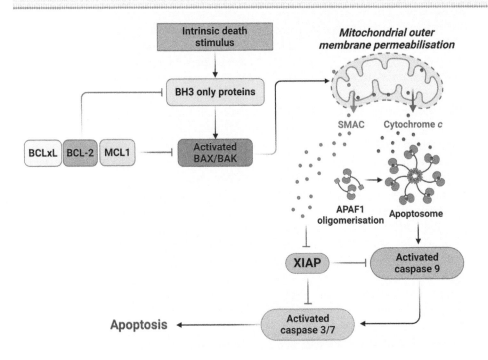

Figure 3.10 Intrinsic apoptosis. BH3-only proteins sense cellular stress and play a pivotal role in initiating the intrinsic apoptosis signalling cascade. They activate the pro-apoptotic effectors, BAX and BAK, leading to the permeabilisation of the mitochondrial outer membrane. This event facilitates the release of apoptotic factors such as cytochrome *c* and the second mitochondria-derived activator of caspases (SMAC) protein. The release of cytochrome *c* into the cytosol promotes the assembly of the apoptosome, which subsequently activates caspase-9. This cascade continues with the activation of effector caspases, allowing apoptosis to proceed. Simultaneously, SMAC supports this process by binding to inhibitor of apoptosis proteins (IAPs), such as XIAP, neutralising their inhibitory activity on caspases. Anti-apoptotic proteins like BCL2, BCL-xL and MCL1 are members of the BCL2 family of proteins and have critical roles in the regulation of the intrinsic apoptosis pathway. Their primary effect is to counteract the actions of the pro-apoptotic members of the BCL2 family, such as BAX, BAK and the BH3-only proteins (e.g. BID, BAD and BIM, not shown). *Source:* Created with BioRender.com.

caspases 3 and 7) and are also able to induce NF-κB activation, ultimately leading to cell survival rather than apoptosis.

Other Forms of Cell Death

Necroptosis is a type of cell death in which the morphological changes resemble necrosis. Necroptosis is regulated by caspase-independent signalling pathways. A key mediator is TNF-α which after binding to TNFR1 can activate receptor-interacting protein kinase 3 (RIPK3). Upon activation, RIPK3 forms a complex with another kinase, RIPK1 (receptor-interacting

protein kinase 1). This complex, known as the *necrosome*, leads to the phosphorylation and activation of a downstream protein called MLKL (mixed lineage kinase domain-like protein) which causes disruption of the cell membrane and cell death. Necroptosis is a mediator of cell death after particular forms of cell injury, e.g. ischaemia reperfusion injury, and is also important in overcoming resistance to some pathogens such as Cytomegalovirus, which can encode caspase inhibitors.

Ferroptosis is, as the name suggests, a form of cell death regulated by iron (and blocked by iron chelators) and is characterised by

Table 3.1 Members of the Bcl-2 protein family.

Subfamily	Activity	BH domain status	Members
Anti-apoptotic	Anti-apoptotic	BH4 domain	BCL-2
			BCL-X$_L$
			BCL-W
			BCL-B (BCL2L10)
			MCL-1L
		No BH4 domain	MCL-1
			BFL-1/A1
			BCL2L12
Pore-forming executioners	Pro-apoptotic	Multi-domain	BAX
			BAK
			BOK
BH3-only	Pro-apoptotic	Activator binds to pro-apoptotic and anti-apoptotic Bcl-2 multi-region proteins	BIM
			BID
			Puma
			Mule
		Sensitizer displaces activator BH3-only proteins from anti-apoptotic proteins	BAD
			Noxa
			BIK./BLK
			BMF
			HRK/DP5
			Beclin-1
	Potentially pro-apoptotic		BCL-Rambo (BCL2L13)
			BCL-G (BCL2L14)
			MCL-1S
			MCL-1ES

Source: Chloe et al. (2019) / Springer Nature / CC by 4.0.

increased lipid peroxidation resulting from lack of activity of the lipid repair enzyme, glutathione peroxidase-4 (GPX4), leading to increased levels of lipid-based ROS, e.g. lipid hydroxyperoxides.

Pyroptosis is a pro-inflammatory form of cell death mediated by inflammatory caspases and mainly manifested by plasma membrane rupture and the release of pro-inflammatory intracellular contents including the *alarmins* IL-1β and IL-18. Pyroptosis is therefore a potent weapon in the host armoury against bacteria, viruses, protozoans and fungi. A key mediator is Gasdermin D which is cleaved and activated by caspases and triggers the pyroptosis process.

The N-terminal cleavage product of Gasdermin has pore-forming activity, which causes cell swelling and death.

Autophagy

Autophagy refers to the 'eating of self', literally meaning that cells consume themselves. It was originally described as a process designed to allow cell survival under conditions of low nutrient supply, but it is now also clear that autophagy can result in cell death if the stress becomes too much for the cell to manage. In autophagy, there is formation of *autophagosomes* into which cell organelles are sequestered. The autophagosomes fuse with lysosomes resulting in the formation of *autophagolysosomes*, in which organelles are dissolved and their constituents recycled. Aberrant autophagy is implicated in many disorders, for example, cancer and neurodegenerative diseases, such as Alzheimer disease.

Summary

This chapter has described the important molecular events in cell reproduction, senescence and death. Perturbations of these key cellular processes frequently lead to the development of disease, examples of which are described in the later chapters.

Key Points

1) The cell cycle is a sequence of events that leads to cell reproduction. It is controlled by the CDKs and their partner cyclins, which allow progress through the cell cycle by phosphorylating proteins such as pRb. The phosphorylation of pRb releases the E2F transcription factors, which in turn allows DNA replication to proceed.
2) The CDK inhibitors comprise two families – the p21 family and the INK4 family. The CDK inhibitors act at specific parts of the cell cycle, known as 'checkpoints', to arrest progress through the cycle.
3) Cellular senescence occurs in most normal cells, usually after a finite number of cell divisions. Shortening of telomeres is responsible for cell senescence. Telomerase stabilises telomere length.
4) There are two major forms of cell death: necrosis and apoptosis. Necrosis is initiated following noxious injury, whereas apoptosis may be physiological or pathological.
5) Apoptosis is controlled by a complex series of biochemical events involving both positive and negative regulators. The stimulus for apoptosis may be via extracellular signals, such as those delivered by CD95L or by intracellular signals.
6) Other important forms of cell death include necroptosis, ferroptosis and pyroptosis.

Further Reading

Alberts B., Hopkin K., Johnson A.D. *et al.* (2021) *Essential Cell Biology* (5th edn). W.W. Norton & Company.

Chloe F.A. *et al.* (2019) BCL-2 family isoforms in apoptosis and cancer. *Cell Death and Disease* 10: 177. https://doi.org/10.1038/s41419-019-1407-6.

Karp G. (2021) *Cell and Molecular Biology: Concepts and Experiments* (9th edn). Wiley.

Yu P., Zhang X., Liu N., *et al.* (2021) Pyroptosis: mechanisms and diseases. *Signal Transduction and Targeted Therapy* 6: Article number 128.

Zhan C., Huang M., Yang X. & Hou J. (2021) MLKL: functions beyond serving as the executioner of necroptosis. *Theranostics* 11: 4759–4769.

Part 3

Infectious Diseases

4

Infectious Diseases

Introduction

Despite significant advances in the treatment and control of infectious diseases, they remain a major cause of morbidity and mortality in the modern world. The COVID-19 pandemic and the re-emergence of serious diseases such as tuberculosis, caused by drug-resistant forms of mycobacteria, have highlighted the need for an extensive research effort to understand the

The Biology of Disease, Third Edition. Edited by Paul G. Murray, Simon J. Dunmore, and Shantha Perera.
© 2024 John Wiley & Sons Ltd. Published 2024 by John Wiley & Sons Ltd.
Companion website: www.wiley.com/go/murray/biologyofdisease3e

nature of these important diseases and improve strategies for their control. This chapter considers the delicate nature of the host–parasite relationship in infectious diseases. Definitions and basic concepts are considered and features of the parasite and the host important for the establishment of disease are outlined. We also consider the major groups of pathogens separately in detail, including bacteria, viruses, fungi and parasites.

Throughout this chapter and elsewhere in this book, we use the abbreviation, 'spp'. to refer to multiple species within a genus. When referring to a single species, 'sp'. can be used as the abbreviation.

The Relationship Between Humans and Their Microbes

Humans are hosts to a variety of microorganisms and animals, not all of which cause disease. All associations in which one species lives in or on another are grouped under the general heading of *symbiosis* which comes from the Greek words 'sym' (meaning 'together') and 'biosis' (meaning 'living'). Symbiosis encompasses a wide range of relationships between organisms, including *mutualism* (both species benefit), *commensalism* (one species benefits and the other is not affected) and *parasitism* (one species benefits at the expense of the other). Thus, a *parasite* is defined as an organism which lives on or in another organism (the *host*) obtains nutrients directly from it, provides no benefit to the host and is often harmful. Some parasites, termed *obligate parasites,* are entirely dependent on the host for their reproduction and are therefore incapable of independent existence. Viruses are examples of obligate organisms. Organisms that are capable of existing outside the host are called *facultative parasites*. Most bacterial parasites belong to this group. *Intracellular parasites* (e.g. *Mycobacterium tuberculosis*, viruses), are adapted for life within cells, whereas those that exist outside cells are *extracellular* parasites (e.g. *Streptococcus mutans*). Organisms are aerobic if they have a strict requirement for oxygen (e.g. *Pseudomonas aeruginosa*) or anaerobic when oxygen is not required and for some organisms may be toxic (e.g. *Clostridium tetani*). *Facultative anaerobes* are microorganisms that can survive whether oxygen is present or not (e.g. *Escherichia coli*).

Humans support a wide range of microorganisms, mostly bacteria living on the skin, in the mouth and in the gastrointestinal tract. Most are harmless and are referred to as *commensals* or 'normal flora'. Commensals may also provide some benefit to the host, for example, the presence of gut commensals can prevent colonisation from other, potentially harmful, organisms. However, if the host becomes immunosuppressed, or if commensals colonise inappropriate sites, then disease may result. The *microbiome* is the general term given to a wide range of organisms that live in or on humans.

An organism living in or on a host can cause an infection but when the infection leads to a disease state, the organism is referred to as a *pathogen. Pathogenicity*, therefore, is defined as the ability to cause disease. While infectious diseases are responsible for significant morbidity and mortality, in biological terms, it may not be in the best interest of the parasite to kill the host. Survival of the host enables the maintenance of a reservoir of potentially infectious organisms, which may be transmitted to other susceptible hosts. The adaptation of an organism and host progresses through several stages. A pathogenic organism entering a new population previously not exposed to it often causes acute disease in people of all ages – this represents the classic picture of an *epidemic*. People who have not yet been exposed to an organism are described as *naïve*. As well as being caused by the introduction of a new pathogen to a population, epidemics can also be caused by other factors, for example, a change in the virulence or transmission of a known pathogen or a breakdown in public health measures that would otherwise prevent the spread of disease. The term epidemic is often used interchangeably with *outbreak*, which refers to the occurrence of a larger number of cases of a particular disease than would normally be expected in a specific area over a short period of time. After a new pathogen is introduced to a population, over generations, the population develops some resistance to the organism and the disease will become *endemic* – this is characterised by a widespread, lower-grade disease or routine childhood illness. With further adaptation by the host and parasite, a symbiotic relationship may develop and the host may not be harmed. Syphilis is a good example of the effect of a new disease on a population. When the causative organism of syphilis, *Treponema pallidum*, first appeared in Europe in the fifteenth century, it caused a very severe disease. Half a century later, symptoms were generally limited to the genitals, face and nervous system and death was less frequent.

Koch's Postulates

The 19th century microbiologist, Robert Koch, devised a set of postulates that must be satisfied before an organism can be designated as the aetiological agent for a given disease. These are as follows:

1) The pathogen must always be present when there is disease and absent when there is no disease.
2) The pathogen must be isolated from an infected host and grown in pure culture.
3) Disease must result from reintroduction of the pathogen into a healthy host.
4) Subsequently, the same pathogen must be isolated from the second host.

Meeting all of Koch's postulates can be challenging. For instance, sometimes the only evidence of an infection is the presence of specific antibodies to a particular microbe. Moreover, for pathogens that exclusively infect humans, fulfilling the third and fourth postulates can be especially difficult.

Transmission of Infectious Diseases

To cause infectious disease, pathogens must be transmitted to the host. Transmission can take place by different routes that include

- In air droplets or droplet nuclei (e.g. *M. tuberculosis,* SARS-CoV-2).
- By ingestion of contaminated water or food (e.g. *Vibrio cholera, Salmonella typhi*).
- By person–person contact, for instance, through sexual contact (e.g. *T. pallidum,* Human immunodeficiency virus, HIV)
- Via *fomites* (that is inanimate objects or materials through which pathogens may be conveyed) such as dust particles (e.g. *M. tuberculosis*).
- By inoculation through the skin by animal (e.g. Rabies virus) or insect bites (e.g. *Plasmodium falciparum*). Where, and in which groups, infection occurs are partly determined by environmental conditions (housing, water supply and sanitation) and climate (rainfall and temperature).

Factors Contributing to the Establishment of an Infectious Disease

Features of the Pathogen

Virulence is the degree of pathogenicity and refers to the ability of the pathogen to invade and multiply within the host and cause damage. Pathogenicity is also determined by the extent to which the pathogen can evade host immunity. For example, African trypanosomes can rapidly change their surface antigens to evade attack by specific antibodies and T-cells. Other pathogens have devised mechanisms that allow intracellular survival, for example *M. tuberculosis* is adapted for survival within phagocytes. The number of infectious organisms to which an individual is exposed is also important in determining whether disease will result. In general, the larger the infective dose, the greater the probability of disease.

The *minimum infective dose* refers to the minimum number of organisms required for infection and varies for different pathogens. In this context, the ID50 (*infective dose 50*) and the LD50 (*lethal dose 50*) are important laboratory measures of the virulence of a pathogen. They refer to the number of organisms required to cause an infection in (ID50), or kill (LD50), 50% of animals within a defined time period.

Features of the Host

Host factors, such as age, nutritional status and genetic constitution, are important in determining whether infection will occur following exposure to a potential pathogen. The very young and the very old are more susceptible to infection – a reflection of developing and gradually failing immune systems, respectively. Malnutrition also increases disease susceptibility. Some populations are more susceptible to certain types of infection because of genetic differences. The status of the host's immune system is also crucial; alterations to the natural barriers to infection may enable entry of potential pathogens. For example, breaks in the integrity of the skin or decreased levels of mucosal antimicrobial proteins, such as immunoglobulin A (IgA, see also Chapter 6), can predispose individuals to infection.

Emerging Resistance to Anti-microbials

Antimicrobial resistance usually occurs when microorganisms undergo genetic changes that render them less responsive to the effects of antimicrobial drugs. As a result, infections caused by these resistant microorganisms become more challenging to treat, leading to increased transmission, more severe illness and higher mortality rates. Anti-microbial resistance is a significant global concern as it threatens our ability to effectively manage common infections and potentially undermines many of the successes achieved so far.

One of the alarming trends is the emergence and spread of multi- and pan-resistant bacteria, often referred to as 'superbugs'. These bacteria have acquired new resistance mechanisms, making them resistant to multiple antimicrobial medicines, including antibiotics (strictly defined as microbial products that inhibit the growth of or destroy other microorganisms. However, in common usage, antibiotics often refer to a broader range of antimicrobial substances). The rapid global spread of these superbugs poses a grave threat as it limits the treatment options available, leaving patients vulnerable to untreatable infections. Several factors have contributed to the acceleration of antimicrobial resistance, for example, the misuse, and overuse, of antimicrobials, in both human healthcare and animal agriculture.

The prevalence of drug-resistant infections is evident across different infectious agents. For example, high rates of resistance have been observed in infectious diseases caused by *E. coli* and *Klebsiella pneumoniae*. Last resort treatments for infections caused by carbapenem-resistant bacteria, including *K. pneumoniae* and *Enterobacteriaceae*, are increasingly ineffective. The rise of drug-resistant strains of *Staphylococcus aureus*, including methicillin-resistant *S. aureus* (MRSA), has led to increased mortality rates compared to drug-sensitive infections. Resistance in tuberculosis and malaria, as well as in infections caused by HIV and other viruses, is also becoming more common, posing severe challenges to infection control.

Types of Infectious Disease

Infectious diseases are *acute* when they are of short duration. The signs and symptoms of acute infections are often severe and usually appear suddenly. In contrast, a *chronic* infection usually develops slowly or insidiously and is long-lasting. There may also be periods of *latency* when overt symptoms are absent. This period is significant in infection control, since the patient may be 'infectious', in other words, capable of spreading the disease to other people, although disease symptoms are not apparent. This can be the case with some viral infections, particularly of the herpesvirus family. *Systemic* infections involve the whole body, whereas *localised* infections are restricted to a particular body site. *Opportunistic* infections are those caused by organisms that are non-pathogenic under normal circumstances. However, they can cause disease when there is impaired immunity or loss of normal flora.

Many different types of organisms can cause disease. The most important groups are bacteria, viruses, fungi, protozoa and higher eukaryotic organisms such as helminths. In addition, very simple agents such as *prions* (proteinaceous infective particles) can cause important diseases such as *scrapie* in sheep but are yet to be fully characterised. We now consider each of these groups in more detail.

General Features of Bacteria

Bacteria are amongst the most successful living organisms. Their ubiquity ensures that humans are obliged to live in constant and intimate contact with a wide variety of species and to encounter, if only briefly, many more. Fortunately, relatively few species routinely cause disease, but many others have the potential to do so, given appropriate conditions. Whether or not a bacterial encounter leads to disease is determined by the balance of two principal factors – host factors, including the state of the individual's immune system and features of the bacterium that enable it to cause disease. These bacterial features are termed *virulence determinants*. Virulence determinants enable bacteria to compete successfully with the normal microflora, survive in adverse conditions, adhere to or enter their target cells and evade host defence mechanisms.

Bacteria are prokaryotes, that is, they lack an organised nucleus. Their genetic information is carried in a double-stranded circular molecule of DNA which is often referred to as a

chromosome although it differs from eukaryotic chromosomes in that no introns (non-coding sequences of DNA) are present. Some bacteria possess small circular extrachromosomal DNA fragments known as *plasmids* which replicate independently of the chromosomal DNA. Plasmids may contain genes that encode virulence factors or antibiotic resistance and may be transferred from one bacterium to another. The cytoplasm of bacteria contains many ribosomes but no mitochondria or other organelles. In all bacteria, the cell is surrounded by a complex cell wall. The nature of the cell wall is important in the classification of bacteria (see below) and in determining virulence.

The Importance of the Bacterial Cell Wall

In 1884, Christian Gram observed that most bacteria could be classified into two broad groups, depending upon their ability to retain crystal violet dye after decolourisation. Those retaining dye were termed Gram-positive and those failing to do so, Gram-negative. This staining phenomenon is a consequence of fundamental differences in the cell walls of the two types of organisms (Fig. 4.1).

All bacteria are bounded by a cytoplasmic membrane, composed of a typical phospholipid bilayer, the function of which is to supply the cell with energy via its associated enzyme systems and to regulate the passage of metabolites in and out of the cell. Surrounding the cytoplasmic membrane is a layer of peptidoglycan, a complex polymer of polysaccharide chains linked by short peptides. This layer gives the cell its strength and shape and is much thicker in Gram-positive cells (accounting for more than 40% of the dry weight of the cell wall) than in Gram-negative cells (where it accounts for only around 10%). In Gram-positive organisms, numerous surface proteins and polymeric molecules other than peptidoglycan are also found closely associated with the peptidoglycan layer. A second outer membrane is present in

(A)

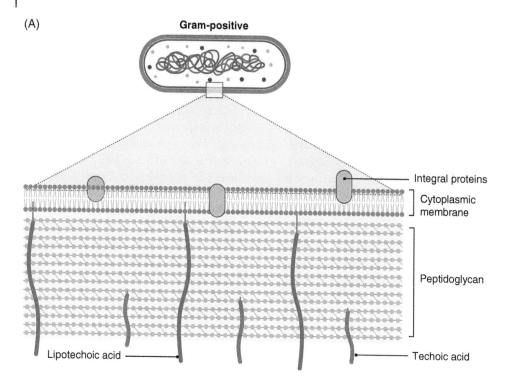

Gram-positive

Integral proteins

Cytoplasmic membrane

Peptidoglycan

Lipotechoic acid

Techoic acid

(B)

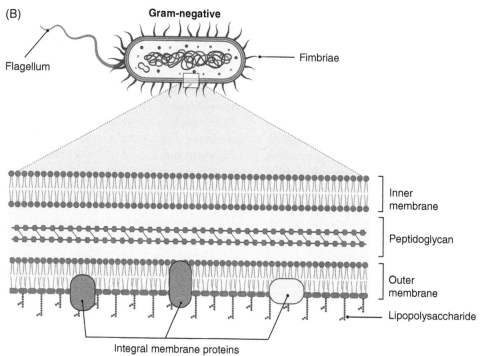

Gram-negative

Flagellum

Fimbriae

Inner membrane

Peptidoglycan

Outer membrane

Lipopolysaccharide

Integral membrane proteins

Figure 4.1 Schematic illustration of the structure of the cell wall of Gram-positive (A) and Gram-negative (B) bacteria. The cytoplasmic membrane in both Gram-positive and Gram-negative organisms is surrounded by a layer of peptidoglycan which is much thicker in Gram-positive cells. In Gram-positive organisms, numerous surface proteins and polymers, including teichoic acid and lipoteichoic acid, are also present. Gram-negative organisms possess a second outer membrane containing lipopolysaccharide and proteins. Flagella and fimbriae are also present in some Gram-negative cells. *Source:* Created with BioRender.com.

Gram-negative organisms which contains lipo-polysaccharide (LPS) and protein molecules. *Flagella* and *fimbriae* are cell appendages, composed of tubular filaments of polymerised protein that project from the cell wall of mainly Gram-negative bacteria (Fig. 4.2). Flagella are much longer than most fimbriae and generate propulsive forces which enable the bacterium to move within a fluid medium. Fimbriae, often also referred to as *pili*, are mainly involved in the adherence of bacterial cells to other bacteria and to host tissues. The notable exceptions are the *sex pili* which are important in the transfer of bacterial DNA, usually plasmids, from one bacterium to another. Finally, external to the cell wall, most pathogenic bacteria, whether Gram-positive or -negative, are covered with a

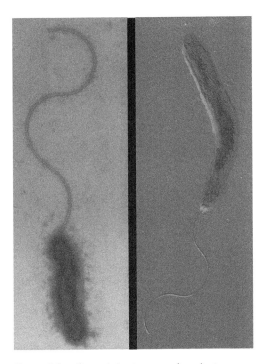

Figure 4.2   Campylobacter, scanning electron micrograph. This genus of Gram-negative bacteria is a common cause of food poisoning, causing diarrhoea, nausea, vomiting and headaches. They are characterised by their curved or comma-shaped appearance. They are also motile, typically possessing one or more flagella that enable them to move. *Source:* Wellcome Collection / https://wellcomecollection.org/works/jjv6j83s/items / last accessed November 13, 2023.

protective layer of carbohydrate known as *capsular polysaccharide.*

Biofilms and Quorum Sensing

Biofilms are complex communities of microorganisms that adhere to surfaces and form a slimy, protective matrix called *extracellular polymeric substances* (EPS), composed of a variety of molecules, including polysaccharides, proteins, lipids and DNA, produced by the cells of the biofilm. The process of biofilm formation begins when individual microbial cells attach to a surface (Fig. 4.3). Once attached, they multiply and secrete EPS, which serves as a glue-like substance that holds the biofilm together. As more cells join the community, the biofilm grows and develops a three-dimensional structure. Biofilms protect the microbes from external threats such as antimicrobial agents, host immune responses and environmental stresses. They also facilitate communication and cooperation between the cells, allowing them to coordinate their activities and share resources. Around two-thirds of bacterial infections in humans, and some fungal pathogens, e.g. *Candida albicans*, utilise biofilms.

Successful pathogens must have the ability to activate or repress their genes to obtain maximum advantage in a given situation. The expression of bacterial genes depends upon a complex interaction between bacteria and their immediate environment. Environmental cues can be many and varied, examples being temperature, nutrient availability, pH and the presence of antibiotics. These signals may act directly on the gene or indirectly through an intermediate response regulator. Closely related to these mechanisms are the effects of cellular communication processes, both between bacteria and between host cells and bacteria; the former is mediated principally by diffusible molecules within bacterial populations; the latter involving cellular interactions that occur after binding of bacteria to host cells.

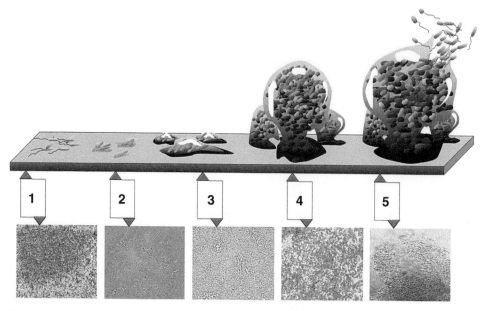

Figure 4.3   Five stages of biofilm development. Stage 1, initial attachment; stage 2, irreversible attachment; stage 3, maturation I; stage 4, maturation II; stage 5, dispersion. Each stage of development in the diagram is paired with a photomicrograph of a developing *P. aeruginosa* biofilm. All photomicrographs are shown to the same scale. *Source:* PUBLIC LIBRARY OF SCIENCE (PLOS) / Wikimedia Commons / CC BY 2.5.

Quorum sensing is a regulatory mechanism employed by some bacteria to coordinate gene expression in response to changes in population density. It enables bacteria to communicate and synchronise their behaviour by producing and sensing specific signalling molecules called *autoinducers*. Autoinducers diffuse across the bacterial cell membrane and accumulate in the extracellular environment as the bacterial population grows. Autoinducers are typically small, diffusible molecules such as *acyl-homoserine lactones* (AHL) in Gram-negative bacteria or oligopeptides in Gram-positive bacteria. As the bacterial population density increases, the concentration of autoinducer molecules also rises. When the concentration of autoinducers reaches a threshold level, they bind to specific receptor proteins located on the bacterial cell surface, triggering intracellular signalling that ultimately leads to changes in gene expression.

Classification of Bacteria

It is beyond the scope of this chapter to detail the classification systems used to separate different groups of bacteria. However, it is useful to acknowledge, in general terms, how this classification is derived. The simplest classification is based on staining characteristics (Gram-positive or Gram-negative) and morphology (Fig. 4.4). Descriptions of the colony types produced when bacteria are grown on simple artificial media improve differentiation, but are not reliable enough for routine diagnostic use. For this reason, a range of biochemical properties, e.g. the ability to ferment certain sugars, can be examined; the wider the range, the more accurate the designation. In practice, a combination of all these methods, together with the more modern approach of genomic sequencing, allow bacteria to be characterised into families, genera, species and

(A)

(B)

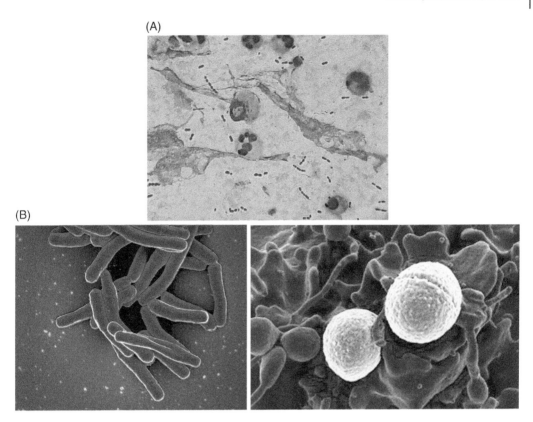

Figure 4.4   (A) Simple classification of bacteria of clinical significance. Bacteria may be classified into broad groups depending upon their ability to retain crystal violet dye after decolorisation. Those retaining dye, shown here, are termed Gram-positive and those failing to do so, Gram-negative. Image shows bacteria (blue) and pus cells (neutrophils; red) in a Gram-stained smear. *Source:* Own work / Wikimedia Commons / CC BY SA 4.0. (B) Microscopic examination of the morphology of bacteria. Shown here are rod-shaped bacilli of *M. tuberculosis* (left panel) and rounded cocci of *S. aureus* (right panel); bacteria (gold) outside a white blood cell (blue). *Sources:* (b1) NATIONAL INSTITUTE OF ALLERGY AND INFECTIOUS DISEASES / Wikimedia Commons / CC BY 2.0. (b2) NATIONAL INSTITUTE OF ALLERGY AND INFECTIOUS DISEASES / Wikimedia Commons / Public Domain.

strains. A *strain* within a bacterial species represents individual bacteria that differ in a certain way from the original species description. Strains can differ in various ways including, but not limited to, morphology, biochemistry or pathogenicity. Traditionally, bacterial *species* were defined phenotypically by their structure and function. However, with the advent of molecular techniques, the definition has included genetic criteria. One widely accepted genetic threshold is based on 16S rRNA gene sequence similarity; strains sharing more than 97% similarity in their 16S rRNA gene sequence are generally considered to be of the same species.

Routes of Acquisition of Bacteria

Bacteria causing infection are acquired from two principal sources: either from amongst the patient's own normal flora (*endogenous infection*) or from external sources (*exogenous infection*). Exogenous infections may be acquired by one of the following four principal routes:

- Ingestion, e.g. food poisoning caused by consumption of foods contaminated with *Salmonella* spp.
- Inhalation, e.g. air droplets containing *M. tuberculosis*, causing pulmonary tuberculosis.
- Inoculation, e.g. rose-thorn punctures introducing *C. tetani* and leading to clinical tetanus.
- Direct contact, e.g. *Neisseria gonorrhoeae*, acquired by intimate person-to-person contact.

Many pathogens, including some bacteria, are inoculated into the host by other animals, principally insects. For example, the bacterium *Borrelia burgdorferi*, which causes Lyme disease, is transmitted by ticks. Members of the bacterial genus, *Rickettsia*, are transmitted by lice, dog ticks or fleas. For example, the human body louse transmits the bacterium *Rickettsia prowazekii*, which causes epidemic typhus. *Yersinia pestis* is the causative agent of the plague and is transmitted by fleas. *Y. pestis* can survive and reproduce inside cells, even if phagocytosed, entering the lymphatic system and causing acute lymphadenitis; the swollen lymph nodes forming the characteristic 'buboes'.

Transmission of Bacterial Infections

The transmission of a bacterial infection is dependent upon several factors:

- *Host factors*, including the degree of immunity to a particular pathogen within the population, the proximity of individuals to each other, and the general state of health and hygiene. Some individuals, while apparently healthy, may harbour and transmit pathogenic bacteria – these individuals are referred to as *carriers*. For example, healthy individuals can excrete *Salmonella* spp. for prolonged periods, causing outbreaks of food poisoning if they are involved in the preparation of food.
- *Bacterial factors* influencing transmission include the organism's virulence, its ability to survive in the environment, the size of the infecting dose and the route through which the bacterium is acquired.
- *Environmental factors* affecting transmission include climate (bacterial growth generally being favoured by warm humid conditions), the standard of sanitation and the presence of non-human 'vectors', e.g. ticks, which transmit bacteria whilst feeding on human or animal blood.
- *Herd immunity*, also known as *population immunity*, is a form of indirect protection from infectious disease that occurs when a large percentage of a population are immune to an infection, whether through vaccination or previous infections, thereby providing a measure of protection for individuals who are not immune. In populations with a high proportion of immune individuals, it becomes more difficult for the disease to spread from person to person. As a result, even individuals who are not vaccinated or have not previously been infected are indirectly protected because the likelihood of them coming into contact with an infected individual is reduced.

Bacteria can be transmitted between individuals (*horizontally*, e.g. *M. tuberculosis*, spread by respiratory droplets) or from mother to baby (*vertically*, e.g. *Listeria monocytogenes*, which can be transmitted from mother to child *in utero*, causing sepsis in the foetus or newborn).

Bacterial Virulence Determinants

As described earlier, pathogenic bacteria possess so-called virulence determinants, which are responsible for their ability to cause disease. Many of these virulence determinants are cell wall constituents. An understanding of the nature and mode of action of virulence determinants is essential if we are to appreciate the mechanisms underlying the pathogenesis of bacterial diseases.

Virulence Determinants Specific to Gram-positive Bacteria

Non-peptidoglycan Polymers

These are a heterogeneous group of teichoic acid-like polymers containing sugar alcohols and phosphodiester linkages, which are found on the surface of Gram-positive cells, bound covalently to peptidoglycan. They are strongly immunogenic and form the identifying group antigens of many species of streptococci. Unlike these 'secondary' cell wall polymers, the closely related molecule, lipoteichoic acid, lies in contact with the cytoplasmic membrane and protrudes through the peptidoglycan layer. It is important in the adherence of bacteria to surfaces, for example, the binding of *S. mutans* to tooth enamel.

Surface Proteins

Many different cell surface proteins have been identified, the majority of which do not appear to be virulence factors. One notable exception, however, is the 'M' protein of group A β-haemolytic streptococci (also known as Group A Streptococcus [GAS] or *S. pyogenes*). The M protein promotes bacterial resistance to phagocytosis by interfering with opsonisation, a process in which bacteria are coated by antibodies that are recognised by phagocytes, facilitating their engulfment. The M protein does this by binding to host proteins, such as fibrinogen and complement regulatory proteins, inhibiting the deposition of opsonins and preventing efficient phagocytosis. The M protein also triggers the release of enzymes, such as streptokinase and hyaluronidase, which facilitate tissue invasion and bacterial dissemination.

Virulence Determinants Specific to Gram-negative Bacteria

Lipopolysaccharide

Lipopolysaccharide (LPS) is one of the most important bacterial virulence factors and is often referred to as *endotoxin*. It is present in the outer membrane of Gram-negative cell walls and consists of an inner glycolipid (lipid A) attached to a 'core' oligosaccharide, with or without a variable length, outer, 'O' polysaccharide. Lipid A is a very potent toxin and is responsible for all the toxic properties attributed to endotoxin, though these are enhanced when the lipid molecule is associated with 'O' polysaccharide. Endotoxin exerts a profound effect when introduced into the host, producing widespread stimulation of the immune system and activation of the complement system (Chapter 5) and the coagulation cascade (Chapter 8). This results in generalised damage to the host, manifested in features collectively referred to as *endotoxic shock*, which may result in death. The 'O' polysaccharide chain of LPS additionally confers resistance to the bacteriolytic effects of serum and protects the bacterial cell from phagocytosis.

Outer Membrane Proteins

Numerous protein molecules can be found within the outer bacterial membrane. They are closely associated with LPS and have functions in cell transport and ion binding. However, in some bacterial species, these proteins are also major virulence factors, enabling bacterial cells to adhere to their target tissues. For example, the pathogen, enterohaemorrhagic *E. coli* (EHEC) – the commonest cause of dysentery-like illness – uses outer membrane proteins, such as *intimin*, that help the bacteria to adhere to the intestinal epithelium. Once attached, the bacteria can produce potent toxins, such as *Shiga toxins*, a family of bacterial toxins produced by certain strains of bacteria, primarily *Shigella dysenteriae* serotype 1 and some strains of *E. coli*.

Flagella and Fimbriae

Flagellar proteins are strong immunogens and represent the 'H' antigens used in typing many Gram-negative bacteria, notably the Salmonellae. However, apart from conferring active motility, which may be a useful attribute in certain circumstances, it is not thought that flagella are of major importance as far as virulence is concerned. Conversely, fimbriae are very significant virulence factors. Examples include *Type I fimbriae* which help bacteria adhere to host cells, specifically those with high mannose structures. Type I fimbriae are commonly found in the family *Enterobacteriaceae*. Fimbrial adherence to epithelium allows a direct interaction between the potent toxins produced by the bacteria and the epithelial cells. *Type III fimbriae* also help the bacteria to adhere to *abiotic* surfaces (non-living surfaces or materials) and promote the formation of biofilms. *Curli fimbriae* (so-named because of their characteristic curly or wavy appearance) are also important for biofilm formation and enable bacteria to adhere to surfaces such as stainless steel.

Virulence Determinants Common to Gram-negative and Gram-positive Bacteria

Capsular Polysaccharides

Capsular polysaccharides enable the bacterium to adhere by forming a sticky layer on surfaces and are important, for example, in the formation of dental plaque and the colonisation of implanted medical devices and intravenous cannulae. They also render the bacterial cell wall inaccessible to the action of complement and to phagocytosis. Some capsular polysaccharides have the added advantage of mimicking host tissue antigens and so are not recognised as foreign by the immune system. For example, certain strains of *E. coli* can cause meningitis in newborn infants. These organisms possess the so-called K1 capsule, which is structurally similar to proteins found in the central nervous system (CNS) of newborn infants. The immune system sees the K1 capsule as 'self' and the bacteria are therefore not destroyed.

Toxins and Enzymes

Large numbers of toxins are known to be produced by bacteria. They are usually proteins of varying molecular weight and are traditionally referred to as *exotoxins* to differentiate them from the endotoxin of Gram-negative bacteria. They are numerous and wide ranging in their effects and are grouped based on the following three main characteristics:

- *Site of action of the toxin.* Some exotoxins act only at the site of release. For example, the enterotoxin of *Clostridium perfringens* acts locally on intestinal epithelial cells to cause diarrhoea. Conversely, certain toxins may have more generalised systemic effects. Diphtheria toxin, for example, acts systemically, inhibiting host cell protein synthesis and resulting in damage to major organs.
- *Mode of action.* Exotoxins may either act directly to cause their effects or their effect may be mediated through other agents. Tetanus toxin, for example, acts directly by blocking the release of neurotransmitters, leading to *tetanus* (painful contractions of muscles resulting in 'lockjaw', in other words when an individual's neck and jaw muscles go into spasm and become rigid, making it hard to open the mouth or swallow).
- *Structure of the toxin.* The toxin of *S. pyogenes*, streptolysin 'O', is a single molecule which binds to cell membranes causing lysis, whereas diphtheria toxin, after binding to a cell, requires cleavage by proteolytic enzymes before its active component can enter the cytoplasm.

Superantigens are a class of toxins produced by certain bacteria, including *S. aureus* and *S. pyogenes*, which cause *toxic shock syndrome*. These toxins can bridge the MHC class II proteins present on antigen-presenting cells, with the T-cell receptor expressed on the surface of T-cells that possess a particular Vβ chain. As a result, a large proportion of T-cells are activated, leading to a massive release of proinflammatory cytokines. This excessive immune response contributes to the symptoms of toxic shock syndrome, which include fever, myalgia, hypotension, rash, confusion/altered mental state and in severe cases, multi-organ dysfunction, including kidney failure, liver impairment, respiratory distress and cardiac complications.

Membrane-damaging toxins are categorised into two main types: channel-forming toxins and toxins that function as enzymes acting on the cell membrane. *Channel-forming toxins*, such as cholesterol-dependent cytolysins (CDCs), form pores in the target cell membrane. CDC requires the presence of cholesterol in the target cell for pore formation. Examples of CDC include pneumolysin from *Streptococcus pneumoniae* and perfringolysin 'O' from *C. perfringens*. CDC also causes modifications of histone proteins in the host cell nucleus, causing the downregulation of genes encoding proteins involved in the inflammatory response. Enzymatically active toxins, such as the α-toxin of *C. perfringens*, possess specific enzymatic activities. For example, *C. perfringens* possesses phospholipase activity, which plays a role in the development of conditions such as gas gangrene caused by this bacterium.

Some exotoxins function intracellularly and are classified on their mode of entry into the target cell or their mechanism of action once inside. Some bacteria, for example *Yersinia* spp., deliver toxins directly from their cytoplasm to the cytoplasm of the target cell through specialised needle-like structures, called type III secretion systems (T3SS). AB toxins are a category of intracellular toxins, in which the 'B' (binding)-subunit attaches to target cell membranes, allowing the 'A' (active)-subunit to enter the cell and exert its enzymatic activity. Examples of AB toxins include cholera toxin, pertussis toxin and heat-labile enterotoxin from *E. coli*.

Other important enzymes, which cannot be classed as true virulence factors but are nevertheless important in human disease, are the enzymes produced by bacteria to counter the effects of antibiotics used to treat infections. Examples are the *β-lactamase enzymes* produced by bacteria that inactivate penicillin-like compounds, *aminoglycoside-modifying enzymes*, which modify aminoglycoside antibiotics, such as gentamicin and streptomycin, and *chloramphenicol acetyltransferase*, which acetylates and inactivates chloramphenicol.

Factors Influencing Bacterial Virulence

Many bacteria cannot express virulence factors from their own genome and are only able to do so if they acquire the necessary genetic material from plasmids or bacteriophages. Plasmid-mediated virulence factors are important in infections caused by some Gram-negative species. As transmissible units of genetic material, plasmids offer enormous potential for the exchange and recombination of gene sequences coding for virulence.

Bacteriophages are viruses capable of infecting bacterial cells and may also mediate the transfer of genetic material from one bacterial cell to another. The best example of bacteriophage-mediated virulence is *Corynebacterium diphtheriae*, which requires the β-phage genome to produce its toxin. In some bacteria, genetic material coding for virulence factors is confined to a single region of the genome. However, other strains may acquire numerous virulence genes distributed around the chromosome, known as *pathogenicity islands*. Pathogenicity islands are defined as gene clusters incorporated in the genome, chromosomally or extrachromosomally, of pathogenic organisms, but which are absent from non-pathogenic organisms of the same or closely related species. Pathogenicity islands are found in both Gram-positive and Gram-negative bacteria and play a significant role in the evolution and pathogenicity of microorganisms. They can be transferred between bacteria through horizontal gene transfer mechanisms, including via plasmids and bacteriophages. Thus, pathogenicity islands are considered key players in bacterial evolution and the development of virulence traits.

Virus Infections

A pandemic caused by the severe acute respiratory syndrome coronavirus 2 (SARS-CoV-2), known as coronavirus disease 2019 (COVID-19), has highlighted the importance of viruses as a cause of morbidity and mortality among humans. First identified in the city of Wuhan, Hubei, China, the WHO declared COVID-19 a pandemic on 11 March 2020. This section discusses the general nature of viruses, how they infect and replicate within cells and how they cause disease. It also discusses how the world is preparing for the next viral pandemic.

Viruses are resistant to antibiotics, are smaller than bacteria (30–200 nm) and are composed of a single type of nucleic acid (either DNA or RNA) surrounded by a protein coat (capsid) which may be further enclosed in an envelope. They are obligate intracellular parasites, i.e. they can only replicate within a living host cell. Virus replication is a stepwise series of metabolic events involving both virus-encoded proteins and the metabolic machinery of the cell. During evolution, viruses have developed exquisite molecular features which allow them to replicate at the expense of the cell. In most cases, the pathology is a result of both the death of infected cells (associated with the release of new virus particles – known as *virions*) and the immune response to the virus. In some cases, the immune system is itself a target for virus infection. This can lead to the destruction of important immune cells, which in turn can lead to immunosuppression. An example of this is the immunosuppression that results from HIV infection of CD4+ T lymphocytes (see also Case Study CS6). Virus infection may also contribute to the development of neoplasia (Chapter 11, discussed in Case Studies CS55, CS57, CS63 and CS67).

Classification of Viruses

Viruses are classified according to the International Committee on Taxonomy of Viruses (ICTV) which publishes annual updates. The classification is based upon the molecular composition of the genome; structure of the virus, including the presence or absence of an envelope; viral gene expression programme; host range; pathogenicity and sequence similarity. While all characteristics are useful, pairwise sequence similarity and phylogenetic relationships have become the most important features in distinguishing virus taxa. Table 4.1 summarises the properties of some virus families, and Fig. 4.5 shows the structures of selected virus particles.

Virus Replication

Viral life cycles are diverse but have in common replication within host cells with eventual release of progeny virions. We use HIV infection as an example here to highlight stages in virus life cycles that can be targeted by antiviral drugs.

First, the virus attaches, via the viral capsid or envelope, to a specific receptor on the surface of a target cell, triggering a series of events that allow the virus to penetrate the cell. It is this initial interaction which accounts for the predisposition of certain viruses to infect specific cell types, and which ultimately confers species-specificity. For example, HIV attaches to one main receptor, the CD4 receptor found on T-helper cells and one of two co-receptors, either the C-C chemokine receptor type 5 (CCR5) or C-X-C chemokine receptor type 4 (CXCR4). Attachment to CD4 occurs via a 120-kDa virus envelope protein (gp120) and fusion/penetration is achieved via a 41-kDa envelope glycoprotein (gp41). Absence of the co-receptor in some individuals appears to make them genetically unsusceptible to HIV infection. Some antiviral drugs work by blocking initial attachment or fusion; approved drugs that inhibit HIV entry include the membrane fusion-inhibitor enfuvirtide (brand name, Fuzeon) and the CCR5 blocker, maraviroc (Selzentry). Ibalizumab (Trogarzo) is an anti-CD4 monoclonal antibody that blocks the

Table 4.1 Taxonomic chart of selected virus families.

Family	Characteristics	Typical members	Diseases caused
Poxviridae	Double-stranded DNA, 'brick'-shaped particles; largest virus	Vaccinia virus (*vaccinia virus*)	Laboratory virus
		Smallpox virus (*variola virus*)	Smallpox (now eradicated)
Herpesviridae	Double-stranded DNA, icosahedron capsid enclosed in an envelope, latency in host common	Herpes simplex virus Type-1 and -2 (*Human alphaherpesvirus 1, 2*)	'Cold' sores, genital infections
		Varicella-zoster virus (*Human alphaherpesvirus 3*)	Chicken pox, shingles
		Human cytomegalovirus (*Human betaherpesvirus 5*)	Febrile illness or disseminated disease in immunosuppression
		Epstein-Barr virus (*Human gammaherpesvirus 4*)	Glandular fever. EBV is also a causative agent in some cancers, e.g. Burkitt lymphoma
Adenoviridae	Double-stranded DNA, icosahedron with fibre structures, non-enveloped	Human adenoviruses, multiple types (*Human mastadenoviruses*)	Respiratory and eye infections
Papillomaviridae	Double-stranded circular DNA, 72 capsomeres in capsid, non-enveloped	Human papillomaviruses, e.g. types 16, 31, 33 (e.g. *Alphapapillomavirus 9*)	Warts, causative agent in epithelial cancers (e.g. cervical cancer)
Hepadnaviridae	One complete DNA minus strand with 5′ terminal protein, DNA circularized by an incomplete plus strand, 42 nm enveloped particle	Hepatitis B virus (*Hepatitis B virus*)	Hepatitis B ('serum' hepatitis), causative agent in hepatocellular carcinoma
Paramyxoviridae	Single-stranded negative-sense RNA, enveloped	Measles virus (*Measles morbillivirus*)	Measles
Pneumoviridae	Single-stranded negative-sense RNA, enveloped	Respiratory syncytial virus (*Human orthopneumovirus*)	Bronchiolitis
Orthomyxoviridae	Single-stranded negative-sense RNA, enveloped, replicates in the nucleus	Influenza virus, e.g. Type A (*Alphainfluenzavirus influenzae*)	Influenza (pandemic virus)
Picornaviridae	Single-stranded positive-sense RNA, 22–30 nm particle of cubic symmetry, non-enveloped	Poliovirus (*Enterovirus C*)	Poliomyelitis
		Hepatitis A virus (*Hepatovirus A*)	'Infectious' hepatitis
Flaviviridae	Single-stranded positive-sense RNA, enveloped	Hepatitis C virus (*Hepacivirus C*)	Hepatitis C
Matonaviridae	Single-stranded positive-sense RNA virus, enveloped	Rubella virus (*Rubivirus rubellae*)	German measles
Rhabdoviridae	Single-stranded RNA, negative-sense, bullet-shaped, enveloped	Rabies virus (*Lyssavirus rabies*)	Rabies
Retroviridae	Single-stranded RNA, enveloped particles with icosahedral nucleocapsid, employ reverse transcriptase enzyme to make DNA copy of genome on infection	Human T-cell leukaemia virus-1 (*Primate T-lymphotropic virus-1*)	Adult T-cell leukaemia/lymphoma
		Human immunodeficiency virus (HIV-1, 2)	Acquired immunodeficiency syndrome (AIDS)

interaction with gp120 and is used to treat adults with multidrug-resistant HIV-1 infection.

Following penetration of the cell, the virus is uncoated releasing the virus genome. In the case of many viruses, the genome migrates rapidly to the host cell nucleus where it may integrate within the host cell DNA or exist as a separate extrachromosomal entity (known as an *episome*). In the case of other viruses, the genome remains within the cytoplasm. Integration is often only achieved with the help of a virus-encoded protein. In the case of HIV, integration is essential for virus replication, and the virus uses its own 'integrase' enzyme to do this. Dolutegravir (Tivicay) is an example of a viral integrase inhibitor used to treat people with HIV/AIDS.

Next, there is a synthesis of virus structural proteins (those proteins that will make up new virions) and non-structural proteins (i.e. the virus enzymes required for the manufacture of the structural proteins). The nucleic acid also replicates to provide the genome for progeny virions. The complexity of the different mechanisms used by viruses to reproduce themselves is well illustrated by the RNA viruses. In the case of 'positive-strand' RNA viruses, the viral RNA genome is the mRNA and is translated directly. On the other hand, 'negative strand' RNA viruses make mirror image copies of their genome, which then act as the mRNA. The retroviruses have a more complex replication mechanism (Fig. 4.6), in which a viral reverse transcriptase enzyme makes a DNA copy of the viral RNA. This DNA copy is integrated into the host cell DNA and in this form is known as a *provirus*. The provirus utilises the cellular machinery to transcribe the viral genetic information into RNA, which is then translated into viral proteins. Viral protease enzymes cleave larger viral protein precursors into their functional forms, facilitating viral assembly. In the case of HIV, reverse transcriptase inhibitors (e.g. tenofovir and emtricitabine) and protease inhibitors, (e.g. darunavir and atazanavir) are important drugs that can suppress virus replication within infected individuals.

Finally, the newly formed viral proteins enclose the viral nucleic acid to form mature virus particles. These particles leave, either by lysing the infected cell, or by budding from the host cell plasma membrane. In some cases, host-derived membrane proteins form part of the virus envelope.

Escape from Apoptosis

In many cases, virus infection triggers a series of events resulting in cell death by apoptosis, meaning that virus replication and the production of progeny virions may be inhibited. However, viral genes can delay the onset of apoptosis to allow sufficient time for the virions to be produced and leave the cell. Cowpox virus, for example, expresses CrmA, an inhibitor of interleukin 1β-converting enzyme (ICE, also known as caspase 1), as well as caspase-8 and caspase-10. In this way, CrmA can prevent apoptosis via *death receptors* that signal through FADD and caspase 8 (Chapter 3). Because CrmA inhibits the caspase-1 mediated cleavage, and thereby the activation, of interleukin 1β, a pro-inflammatory cytokine, it also reduces the potentially beneficial inflammation that should follow infection with cowpox virus. Genes that counteract apoptosis are also found in many other viruses, e.g. Kaposi sarcoma-associated herpesvirus (KSHV; viral homologue of cellular FLIP known as v-FLIP, see also Chapter 3) and Epstein-Barr virus (EBV; viral homologue of cellular BCL2, known as BHRF1). Alternatively, viruses may induce expression of cellular anti-apoptosis genes, for example the EBV gene, latent membrane protein-1 (LMP1), induces expression of the cellular anti-apoptosis genes, BCL2 and MCL1.

Virus Evasion of Host Immunity

The immune response to virus infections can be mediated by both the innate and adaptive arms of the immune system. It is important to emphasise that while the replication of viruses can cause cell death and therefore tissue damage, this can be further exacerbated by the immune response to the infection which will also lead to significant tissue destruction, especially in the case of the innate response. This immune-mediated damage is referred to as *immunopathology* and is discussed more fully in Chapter 6. During co-evolution with their hosts, viruses have developed sophisticated strategies to evade host innate and adaptive immune responses:

Escape from Innate Immunity

Viral *pathogen-associated molecular patterns* (PAMPs) are sensed by host *pattern recognition receptors* (PRRs; see also Chapter 5). PRRs include Toll-like receptors (TLRs), RIG-I-like receptors (RLRs) and DNA sensors. Activation of PRRs induces intracellular signalling via key adaptor proteins, such as mitochondrial antiviral-signalling protein (MAVS) and stimulator of interferon genes (STING). MAVS and STING activate other kinases and transcription factors leading to expression of type I and III interferons (IFNs). Viruses have evolved numerous strategies to evade these innate immune responses:

- *Inhibition of PRR signalling:* For example, the Hepatitis C virus (HCV) produces a protease, NS3/4A, which can cleave and inactivate MAVS.
- *Antagonising type I interferon production:* The NS1 protein of Influenza A virus and the E3L and K3L proteins of Vaccinia virus inhibit type I IFN production.
- *Counteracting the antiviral actions of type I interferons:* For example, the Ebola virus VP24 protein can bind to and inhibit STAT1, a key transcription factor activated by IFNs. In doing so, it prevents the expression of IFN-stimulated genes (ISGs).
- *Blocking complement pathways:* For example, Cytomegalovirus (CMV) encodes a protein known as pUL55 that binds to the host complement component C8,

preventing its incorporation into the membrane attack complex (Chapter 5) and thereby inhibiting cell lysis. ORF4 of KSHV encodes a viral protein that can bind to complement components C3b and C4b, inhibiting their ability to form C3 and C5 convertases.

Escape from Adaptive Immune Responses

Viruses have also evolved diverse mechanisms to escape adaptive immunity;

- *Evolving antigenic variability:* Many viruses can rapidly mutate their surface antigens, such as viral glycoproteins, to produce slightly different versions of these proteins. This makes it difficult for the host immune system to recognise the virus. Influenza virus, HIV and HCV are examples of viruses that exhibit high levels of antigenic variation.
- *Suppressing antigen presentation:* Virus proteins can be processed by the host cell into peptides and presented on the surface of the cell in association with major histocompatibility complex (MHC) class I molecules (see Chapter 6). In this way, infected cells can be recognised and destroyed by cytotoxic T-lymphocytes (CTLs) carrying the CD8 surface receptor. Some viruses inhibit the intracellular transport mechanism by which viral peptide fragments are presented to CTLs. For example, both Herpes simplex virus 1 (HSV1) and Herpes simplex virus 2 (HSV2) express ICP47, which inhibits the transporter associated with antigen processing (TAP) protein, impairing the loading of viral peptides onto MHC class I molecules and reducing recognition by CTLs. CMV expresses US2 and US11 proteins that target MHC class I molecules for degradation. The EBV-encoded BILF1 protein downregulates the expression of MHC class I in infected cells.
- *Persistent latency:* By establishing persistent latent infections, viruses can effectively

'hide' from the immune system. Herpes-viruses, such as EBV, downregulate viral gene expression in infected cells, allowing them to persist long-term unseen by the immune system, with the potential for inter-mittent reactivation during the lifetime of the host.

- *Induction of immune tolerance:* For example, the Hepatitis B virus (HBV) produces a pro-tein (HBx) that can promote the differentia-tion of naive T cells into Tregs, which then suppress antiviral immune responses. This contributes to the establishment of immune tolerance and chronic infection.

(A)

(B)   (C)

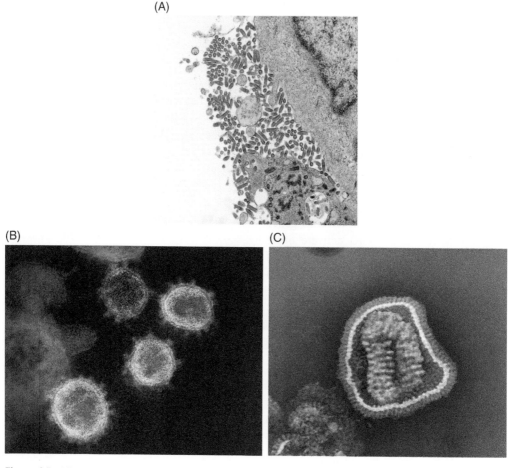

Figure 4.5 Virus particles. (A) Rabies virus. Colorized transmission electron micrograph of BSK cells in culture, heavily infected with Rabies virus (red). Image captured and colour-enhanced at the NIAID Integrated Research Facility (IRF) in Fort Detrick, Maryland. *Source:* NATIONAL INSTITUTE OF ALLERGY AND INFECTIOUS DISEASES / Wikimedia Commons / CC BY 2.0. (B) SARS-CoV-2. This colorized transmission electron microscope image shows SARS-CoV-2, the virus that causes COVID-19, isolated from a patient in the United States. Virus particles are shown emerging from the surface of cells cultured in the lab. The protruding capsid structures (the so-called spike protein) of the virus give these viruses their name, corona being the Latin term for crown. *Source:* NATIONAL INSTITUTE OF ALLERGY AND INFECTIOUS DISEASES / Wikimedia Commons / CC BY 2.0. (C) Pseudocoloured negative-stained transmission electron micrograph depicting the ultrastructural details of an Influenza virus particle. https://commons.wikimedia.org/wiki/File:Influenza_virus_particle_color.jpg. *Source:* Centers for Disease Control and Prevention / Wikimedia Commons / Public Domain.

- *Expression of immunoevasins:* Viruses can encode specific immunoevasion molecules that can directly interfere with host immune responses. Many of these target MHC functions, as described above. Other immunoevasins target different cell functions. For example, HIV encodes the *viral infectivity factor* (Vif) protein, which counteracts the host restriction factor, APOBEC3G. APOBEC3G inactivates HIV by inducing mutations in the viral genome.

Virus Spread

In order to survive, viruses need a large population of susceptible hosts and an efficient means of spread between them. The main routes for transmission are as follows:

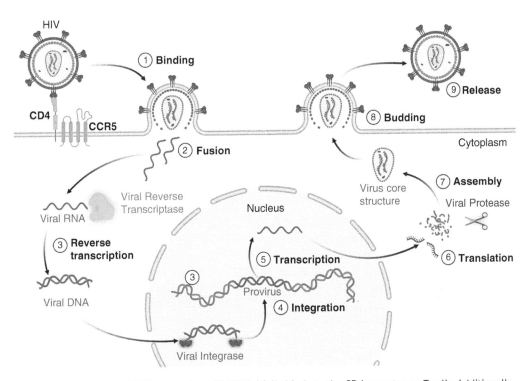

Figure 4.6 Life cycle of HIV, a retrovirus: (1) HIV initially binds to the CD4 receptor on T cells. Additionally, HIV requires co-receptors, such as CCR5, shown here. (2) Once bound to these receptors, the virus fuses with the host cell membrane and enters the cell. After entry, the viral capsid, which protects the viral RNA, is uncoated, releasing the viral RNA into the host cell's cytoplasm. (3) The virus utilises a viral enzyme, known as reverse transcriptase, to convert its single-stranded RNA genome into a double-stranded DNA copy. This newly formed viral DNA is called complementary DNA (cDNA). (4) The cDNA is then integrated into the host cell's genome with the help of another enzyme called integrase. This integrated viral DNA becomes a permanent part of the host cell's DNA and is referred to as a provirus. The provirus serves as a template for the production of genomic RNA and messenger RNA. (5) These viral RNAs are transcribed and (6) are translated by the host cell's machinery to produce viral proteins, including structural proteins and enzymes. HIV relies on the action of a protease enzyme that cleaves long precursor proteins into smaller, functional pieces. (7) The newly synthesised viral proteins and RNA are transported to the cell membrane, where they assemble into new virus particles. (8) The assembled virus particles bud from the host cell's membrane, acquiring an envelope derived from the host cell membrane. (9) The virus particles, now mature and infectious, are released from the host cell, ready to infect other target cells and continue the infection cycle. *Source:* Created with BioRender.com.

The Respiratory Route

This is the most common pathway for virus entry and exit. Following inhalation, viruses can infect and replicate within epithelial cells of the upper or lower respiratory tract. Following release from these cells, progeny virions enter surrounding airways and exit via sneezing and coughing. Examples of viruses that are transmitted in this way include SARS-CoV-2, influenza viruses, common cold viruses, adenoviruses and the Respiratory syncytial virus (RSV). Although the entry and exit of many viruses are via the respiratory tract, some viruses do not remain within the respiratory tract but may enter the bloodstream causing *viraemia* (defined as the presence of infectious virions in the bloodstream) and subsequent infection of other target organs. Examples of such viruses include the Varicella-zoster virus, which is responsible for chicken pox, the Measles virus and the Rubella virus, the causative agent of German measles.

The Oral-Gastrointestinal Route

This route is used by viruses responsible for GI-tract infections (e.g. Rotavirus, enteroviruses). A number of these viruses (e.g. Poliovirus) can spread to other tissues, causing disease elsewhere, for example in the CNS. Vast numbers of infectious virus particles can be excreted in faecal material (e.g. of the order of 10^{12} particles/g) facilitating the easy spread of these viruses in conditions of poor sanitation. The drinking of faecally contaminated water, consumption of contaminated shellfish and food prepared by unhygienic food handlers are ways in which these viruses are spread between susceptible hosts. Infections caused by these viruses are widespread in resource-poor countries where sanitation is a particular problem.

Skin Inoculation

While the skin normally provides an impenetrable barrier to virus invasion, infectious viruses can enter following trauma. This may be a bite from an animal vector (e.g. transmission of the Rabies virus from the bite of an infected canine) or an arthropod vector (e.g. the transmission of the Yellow fever virus by mosquitoes). HIV, HBV and HCV can be transmitted by the injection of blood or blood products, either in the form of a blood transfusion, a needle-stick injury, or by intravenous drug abuse.

Sexual Transmission

This is an important route for the spread of viruses such as HSV2 (genital herpes) the papillomaviruses (genital warts) and HIV. Unprotected penetrative anal intercourse among men who have sex with men (MSM) remains a high-risk behaviour for HIV transmission. However, there have been advances in HIV prevention, including pre-exposure prophylaxis (PrEP; see also Case Study CS6), which reduces the risk of HIV acquisition among high-risk populations. While heterosexual transmission continues to be a significant mode of HIV spread in some populations, for example, in sub-Saharan Africa and the Caribbean, there have been improvements in prevention and treatment efforts. Initiatives to increase awareness, provide access to antiretroviral therapy (ART) and promote safe practices have helped to curb the epidemic's impact. However, challenges remain, including addressing stigma and discrimination, reaching key at-risk populations and ensuring that treatment and prevention services are accessible to all who need them.

Vertical Transmission

Viruses may be transmitted *vertically* that is from mother to offspring, via the placenta, during childbirth or in breast milk. For example, Rubella virus and CMV infections acquired by the mother during pregnancy may be transmitted to the developing embryo, leading to severe congenital abnormalities and/or spontaneous abortion. Some virus infections, e.g. HSV infections, if acquired *in utero* or during birth can present as an acute disease syndrome at birth. Alternatively, the child may be born without any overt signs of infection but carries the virus asymptomatically. The development of this 'carrier state' follows congenital infections with viruses like HIV or HBV. The developing foetus of an HIV-positive mother has an estimated 40–60% chance of becoming infected with HIV, but the risk is reduced if prophylactic chemotherapy is given during pregnancy.

Virus Pathogenesis and the Clinical Effects of Virus Infections

The outcome of a virus infection is dependent on several factors that include age, immune status and nutritional state of the host, route of infection, viral load and viral strain. HSV1 and HSV2 infection puts newborns at high risk of developing severe and life-threatening symptoms, including fatal organ damage (e.g. of liver, lungs and heart), but may cause inapparent infections or minor epithelial lesions (e.g. 'cold sores') in older children and adults. EBV causes a very mild febrile illness in infants, but infectious mononucleosis (glandular fever) in teenagers; the same virus in parts of tropical Africa and in southeast Asia contributes to the development of Burkitt lymphoma and nasopharyngeal carcinoma, respectively. EBV is also a causative agent in the development of other cancers, including diffuse large B-cell lymphoma (Case Study CS55), classical Hodgkin lymphoma (Case Study CS57) and extranodal NK/T-cell lymphoma (case Study CS63).

CMV in a healthy individual may cause a mild febrile illness, but in immunosuppressed individuals can lead to fatal pneumonia. Measles only rarely has side effects in healthy children, but can cause severe disease, and even death, in others. Rarely, measles sufferers may develop subacute sclerosing panencephalitis (SSPE), a usually fatal disease of the CNS that develops 7–10 years after initial infection.

There are also many 'unknowns' in studies on virus infections. Why, for example, even if not vaccinated, do most people infected with Poliovirus show no signs of ill health, and in those with symptoms, why do only about 10% suffer CNS damage? Likewise, why do a very small proportion of cold sore sufferers develop fatal HSV encephalitis?

Pathogenic viruses must be able to spread from host to host, invading tissues (often mucous membranes), replicating inside cells, interfering with host defences and damaging the host. Cellular changes are both biochemical and morphological (cytopathic effect of viruses) with a shutdown of host cell macromolecular synthesis, chromosome alterations and inclusion body formation, resulting in direct cytotoxicity. The resultant disease syndrome is often a combination of cell damage and host response to infection. Influenza virus, which kills endothelial cells lining the respiratory tract, induces very high levels of interferon-alpha (IFN-α) and interferon-beta (IFN-β), which can give rise to symptoms of lethargy, depression etc. Pathogenesis is therefore a multifactorial phenomenon which can lead to different clinical outcomes.

Inapparent (Asymptomatic) Infection

Many virus infections are *subclinical*, there being no apparent outward symptoms of disease. This is virtually always true in the immune host, in which recovery from a previous infection or vaccination protects the host from virus replication following re-infection by the wild-type virus. However, several viruses, including some respiratory viruses and enteroviruses do not produce clinical symptoms even in non-immune individuals. Thus, the Poliovirus, in 80% of infected individuals, replicates in the epithelial cells of the gastrointestinal tract, is excreted in the faeces, but causes no symptoms. After a few days, the virus is eradicated, and the host acquires life-long immunity.

Disease Syndrome, Virus Eradication and Recovery

The pattern we expect following most virus infections in otherwise healthy individuals is that of clinical symptoms of varying degrees of severity, i.e. *disease syndrome*, followed by virus eradication by the immune system, recovery and often lifelong immunity. This applies to the majority of childhood viral infections, such as measles, mumps, and rubella, as well as numerous respiratory infections caused by a wide variety of viruses, along with a diverse range of other viral infections, including those resulting from Hepatitis A virus (infectious jaundice; Case Study CS1), Rotavirus (gut infections) and Coxsackie virus (myocarditis, pericarditis, conjunctivitis). The following section outlines the typical patterns of infection seen in local and systemic virus infections.

Local Virus Infection

RSV infections are confined to the respiratory tract and are responsible for severe respiratory distress in children less than 12 months old. RSV infection causes necrosis of the bronchiolar epithelium, which sloughs off, blocking the small airways. This in turn leads to obstruction of airflow and respiratory distress. Obstruction is compounded by the increased secretion of mucus and the presence of inflammatory exudates within the airways. This condition, known as *bronchiolitis*, may progress to pneumonia, in which oedema and necrosis of the lung parenchyma result in the filling and collapse of the alveoli. Children recovering from acute RSV bronchiolitis are sometimes left with a weakened and vulnerable respiratory system, predisposing them to a lifetime of chronic lung disease.

Systemic Virus Infection

Following respiratory infection with Measles virus, the virus replicates in the lymph nodes that drain from the infected tissue. The virus spreads to the rest of the lymphoid system and respiratory tract through the blood in a primary viraemia. Giant cells (multinucleated virus-infected cells) are formed in lymphoid tissues and on epithelial surfaces, giving rise to free virus which again enters the bloodstream in a secondary viraemia. Subsequently, the virus infects the skin and viscera, kidneys and bladder. At this stage, the patient is highly infectious and may present with fever, malaise, sneezing, rhinitis, congestion, conjunctivitis and a cough. The distinctive measles rash appears about 14 days post-infection.

Latency

A restricted range of viruses, most being of the herpesvirus family (HSV, Varicella-zoster, EBV and CMV), are not eradicated from the body following recovery, but instead become 'latent' within the host. Virus replication may be initiated some time later (reactivation) and cause clinical symptoms which are similar to those observed in primary infection.

HSV can reactivate many times during the life of an individual and produces the typical painful 'cold-sore' lesions on the mouth or genitals. Following primary infection, HSV travels via the nerve axons to the CNS where it lays dormant in the trigeminal or sacral ganglia. During dormancy, there is no viral replication, indeed no proteins are produced. One mRNA species, the LAT (latency) transcript, is produced but is not translated. HSV may remain dormant for the lifetime of the patient, for several years, or occasionally only for a few months before reactivating. A variety of stimuli, including menstruation, exposure to ultraviolet light and stress, induce the virus to reactivate and to travel back down the axon to the periphery, where it causes the characteristic lesions. HSV can spread by cell-to-cell fusion, thus protecting it from the high concentrations of HSV antibody found in the sera of infected patients. Cytotoxic T-cells are eventually responsible for abrogating clinical infection.

Alternatively, a very different clinical syndrome may result following virus reactivation. Thus, primary infection with the Varicella-zoster virus causes chicken pox, whereas reactivation is associated with the development of shingles. Shingles is characterised by a localised area of extremely painful vesicles from which the virus can be isolated. The lesions may clear after 1–2 weeks but often result in an aftermath of severe neuralgia, which may persist for months or years. However, it is unusual for patients to present with further attacks of shingles unless they are immunocompromised.

It is important to make the distinction here between clinical latency and microbiological latency. *Microbiological latency* defines a situation in which the virus exists within a cell but does not replicate, usually producing only those virus proteins required to maintain the virus within the cell. Microbiological latency usually also results in *clinical latency*, there being no outward signs of infection. The difficulty in accurately defining latency is well illustrated in the case of EBV. Following primary infection, EBV persists by infecting memory B-lymphocytes, maintaining itself in a latent state characterised by the production of only those viral proteins required for the maintenance of the viral genome. However, free virus can be isolated from the saliva of normal individuals during this clinically latent infection, indicating that virus is replicated at low levels in a few cells of the oropharynx. Clinical reactivation of EBV can occur during immunosuppression and is characterised by the onset of the symptoms typical of infectious mononucleosis with or without the development of lymphoproliferative disease.

Carrier/Persistent State

Although a rare event after virus infection, the virus carrier state is often seen following HBV, HCV or HIV infection. Around 5–10% of individuals infected with HBV will carry the infective particles in their blood for months or years (see Case Study CS4). Worldwide, between 240 and 350 million people live with chronic HBV infection and have a greatly increased risk of developing cirrhosis and hepatocellular carcinoma. The immediate clinical consequences of infection with HCV are usually very mild; 70–80% of infected individuals are asymptomatic. However, most primary infections become established as long-term, usually lifelong, chronic persistent infections. HCV evades the immune system through rapid mutation, producing a myriad of viral variants, and by interfering with the host's antiviral defenses, including impairing key immune cells and inducing exhaustion of virus-specific T-cells. These combined strategies enable HCV to establish and maintain chronic infections in the host. Globally, an estimated 58 million people have chronic HCV infection, with about 1.5 million new infections occurring per year. As with HBV, there is an increased risk of developing cirrhosis and hepatocellular carcinoma. However, the introduction of direct-acting antivirals (DAAs) has revolutionised the treatment landscape for individuals infected with HCV, offering cure rates exceeding 90%. Unlike previous treatments, which often had severe side effects and lower efficacy, DAAs specifically target steps in the HCV life cycle, ensuring minimal adverse effects and shorter treatment durations. As a result, many patients previously ineligible or unresponsive to treatment can now achieve sustained viral clearance, drastically reducing the risk of liver-related complications and improving overall quality of life.

As described earlier, following exposure to HIV, the provirus becomes integrated into the host genome. Primary infection results in a significant viraemia, which usually subsides to extremely low levels. However, HIV continues to replicate at a low level. During this time, the virus undergoes mutations, some of which result in more pathogenic and less immunogenic forms. The infected individual (who is antibody positive) will carry infectious virus and virus-infected cells in their blood, semen vaginal fluid or saliva. As the disease progresses, several key events occur that contribute to the worsening of the condition. These include the eventual depletion of CD4+ T cells, increased viral replication and the development of opportunistic infections and malignancies. However, the progression of HIV can be managed effectively using ART. Modern ART regimens are more potent than before, have fewer side effects and often involve a combination of drugs in a single tablet taken once daily. As a result, individuals with HIV, when adhering to their prescribed ART, can achieve and maintain an undetectable viral load, reducing the risk of HIV transmission and allowing them to lead longer, healthier, lives.

Host Immunosuppression

Although mild immunosuppression can often result from virus infections such as measles and glandular fever, the advent of AIDS has highlighted how dramatic virus invasion of the immune system can be. HIV infects and grows best in cells that express CD4 receptors (i.e. T-helper cells). The progression from HIV positivity to a pre-AIDS syndrome and ultimately to AIDS is usually best indicated by a rise in viral load (the number of viruses present within an infected individual) and a dramatic drop in the CD4+ T-cell count, extremely low levels demonstrating the extent to which the patient may become immunosuppressed. This drop in CD4+ T-cells is brought about by the virus, which lyses the infected cell, releasing hundreds of progeny virions, which infect and kill other CD4+ T-cells. T-helper cells are of central importance in the immune system (Chapter 6) and their depletion leads rapidly to infections with opportunistic pathogens, including HSV, CMV, *Pneumocystis jirovecii* and *Candida albicans* (see Case Study CS6).

Death

While many virus infections are fatal in distinct circumstances and others in a percentage of victims, some are almost always fatal. For example, almost all patients who develop symptoms of Rabies virus infection will die within 7–12 days. Following a bite, the virus enters the peripheral nerves and moves centripetally to the spinal cord and brain where it replicates. The virus then leaves the CNS and spreads centrifugally to virtually all tissues of the body, including the salivary glands where it is excreted in saliva. The patient develops a variety of abnormalities including *hydrophobia* (aversion to water), rigidity, *photophobia* (aversion to light), *fasciculations* (muscle twitches) and *paresis* (motor weakness), cerebellar signs, cranial nerve palsies, hypo- or hyperreflexia, focal or generalised convulsions and a variety of autonomic disturbances. Development of a flaccid paralysis and onset of coma precede fatal complications. Treatment for Rabies virus infection can be effective if given early after infection and before the development of symptoms and is referred to as *post-exposure prophylaxis* (PEP). Rabies vaccinations are the cornerstone of PEP and are administered in a series of doses over a period of weeks. Anti-rabies immunoglobulin, a concentrated solution of antibodies directed against the Rabies virus, can also be included and provides immediate passive immunity while the body's own immune response is being stimulated by the vaccines. PEP can also be used to prevent disease caused by other virus infections, for example, HIV and HBV.

Virus Evolution and Pandemics

Viruses mutate over time, and this can account for changes in virus behaviour. Genomic mutations occurring in a viral sequence can be beneficial, neutral or deleterious. Beneficial mutations increase the fitness of the virus by enhancing replicative efficiency, transmissibility or immune escape. On the contrary, deleterious mutations hamper efficient replication and transmission, and viruses with such mutations will likely eventually disappear.

Of all the mutations in different regions of the SARS-CoV-2 genome, the spike region is a major focus because the spike protein interacts with the virus receptor, angiotensin-converting enzyme 2 (ACE2) on target cells. Several SARS-CoV-2 variants of interest (VOIs) and variants of concern (VOCs) with mutations in spike regions have been characterised (https://www.who.int/en/activities/tracking-SARS-CoV-2-variants/). To date several VOCs [B.1.1.7 (Alpha), B.1.351 (Beta), P.1 (Gamma), B.1.617.1 (Kappa), B.1.617.2 (Delta), B.1.525 (Eta), B.1.526 (Iota), C.37 (Lambda), B.1.621 (Mu) and at the time of writing B.1.1.529 and BA.1 and BA.2 (Omicron)] have been reported to influence viral transmission, disease severity, binding ability and vaccine efficacy. The Omicron variant carries >46 mutations in the spike protein with higher transmissibility and infectivity, allowing it to outcompete other variants on the global stage.

In general, more complex organisms have a lower mutation frequency. John Drake showed that DNA-based life forms mutate at a roughly constant rate of 0.003 mutations per genome/multiplication round, known as Drake's rule. Mutation rates for RNA viruses, which have generally smaller and less complex genomes, are higher (as they are for smaller bacterial DNA genomes). For larger genomes, a proofreading-repair function helps to ensure genome integrity. Thus, for all organisms, including viruses, there is a trade-off between maintaining viable genomes (reduced by excessive mutation) and rapid evolution of potentially beneficial variants (enhanced by increased mutational frequency).

Another example of the importance of mutations in virus genomes is provided by the Influenza virus. In this case, mutations resulting in amino acid substitutions in the surface antigens, haemagglutinin and neuraminidase, are responsible for '*antigenic drift*'. On the other hand, the exchange of segments of genome encoding the surface antigens, particularly between human and animal influenza viruses, known as '*antigenic shift*', is responsible for influenza pandemics. The ability of influenza viruses to evade host immunity

is the reason why influenza vaccines must be updated annually. However, at the time of writing, there is considerable promise that an mRNA-based vaccine (similar to some of the successful vaccines deployed against SARS-CoV-2) could provide protection against all known subtypes of influenza virus.

Pathogenic Fungi

Fungal infections (*mycoses*) contribute significantly to the global burden of disease, particularly in individuals with weakened immune systems. While precise data on the global burden of fungal infections is challenging to determine due to underdiagnosis and lack of reporting in some regions, estimates suggest that several hundred million people worldwide are affected by fungal infections each year. The burden is particularly high in certain geographical regions, for example, in areas with a high prevalence of HIV infection, in some resource-limited populations and in areas with specific environmental conditions conducive to fungal growth.

Superficial mycoses are fungal infections that primarily affect the outermost layers of the skin, hair and nails. They are usually limited to the surface of the body and do not usually invade deeper tissues or organs. *Deep mycoses*, on the other hand, occur when fungi invade and spread to internal organs and deeper tissues. These infections can be severe and may pose a risk to life. Deep mycoses are usually caused by opportunistic fungi, which take advantage of weakened immune systems. Some usually superficial fungal infections, for example *Candida albicans*, can become invasive under certain conditions.

Growth Characteristics of Pathogenic Fungi

Pathogenic fungi exist as either filamentous forms (moulds, or 'mold' in the United States) or yeasts. Some fungi exhibit both growth forms, depending on environmental conditions, and are known as *dimorphic fungi*. *Filamentous forms*, or molds, consist of branching hyphae that form a network known as a *mycelium*. The hyphae can extend through transverse divisions, allowing the fungus to grow and spread. Asexual reproduction in molds occurs through the production of spores called *conidia*, which are responsible for the dissemination of the fungus. *Yeasts*, on the other hand, are single-celled organisms that reproduce by simple division. They can undergo budding, in which a smaller daughter cell forms as an outgrowth of the parent cell. In some cases, yeasts may form chains of cells known as *pseudohyphae*.

Among the fungi causing human infections, many are typically found in the environment as free-living *saprophytes* (organisms that obtain nutrients from decaying organic matter) or as plant parasites. Infection occurs when these fungi are accidentally introduced into host tissues. Many of these pathogenic fungi have adapted to warm habitats, such as self-heating accumulations of decomposing vegetation, resulting in an optimum growth temperature close to that of the human body.

Some commonly encountered pathogenic fungi are described below. They include *Candida* spp. (causing candidiasis), *Aspergillus* spp. (causing aspergillosis; see also Case Study CS2), *Cryptococcus neoformans* (causing cryptococcosis), *Histoplasma capsulatum* (causing histoplasmosis) and *Coccidioides* spp. (causing coccidioidomycosis). *Pneumocystis jirovecii* is a yeast-like fungus that is discussed in detail in Case Study CS6.

Examples of Medically Important Fungi

Candida Species

Candida albicans is a commensal organism found in the gut of many species of mammals and birds, particularly those which habitually have a diet high in sugar and other carbohydrates. Approximately 60–70% of healthy people carry *C. albicans* in the intestine and 40–50% also in the mouth. Around 40% of women have *C. albicans* in the vagina. Infections by *C. albicans* (and, less commonly, by other species of *Candida* such as *C. tropicalis*, *C. krusei* and *C. parapsilosis*) can take many forms depending upon the part of the body involved and range from superficial skin rashes or diseases of mucous membranes, to fatal infections of deep organs.

Candida auris is a newly emergent fungus that poses a global health threat. This is because it has been shown to be resistant to multiple anti-fungal drugs used to treat *Candida* infections. Moreover, standard laboratory methods used to identify *Candida* spp. may not accurately detect *C. auris*. It is responsible for outbreaks in hospitals and long-term care facilities and can persist on surfaces and equipment, making it difficult to control. Adhesion of *Candida* spp. is mediated by various fungal adhesins that interact with specific receptors on host cells or on the extracellular matrix.

Superficial Infections

Candida infection of the skin of the napkin area is a common problem in infants, producing *macular erythema* (discrete patches of redness, especially in the skin folds and inguinal fold). Smaller 'satellite' lesions also occur outside the main area of reddening. Similar changes are seen in adults, especially in diabetics who are particularly prone to skin infections. *Candida* infection of the mouth and vagina (often referred to as 'thrush') represents another common form of superficial infection. White plaque-like lesions are produced on the epithelial surface which are seen as reddened areas if the plaque is removed. The plaque consists of a mixture of keratinized epithelial cells and fungal cells. Oral thrush is common in infants, in diabetic patients and under denture plates. It is also commonly encountered in patients with HIV/AIDS. Vaginal thrush causes itching, a burning sensation and discharge. *Chronic mucocutaneous candidiasis* (CMC) is a special type of superficial infection that occurs in children. CMC is often indicative of an underlying immune deficiency, specifically in the T-cell mediated immune response. It is characterised by the development of extensive and persistent skin infection and oral thrush at a very early age. As the child grows, the infection causes disfiguring scaly swellings.

Deep Infections

Serious, often fatal, deep infection occurs in several groups of patients. First, among patients undergoing immunosuppressive therapy for organ transplantation, *Candida* infections of deep organs such as liver, kidney and brain represent a common cause of morbidity and mortality. A second major group of cases occurs among surgical patients, often without severe immunosuppression but in whom physical barriers to infection have been removed. A smaller third group is represented by intravenous drug abusers who may inject *Candida* organisms directly into the bloodstream. *Candida* septicaemia (the presence of *Candida* organisms in the bloodstream) is a serious complication in immunosuppressed patients. Clinical indicators of systematically spread *Candida* are the development of discrete cutaneous lesions and *endophthalmitis* (infection within the eye). *Candida* septicaemia in immunocompetent individuals may result in colonisation of the endocardium leading to endocarditis, particularly if there is already some abnormality of heart structure.

Aspergillus Species

Aspergillosis is the clinical term given to disease produced by members of the genus *Aspergillus*. These saprophytic filamentous fungi are ubiquitous in the environment, for example, in decaying organic debris. *Aspergillus fumigatus* is the most common

Figure 4.7 Scanning electron micrograph of Aspergillus producing spores, computer-coloured green. The picture shows an *Aspergillus* conidiophore with conidia (spores) budding off. *Source:* Wellcome Collection / https://wellcomecollection.org/works/xpmvf84z / last accessed November 13, 2023.

pathogen in this genus, although *A. flavus*, *A. terreus*, *A. nidulans* and *A. niger* can cause disease. Asexual spores (conidia) are produced in vast numbers from the heads of maturing conidiophores, from which they are liberated into the environment (Fig. 4.7). *Aspergillus* spp. are opportunistic pathogens that are unable to invade or colonise healthy lungs but can cause devastating systemic infection in immunocompromised patients. The two most common forms of aspergillosis that occur in immunocompetent people are allergic aspergillosis and aspergilloma; in neither of these is the fungus invasive.

Allergic aspergillosis is a localised reaction to inhaled spores and the mycelium produced by their subsequent germination in the bronchi. It is primarily a type 1 (IgE-mediated) hypersensitivity response (Chapter 6) and may be particularly severe in atopic individuals. The production of circulating aspergillus-specific IgG antibody can also result in the formation of antibody–antigen complexes leading to complement activation and localised tissue damage.

Aspergilloma is the name given to the fungus ball that can colonise old tuberculosis lesions or other pre-existing lung cavities. The fungus grows as a compact mass of mycelium and normally does not invade adjacent tissue. This infection is only serious if the patient becomes immunosuppressed or if the fungus ball impinges on a blood vessel. Damage to the blood vessel can lead to bleeding in the lungs which is often clinically apparent as haemoptysis (see also Case Study CS2).

The most serious form of the disease is *invasive aspergillosis*. This is seen in immunocompromised patients, especially those with low neutrophil counts. In these patients, inhalation of spores may be followed by extensive growth of fungal forms within the lungs and frequent dissemination to other organs, especially the brain. Invasive aspergillosis is rapidly fatal if left untreated.

It is noteworthy that while *Aspergillus* is not directly carcinogenic, it can produce cancer-causing substances, known as aflatoxins. Aflatoxins are naturally occurring mycotoxins produced by certain species of *Aspergillus*, primarily *Aspergillus flavus* and *Aspergillus parasiticus*. These toxins are formed by fungal contamination of various food commodities, such as grains, nuts and spices. Prolonged consumption of aflatoxin-contaminated food increases the risk of developing hepatocellular carcinoma.

Cryptococcus neoformans

Cryptococcus neoformans is a fungal pathogen that causes cryptococcosis. Most infections occur in the lungs. However, it can also cause meningitis and encephalitis, particularly in individuals with weakened immune systems. Infections with *C. neoformans* are rare in people with intact immune systems. The fungus is a facultative intracellular pathogen and can exist in macrophages, providing a means for its spread within the body; the intracellular lifestyle contributing to its ability to evade immune responses and establish persistent infections. In 2006, *Cryptococcus neoformans* was shown by two independent research groups – one in the United Kingdom and the

other in the United States – to exhibit *vomocytosis*, a process in which intact fungal organisms are ejected from infected macrophages. Following its initial discovery, vomocytosis was found to be utilised by other fungal pathogens, including *Candida albicans* and *Candida krusei*. By escaping in this way from infected immune cells, the fungal cells can disseminate and establish infections in other tissues or infect new host cells. *Cryptococcus* spp. also produce extracellular enzymes, for example, proteases that allow tissue invasion, facilitating the spread of the infection.

Histoplasma capsulatum

Histoplasma capsulatum is a dimorphic fungus. In its natural environment, such as soil enriched with bird or bat droppings, it exists as a mycelial form, consisting of hyphae. At ambient temperatures, the fungus remains in this mycelial form. However, when it infects mammals, including humans, and encounters the higher body temperature, it undergoes transformation to a yeast form (Fig. 4.8). *Histoplasma capsulatum* is adapted to survive within macrophages. The fungus is globally distributed but is endemic in North and

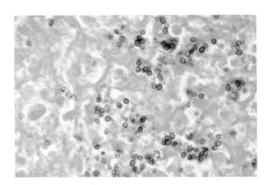

Figure 4.8 *Histoplasma capsulatum* yeasts. Grocott silver stain. This is a high-power view of a lung lesion showing several small black-staining yeasts within necrotic material. Creative Commons Zero (CC0) terms and conditions https://creativecommons.org/publicdomain/zero/1.0. *Source:* Wellcome Collection / https://wellcomecollection.org/works/pj5584yp/items / last accessed November 13, 2023.

Central America, parts of Europe (especially eastern and southern Europe), Africa, eastern Asia and Australia, the soil in these endemic areas provides an ideal environment for the growth of the organism in its mycelial form. In some cases, histoplasmosis may remain asymptomatic or cause only mild flu'-like symptoms. However, in individuals with weakened immune systems, infection can spread beyond the lungs and involve other organs, leading to severe illness.

Coccidioides Species

The name 'Coccidioides' derives from its resemblance to the protozoan Coccidia. *Coccidioides* spp. are dimorphic fungi and the causative agents of coccidioidomycosis, also known as San Joaquin Valley fever, a fungal infection primarily found in the Western Hemisphere, particularly in the Southwestern United States. The disease is acquired by inhaling spores of *Coccidioides* spp. The two main species responsible are *Coccidioides immitis* (the term 'immitis', referring to its harsh clinical course) and *Coccidioides posadasii*. Approximately 60% of infected individuals show no symptoms, while the remaining 40% have an acute respiratory illness characterised by fever, cough and pleuritic pain; in some cases, this can lead to acute respiratory distress syndrome or fatal multilobar pneumonia. Coccidioides infection can also develop into a severe form of meningitis.

Dermatophytosis (Ringworm)

Ringworm is caused by a group of fungi known as *dermatophytes*. These fungi live on the keratinized tissues of the skin, hair and nails and although closely related to certain soil fungi, many have become adapted to a purely parasitic existence. Dermatophyte fungi produce enzymes, such as keratinases, that allow them to degrade keratin, enabling invasion and colonisation of the keratinized layers. Human infections can be caused by

- *Anthropophilic dermatophytes* which are entirely dependent on passage from one human host to another. Infections can occur in any body site. They are passed on to susceptible hosts by direct contact or by contact with shed infected skin cells in which the fungus can lie dormant for several years.
- *Zoophilic species* which are adapted to life on animal hosts. When they are passed to humans, infection most often occurs in the exposed sites of the head, neck, trunk and limbs.
- *Geophilic species* are more suited to a saprophytic existence in the soil but can cause infection following contact of the host with contaminated soil.

There are three main genera of dermatophyte fungi: *Trichophyton, Microsporum* and *Epidermophyton*. The three genera are indistinguishable in tissue. In culture, all three genera produce distinctive multicellular macroconidia or large spores. In addition, *Trichophyton* and *Microsporum* species produce unicellular macroconidia. Hair infections are characterised by the production of large numbers of athroconidia (spores formed from fragmentation of hyphae). These may be *ectothrix*, forming a sheath around the hair, or, *endothrix* in which hair penetration is followed by spore production within the hair shaft.

The most common, and therefore the best known, anthropophilic dermatophytosis is athlete's foot or *tinea pedis*, caused most often by *Trichophyton rubrum* or *T. interdigitale*. Athlete's foot is common in affluent countries in which there is ready access to swimming baths and sports facilities where the use of communal changing rooms facilitates transfer from host to host. Occlusive footwear also provides the warm, moist conditions in which the fungus thrives. Athlete's foot is characterised by itching, skin peeling, maceration and fissuring of the toe webs, often with concomitant nail infection in which the nails may become discoloured and crumbly. Lesions on smooth skin are itchy, erythematous and flaky and may take on the circular appearance from which the term 'ringworm' was derived. Infections with zoophilic species often induce a more marked inflammatory response. Occasionally, suppurating highly inflammatory lesions known as *kerions* may result.

Mycetoma

This exemplifies a range of mycoses which are mainly subcutaneous in nature, resulting from traumatic inoculation of a saprophytic fungus through the skin. Many fungal species and aerobic actinomycetes (bacteria capable of forming hyphal forms) can cause mycetoma and, whilst they show minor differences in their histological appearance, there are several unifying features. The inoculated fungi grow slowly in the presence of a massive and intense host inflammatory response. The resulting mycelial microcolonies appear as tightly packed 'grains' just visible to the naked eye. These microcolonies break apart and re-form, enabling a slowly progressive, localised infection to develop. There is often massive swelling of the affected area, which may penetrate and erode underlying bone. Sinus tracts filled with pus develop around the grains and these may drain out onto the surface of the skin. Left untreated, many mycetomas will continue an inexorable spread further and further from the original site and cause death if vital viscera become involved. Initial infections most often occur on the feet and render amputation necessary in severe cases.

Parasitic Infections

Earlier in this chapter, a parasite was defined as an organism which lives on or in another organism, takes nutriment directly from it and which, under certain circumstances, may be harmful. However, in the study of microbiology, the term 'parasite' is often used to designate protozoa and members of the animal kingdom that infect and cause disease in other animals, including humans.

The life cycles of parasites are often complex, but may be separated into two broad types: *direct* (monoxenous) and *indirect* (heteroxenous). Parasites with direct life cycles spend most of their adult lives in one host (known as the *parasitic stage*), with their progeny transmitted from one host to another. Direct parasites often lack an intermediate stage and must leave their host. To do this, they must be able to survive in an environment outside their original host (known as the *free-living stage*) and then locate and establish in a new host. *Entamoeba histolytica* is an example of a parasite with a direct life cycle (Figs. 4.9 and 4.10).

Parasites with indirect life cycles are characterised by two host stages; the *definitive* (primary) host stage required for reproduction and the adult life phase and an *intermediate* (secondary) host within which parasite development occurs. Multiple developmental stages may take place in an intermediate host. *Plasmodium* and *Leishmania* species are examples of parasites with indirect life cycles (see Case Study CS8 and Fig. 4.11).

The so-called *reservoir hosts* typically tolerate parasites with no ill-effects. However, reservoir hosts serve as a source of infection for other susceptible species, which has significant implications for disease control. Reservoir hosts can be infected multiple times and play a crucial role in maintaining and spreading the pathogen to other species. *Paratenic* ('transport') *host* is a term given to an organism that carries the sexually immature form of a parasite but is not necessary for the parasite's development. Instead, it acts as a temporary 'dump', in which the non-mature stages of the parasite can accumulate in high numbers. *Dead-end host* (also called incidental or accidental host) is a term used to describe a host that does not allow the transmission of the parasite to the definitive host, preventing the parasite from completing its development.

Parasites are conveniently separated into three major groups: *the protozoa, helminths and arthropods.*

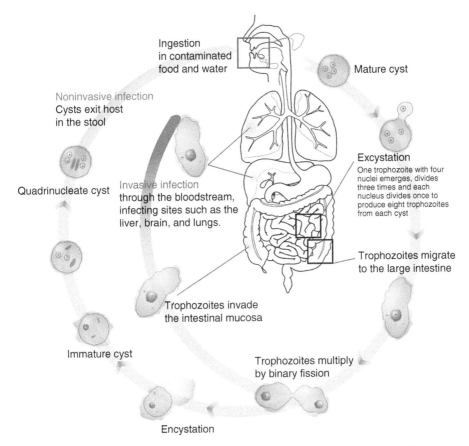

Ingestion
in contaminated
food and water

Mature cyst

Noninvasive infection
Cysts exit host
in the stool

Excystation
One trophozoite with four
nuclei emerges, divides
three times and each
nucleus divides once to
produce eight trophozoites
from each cyst

Quadrinucleate cyst

Invasive infection
through the bloodstream,
infecting sites such as the
liver, brain, and lungs.

Trophozoites migrate
to the large intestine

Trophozoites invade
the intestinal mucosa

Immature cyst

Trophozoites multiply
by binary fission

Encystation

Figure 4.9 *Entamoeba histolytica* life cycle: *E. histolytica* is a parasitic protozoan which infects predominantly humans and other primates. The life cycle begins when the mature cysts of *E. histolytica* are ingested in contaminated food or water. In the host's small intestine, the cysts release infective trophozoites (excystation) due to the action of gastric acid and enzymes. The trophozoites migrate to the large intestine, where they multiply by binary fission. In some cases, trophozoites can invade through the intestinal wall and enter the bloodstream to reach various organs, most commonly the liver. This can lead to extraintestinal infections, including amoebic liver abscesses. Alternatively, trophozoites can differentiate into immature, and then mature, cysts (encystation). These cysts are resistant to the harsh external environment and can survive outside the host's body. https://commons.wikimedia.org/wiki/File:Entamoeba_histolytica_life_cycle-en.svg. *Source:* LadyofHats, Public domain, via Wikimedia Commons.

Protozoa

Protozoan infections are parasitic diseases caused by organisms formerly classified in the kingdom Protozoa. These single-celled eukaryotic organisms can be found in diverse habitats and exhibit a wide range of characteristics and behaviours. We consider some of the important diseases caused by these organisms below. *Plasmodium* spp. which cause malaria, and

Toxoplasma gondii, responsible for toxoplasmosis, are considered elsewhere (Case Studies CS8 and CS5, respectively).

Amoebiasis

Amoebiasis is caused by *Entamoeba histolytica*. Cysts of *Entamoeba histolytica* are transmitted through the faecal-oral route, usually from contaminated food or water. The parasite

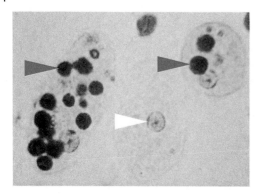

Figure 4.10 Trophozoites of *Entamoeba histolytica* with ingested erythrocytes (trichrome stain). The ingested erythrocytes appear as dark inclusions (red arrowheads). Erythrophagocytosis is the only characteristic that can be used to morphologically differentiate *E. histolytica* from the non-pathogenic *E. dispar*. In these specimens, the parasite nuclei have the typical small, centrally located karyosome (white arrowhead), and thin, uniform peripheral chromatin. *Source:* Centers for Disease Control and Prevention / Wikimedia Commons / Public Domain.

usually colonises the gastrointestinal tract but may also invade other tissues. Most infected people will show no symptoms; in these cases, *Entamoeba histolytica* primarily feeds on bacteria and food particles in the gut. However, in some cases, it can invade through the epithelium of the gastrointestinal tract, utilising enzymes, such as cysteine proteases, to do so. As a result of the tissue damage caused by the parasites, flask-shaped ulcerations may develop in the intestine. When observed in stool samples, the amoeba can often be seen with red blood cells inside (erythrophagocytosis; Fig. 4.10). In some cases, a granulomatous mass known as an *amoeboma* may form in the wall of the colon.

Trypanosomiasis

The Old-World trypanosomes, *Trypanosoma brucei gambiense* and *Trypanosoma brucei rhodesiense* cause chronic African trypanosomiasis (also known as West African sleeping sickness) and acute African trypanosomiasis (also known as East African sleeping sickness), respectively. The life cycle of African trypanosomes involves both a mammalian host and a tsetse vector. As the parasites proliferate in the bloodstream, the host immune system, mediated by antibodies to variant surface glycoproteins (VSGs) and accompanied by host expression of inflammatory cytokines and the recruitment of immune cells, eliminates most of the pathogens. However, a few variants escape by switching their VSG genes. The host mounts a second immune response directed against the altered VSG, before new escape variants emerge, and the cycle is repeated. This explains the cyclical nature of the fever observed in people infected with these organisms.

Chagas disease (AKA American trypanosomiasis), caused by the parasite *Trypanosoma cruzi*, is transmitted by insects known as 'kissing bugs' that belong to the subfamily *Triatominae*. In the early stages of infection, symptoms may be absent or mild (e.g. fever, lymphadenopathy, headaches and localised swelling at the site of the bite). However, if left untreated, the infection usually progresses to a long-term asymptomatic chronic phase. Some 10–20 years after initial infection, around 20–30% of chronically infected people will develop complications that include heart disease, known as *Chagasic cardiomyopathy* or enlarged oesophagus (*megaesophagus*) and/or colon (*megacolon*).

Leishmaniasis

Leishmaniasis, a neglected tropical disease that affects millions globally, is caused by the *Leishmania* genus of parasites. These parasites are transmitted to humans through the bite of infected phlebotomine sandflies (Fig. 4.11). The disease is common in Asia, Africa, South and Central America, as well as in southern Europe. Leishmaniasis primarily manifests in three forms: cutaneous, mucocutaneous and visceral. *Cutaneous leishmaniasis* typically results in skin ulcers, leading to disfiguring scars, although it

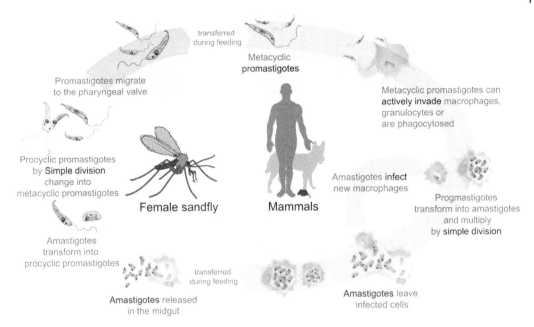

Figure 4.11 Life cycle of parasites from the genus *Leishmania*. Leishmaniasis is a vector-borne parasitic disease caused by various species of *Leishmania* protozoa. Leishmaniasis is transmitted through the bite of female phlebotomine sandflies, which deliver the infective stage known as promastigotes into the host's skin during blood feeding. The promastigote form is found in the alimentary tract of sandflies. It is an extracellular and motile form with a long flagellum projected externally at the anterior end. There are two kinds of promastigote, procyclic forms which are adapted to survival in the sandfly and the metacyclic forms which can infect mammals. Upon entering the host, the metacyclic promastigotes invade or are phagocytosed by macrophages in the vicinity of the puncture wound and undergo transformation into amastigotes, the intracellular and non-motile form, being devoid of external flagella. Amastigotes undergo multiplication within host cells. Sandflies become infected with *Leishmania* parasites while feeding on an infected host, including consuming macrophages that contain amastigotes. Within the sandfly, these parasites transform into promastigotes in the midgut, where they replicate and eventually migrate to the proboscis, preparing for transmission to another host. https://commons.wikimedia.org/wiki/File:Leishmaniasis_life_cycle_diagram_en.svg. *Source:* LadyofHats, Public domain, via Wikimedia Commons.

does not generally result in death. *Mucocutaneous leishmaniasis* causes ulcerations, not only on the skin but also in the mouth and nose. *Visceral leishmaniasis*, also known as *kala-azar*, is a severe and potentially life-threatening form of leishmaniasis. It primarily affects the internal organs, including the spleen, liver and bone marrow. The disease initially presents with skin ulcers. However, unlike cutaneous leishmaniasis, visceral leishmaniasis progresses beyond the skin and affects the entire body. As the disease progresses, individuals experience fever, hepatomegaly, splenomegaly, anaemia, thrombocytopenia and lymphadenopathy. If left untreated, visceral leishmaniasis can be life-threatening.

Helminths

The term 'helminth' is a broad and somewhat artificial term used to describe a group of organisms that share a similar body form but are not necessarily closely related in evolutionary terms. The taxonomy of helminths is still the subject of debate. The major groups are briefly outlined below:

Tapeworm Infections

Tapeworm infections also known as cestodiasis, are caused by flat, segmented worms belonging to the class *Cestoda*. These parasites typically reside in the intestines. Examples of disease-causing tapeworms include *Taenia solium*, the pork tapeworm, which is acquired by consuming undercooked pork contaminated with cysts. Infection can lead to neurocysticercosis, a condition in which the larval forms of the tapeworm infect the brain, muscles and other tissues, causing seizures and neurological symptoms. *Taenia saginata* is the beef tapeworm and is transmitted via the consumption of raw or undercooked beef containing tapeworm larvae. Infection with *T. saginata* usually causes only mild or asymptomatic intestinal infection. *Echinococcus granulosus*, the dog tapeworm, is considered in detail in Case Study CS7.

Fluke Infections

Fluke infections or trematodiases are caused by parasitic flatworms, known as flukes. These worms have complex life cycles involving intermediate hosts, such as snails or fish. Examples of disease-causing flukes include *Schistosoma* spp., the blood flukes that cause schistosomiasis. Among the six species implicated in human schistosomiasis, *S. haematobium* and *S. mansoni* are the most prevalent, affecting millions of individuals worldwide. Mortality rates are estimated to be around 280 000 deaths per year, and a health burden estimated at over 70 million disability-adjusted life years. Chronic infection can result in portal hypertension, oesophageal varices, hepatic periportal fibrosis and ascites. Urinary bladder cancer is recognised as a well-known late complication of schistosomiasis, with infected individuals having a 2–14 fold increased risk (Fig. 4.12). The transmission of schistosomiasis is closely linked to poverty and the use of contaminated river water.

Fasciola hepatica, commonly known as the liver fluke (Fig. 4.13), causes the disease known as fascioliasis. The life cycle of *Fasciola hepatica* involves two primary hosts: freshwater snails and mammals, including humans and livestock. The transmission of *Fasciola hepatica* occurs through the ingestion of

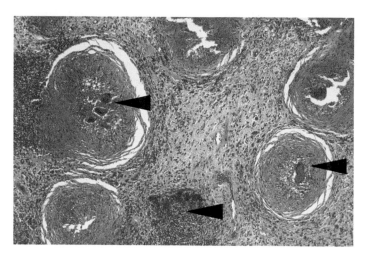

Figure 4.12 Schistosomiasis due to *Schistosoma haematobium*: histopathology of the urinary bladder. This is a medium power view of biopsy tissue taken from a patient 1 month after starting anti-schistosomal chemotherapy. H&E stain. There are numerous well-defined granulomas (examples shown by black arrowheads) within the fibrotic suburothelial tissue. Some dead schistosome eggs can also be seen. The inflammation of acute schistosomal cystitis may be granulomatous but is often more diffuse and flooded with eosinophils. After therapy, a more granulomatous appearance, such as shown here, is often seen. Creative Commons Zero (CC0) terms and conditions. *Source:* Wellcome Collection / https://wellcomecollection.org/works/m56du77e / last accessed November 13, 2023.

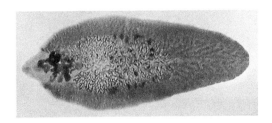

Figure 4.13 *Fasciola hepatica.* Slide of *Fasciola hepatica* (from teaching slides at the University of Edinburgh). https://commons.wikimedia.org/wiki/File:Fasciola_hepatica2.jpg. *Source:* Own work / Wikimedia Commons / Public Domain.

aquatic vegetation contaminated with metacercariae, the infective stage of the parasite. The metacercariae attach to the vegetation and can survive in water for extended periods. When humans or animals consume the contaminated vegetation, the metacercariae are released into the intestines and penetrate the gut wall to enter the abdominal cavity. The immature flukes migrate to the liver, causing inflammation (which can become granulomatous), with fibrosis and bile duct obstruction. As the flukes mature, they eventually reach the bile ducts, where they release eggs that are passed into the intestine and excreted in the faeces, completing the life cycle.

Clonorchis sinensis is another species of liver fluke that infects humans through the consumption of raw or undercooked fish. The parasites reside in the bile ducts, and chronic infection increases the risk of cholangiocarcinoma. The life cycle of *C. sinensis* involves multiple stages and hosts, beginning with freshwater snails, which are the first intermediate host (Fig. 4.14). The snails ingest the embryonated eggs of *C. sinensis* from contaminated water sources. Inside the snail, the eggs release miracidia, which undergo further development into sporocysts, rediae and cercariae. Cercariae, the free-swimming larvae, are released from the snail and enter freshwater fish or shrimps, which act as the second intermediate host. Inside the fish or shrimp, the cercariae encyst and become metacercariae. Definitive hosts, including humans and

other carnivores, such as cats and dogs, become infected by consuming raw or undercooked fish or shrimp containing the infectious metacercariae. Upon ingestion, the metacercariae *excyst* (the process by which a parasite, enclosed within a protective cyst, emerges or is released from the cyst) in the duodenum and migrate to the hepatobiliary system, primarily the bile ducts. Within the hepatobiliary system, the parasites mature into adult trematodes and release eggs. These eggs are passed from the bile into the gut and are excreted via the stool of infected individuals. If the excreted eggs reach freshwater environments, they can be ingested by snails, restarting the life cycle.

Paragonimiasis is a food-borne parasitic disease caused by various species of lung flukes belonging to the genus *Paragonimus*. The primary mode of transmission is through the consumption of undercooked or raw crustaceans, such as crabs and crayfish, that harbour the infective metacercariae stage. Paragonimiasis can also be acquired by consuming the meat of infected mammals that have ingested the metacercariae from crustaceans. There are more than 40 identified species of *Paragonimus*, 10 of which are reported to cause human disease. The most common cause of human paragonimiasis is *Paragonimus westermani*, also known as the oriental lung fluke, which is common in parts of Asia, including China, Japan and Korea. An estimated 22 million people worldwide are affected by paragonimiasis each year. The infection primarily affects the lungs, where the adult flukes reside and lay eggs. The clinical manifestations of paragonimiasis vary, but commonly include chronic cough, chest pain and haemoptysis.

Nematode Infections

Nematode infections, or nematodiases, are caused by roundworms belonging to the phylum *Nematoda*. Examples of disease-causing roundworms include *Ascaris lumbricoides* which infect humans through the faecal-oral route. The life cycle of *Ascaris lumbricoides* is

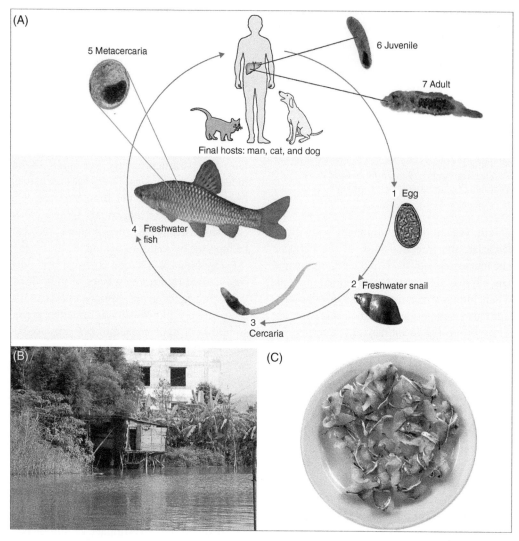

Figure 4.14 Life cycle of *Clonorchis sinensis*. (A) Life cycle of *C. sinensis* (blue colour represents water environment). (1) Eggs enter the water environment from faeces; (2) eggs are ingested by specific freshwater snails, the first intermediate host; (3) cercariae released from the snail; (4) cercariae infect specific freshwater fish, the second intermediate host; (5) cercariae encyst in the subcutaneous tissues or muscles of the fish to form metacercariae; (6) metacercariae are ingested by man (and other hosts) through consumption of raw or undercooked fish, and excyst to become juvenile flukes in the duodenum; (7) hermaphroditic adult *C. sinensis* are predominantly found in intrahepatic bile ducts. (B) A hanging toilet above a river. (C) Raw fish dish. *Source:* Men-Bao Qian et al. 2016 / Reproduced with permission from Elsevier.

complex (Fig. 4.15). Adult females release eggs that are excreted in the faeces. Human infection occurs when individuals ingest food or water contaminated with embryonated eggs. In the duodenum, each ingested egg hatches into larvae. The larvae penetrate the intestinal wall and enter the bloodstream. When they reach the pulmonary circulation, they break through the walls of the pulmonary capillaries to enter the alveoli. From the

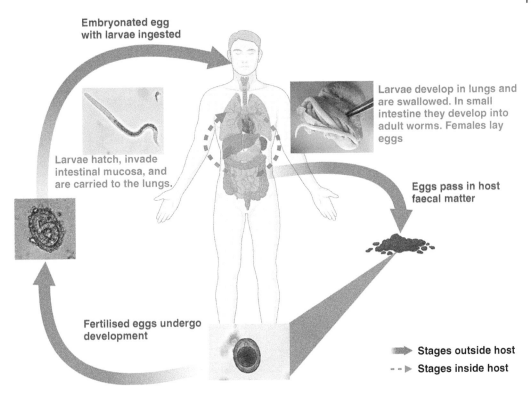

Embryonated egg with larvae ingested

Larvae develop in lungs and are swallowed. In small intestine they develop into adult worms. Females lay eggs

Larvae hatch, invade intestinal mucosa, and are carried to the lungs.

Eggs pass in host faecal matter

Fertilised eggs undergo development

➡ **Stages outside host**

---▶ **Stages inside host**

Figure 4.15 Life cycle of *Ascaris lumbricoides*. Adult worms live in the lumen of the small intestine. A female may produce approximately 200 000 eggs per day, which are passed with the faeces. Unfertilised eggs may be ingested but are not infective. Larvae develop to infectivity within fertile eggs after 18 days to several weeks, depending on the environmental conditions (optimum: moist, warm, shaded soil). After infective eggs are swallowed, the larvae hatch, invade the intestinal mucosa and are carried via the portal, then systemic, circulation to the lungs. The larvae mature further in the lungs (10–14 days), penetrate the alveolar walls, ascend the bronchial tree to the throat and are swallowed. Upon reaching the small intestine, they develop into adult worms. Between 2 and 3 months are required from ingestion of the infective eggs to oviposition by the adult female. Adult worms can live 1–2 years. *Sources:* SuSanA Secretariat / Wikimedia Commons / CC BY 2.0. Department of Pathology, Calicut Medical College / Wikimedia Commons / CC BY-SA 4.0. VlaminckJ / Wikimedia Commons / CC BY-SA 4.0. SuSanA Secretariat / Wikimedia Commons / CC BY 2.0. Sambon, L. W. (Louis Westenra) / Wellcome Collection / CC BY 4.0. Created with BioRender.com.

alveoli, the juvenile worms migrate through the respiratory tract until they reach the trachea. Once in the trachea, the worms are coughed up and swallowed again, finally reaching the small intestine, where they mature into adult worms. The adult female worms produce around 200 000 fertilised eggs per day for a period of 12–18 months. The eggs are excreted in the faeces and can remain viable in soil for many years. The reason for the migration pattern of *Ascaris* worms, from the small intestine to the lungs and back to the small intestine, is not fully understood. One hypothesis suggests that this migration mimics the life cycle of an intermediate host, which may have been necessary for the development of ancestral forms of the worm. Worldwide, an estimated one billion people are infected with *A. lumbricoides*. Löffler syndrome, a rare condition characterised by pulmonary infiltrates and eosinophilia, can occur in some parasitic infections, especially with *A. lumbricoides.*

Enterobius vermicularis, or the pinworm, is transmitted via the ingestion of its eggs, typically through poor hygiene practices or contaminated objects. Pinworm infection often affects children and causes intense anal itching, especially at night when the female worm lays her eggs on peri-anal skin.

Trichinella spiralis is a parasitic nematode that causes trichinellosis, which occurs when humans consume undercooked or raw meat containing the encysted larvae of *Trichinella.* After ingestion, the larvae mature in the intestines and penetrate the intestinal wall, entering the bloodstream. They can migrate to various tissues, particularly skeletal muscles, where they form cysts. Trichinellosis can cause muscle pain, fever, swelling and gastrointestinal disturbances. In severe cases, it can lead to myocarditis or encephalitis.

Strongyloides stercoralis is a parasitic nematode that causes strongyloidiasis. Infection occurs when the skin comes into contact with contaminated soil or when individuals consume food or water contaminated with *Strongyloides* larvae. The larvae penetrate the skin and migrate to the lungs, where they are coughed up and swallowed. In the intestines, they mature into adult worms, which can reproduce and release eggs. Strongyloidiasis can cause a range of symptoms, including abdominal pain, diarrhoea and a skin rash. In immunocompromised individuals, such as those with HIV/AIDS or those receiving immunosuppressive therapy, the infection can become chronic and can spread throughout the body (known as disseminated strongyloidosis), which is potentially life-threatening.

Wuchereria bancrofti is a filarial nematode that causes lymphatic filariasis, also known as elephantiasis. The disease is transmitted to humans through the bite of infected mosquitoes. The larvae enter the lymphatic system and mature into adult worms. Lymphatic filariasis can result in chronic, usually unilateral,

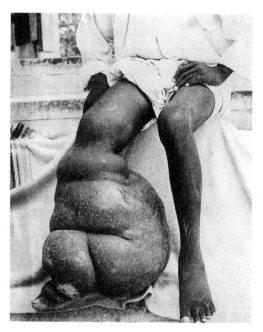

Figure 4.16 Elephantiasis in the right leg of an inhabitant of the West Indies. Photograph. Sambon, L.W. (Louis Westenra), 1865–1931. Date: [between 1900 and 1999]. *Source:* Sambon, L. W. (Louis Westenra) / Wellcome Collection / CC BY 4.0.

swelling and thickening of the skin of limbs, genitals or breasts, leading to severe disfigurement and disability (Fig. 4.16).

Arthropods

Most arthropods which attack humans are blood feeders (e.g. mosquitoes, ticks and fleas) and reside only briefly on the host. Many of these arthropods are vectors for the transmission of other infectious agents. Some arthropods, notably the louse, *Pediculus humanus,* and the crab louse, *Phthirus pubis,* can reproduce on humans and produce inflammatory reactions following penetration of the skin during feeding.

Scabies occurs when the microscopic mite, *Sarcoptes scabei,* burrows into the skin and lays eggs, leading to intense itching and rash. Scabies is most prevalent in hot tropical countries and areas with high population density. It is estimated that up to 200 million people are affected by scabies at any given time. The burden of scabies is particularly high in resource-poor settings. Children in these areas are especially vulnerable, with up to 10% of children infested. Recurrent infestations are common, and the complications of scabies, such as secondary bacterial infections, can lead to more serious complications, including septicaemia. Immunosuppressed people develop an uncommon manifestation, called crusted (Norwegian) scabies, a hyper-infestation with thousands to millions of mites, producing widespread scale and crust. This condition has a high mortality due to secondary sepsis, if left untreated. Scabies is transmitted through close skin contact with an infested individual; the risk of transmission is higher in individuals with a heavy infestation, particularly those with crusted scabies.

The larvae of some species of fly can infest humans, For example, *Cordylobia anthropophaga*, commonly known as the mango fly, or skin maggot fly, is a species of blow-fly. During its larval stage, this fly acts as a parasite, infesting large mammals, including humans. It is primarily found in the subtropical regions of Africa where it is a common cause of myiasis in humans. Myiasis refers to the infestation of living tissue by fly larvae. There are several other insects whose larval stages can infect humans and cause myiasis, including *Dermatobia hominis* (human botfly), *Cochliomyia hominivorax* (New World screw-worm fly) and *Wohlfahrtia magnifica* (flesh fly).

Summary

Despite man's attempts to mitigate the impact of infectious diseases, the constant evolution of pathogens poses an ongoing threat to global health. These challenges include the emergence of drug-resistant pathogens, such as MRSA, multidrug-resistant tuberculosis (MDR-TB) and carbapenem-resistant *Enterobacteriaceae* (CRE), as well as new infectious agents. Recent outbreaks of diseases like Zika, Ebola, Middle East respiratory syndrome (MERS), severe acute respiratory syndrome and COVID-19 have demonstrated the potential for new pathogens to cause widespread harm. To address this, early identification and surveillance of potential zoonotic pathogens, which can 'jump' from animals to humans, is essential. This will involve rigorous monitoring, enhanced diagnostics and research into the reservoirs and transmission dynamics of potential pathogens. Preparing for future pandemics requires a global effort. It involves strengthening public health systems, investing in research and development and establishing robust surveillance networks. Timely detection, rapid response and effective communication are critical in containing and mitigating the impact of emerging infectious diseases.

We must also redress the global inequalities in the incidence and treatment of infectious diseases. To this end, the World Health Organization has identified the so-called neglected tropical diseases (NTDs). NTDs include Buruli ulcer; Chagas disease; dengue and chikungunya; dracunculiasis; echinococcosis; foodborne trematodiases; human African trypanosomiasis; leishmaniasis; leprosy; lymphatic filariasis; mycetoma, chromoblastomycosis and other deep mycoses; onchocerciasis; rabies; scabies and other ectoparasitoses; schistosomiasis; soil-transmitted helminthiases; taeniasis/cysticercosis; trachoma; and yaws. The impact of NTDs on individuals and communities is profound, causing severe health, social and economic impacts. Many NTDs are transmitted through vectors, have animal reservoirs and exhibit intricate life cycles, posing challenges for public health control efforts. NTDs thrive in rural areas, conflict zones and hard-to-reach regions. They find fertile ground in areas lacking access to clean water and sanitation, with the situation further exacerbated by the effects of climate change and military conflicts. Moreover, NTDs disproportionately affect regions with limited healthcare resources, leaving impoverished populations susceptible to these debilitating diseases.

At the time of writing (May 2023), the only significant human pathogen to be successfully eliminated is the smallpox Variola virus. The fact that this virus only infected humans made it a good target for eradication; human-to-human transmission could be interrupted through targeted vaccination campaigns. Similarly, Poliovirus, which also only infects humans, has been eliminated in most countries. Of the three strains of wild Poliovirus (type 1, type 2 and type 3), wild Poliovirus type 2 was eradicated in 1999 and wild Poliovirus type 3 in 2020. As of 2022, endemic wild Poliovirus type 1 remains in only two countries, Pakistan and Afghanistan.

Key Points

1) Pathogens are parasites that cause disease. In general, Koch's postulates must be satisfied before an organism can be identified as the aetiological agent for a given disease.
2) Parasite and host factors determine whether infectious disease will occur. Parasite factors include virulence factors, the ability of the parasite to evade host immunity and the size of the infective dose. Host factors are age, nutritional status, genetic constitution and the integrity of the immune system.
3) Bacteria are prokaryotes that are classified on the basis of their staining properties, morphology, biochemical characteristics and more recently, genome sequence. Bacterial infections may be acquired from two principal sources: from amongst the patients' own flora or from external sources.
4) Bacteria capable of causing disease possess virulence determinants. Most virulence determinants are found in the cell wall.
5) Viruses are intracellular parasites composed of either DNA or RNA genomes, surrounded by a protein coat (capsid) and in some cases by an outer envelope.
6) Viruses can only survive and replicate inside live host cells.
7) The tissue damage caused by virus infections is the result of the direct cytopathic effects of the virus or the immune response to the virus, or a combination of both.
8) Medically important fungi include those that cause only superficial infections and those that can cause potentially life-threatening systemic disease; the latter are usually the so-called opportunistic infections that arise more commonly in immunosuppressed individuals.
9) Parasitic infections are caused by a wide variety of different types of organisms, ranging from single cell protozoa to large complex worms, flukes and arthropods.

Further Reading

Riedel S., Morse S.A., Mietzner T. & Miller S. (eds) (2019) *Jawetz, Melnick and Adelberg's Medical Microbiology* (28th edn). McGraw Hill/Medical.

Ryan K. (ed) (2022) *Sherris & Ryan's Medical Microbiology* (8th edn). McGraw Hill/Medical.

Arevalo C.P., Bolton M.J., Le Sage V. *et al.* (2022) A multivalent nucleoside-modified mRNA vaccine against all known influenza virus subtypes. *Science* **378**: 899–904.

Sridhar D. (2022) Five ways to prepare for the next pandemic. *Nature* **610**: S50. https://doi.org/10.1038/d41586-022-03362-8.

Part 4

Inflammation and Disorders of Immunity

5

Inflammation and Repair

Introduction

Injury, trauma or infections induce a series of complex and interconnected reactions initiated at the site of tissue damage. These serve to contain and destroy the infection or damaging agent, prevent continued tissue damage and initiate repair processes to restore normal function. *Innate immunity* provides an immediate but general defence against pathogens using pre-existing mechanisms, without requiring prior exposure to the specific invader. In contrast, *adaptive immunity* is a delayed response that recognises and targets specific pathogens based on prior exposure, generating memory cells for long-term protection against subsequent encounters with the same pathogen. The rapid innate response, employing a variety of components of the innate immune system, is known as *acute inflammation*. Adaptive immunity is discussed in detail in Chapter 6.

The toxic reactions induced during acute inflammation that are designed to destroy invading pathogens also have the capacity, paradoxically, to injure host tissues. Thus, negative feedback mechanisms are required to prevent excessive tissue damage. If these toxic responses are not tightly regulated, then tissue injury may predominate over tissue protection and repair, leading to inflammatory disease. Similarly, if the infectious agent has evolved mechanisms to protect itself from attack by the immune system, it may persist, giving rise to *chronic inflammation*.

The characteristics of the acute inflammatory response have long been identified and are part of everyday experience. The effects of even a minor wound, such as a small splinter in the finger, allow us to confirm the observations of the Roman physician Celsus who described the four cardinal signs of inflammation: *rubor* (redness), *tumour* (swelling), *calor* (heat) and *dolor* (pain). The challenge is to explain the physiological events that generate this response.

The Biology of Disease, Third Edition. Edited by Paul G. Murray, Simon J. Dunmore, and Shantha Perera.
© 2024 John Wiley & Sons Ltd. Published 2024 by John Wiley & Sons Ltd.
Companion website: www.wiley.com/go/murray/biologyofdisease3e

Acute Inflammation

The initial stimulus to the inflammatory response is tissue damage or the presence of infectious agents and their products. Within the damaged tissue, localised changes occur which include the following:

- The release of pre-formed inflammatory mediators from intracellular stores (e.g. the degranulation of mast cells).
- The initiation of reaction cascades through the activation of soluble plasma components produced by the liver.
- The new synthesis of inflammatory mediators.

Release of Preformed Inflammatory Mediators

The release of preformed inflammatory mediators from cells is one of the most rapid responses to tissue injury. The immediate response to damage of blood vessels is the aggregation of platelets, which adhere to exposed extracellular surface structures such as collagen. There is a release of soluble factors from platelet stores, which helps to attract further platelets and promote aggregation. Vasoconstriction reduces blood flow in muscular blood vessels such as arterioles and facilitates the formation of a platelet plug. Subsequently, in conjunction with products of the fibrin cascade, a clot is formed.

Histamine is released by the degranulation of mast cells. Histamine, in contrast, is a *vasodilator* that increases the volume of blood reaching the area of the damaged tissue. *Lysosomal enzymes* are released from damaged cells and help to breakdown cell debris. *Serotonin* is also a pre-formed vasodilator present mainly in platelets that is released very early in the response.

Initiation of Plasma Protein Cascades Generating Inflammatory Mediators

Damage to vascular endothelial cells results in the activation of Factor XII. This in turn activates cascades of reactions that generate plasma mediators with different roles in inflammation.

Fibrin and Fibrinolytic Cascades

Activated Factor XII initiates a cascade of reactions that result in the conversion of prothrombin to thrombin, which converts fibrinogen to fibrin to form a mesh of fibres that trap a variety of circulating cells in a clot. This not only prevents further loss of blood but also contains the damaged tissue and serves as a barrier to infection. Another cascade of reactions triggered by activated Factor XII brings about the conversion of plasminogen to the proteolytic enzyme plasmin, which breaks down clots releasing products that play a role in the inflammatory process. Chapter 8 gives a fuller account of the arrest of bleeding and the coagulation cascade.

Kinin Cascade

Activated Factor XII has a further role in triggering the kallikrein–bradykinin system (Fig. 5.1). Bradykinin is a potent pro-inflammatory nonapeptide (a peptide containing nine amino acid residues) that can;

- Induce vasodilation to increase blood flow.
- Increase vascular permeability, allowing fluids and cells to pass from the bloodstream into tissues.
- Stimulate nerve endings to cause pain.
- Cause prostaglandin release, further amplifying its pro-inflammatory effects.

High-molecular-weight kininogen (HMWK) is a precursor of bradykinin. It is converted to bradykinin by the action of kallikrein, which itself is formed from pre-kallikrein, an inactive precursor. Various stimuli cause the conversion of pre-kallikrein to kallikrein, including factors from the coagulation system (Factor XII), or contact with negatively charged surfaces. Bradykinin is negatively regulated by enzymes, including angiotensin-converting enzyme (ACE).

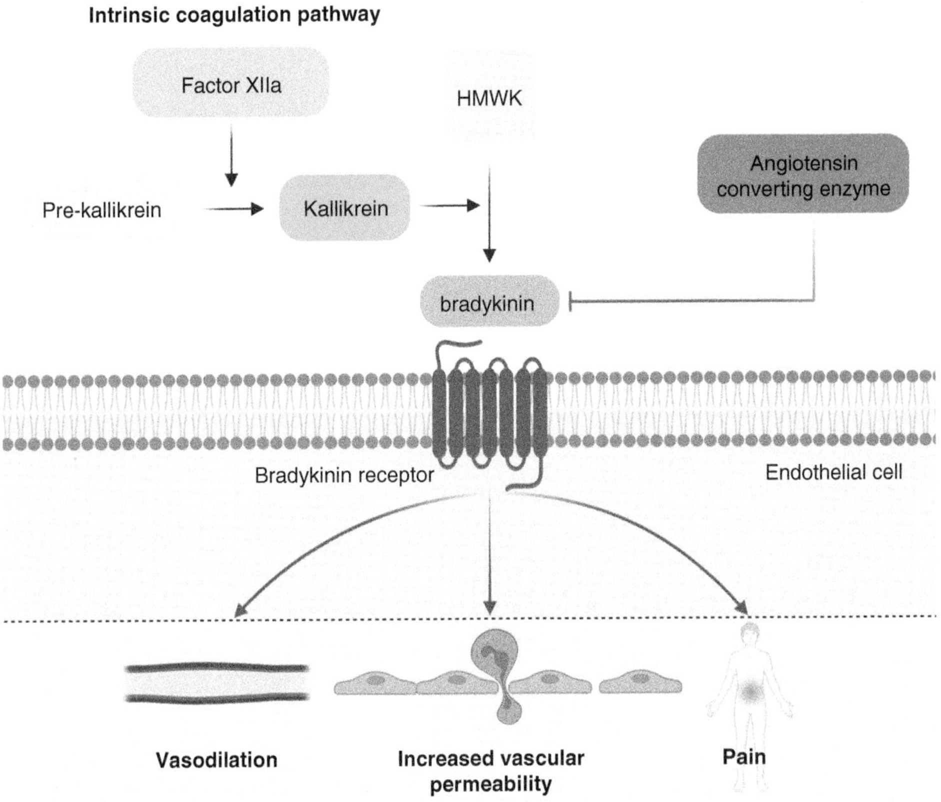

Figure 5.1 The kallikrein–bradykinin system. This pathway plays a pivotal role in mediating inflammation. Activation results in the production of bradykinin, a nonapeptide with potent pro-inflammatory properties. Originating from its precursor, high-molecular-weight kininogen (HMWK), bradykinin is formed via the enzymatic activity of kallikrein. The conversion of pre-kallikrein, an inert precursor, to its active form, kallikrein, can be triggered by various factors, notably Factor XII from the coagulation system or through contact with negatively charged surfaces. Once produced, the actions of bradykinin include vasodilation, increased vascular permeability and stimulation of pain. The effects of bradykinin are counteracted by enzymes such as angiotensin-converting enzyme (ACE), which helps maintain a balance in the inflammatory and blood pressure response. *Source:* Created with BioRender.com.

Complement Cascades

There are three complement pathways that interact through common intermediates and form a key part of the immune system (Fig. 5.2). All three pathways involve a cascade of reactions in which precursor proteins are cleaved to generate major and minor fragments. The major fragments decay rapidly unless they bind to other components or surfaces and retain their activity. In turn, they enzymatically cleave and activate the next component in the pathway. The minor fragments frequently have functions that mediate the process of inflammation. The assembly of complement components on antibody molecules that are complexed with one another or bound to target antigen surfaces initiates the *classical pathway*. This pathway is therefore principally dependent for its activation on the presence of antibody generated by an acquired immune response. It is not therefore involved in the early stages of the inflammatory response. The *alternative pathway* is initiated by a variety of foreign cell surface components such as constituents of bacterial or fungal cell walls. It is an essential element of the innate

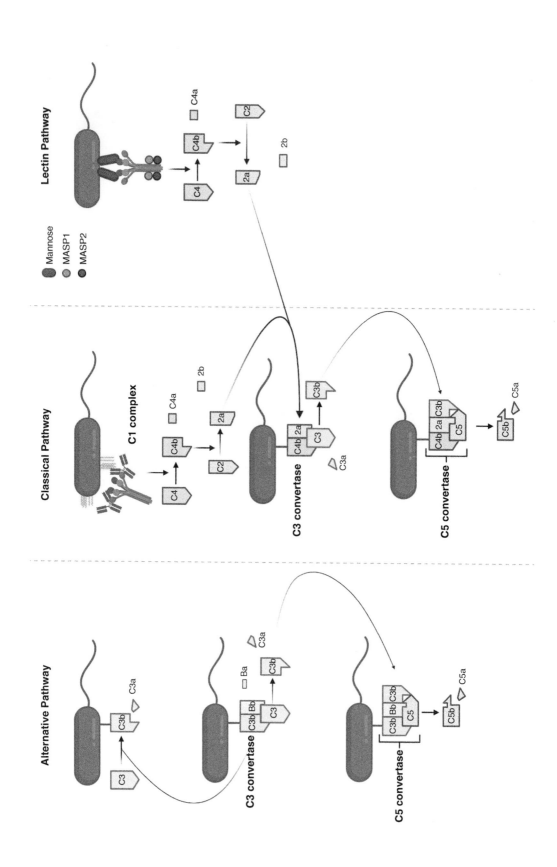

Figure 5.2 The complement cascades. The complement cascades form a key part of the innate immune system and complement components are important mediators of inflammation. The *alternative pathway* is initiated by a variety of foreign cell surface components such as constituents of bacterial or fungal cell walls. It is an essential element of the innate immune response. The complement component C3 undergoes spontaneous hydrolysis to C3a and C3b. Binding of C3b to foreign antigenic surfaces prevents its inactivation. In this environment, C3b binds to factor B. This alteration in factor B exposes a specific site or region that can then be cleaved, releasing a small fragment Ba and leaving the active complex C3bBb, which has C3 convertase activity. The C3 convertase initiates the remaining stages of the complement cascade. The C3 convertase establishes a positive feedback loop converting further C3 molecules to C3a and C3b. This amplifies the response and leads to the production of C3bBbC3b complexes. These have C5 convertase activity and cleave C5 to generate C5a and C5b. The *classical pathway* is initiated by the assembly of C1 complement components on antibody molecules bound to target antigenic surfaces. The active C1 complex cleaves C4 to release C4a and C4b. C4b is stabilised by binding to the antigenic surface and, in turn, acts as a binding site for C2, which is cleaved to release C2b. The remaining complex, C4b2a, has C3 convertase activity. The binding of C3b generates the C5 convertase of the classical pathway (C4bC2aC3b). The *lectin pathway* is initiated when pattern recognition molecules, known as mannose-binding lectins (MBLs) or ficolins recognise specific patterns on the surface of pathogens. These patterns typically consist of carbohydrate structures, such as mannose, on the pathogen's surface. Upon binding to the pathogen, MBL or ficolins form complexes with serine proteases called MBL-associated serine proteases (MASPs). These complexes are responsible for triggering the subsequent cascade of complement activation. Once activated, MASPs, particularly MASP-2, cleave complement component C4 into C4a and C4b. *Source:* Created with BioRender.com.

immune response. The *lectin pathway* is initiated when mannose-binding lectin (MBL) binds to carbohydrates in microbes.

The complement components C3a and C5a have particularly important roles in inflammation (Fig. 5.3). Both induce the release of histamine from mast cells and basophils, which increases vascular permeability and causes vasodilation, facilitating the recruitment of immune cells to the site of inflammation. C5a, and to a lesser extent C3a, are also chemoattractants for various immune cells, including neutrophils, basophils, eosinophils and monocytes. C5a also stimulates the release of inflammatory mediators such as cytokines and reactive oxygen species (ROS; see also Chapter 3), contributing to the amplification of the inflammatory response. Additionally, C5b initiates the assembly of the membrane attack complex (MAC), which is a cytolytic complex that forms on the surface of pathogens, leading to their destruction by lysis (Fig. 5.4).

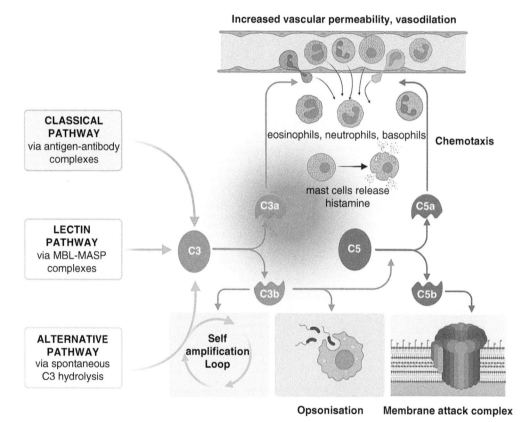

Figure 5.3 Role of complement components C3 and C5 in inflammation. C3 is split into two fragments, C3a and C3b, by C3 convertase (not shown). C3b is the larger fragment generated from the cleavage of C3. It covalently attaches to the surface of pathogens, marking them for recognition by phagocytic cells (opsonisation), accelerated by the amplification loop. C3b, along with other complement components, contributes to the formation of the C5 convertase enzyme, which cleaves C5 into C5a and C5b. C3a and C5a promote histamine release from mast cells and basophils, increasing vascular permeability and vasodilation and facilitating the recruitment of immune cells to the site of inflammation. C5a and C3a are also chemoattractants for various immune cells, including neutrophils, basophils, eosinophils and monocytes. C5a also stimulates the release of inflammatory mediators, such as cytokines and ROS, contributing to the amplification of the inflammatory response. C5b also serves as the initial component that triggers the assembly of the membrane attack complex, a cytolytic complex that forms on the surface of pathogens, leading to their lysis and destruction. *Source:* Created with BioRender.com.

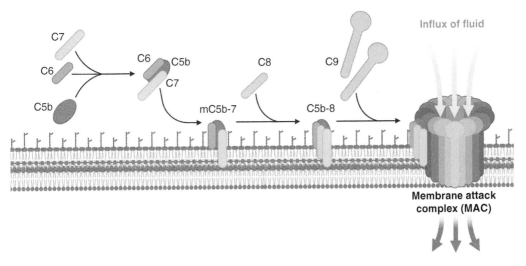

Figure 5.4 The membrane attack complex (MAC) is a cytolytic complex formed by the complement system. Once C5 is activated, it is cleaved into two fragments: C5a and C5b. C5b initiates the assembly of the MAC by binding to nearby cell membranes or surfaces. Binding of C5b promotes the recruitment and sequential binding of complement components C6, C7, C8 and multiple copies of C9. The C9 molecules form a pore-like structure called the MAC, which spans the lipid bilayer of the target cell membrane. The MAC disrupts the integrity of the cell membrane allowing the unregulated influx of ions and fluids into the cell. This influx causes osmotic imbalance and cell swelling ultimately leading to cell lysis and destruction. The MAC can directly kill pathogens, such as bacteria, by lysing their cell membranes. It can also participate in the destruction of virus-infected cells or cancer cells. *Source:* Created with BioRender.com.

Synthesis of Inflammatory Mediators

Membrane Phospholipid-Derived Eicosanoids

Inflammatory stimuli can initiate the production of a family of compounds known as *eicosanoids*. The initial trigger can be the activation of cell surface receptors (e.g. receptors for histamine, bradykinin, thrombin, TNF-α and IL-1), which in turn activate phospholipases (PLA), including PLA$_2$ and trigger arachidonic acid release. Arachidonic acid is further metabolised by either the cyclooxygenase or lipoxygenase pathways (Fig. 5.5). These reactions take place in a variety of cells involved in inflammation including monocytes/macrophages, neutrophils, mast cells, platelets and endothelial cells. The pattern of metabolites generated depends on the cell type. The *cyclooxygenase pathway* converts arachidonic acid to *prostaglandins*, including *prostacyclin* (prostaglandin I$_2$ or PGI$_2$) and *thromboxanes*.

Platelets mainly produce thromboxane A$_2$ which causes vasoconstriction and platelet aggregation. Conversely, vascular endothelial cells mainly produce prostacyclin, which stimulates vasodilation and inhibits platelet aggregation. The *lipoxygenase pathway* produces *leukotrienes* that are chemotactic, induce smooth muscle contraction and increase vascular permeability, as well as *lipoxins*, which inhibit neutrophil adhesion and chemotaxis.

Cytokines and Chemokines

Macrophages present in damaged tissues become activated and synthesise and secrete cytokines of which *interleukin 1* (IL-1) and *tumour necrosis factor alpha* (TNF-α) are the most important. One of the effects of these cytokines is to induce the expression of adhesion molecules on the surface of vascular endothelial cells. They also act on endothelial cells and other cells such as fibroblasts to induce the synthesis of cytokines such as IL-8

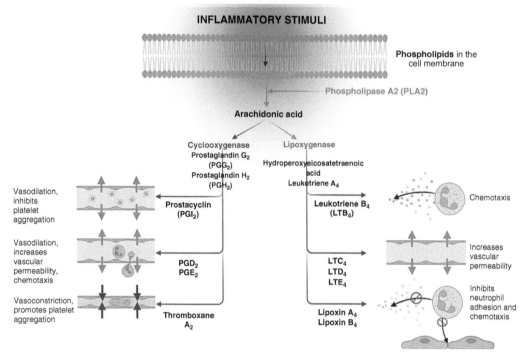

Figure 5.5 The arachidonic acid pathway. The arachidonic acid pathway, also known as the eicosanoid pathway, plays a significant role in inflammation. It involves the conversion of arachidonic acid, a fatty acid found in cell membranes, into various inflammatory mediators called eicosanoids. In response to an inflammatory stimulus, phospholipase enzymes, particularly phospholipase A_2 (PLA2), are activated. PLA2 acts on cell membrane phospholipids, specifically phospholipids containing arachidonic acid, releasing arachidonic acid into the cytoplasm. Arachidonic acid can be metabolised through two major pathways: the cyclooxygenase (COX) pathway and the lipoxygenase (LOX) pathway. These pathways give rise to different classes of eicosanoids. In the COX pathway, arachidonic acid is converted by cyclooxygenase enzymes (COX-1 and COX-2) into prostaglandin G_2 (PGG$_2$). PGG$_2$ is further converted into various prostaglandins, including prostaglandin E_2 (PGE$_2$), prostaglandin D_2 (PGD$_2$), prostaglandin I_2 (PGI$_2$) and thromboxane A_2. Prostaglandins exert diverse effects, including vasodilation, increased vascular permeability and leucocyte chemotaxis. In the LOX pathway, arachidonic acid is converted by different lipoxygenase enzymes into leukotrienes and lipoxins. Leukotriene A4 (LTA$_4$) is a precursor of various leukotrienes, including leukotriene B4 (LTB$_4$), leukotriene C4 (LTC$_4$), leukotriene D4 (LTD$_4$) and leukotriene E4 (LTE$_4$). Leukotrienes are potent mediators of inflammation, contributing to the recruitment and activation of immune cells, bronchoconstriction and increased vascular permeability. Lipoxins inhibit neutrophil adhesion and chemotaxis. *Source:* Created with BioRender.com.

and *monocyte chemotactic protein-1* (MCP-1/ CCL2). These molecules, which are also produced by monocytes and macrophages, are examples of a family of pro-inflammatory molecules known as *chemokines*. They attract different types of leucocytes, regulating their expression of cell surface adhesion molecules and facilitating their attachment to the vascular endothelial surface and their attraction towards localised concentrations of chemokine. These interactions are critical in

bringing cells of the immune system such as neutrophils, and subsequently monocytes and lymphocytes, to the site of injury.

Effects of Inflammatory Mediators

The interconnected reaction sequences described above generate a wide range of inflammatory mediators (see Table 5.1 for a summary). These immediate responses isolate the area of tissue damage and restrict the loss

Table 5.1   The origin and functions of mediators of acute inflammation.

Type	Mediator	Origin	Functions
Released from intracellular stores	Serotonin	Platelets	Vasoconstriction, platelet aggregation
	Histamine	Mast cells	Vasodilation, increased permeability of vascular endothelium, smooth muscle contraction
Generated by plasma protein cascades	Thrombin	Plasma protein coagulation cascade	Promotion of platelet aggregation
	Fibrin and fibrinopeptides	Plasma protein coagulation cascade	Increased permeability of vascular endothelium, neutrophil chemotaxis
	Plasmin	Plasma protein fibrinolytic cascade	Proteolytic degradation of clots to release chemotactic factors, activation of classical complement pathway
	Kallikrein	Plasma protein kinin cascade	Cleavage of complement component C5 to generate C5a and C5b
	Bradykinin	Plasma protein kinin cascade	Vasodilation, increased permeability of vascular endothelium, smooth muscle contraction, stimulation of pain receptors
	C3a	Plasma protein complement pathway	Degranulation of mast cells, basophils and eosinophils, platelet aggregation
	C5a	Plasma protein complement pathway	Degranulation of mast cells, basophils and eosinophils, platelet aggregation, chemotaxis of leucocytes, neutrophil activation
Products of membrane phospholipid degradation	Prostaglandin	Cyclooxygenase breakdown of membrane phospholipids	Vasodilation, increased permeability of vascular endothelium, neutrophil chemotaxis
	Thromboxane	Cyclooxygenase breakdown of membrane phospholipids	Vasoconstriction, platelet aggregation
	Leukotriene B4	Lipoxygenase breakdown of membrane phospholipids	Neutrophil chemotaxis
	PAF	Conversion from lyso-PAF	Platelet aggregation, increased permeability of vascular endothelium, neutrophil activation, chemotaxis of eosinophils
Chemokines synthesised at the site of inflammation	IL-8	Neutrophils, monocytes, macrophages, fibroblasts, endothelial cells	Neutrophil chemotaxis, activation and degranulation of neutrophils
	CCL3, CCL4	Neutrophils	Monocyte chemotaxis, macrophage activation

of blood, thus containing any infection to a localised area and forming a physical barrier to the entry of further infectious organisms. The flow of blood to the damaged tissue is increased by vasodilation. The influx of plasma and plasma proteins to the area dilutes any toxic products from infectious agents, ensures a ready supply of the components of the plasma protein cascades and carries leucocytes to the site of injury. The increased permeability of vascular endothelium allows the damaged tissue to be flooded with fluid and plasma proteins. This fluid drains eventually via the afferent lymphatics to local lymph nodes. Foreign antigens carried to the lymph nodes will initiate an acquired immune response, which will, in time, generate antigen-specific effector lymphocytes of the adaptive immune system (Chapter 6) that will return in the blood to the site of infection and supplement the activities of the innate immune system.

The fluid 'leaking' into the damaged tissue causes localised *swelling* (oedema) and increased pressure. This fluid is known as *exudate* and is protein rich. It should not be confused with a *transudate* which has a low protein concentration and is caused by either increased hydrostatic pressure (e.g. in congestive heart failure) or decreased colloid osmotic pressure (e.g. in liver failure). The increased pressure within the tissue caused by an exudate, along with the release of mediators such as bradykinin, triggers *pain* receptors in the tissue. The increased flow of blood to the wound causes *redness* and generates *heat*, the other cardinal signs of inflammation. Cells of the innate immune system present at the site of injury will begin to attack any foreign microorganisms. However, they may be few in numbers and an effective response will depend on the recruitment of many more leucocytes to the damaged tissue.

Systemic Effects of Cytokines

Systemic physiological responses are also associated with the release of cytokines during acute inflammation. IL-1 and TNF-α generated by activated macrophages at the site of infection act on other cells in the locality, such as fibroblasts, to stimulate their production of IL-1, TNF-α and IL-6, which alter the temperature set-point in the hypothalamus, mediating fever through the stimulation of prostaglandin synthesis (PGE_2). The same cytokines act on the adrenal pituitary axis to generate adrenocorticotrophic hormone (ACTH) and induce the production of glucocorticoids that act cooperatively with these cytokines to induce the synthesis of acute phase proteins in the liver. Although glucocorticoids induce the synthesis of acute phase proteins, they provide a negative feedback loop by downregulating further synthesis of cytokines by macrophages at the site of infection. The acute phase proteins include

- Components of the coagulation and complement cascades. Elevated levels of fibrinogen are the basis of a test that can indicate inflammation – the erythrocyte sedimentation rate (ESR) – fibrinogen causes red blood cells to stick together, meaning that they sediment more rapidly.
- Proteinase inhibitors that control the effects of lysosomal hydrolases released at the site of infections.
- Metal binding proteins that prevent iron loss during injury and infection.
- The major acute phase proteins, serum amyloid A, serum amyloid P and C-reactive protein (CRP), all of which show massive induction (for example, as much as 1000-fold increase on normal levels). The measurement of CRP in the blood of patients is a useful indicator of inflammation. Amyloid proteins can cause secondary amyloidosis if the inflammation is not resolved.

Recruitment of Leucocytes

Vascular endothelium separates the site of tissue injury from the circulating lymphocytes in the blood system. Vasodilation not only increases the volume of blood brought to the area but also reduces the rate of blood flow in capillaries adjacent to the site of inflammation. This increases the number of leucocytes arriving in the area and provides a better microenvironment for their interaction with vascular endothelium. This interaction is initially of low affinity and is mediated by adhesion molecules that become expressed on the endothelial cell surfaces, and which bind to their ligands present on leucocytes, principally neutrophils in the first instance. The flow of blood within the vessel tends to detach the neutrophils, but new interactions may form to reattach the cell. This gives the appearance of the neutrophil 'rolling' along the endothelial surface (Fig. 5.6). Attachment brings the neutrophil into contact with localised concentrations of chemokines secreted by the endothelial cells. The chemokines bind to specific chemokine receptors on the surface of the neutrophils, which triggers the activation of adhesion molecules. Cytokines induce the expression of additional adhesion molecules on endothelial cells. Interaction between these adhesion molecules brings about firm adhesion of the neutrophils to the endothelial surface. The neutrophils then pass through the intercellular junction between endothelial cells and enter the damaged tissue. This process is termed *extravasation*. Although the process of extravasation is similar in outline for all leucocytes, there are essential differences in the nature of the specific adhesion molecules and chemokines involved. Concentration gradients of chemoattractants guide the migration (chemotaxis) of the neutrophils to the site of infection. In essence then, tissue damage triggers the generation of inflammatory mediators, which cause changes at the vascular endothelial surface and provide the necessary signals for the recruitment of the neutrophils (Fig. 5.7). The recruitment of neutrophils is followed by the migration of monocytes into the site of inflammation, which mature to form tissue macrophages. Depending on the nature of the response, other cells of the innate system may also be recruited. The cells of the innate immune system have much broader functions

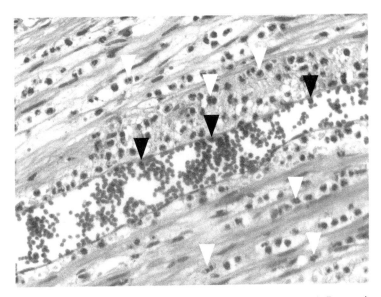

Figure 5.6   Neutrophil margination and extravasation in acute inflammation. A blood vessel is shown, which contains red blood cells and neutrophils which are attached to the endothelial surface (black arrowheads). Some have spilled out in the surrounding tissue (examples shown by white arrowheads). *Source:* Department of Pathology, Calicut Medical College / wikimedia Commons / CC BY-SA 4.0.

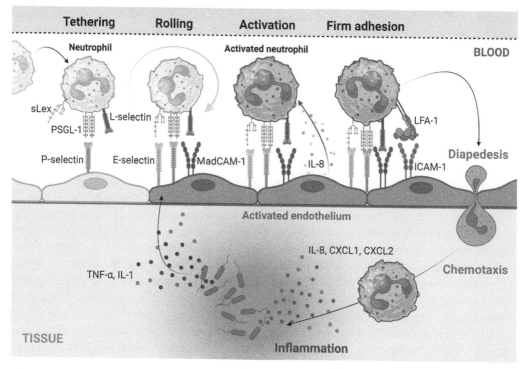

Figure 5.7 Neutrophil recruitment to sites of acute inflammation. Neutrophils play a crucial role during inflammation, as they are usually the first immune cells to arrive at the site of infection or tissue injury. Neutrophils initially tether to the activated endothelial cells lining the blood vessels near the inflamed tissue. This step is mediated by the interaction between selectins, a family of adhesion molecules, and their ligands. There are three types of selectins involved: P-selectin: Expressed on the surface of activated endothelial cells and platelets, it interacts with P-selectin glycoprotein ligand-1 (PSGL-1) on neutrophils. E-selectin is induced on the surface of endothelial cells in response to inflammatory cytokines, such as tumour necrosis factor-alpha (TNF-α) and interleukin-1 (IL-1). It binds to sialyl-Lewis X (sLex) and sialyl-Lewis A (sLea; not shown) expressed on neutrophils. L-selectin is expressed on the surface of neutrophils, it interacts with various ligands on endothelial cells, including peripheral node addressin (PNAd; not shown) and mucin-like addressin cell adhesion molecule-1 (MadCAM-1). Once tethered, neutrophils undergo a 'rolling' motion along the endothelial surface. This rolling is mediated by weak interactions between selectins and their ligands. The rolling allows the neutrophils to survey the endothelial surface for signals of inflammation. Endothelial cells produce chemokines, such as IL-8, in response to various inflammatory stimuli. As neutrophils roll along the endothelium, they encounter these chemokines. Upon binding of the chemokines to their respective receptors on neutrophils, signalling events are triggered, activating the neutrophils. After activation, neutrophils transition from rolling to firm adhesion, sticking more firmly to the endothelium. This strong attachment is mediated by the β2 integrin, LFA-1, which binds to ICAM-1 on endothelial cells. Upon firm adhesion, neutrophils undergo transmigration, crossing the endothelial cell barrier to enter the inflamed tissue. Transmigration occurs either through paracellular routes (between endothelial cells) or transcellular routes (through the endothelial cell itself). Once neutrophils have successfully transmigrated into the tissue, they can move to the site of injury, guided by chemokines. This process is called chemotaxis. Once at the site of injury, they engage in various immune functions, including phagocytosis of pathogens, release of antimicrobial factors and interaction with other immune cells. *Source:* Created with BioRender.com.

than their simple participation in the inflammatory response and are discussed in more detail in Chapter 6. Of relevance here is that several of these cell types, including macrophages and dendritic cells, can present antigen, principally to T-cells, to initiate adaptive immunity. Thus, recruitment of innate cells is eventually followed, some days later, by the activation and clonal expansion of antigen-specific lymphocytes.

Role of Adhesion Molecules in Endothelial Cell–Leucocyte Interactions

There are several families of adhesion molecules that play a role in inflammation. They include

- Selectins.
- Immunoglobulin-like cell adhesion molecules.
- Integrins.

Selectins are glycoproteins that consist of a lectin-like domain, an epidermal growth factor repeat, a variable number of domains resembling those found in complement binding proteins (containing approximately 60 amino acids), a transmembrane domain and a cytoplasmic tail. There are three main types – selectins P, E and L. *P-selectin* is stored in granules in platelets and endothelial cells and is rapidly redistributed to the cell surface in response to histamine or components of the complement pathway. *E-selectin* is synthesised and expressed by endothelial cells in response to IL-1, TNF-α or bacterial endotoxins. Both P- and E-selectins bind to their ligands on the surface of neutrophils, monocytes and some memory T lymphocytes. *L-selectin* is constitutively expressed on the surface of circulating lymphocytes, monocytes and neutrophils and plays a role in their adhesion to vascular endothelium and to high endothelial venules (specialised post-capillary venules found in lymph nodes and Peyer's patches that facilitate the extravasation of lymphocytes from the bloodstream into lymphoid tissues). Selectins bind to specific carbohydrate components, including sialylated, sulphated and phosphated polysaccharides.

Members of the *immunoglobulin supergene family* contain the characteristic immunoglobulin structural domain (consisting of approximately 110 amino acids in antiparallel β-pleated sheet conformation, with an intrachain disulphide bridge). Examples are intercellular adhesion molecule-1 (ICAM-1) and ICAM-2, both of which are expressed on the surface of endothelial cells. While ICAM-2 is constitutively expressed, ICAM-1 is typically present only at low levels. However, the synthesis and expression of ICAM-1 are significantly amplified by IL-1 and TNF-α. ICAM-1 is synthesised more slowly than E-selectin, but its expression lasts longer. VCAM-1 is another member of this adhesion molecule family. Its expression is upregulated similarly to ICAM-1. Notably, VCAM-1 primarily supports the adhesion of monocytes, as opposed to neutrophils which are associated with ICAM-1 and ICAM-2.

The molecules of the immunoglobulin superfamily interact with a further group of adhesion molecules known as *integrins*. Integrins are protein heterodimers (consisting of α and β-chains). The integrin, leucocyte function-associated antigen-1 (also known as LFA-1 or αLβ2), is present on the surface of neutrophils, monocytes and lymphocytes and binds to ICAM-1 and ICAM-2. Integrin molecules are stimulated by chemokines such as IL-8 and platelet-activating factor (PAF), secreted by vascular endothelial cells. Stimulation increases the affinity of integrins for ICAMs and promotes firm adhesion between leucocytes and the endothelial surface.

The importance of adhesion molecules in the recruitment of leucocytes is illustrated by immunodeficiency diseases such as *leucocyte adhesion deficiency* (LAD) syndromes in which defective cell adhesion results in recurrent infections (Chapter 6).

Leucocyte Functions in Inflammation

Neutrophils are the first cells to infiltrate the site of infection in significant numbers. Microbial products such as lipopolysaccharide (LPS) and *N*-formyl methionyl peptide (sometimes simply referred to as 'chemotactic peptide') and the products of other inflammatory cells such as PAF and TNF-α are potent activators of neutrophils. Activated neutrophils show increased expression of chemokine receptors and secrete chemokines that attract monocytes and macrophages.

Certain cells are specialised to accomplish phagocytosis with high efficiency. These include macrophages, neutrophils, monocytes, dendritic cells and osteoclasts, collectively referred to as *professional phagocytes*. Their primary roles encompass the elimination of microorganisms and the presentation of antigens to lymphocytes to initiate an adaptive immune response (see Chapter 6). Conversely, fibroblasts, epithelial cells and endothelial cells, though capable of phagocytosis, do so with lower efficiency. Termed *non-professional phagocytes*, these cells primarily focus on the removal of dead cells, playing a pivotal role in maintaining tissue homeostasis.

Upon their activation, neutrophils show increased expression of complement receptors and Fc receptors which recognise complement components and antibodies, respectively. The increased receptor expression on neutrophils enhances the phagocytosis of microorganisms coated with complement and at a later stage with specific antibodies (opsonisation). In the early stages of pathogen invasion, phagocytosis is a crucial mechanism that eliminates foreign particles. Phagocytosis involves engulfment of the foreign material into a specialised organelle called the *phagosome*. As part of the phagosome maturation process, this organelle undergoes structural membrane alterations and content modifications. Eventually, the maturing phagosome fuses with lysosomes, forming a *phagolysosome*. This resultant organelle houses enzymes designed to break down and degrade the internalised foreign material. These enzymes work within the confines of the vesicle ensuring the invaders are neutralised without harming the host cells. These enzymes include *hydrolytic enzymes* such as proteases, elastases, collagenases and phospholipases.

Neutrophils also kill pathogens by extracellular mechanisms. Thus, their activation increases membrane-bound reduced nicotinamide adenine dinucleotide phosphate (NADPH) oxidase activity, generating reactive oxygen species (ROS), including superoxide anions, hydroxyl radicals, hydrogen peroxide and hypochlorous acid. These are released from neutrophils and are highly toxic to pathogens. However, this potent defence mechanism is a 'double-edged' sword. While ROS are crucial for pathogen clearance, their excessive or prolonged production can result in collateral damage to surrounding host tissues. The oxidative stress induced by ROS can damage cellular membranes, proteins and DNA, potentially leading to tissue injury and chronic inflammation. Therefore, a delicate balance in ROS generation and regulation is essential to ensure effective immune responses without compromising host tissue integrity. Endogenous antioxidant systems play a crucial role in maintaining ROS homeostasis. Enzymatic antioxidants such as superoxide dismutase (SOD), catalase and glutathione peroxidase inactivate harmful ROS species. Non-enzymatic antioxidants, like glutathione, and vitamins C and E, provide an added layer of defence by directly scavenging ROS or supporting enzymatic antioxidant functions. Individuals with chronic granulomatous disease suffer from recurrent infections because of defects in the generation of ROS (Chapter 6).

Neutrophils can also kill extracellularly by forming *neutrophil extracellular traps* (*NETs*) produced in response to microbes and consisting of a network of neutrophil-derived chromatin that can entrap the offending organism. NETs can be generated by a form of cell death known as *suicidal NETosis*, in which the neutrophil undergoes disintegration of its nuclear membrane, chromatin decondensation and loss of plasma membrane integrity resulting in the spillage of NETs into the extracellular space. *Vital NETosis* can also occur when the neutrophil releases some of its nuclear or mitochondrial DNA but remains alive.

Morphological Appearance of Inflammation

Although the classical morphological appearances of acute inflammation are well described, there are situations in which these can vary. For example, in *fibrinous inflammation*, a large amount of exudate (containing fibrinogen) and/or a strong procoagulant response can result in fibrin deposition.

Fibrinous inflammation can be seen, for example, in severe endothelial injury and results from the leakage of fluid from damaged vessels and concomitant activation of the coagulation cascade. Another form of inflammation is known as *serous inflammation* and occurs when there is a large collection of exudate that is relatively free of cells. A blister is a form of serous inflammation.

Resolution of the Acute Inflammatory Response

Resolution of the acute inflammatory response is in part due to the short half-lives of the cytokines involved in the response. Glucocorticosteroids involved in the systemic response in the liver provide a negative feedback loop by inhibiting the production of cytokines by tissue macrophages. In addition, naturally occurring antagonists interfere with the ongoing cascades of acute inflammation. These include IL-1 and TNF-α receptor antagonists, and cytokines such as IL-4 and IL-10. IL-4 and IL-10 are produced by helper T lymphocytes and downregulate IL-1, TNF-α and IL-8 production and upregulate production of IL-1 receptor antagonist (IL-1RA). IL-4 also reduces the release of PGE_2 and superoxide anions. The release of IL-4 and IL-10 from cells close to the site of infection plays a large part in dampening acute inflammation. The remaining inflammatory cells undergo apoptosis and are phagocytosed by macrophages.

Tissue Repair

Tissue repair, sometimes also referred to as healing, occurs by two processes – *regeneration* and *scarring*. Cells and tissues have different capacities for regeneration. In general, tissues can be classified as; (i) *labile*, in which cells are continuously being replaced, e.g. most surface epithelia, including skin and gastrointestinal tract; (ii) *stable*, comprising cells that are normally quiescent, but which are capable of proliferation following tissue injury.

Examples here include most solid organs, such as the liver and kidneys, and; (iii) *permanent*, in other words, tissues that do not usually have the capacity to regenerate, e.g. brain and muscle. In cases where injury occurs in permanent tissues or when repair cannot be accomplished by regeneration alone, fibrotic *scar* tissue is formed, consisting largely of connective tissue proteins such as collagens. Although scars are adequate to maintain the structural integrity of the tissue, they are not able to restore lost function.

Let us take the example of a skin wound. After the acute inflammation has resolved, there is proliferation and migration of epithelial cells. The epithelial cells become motile by a process known as *epithelial–mesenchyme transition*, in which they become spindle shaped. In normal wound healing, this is a reversible process so that once the skin surface has been covered with new epithelial cells they revert to their normal morphology and stop proliferating. Endothelial cells and pericytes proliferate to form new blood vessels. This process is known as *angiogenesis*. Angiogenesis is induced by soluble factors, mainly vascular endothelial growth factors (VEGFs) and fibroblast growth factors (FGFs). At the same time, fibroblasts and macrophages (mainly the so-called M2 macrophages) migrate into the wound. TGF-β produced by M2 macrophages stimulates fibroblasts to deposit connective tissue. At this stage, the tissue has a pink granular appearance and is known as *granulation tissue* (not to be confused with granulomas, see later). If the wound to the skin is superficial, then the predominant mechanism of repair is regeneration and there is little or no scarring. This is known as 'healing by first intention'. However, if the wound creates a larger tissue deficit, then more granulation tissue is required to fill in the missing tissue and eventually there is more marked fibrosis in the form of a large scar. This is known as 'healing by second intention'. In both cases, after scar tissue is formed, there is ongoing extracellular matrix remodelling mediated by enzymes, including matrix metalloproteinases (MMPs).

Chronic Inflammation

Infectious agents that have evolved mechanisms to avoid or counteract the activities of the immune system (e.g. intracellular pathogens such as *Mycobacteria* spp.) may persist at the site of infection. This induces *chronic inflammation* in which the mechanisms of inflammation, invoked in a vain attempt to destroy the invading microorganism, destroy the host tissue instead. Chronic inflammation is characterised by the accumulation of macrophages that are activated by interferon γ (IFN-γ) and have increased cytokine production and microbicidal activities. Fibroblast proliferation and collagen synthesis lead to the formation of fibrous scar tissue. Often granulomata form (e.g. in *M. tuberculosis* infection) in which a mass of macrophages, many of which have changed their morphology to form 'epithelioid' cells or have fused to form multinucleate giant cells, are surrounded by effector lymphocytes. Exposure to, and the intracellular accumulation of, foreign insoluble agents (e.g. silica or asbestos particles) can also have a similar effect. This is sometimes described as a 'foreign body reaction'. Fig. 5.8 shows the example of crospovidone, an inert, water-insoluble polymer utilised in pharmaceutical tablets as a disintegrant. Some individuals misuse oral tablets by crushing them, forming an aqueous suspension and then injecting them intravenously. Because crospovidone is water-insoluble, it is carried to the lungs where granulomas form in an attempt to remove the particles.

Chronic inflammation may also result from loss of control of immunity, for example, in autoimmunity. Thus, many autoimmune diseases are characterised by altered endothelium with increased expression of adhesion molecules and increased leucocyte infiltration (e.g. Crohn's disease, type 1 diabetes, rheumatoid arthritis and ulcerative colitis). Autoimmunity is discussed in more detail in Chapter 6.

In other cases, there is a hyper-reactivity of the chronic inflammatory response, in which there is excessive scarring, leading to the formation of the so-called *keloids* (Fig. 5.9).

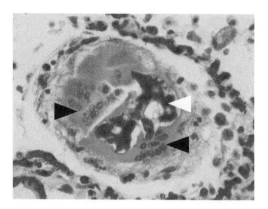

Figure 5.8 Extravascular embolised crospovidone with a foreign body giant cell reaction. Crospovidone has a coral-like appearance (white arrowhead) and is an insoluble polymer of *N*-vinyl-2-pyrrolidone that is used as a disintegrant in pharmaceutical tablets. It embolises the lungs when aqueous suspensions of pulverised tablets are injected intravenously. A giant cell is shown that has phagocytosed the crospovidone. Black arrowheads indicate the multiple nuclei present in the giant cell. *Source:* Y. Rosen, MD / Wikimedia Commons / CC BY-SA 2.0.

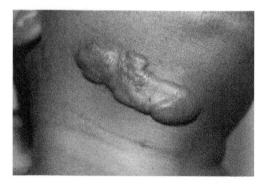

Figure 5.9 Keloid on the neck. Keloids are raised overgrowths of scar tissue that occur commonly at the site of skin injuries. They form when the body overreacts to a wound and produces excessive amounts of collagen during the healing process. Keloids can develop from various types of skin injuries, including surgical incisions, piercings, burns and even minor abrasions. Unlike typical scars, keloids do not diminish in size over time and can continue to grow outside the original boundaries of the wound. Treatment can be challenging, as they often recur after removal. https://commons.wikimedia.org/wiki/File:Keloid-Neck,_Bulky_Keloid.JPG. *Source:* Htirgan / wikimedia Commons / CC BY-SA 3.0.

Summary

Inflammation is the physiological response to tissue injury and infection. It is a rapidly induced complex network of molecular and cellular interactions that is carefully regulated and resolves once the infection or the noxious stimulus is removed. If inappropriately controlled, the toxic mechanisms that lead to the destruction of invading microorganisms can persist, resulting in extensive damage to host tissues.

Key Points

1) Inflammation results from a complex and interconnected network of reactions, initiated at sites of tissue damage, which serve to contain and destroy foreign agents, prevent continued tissue damage and initiate repair processes to restore normal function.

2) A wide range of inflammatory mediators released or formed rapidly in response to infection, or tissue damage, initiate acute inflammation. These mediators are released from granular stores within cells (e.g. histamine), generated by the activation of plasma protein reaction pathways (e.g. coagulation and complement cascades), derived from breakdown of membrane phospholipids (e.g. prostaglandins and leukotrienes) or are synthesised de novo (e.g. cytokines).

3) The destruction of invading microorganisms depends on the recruitment of circulating leucocytes (particularly neutrophils, monocytes and lymphocytes), which is regulated by chemokines and adhesion molecules produced at the vascular endothelial surface.

4) The effector functions of cells of the innate immune system, including phagocytosis, the generation of toxic reactive oxygen intermediates and the action of hydrolytic enzymes, provide the initial attack on invading microorganisms. These are later supplemented by an acquired immune response from lymphocytes that are specific to individual antigens.

5) The toxic reactions employed to destroy infectious agents also have the capacity to damage host tissues. Chronic inflammation results when microorganisms or foreign bodies are resistant to destruction or where the inflammation process is inadequately regulated.

Further Reading

Meizlish M.L., Franklin R.A., Zhou X. & Medzhitov R. (2021) Tissue homeostasis and inflammation. *Annual Review of Immunology* **39**: 557–581.

6

Disorders of the Immune System

The Biology of Disease, Third Edition. Edited by Paul G. Murray, Simon J. Dunmore, and Shantha Perera.
© 2024 John Wiley & Sons Ltd. Published 2024 by John Wiley & Sons Ltd.
Companion website: www.wiley.com/go/murray/biologyofdisease3e

Introduction

The immune system functions as a network of cellular interactions which take place in a wide range of different microenvironments throughout the body. These interactions are facilitated by receptor molecules on the outer surface of cells which bind to molecules in the microenvironment, often on the surface of other cells, initiating biochemical changes within the cells which change gene expression and consequently cell function. In this respect, the immune system is just a subset of the cellular interactions which take place within the body and determine its physiology. However, the immune system is extraordinarily complex, operating at both local and systemic levels. This chapter gives an overview of the mechanisms of innate and adaptive immunity, before considering each of the major groups of immune disorders.

Innate Immunity

The innate immune system is a broad collection of mechanisms that work together to protect the host from pathogens and other threats. The innate immune system is rapid but non-specific and is present from birth, providing immediate protection against a wide range of pathogens, including bacteria, viruses and fungi. While inflammation (described in detail in Chapter 5) is one of the key responses of the innate immune system, there are many other important components. For example, the innate immune system includes various physical and chemical barriers that prevent pathogens from entering the body. These include the skin, mucous membranes and the acidic environment of the stomach, providing a first line of defense by physically blocking the entry of pathogens or by creating an inhospitable environment for their survival and growth.

Cellular Components of Innate Immunity

Of central importance to the innate immune system are the immune cells that work together with a variety of soluble molecules to recognise and eliminate pathogens.

Neutrophils

Neutrophils are highly mobile blood cells that can pass through endothelium and migrate into tissues. As such, they are among the first responders in inflammation (see Chapter 5). Once in tissues, they migrate towards the site of inflammation, guided by chemical signals such as IL-8, C5a and leukotriene B4 (LTB_4). They can phagocytose microorganisms, releasing reactive oxygen species (ROS; Chapter 3) and hydrolytic enzymes into the phagosome. Neutrophils can also release antimicrobial proteins into the extracellular environment through degranulation and can form neutrophil extracellular traps (NETs) composed of DNA fibres and serine proteases that primarily function to contain and prevent the dissemination of pathogens.

Macrophages

Macrophages are also phagocytes. They can exist as specialised so-called *resident* macrophages that are permanent occupants of certain tissues, for example, in the liver (where they are known as Kupffer cells), the brain (microglial cells) and the placenta (Hofbauer cells). Blood monocytes can also migrate into tissues and become macrophages, differentiating into one of two major phenotypes;

- *M1 macrophages* which are activated by lipopolysaccharide (LPS) and IFN-γ. They have pro-inflammatory functions and are involved in defense against intracellular pathogens. M1 macrophages produce cytokines such as IL-12 and have enhanced bactericidal and phagocytic capabilities.
- *M2 macrophages* are activated by signals such as IL-4 and IL-13. They have an anti-inflammatory function and are involved in tissue repair, wound healing and defence against extracellular pathogens. M2 macrophages produce IL-10, TGF-β and other factors that promote tissue remodelling, angiogenesis and collagen synthesis.

Macrophages also interact with the adaptive immune response. They present antigens to T-helper cells via Major histocompatibility complex (MHC) class II molecules and provide co-stimulatory signals for T-cell activation (see later). Macrophages can also be activated by T-helper cells, particularly Th1 cells (later), and play a role in antibody production and isotype switching, for example, by secreting B-cell growth factors, such as BAFF (B-cell activating factor) and APRIL (A proliferation-inducing ligand).

Eosinophils

Eosinophils are important for immunity against parasites, particularly helminths (worms; Chapter 4). They release cationic granule proteins, such as major basic protein, eosinophil peroxidase and eosinophil cationic protein, which are toxic to parasites. They also release ROS such as superoxide and peroxide, eicosanoids from the leukotriene (e.g. LTC_4, LTD_4 and LTE_4) and prostaglandin (e.g. PGE_2) families (Chapter 5), as well as tissue-damaging enzymes such as elastases. Eosinophils are also important in the immune response to allergens (see later).

Basophils

Basophils are circulating leucocytes, which, like eosinophils, can be activated through antigen crosslinking of their IgE receptors (FcεRI) bound to IgE, leading to rapid degranulation and the release of soluble factors including heparin, histamine and leukotriene C4 (LTC_4). Basophils also express various cell surface molecules, including cytokine receptors (e.g. IL-3Rα, IL-2Rα and IL-5R), complement receptors and chemokine receptors, allowing them to be recruited to sites of inflammation, where they can be activated by pathogen-associated molecular patterns (PAMPs; see later). Basophils also produce cytokines, particularly IL-4, IL-13 and MIP-1α. As well as being involved in allergic responses, basophils are also important for immunity against parasites, including ticks and filarial worms (Chapter 4).

Mast Cells

Mast cells are tissue-resident cells that share many features in common with basophils. They are triggered to degranulate through various mechanisms, including the cross-linking of their cell surface receptor, FcεRI (Fig. 6.1). Microbial pathogens also activate mast cells through PAMPs. When activated, mast cells release a variety of mediators, including histamine, cytokines and other chemotactic factors that play an important role in allergic reactions and in the recruitment of other immune cells to the site of inflammation.

Natural Killer Cells

Natural killer (NK) cells are a subset of *innate lymphoid cells* (ILCs) that make up around 8–20% of circulating lymphocytes and play a crucial role in immune surveillance and host defence, particularly against virus-infected cells and cancer cells. Unlike other lymphocytes, NK cells do not possess antigen-specific receptors. They detect alterations in the expression of surface molecules on target cells, such as the downregulation of MHC class I molecules, normally present in most healthy cells. This recognition triggers a cytotoxic response, leading to the destruction of the target cell. Humans possess a wide repertoire of NK cell receptors, some of which are activating, and some of which are inhibitory (Fig. 6.2). Ligands for these receptors include MHC class I proteins. When NK cells encounter a target cell, they assess the balance of inhibitory and activating signals received through these receptors. If the inhibitory signals dominate, NK cells remain quiescent and do not initiate an immune response against the target cell. However, if the activating signals prevail, NK cells are triggered to unleash their cytotoxic capabilities. NK cells directly kill target cells through the release of cytotoxic granules containing perforin and granzymes. They can also trigger antibody dependent cell mediated cytotoxicity (ADCC) (Fig. 6.3) by binding to target cells opsonised with antibodies through their Fc receptors. Additionally, NK cells produce various cytokines and chemokines, such

(A)

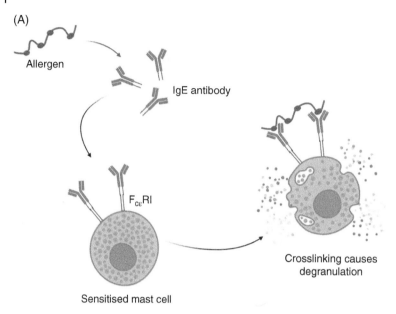

(B)

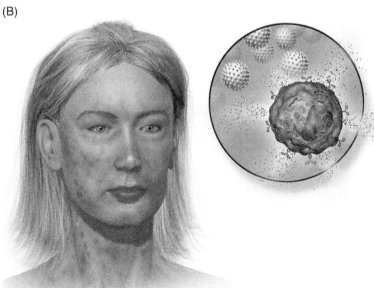

Figure 6.1 (A) Role of IgE in mast cell degranulation. When a person is exposed to an allergen, such as pollen or certain foods, it triggers the production of IgE antibodies by B-cells. These IgE antibodies bind to specific receptors on the surface of mast cells, called FcεRI (Fc epsilon receptor I). This process is known as sensitisation. Upon re-exposure, the allergen binds to the IgE antibodies already attached to the mast cells. This cross-linking of IgE antibodies on the mast cell surface triggers a signalling cascade within the mast cell, leading to degranulation. As a result of mast cell degranulation, histamine and other inflammatory mediators are released into the surrounding tissue. *Source:* Created with BioRender.com. (B) Mast cell degranulation can lead to immediate symptoms, in the affected area. Histamine and other inflammatory mediators cause symptoms including itching, hives, redness and swelling that can occur locally, or which can manifest systemically, affecting multiple organ systems. *Source:* Blausen Medical / Wikimedia Commons / Public domain.

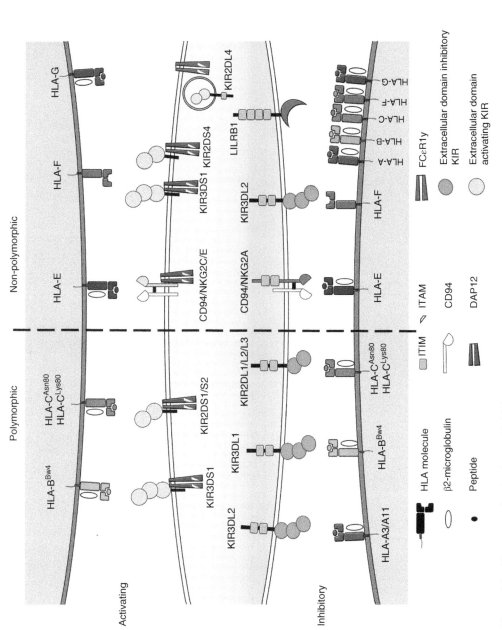

Figure 6.2 The range of activating and inhibitory natural killer (NK) cell receptors and their major histocompatibility complex (MHC) class I ligands. Shown are the different interactions between the killer cell immunoglobulin-like receptors (KIRs) and CD94:NKG2 family of receptors and their MHC class I ligands. LILRB1 interacts with folded HLA class I molecules, HLA-F and HLA-G. KIR2DL4 is thought to interact with HLA-G through an endosomal compartment. HLA-F binds as an open conformer to KIR3DS1, KIR2DS4 and KIR3DL2. *Source*: Adapted from https://onlinelibrary.wiley.com/doi/10.1111/imm.13039 / John Wiley & Sons.

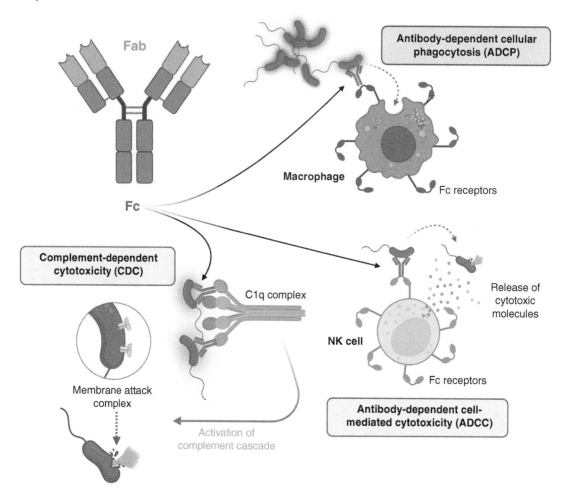

Figure 6.3 Importance of the Fc portion of antibody molecules. An antibody consists of two identical fragment antigen-specific binding (Fab) domains and one fragment crystallisable (Fc) domain, forming a Y-shaped glycoprotein. This structure enables the antibody to recognise and bind to specific antigen epitopes on invading microorganisms. Additionally, antibodies can interact with other components of the immune system via their Fc domains. Fc interactions initiate various antibody effector functions, including complement-dependent cytotoxicity (CDC), antibody-dependent cellular phagocytosis (ADCP) and antibody-dependent cell-mediated cytotoxicity (ADCC). ADCC involves the recognition of the Fc portion of IgG antibodies by the CD16 (FcγRIII) Fc receptor on NK cells, leading to the killing of antibody-coated cells. CDC occurs through the binding of the Fc domain of IgG antibodies to the C1q complex, resulting in the activation of complement components and subsequent destruction of targeted cells. ADCP involves the cross-linking of FcγRIIa and FcγRI Fc receptors on macrophages with IgG antibodies, leading to the internalisation and phagocytosis of target cells. *Source:* Created with BioRender.com.

as IFN-γ and TNF-α, which modulate immune responses and contribute to the recruitment and activation of other immune cells. Overall, the interplay between inhibitory and activating signals mediated by NK cell receptors allows for the discrimination between healthy cells and those that are infected, transformed or otherwise abnormal.

Gamma Delta (γδ) T-cells

Gamma delta (γδ) T-cells are considered part of both the innate and adaptive immune systems. They represent a distinct subset of T-cells that express a unique T-cell receptor composed of gamma and delta chains, in contrast to the conventional alpha and beta chains found in most T-cells (see later). γδ T-cells are often found in epithelial and mucosal tissues, acting as one of the first lines of defence against pathogens. One of the distinguishing features of γδ T-cells is their ability to recognise antigens in an MHC-independent manner. While the antigens recognised by most γδ T-cells are still not fully described, they are thought to respond to markers of cellular stress, such as those found during infection or in tumours. Upon activation, γδ T-cells produce cytokines such as IFN-γ, TNF-α and IL-17, as well as chemokines that attract other immune cells. γδ T-cells can also directly kill infected or transformed cells through cytolysis.

Dendritic Cells

Dendritic cells like macrophages are specialist (sometimes called 'professional') antigen-presenting cells, meaning they can process antigens and present them on the cell surface to T-cells (see later), thus acting as an important link between the innate and adaptive immune systems. Dendritic cells are found in tissues that are in contact with the external environment, such as the skin (where a specialised type of dendritic cell, called a Langerhans cell, resides) and the inner linings of the nose, lungs, stomach and intestines. Once activated, dendritic cells migrate to the lymph nodes, where they interact with T-cells and B-cells to initiate the adaptive immune response. Dendritic cells are classified into conventional dendritic cells and plasmacytoid dendritic cells. Whereas conventional dendritic cells are specialised for antigen presentation, plasmacytoid dendritic cells have evolved to respond to viral infections.

Pattern-Recognition Receptors

The innate immune system relies on pattern-recognition receptors (PRRs) to detect molecular patterns from microorganisms and damaged cells. These PRRs are essential for the recognition of both PAMPs that are specific to microbes, and damage-associated molecular patterns (DAMPs) released by damaged cells. Different types of PRR have been identified, including Toll-like receptors (TLRs), nucleotide-binding oligomerisation domain (NOD)-like receptors (NLRs), retinoic acid-inducible gene-I-like receptors (RLRs), C-type lectin receptors and AIM2 (absent in melanoma-2)-like receptors.

Toll-like Receptors

TLRs are receptors located on the plasma membrane or endosomal membrane. They are involved in pathogen recognition. Different TLRs recognise different types of microbial products (Fig. 6.4). For example,

- TLR1, 2, 4 and 6 recognise bacterial lipids.
- TLR3 and 7 recognise viral RNA.
- TLR9 recognises single-stranded DNA.
- TLR5 recognises bacterial proteins.

Upon ligand binding, TLRs recruit adaptor proteins to their cytoplasmic tails to initiate downstream signalling. The two main adaptors are MYD88 and TRIF. MYD88 is used by most TLRs, while TLR3 and TLR4 use TRIF. These adaptors activate signalling cascades, including the NF-kB and IRF (interferon regulatory factor) pathways, inducing the expression of pro-inflammatory cytokines (e.g. IL-6, TNF-α and IL-12), anti-inflammatory cytokines (e.g. IL-10), type I interferons, chemokines and co-stimulatory molecules for T-cell activation. Thus, TLRs fine-tune the quality, intensity and duration of these signalling cascades to generate a response that is tailored to the pathogen they recognise, allowing for the precise activation of immune mechanisms required to eliminate the invading organism.

Cytosolic PRR

While TLRs primarily detect PAMPs on the cell surface or within intracellular vesicles, there is also a cytosolic system of PRRs. The NLRs are among the best characterised. Some PRR can

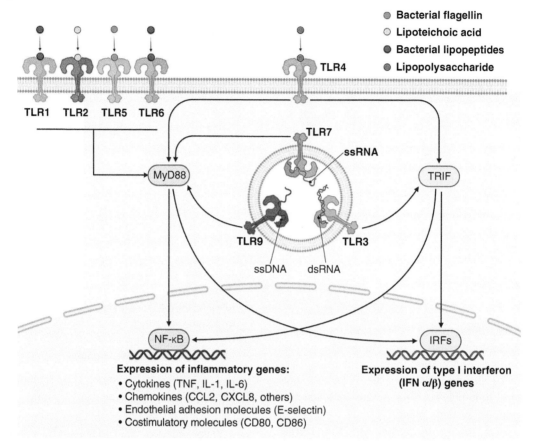

Figure 6.4 Toll-like receptor (TLR) signalling. TLRs are pattern-recognition receptors (PRRs) found on immune cells, including macrophages and dendritic cells. When TLRs recognise pathogen-associated molecular patterns (PAMPs) or damage-associated molecular patterns (DAMPs), they activate signalling cascades that lead to the production of pro-inflammatory cytokines, chemokines and interferons. TLR signalling requires the recruitment of adaptor proteins, such as MyD88 and TRIF, to the cytoplasmic tails of TLRs. This triggers the activation of downstream signalling molecules, including nuclear factor (NF)-kappa-B and interferon regulatory factors (IRFs), which promote the expression of the genes that mediate the response. Modified from Ruslan Medzhitov (Creator), Akiko Iwasaki, Jung-Hee Lee, Abbas *et al. Cellular and Molecular Immunology* 2015 Elsevier. *Source:* Created with BioRender.com.

form the so-called *inflammasome*; multiprotein complexes that initiate and regulate the downstream effects of the PRR. The NLRP3 (NOD-like receptor family, pyrin domain-containing 3) inflammasome is described in Fig. 6.5.

Adaptive Immunity

In higher organisms, adaptive (sometimes also referred to as 'acquired') immunity has evolved

to complement and enhance innate immunity. There are two main components of adaptive immunity: cell (T-cell)-mediated and antibody (B-cell)-mediated. Antibody-mediated immunity is sometimes referred to as *humoral immunity*. Adaptive immunity is characterised by

Specificity

Each B-cell and T-cell is equipped with a unique B-cell receptor, or T-cell receptor, respectively. As these cells mature, the critical

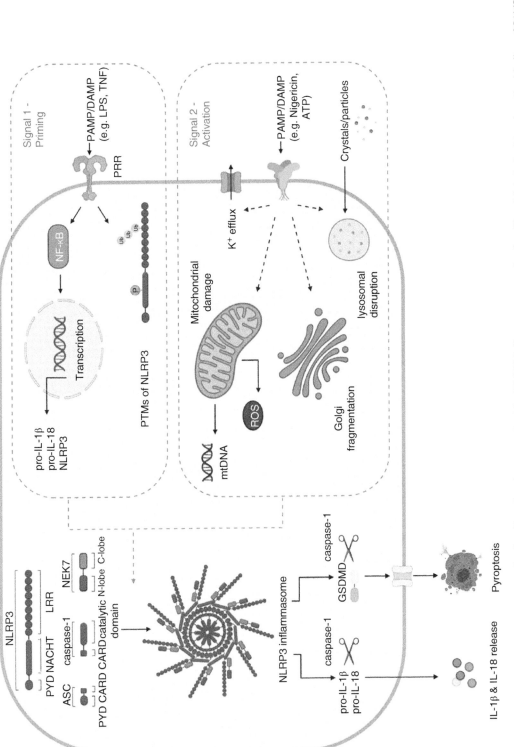

Figure 6.5 NLRP3 inflammasome activation. Two signals are required for NLRP3 inflammasome activation-Signal 1 (priming) involves the binding of PAMPs (e.g. LPS) and/or DAMPs (e.g. TNF-α) to pattern recognition receptors (PRRs) triggering the activation of NF-κB-dependent pathways and the upregulation of

(Continued)

Figure 6.5 (Cont'd) NLRP3, pro-IL-1β and pro-IL-18 transcription. Priming also triggers post-translational modifications (PTMs) of the NLRP3 protein. Signal 2 (activation) triggers various intracellular events, including potassium (K+) efflux, lysosomal disruption, dispersal of the trans-Golgi network and mitochondrial damage leading to release of mitochondrial (mt)DNA and the production of ROS. Nigericin, a potassium ionophore produced by *Streptomyces hygroscopicus*, can act as a PAMP. Extracellular ATP can act as a DAMP; under physiological conditions, ATP is found in the intracellular environment, and its release is a danger signal, alerting the immune system to cellular damage. Crystals and particles, like monosodium urate crystals in gout, or silica or asbestos particles in lung disease (penumoconiosis, see Case Study CS69), can also act as DAMPs. Together, priming and activation stimuli induce the oligomerisation of the NLRP3 protein, which forms a complex called the NLRP3 inflammasome. This oligomerisation recruits ASC (apoptosis-associated speck-like protein containing a CARD), leading to the formation of a large protein structure called the ASC speck. Within the inflammasome complex, the pro-caspase-1 enzyme is also recruited. The NLRP3 inflammasome complex additionally recruits NEK7 (NIMA-related kinase 7), which mediates the formation of interactions between adjacent NLRP3 subunits. The pro-caspase-1 enzyme within the inflammasome undergoes auto-catalytic self-cleavage, resulting in its activation. Once activated, caspase-1 cleaves its effector substrates, including pro-IL-1β (pro-interleukin-1β), pro-IL-18 (pro-interleukin-18) and GSDMD (gasdermin D). Cleavage of pro-IL-1β and pro-IL-18 generates the mature forms of these pro-inflammatory cytokines, which are then released from the cell to initiate immune responses. Cleavage of GSDMD leads to the formation of pores in the plasma membrane, resulting in a pro-inflammatory form of cell death called pyroptosis (see also Chapter 3).
Source: https://jlb.onlinelibrary.wiley.com/doi/full/10.1002/JLB.3MR0720-513R / John Wiley & Sons.

segment of these receptors responsible for recognising pathogens undergoes genetic rearrangement, a process that ensures a vast diversity in receptor sequences. As a result, within the populations of B-cells and T-cells, there exists an immense variety of receptors, each tailored to detect a different antigen, allowing the immune system to recognise and respond to an extensive range of pathogens. Antigens include the components or products of microorganisms such as viruses, bacteria and fungi, other parasites such as helminth worms, and cells from unrelated human donors, as well as commonplace substances such as pollen or animal hair. The B-cell receptor (BCR) and the T-cell receptor (TCR) on the surface of an individual B-cell or T-cell, respectively, are specific for a single molecular cluster or arrangement within the structure of an antigen. This molecular cluster or arrangement is referred to as an *antigenic determinant* or *epitope*. Within a given antigen, some epitopes are more likely to elicit a strong immune response than others. These are often referred to as *dominant epitopes*. *Subdominant epitopes*, on the other hand, can elicit an immune response, but this response is typically weaker than that caused by dominant epitopes. *Cryptic epitopes* are antigenic determinants that are not normally accessible to the immune system but can become accessible under certain conditions, such as changes in the protein conformation, proteolytic cleavage, or during the process of antigen presentation. Unlike subdominant epitopes, cryptic epitopes are typically not recognised by the immune system under normal conditions. Their exposure may lead to new immune responses, which can sometimes be detrimental, as is the case with some autoimmune diseases (see later).

Diversity

The body mounts responses against a wide range of antigenic determinants. Although each B- or T-cell has surface receptors with a single specificity, collectively these cells have a wide range of specificities facilitating the recognition of a similarly wide range of epitopes.

This diversity is generated during B- and T-cell development, by re-arrangement of the genetic information encoding the polypeptides that make up the receptor molecules and is independent of the presence of antigen. The range of receptors (known as the *immunological repertoire*) produced is unique to the individual, may alter with time and will contain receptors specific to epitopes that exist at present, or which may be generated in the future as a consequence of evolutionary changes occurring in infectious organisms.

Memory

Once the adaptive immune system has encountered a particular antigen, it is able to make a quicker and more effective response to the same antigen on subsequent re-exposure. So, B- and T-cells have 'memory'. This is probably the most widely known property of the immune system –every parent knows that an infant contracting measles in childhood is protected from the disease in later life. It is the basis of vaccination, the first demonstration of which by Edward Jenner marked the foundation of immunology as a discipline (Chapter 1).

Discrimination of Self and Non-self

Under normal circumstances, the immune system mounts an effective response against 'foreign' antigens but does not attack self-components. In part, this is due to the removal of cells that carry receptors specific for 'self', before they become fully able to respond to antigen (i.e. before they become immunocompetent). However, cells with self-reactive receptors do survive and so mechanisms must exist to suppress their function.

Generation of Diversity in the Immunological Repertoire

B- and T-cells originate from stem cells that are formed first in the yolk sac, then in the foetal liver or spleen and later in adult bone marrow.

B-cells complete their development in the bone marrow, but T-cell precursors migrate from the bone marrow to the thymus where they mature. The bone marrow and the thymus are responsible for the production of the repertoires of immunocompetent B- and T-cells, respectively, and are the primary lymphoid organs of humans.

B- and T-cell receptors are made up of two types of polypeptide chain – the heavy and light chains of immunoglobulins on B-cells, and the α- and β-chains, which are the most common constituents of the T-cell receptor. In both cases, the polypeptides have a non-polymorphic or *constant* region and a polymorphic or *variable* region which forms the antigen recognition site. *Hypervariable* amino acid residues within the variable regions are involved in antigen recognition and binding.

Within each haploid set of chromosomes, there are different versions of the gene segments (*variable, joining* and *diversity* gene segments) which carry the information for the synthesis of the polymorphic regions of the B- and T-cell receptors, but usually only single versions of the gene segments which encode the constant regions. Rearrangement of the gene segments occurs to produce functional coding information for the diverse repertoire of B- and T-cell receptors (Fig. 6.6). The mechanisms involved have the same essential features in both B- and T-cells:

- Recombined gene segments for a particular polypeptide are contained on only one of the pair of chromosomes that are present in the cell. Although the gene segments can recombine in many ways, a given chromosome in a particular cell can only make one rearrangement to produce a functional coding sequence. This is because the segments are brought together to form a functional sequence by looping out of DNA, involving the irreversible breakage and rejoining of the chromosome. Successful rearrangement on one chromosome prevents rearrangement of the other. This is termed *allelic exclusion*

and ensures that each cell has only a single receptor specificity.
- Gene rearrangement does not depend on the presence of antigen.
- Rearrangements occur at random, so that different precursor cells can rearrange in different ways to produce functional genes that encode receptors with different specificities.
- A functional gene is assembled from a single variable, joining, diversity (in some cases) and constant region segment. The more versions of the gene segments there are on the chromosome, the greater the number of possible combinations.
- When a junction between segments is formed, there is some variation in the precise point of joining, and extra coding information can be added at the junctions.
- Functional receptors are formed by the pairing of the polypeptide products of two different types of rearranged genes (heterodimer formation). The number of possible receptor structures is therefore the product of the number of possible structures for each polypeptide.

Lymphocyte Recirculation

Once mature, the immunocompetent lymphocytes pass from the bone marrow or thymus into the circulation. They are transported around the body in the blood and the lymphatic system and reach the lymph nodes. They pass through the cuboidal endothelial cells of the venules (known as high endothelial venules) into the cortical tissues of the node. From here they are collected by efferent lymphatics draining the lymph nodes and returned to the blood system via the thoracic ducts. Foreign antigens within the peripheral tissues are collected with other extracellular fluids and drained to the nearest lymph node. Thus, the lymph nodes, along with the other secondary lymphoid organs (the spleen and the mucosa-associated lymphoid tissues) provide a microenvironment which facilitates interaction between the lymphocyte surface receptors and

(A)

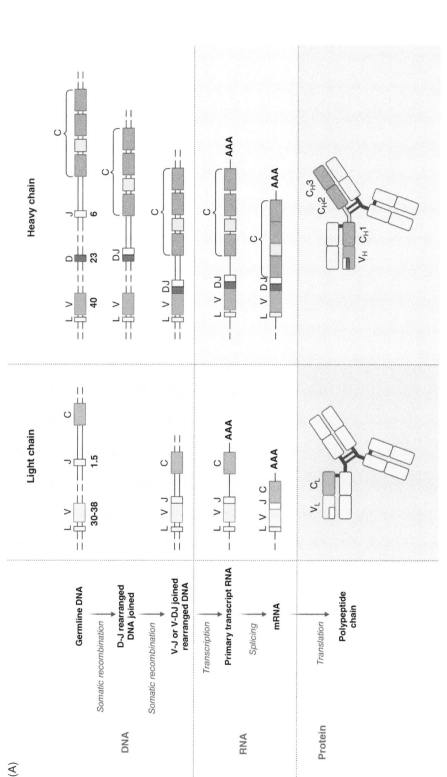

Figure 6.6 Generation of diversity in the B-cell and T-cell receptors. (A) During B-cell development, a process called V(D)J recombination occurs which generates the diversity of immunoglobulin heavy chains and light chains. Right panel: The recombination process begins with the rearrangement of one D (diversity) gene segment and one J (joining) gene segment in the heavy chain locus, resulting in deletion of any DNA between the D and J segments. This process is known as D-J recombination. Next, a V (variable) gene segment from a region upstream of the newly formed DJ complex undergoes recombination with the DJ complex. This rearrangement process joins the V, D and J gene segments, resulting in the formation of a rearranged VDJ gene segment. All other gene segments between the V and D segments are deleted from the genome. A primary (un-spliced) transcript containing the VDJ region of the heavy chain, and the constant regions is formed. A polyadenylated tail is added, and sequences between the VDJ segment and the constant gene segment are removed. The processed mRNA is translated, leading to the production of the IgM heavy chain protein. Left panel: A similar process creates the light chains. The heavy and light chains (either kappa or lambda) combine to create the membrane-bound form of the immunoglobulin IgM that is

(Continued)

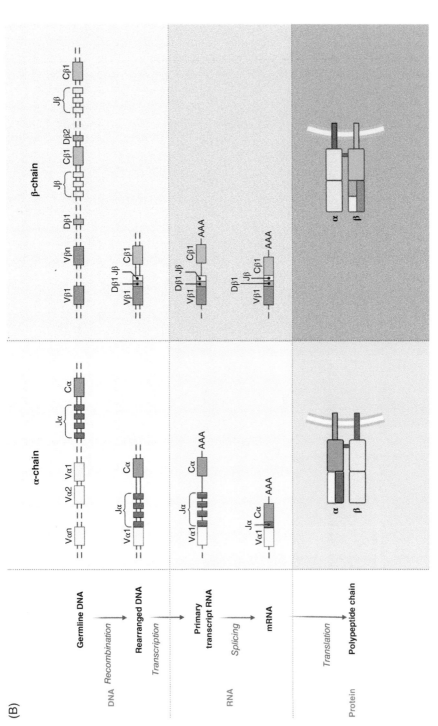

Figure 6.6 (Cont'd) expressed on the surface of immature B-cells. Acknowledgements: Minor changes to the Figure created by Akiko Iwasaki (Creator) Jung-Hee Lee. Janeway's Immunology, 8ed. (© Garland Science 2012). Created by Biorender.com. (B) In T-cells, the TCR genes undergo a series of ordered recombination events, similar to immunoglobulin gene rearrangement. Right panel: The process begins with the Dβ-Jβ recombination event in the β-chain gene. This involves the joining of one of the Dβ (diversity) gene segments with one of the Jβ (joining) gene segments. There are two sets of Dβ and Jβ gene segments available. Following Dβ-Jβ recombination, Vβ-to-DβJβ rearrangements occur. All gene segments between the Vβ, Dβ and Jβ segments are deleted. The rearranged Vβ-Dβ-Jβ gene segment is transcribed into a primary transcript that includes the Vβ-Dβ-Jβ-Cβ segments. The spliced mRNA is translated, allowing the production of the full-length protein for the TCR β-chain. Left panel: After β-chain rearrangement, recombination of the α-chain occurs. The rearranged β-chain and α-chain of the TCR are assembled, resulting in the formation of the αβ-TCR which is expressed on the majority of mature T-cells. Note: Not all V, D and J genes are shown. Acknowledgements: Minor changes to the Figure created by Akiko Iwasaki (Creator) Jung-Hee Lee. Janeway's Immunology, 8ed. (© Garland Science 2012). *Source:* Created with Biorender.com.

foreign antigens. Thus, the secondary lymphoid organs are the location at which an immune response is initiated.

Lymphocyte Activation

Gene rearrangement ensures that the immunological repertoire contains an enormous number of lymphocytes with receptors for different antigens. There is no reliable way to determine the size of the repertoire, but estimates suggest something like 10^{10} different B-cell receptors and rather more, maybe as many as 10^{15}, αβ T-cell receptors. There are likely to be only a few cells with receptors of the same specificity. How then is the immune system able to counteract a particular infection effectively? This is explained by the *clonal selection theory*, which is the central concept of immunology. Clonal selection means that when a B- or T-cell encounters an epitope for which its surface receptor is specific, it is in effect 'selected'; binding of epitope to the receptor triggers biochemical changes inside the cell which lead ultimately to proliferation and the production of a clone of cells all with the same receptor specificity. The antigen is therefore responsible for *clonal selection*, and receptor binding results in *clonal expansion*. Changes occur within the cells as they differentiate to form either effector cells or memory cells. The same process occurs in both B- and T-cells, but the details of the receptor interactions are very different, and are described below.

Recognition by Antigen Receptors and the Major Histocompatibility Complex

B-cell receptors recognise epitopes on the surface of complex antigens (e.g. components of the surface coat of bacteria). Key amino acid residues in the receptor-binding site interact with a particular three-dimensional arrangement of molecules on the antigen. These amino acids are *hypervariable residues*, often also described as *complementarity-determining regions*. The B-cell receptor therefore recognises what is often described as a *conformational epitope*.

The situation in T-cells is more complex. The T-cell receptor can only recognise short peptides displayed on the surface of cells by MHC proteins (Fig. 6.7, Table 6.1). These genes were first identified in the context of transplantation. A graft between genetically identical animals is usually successful, but if there are genetic differences between donor and recipient, the graft is rejected. The MHC genes are mainly responsible. In humans, this gene region is referred to as the *human leucocyte antigen* (HLA) region. There are two sets of genes which produce two different types of protein (Table 6.1); the MHC class I genes (HLA-A, -B and -C in humans) and the MHC class II genes (HLA-DP, -DQ and -DR in humans).

In the population, there are many different alleles for each of the MHC genes. So, there is a high level of polymorphism at the population level. Some of the alleles are common in the population, for example, the HLA-A2 allele is carried by 50% of White Europeans, while other alleles are rare. Similarly, the HLA-B*5703 allele is more common in Africans, particularly in sub-Saharan populations, HLA-A*1101 allele is frequently found in South Asians, including people of Indian, Pakistani and Bangladeshi descent, while the HLA-B*5001 allele is more prevalent in Middle Eastern populations. Each individual carries only two alleles (one from each parent) for each of the three class I genes and two alleles for each of the three class II genes. It is the nature of these alleles which characterises a 'tissue type', a vitally important feature when considering the suitability of donors for transplantation (see later). Different alleles encode MHC molecules which have the same overall structure, but differ in the makeup of their peptide-binding sites and are therefore specific for peptides with different structural motifs.

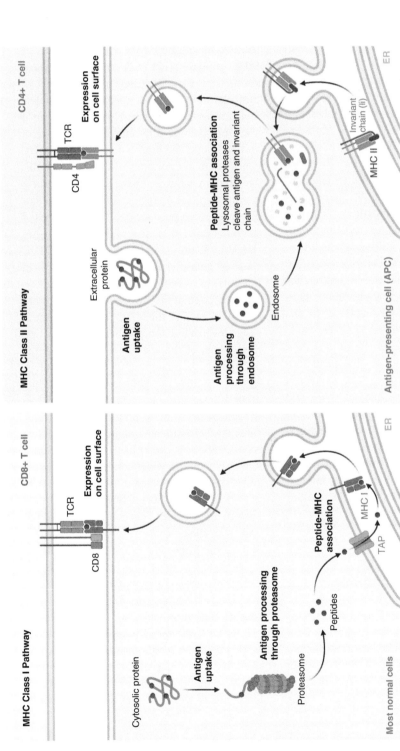

Figure 6.7 Antigen presentation by MHC class I and II. Left panel: MHC class I molecules primarily present endogenous antigens derived from intracellular pathogens (e.g. viruses) or from self-proteins. Intracellular antigens are broken down into peptide fragments by the proteasome. The peptides are transported into the endoplasmic reticulum by the transporter associated with antigen processing (TAP). In the ER, peptides bind to the MHC class I molecule, which consists of a heavy chain (encoded by HLA genes in humans) and a beta-2 microglobulin (β2m) chain. The MHC class I complex is transported to the cell surface and displayed on the cell membrane. Here, it can be recognised by CD8+ T-cells (cytotoxic T-cells), which recognise the complex through their T-cell receptor. Right panel: MHC class II molecules primarily present exogenous antigens derived from extracellular pathogens (e.g. bacteria) or from engulfed particles, such as proteins or fragments of pathogens acquired through endocytosis or phagocytosis. Exogenous antigens are processed in endosomes or lysosomes within antigen-presenting cells. These antigens are broken down into peptide fragments by proteases. In the endosomal or lysosomal compartments, the peptide fragments bind to the MHC class II molecule, consisting of an alpha chain and a beta chain. The invariant chain occupies the peptide-binding groove of MHC class II molecules, preventing premature binding of endogenous peptides during their assembly and transport, ensuring that only exogenous antigens are presented. The peptide–MHC class II complex is transported to the cell surface, where it can interact with CD4+ T-cells (helper T-cells), which recognise the complex through their T-cell receptor. Acknowledgements: Minor changes to the Figure created by Akiko Iwasaki (Creator) Jung-Hee Lee. Janeway's Immunology, 8ed. (© Garland Science 2012). *Source:* Created with Biorender.com.

Table 6.1 Defining features of major histocompatibility complex (MHC) I and II.

MHC class I	MHC class II
Presents peptides to CD8+ T-cells	Presents peptides to CD4+ T-cells
Encoded by HLA-A, -B and -C genes	Encoded by HLA-DP, -DQ and -DR genes
Expressed on the surface of most cells	Expressed on antigen-presenting cells
Presents endogenous peptides	Presents peptides from extracellular antigens

Peptide binding is essential for effective presentation of antigens to T-cells. The MHC therefore exerts a genetic influence on the ability of an individual to respond to particular antigens – it influences *immune responsiveness*.

As T-cells can only recognise peptides presented by MHC molecules, there must be mechanisms by which antigens can be processed to produce peptides which are bound to MHC molecules and presented at the cell surface. There are separate mechanisms for MHC class I and II molecules:

MHC class I molecules are present on the surface of virtually all cells. They are produced in the endoplasmic reticulum and depend on peptide binding at this site for their assembly and transport to the cell surface. They provide a mechanism for sampling the peptides available in the cytoplasm and endoplasmic reticulum. For example, if a cell is infected with a virus, it may produce viral proteins which will be processed and presented at the cell surface where they will generate an immune response. In a similar way, tumour cells may produce antigens which are seen as 'non-self'. Peptides presented by MHC class I are recognised by a subset of T-cells which carry the surface glycoprotein CD8. Most of these cells have a cytotoxic function which brings about lysis of the target cell.

MHC class II molecules are restricted to the surface of antigen-presenting cells, which include dendritic cells, B-cells and macrophages. Like MHC class I molecules, MHC class II molecules are produced in the endoplasmic reticulum but in this case are initially prevented from binding peptides by the action of an additional polypeptide, referred to as the *invariant chain*. The MHC class II molecules are eventually transported to the endosomal compartment where they are dissociated from the invariant chain and can then bind available peptides. In contrast to MHC class I, these peptides are derived from proteins that have been taken up, for example, by endocytosis of exogenous antigens. MHC class II therefore samples and presents antigens from the extracellular environment. In this case, the presented peptides are recognised by a subset of T-cells defined by the presence of the glycoprotein CD4 on their surface. The majority of CD4+ T-cells have a 'helper' function.

Non-classical MHC (HLA) molecules refer to HLA proteins that differ from the 'classical' HLA class I and class II molecules in terms of their expression patterns, functions and molecular characteristics. Non-classical HLA molecules have specialised roles in the immune system, often contributing to immune regulation and tolerance. Some key characteristics and examples of non-classical HLA molecules include the following:

- *Limited polymorphism:* Unlike the highly polymorphic classical HLA molecules, non-classical HLAs show limited allelic variability.
- *Restricted tissue expression:* While classical HLA molecules are broadly expressed on nucleated cells (for class I) or on antigen-presenting cells (for class II), non-classical HLA molecules have more restricted and specialised patterns of expression. For instance, HLA-G is primarily found at the maternal–foetal interface in the placenta.
- *Alternative functions:* Non-classical HLAs often have functions distinct from classical HLA molecules. For example, while classical HLA molecules primarily present peptides to T-cells, non-classical HLAs, such as HLA-G, can inhibit immune cell activity,

contributing to immune tolerance. Another example is HLA-E, which interacts with NK cells to mediate their function.

Effector Functions

T-cell Effectors

When T-cells recognise their cognate antigen, several events occur. First, TCR engagement triggers signalling events, including the activation of protein kinases, phosphorylation cascades and the recruitment of adapter proteins, ultimately leading to the activation of transcription factors. In addition to TCR engagement, co-stimulatory signals are required to fully activate T-cells. For example, co-stimulatory molecules on antigen-presenting cells, such as CD80 and CD86, interact with CD28 receptors on T-cells, providing the necessary co-stimulatory signal. The activated T-cells undergo clonal expansion, resulting in the proliferation of antigen-specific T-cells. Activated T-cells can differentiate into different effector T-cell subsets, depending on the nature of the cytokine milieu and the antigen encountered. Examples of effector T-cell subsets include:

CD4+ T-cells

There are different CD4+ T-cell subsets which include;

- *Th1 cells*, which play a vital role in cellular immunity against intracellular pathogens, particularly bacteria and viruses. They secrete cytokines such as IFN-γ, TNF-α and IL-2, activate macrophages, stimulate the production of opsonising antibodies and promote the development of cytotoxic CD8+ T-cells.
- *Th2 cells*, which are involved in allergic and humoral immune responses, particularly against extracellular parasites and allergens. They produce IL-4, IL-5, IL-9 and IL-13, facilitate class switching (see later) and promote eosinophil recruitment and activation.
- *Th17 cells*, which participate in inflammatory and immune responses against extracellular pathogens, especially bacteria and fungi.

CD8+ T-cells

Cytotoxic T lymphocytes can directly kill infected or abnormal cells by releasing cytotoxic granules containing perforin and granzymes or by inducing Fas-mediated apoptosis of target cells (Chapter 3). They have a critical role in clearing viral infections and eliminating tumour cells.

Regulatory T-cells (Tregs)

Tregs are crucial for preventing excessive immune responses that could lead to autoimmunity. They express the transcription factor FoxP3 and suppress the activation and function of other immune cells, by secreting anti-inflammatory cytokines such as IL-10 and TGF-β.

B-cell Effectors

After naïve B-cells recognise an antigen through their surface B-cell receptors, a series of events occur that lead to B-cell activation, differentiation and the generation of effector B-cells and memory B-cells (Fig. 6.8). Antigen binding to the B-cell receptor triggers intracellular signalling cascades, including phosphorylation events and activation of signalling molecules such as kinases and adapter proteins. B-cells require additional co-stimulatory signals, typically provided by helper T-cells. After antigen binding, the B-cell receptor–antigen complex is internalised into the B-cell through receptor-mediated endocytosis. B-cells present antigen-derived peptides on their MHC class II molecules to CD4+ helper T-cells. This interaction requires the CD40–CD40L interaction. B-cell activation and differentiation in response to helper T-cell interactions are referred to as *T-dependent responses*. Activated B-cells migrate to secondary lymphoid organs, such as lymph nodes. Here, the antigen-specific activated B-cells undergo clonal expansion within germinal centres. Germinal centres are not pre-existing structures; they develop after an immune response is initiated within the follicles of the lymphoid tissue. In the germinal centre,

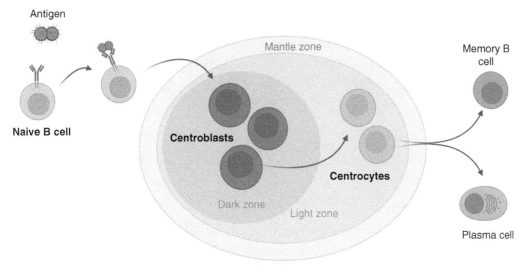

Figure 6.8 Schematic representation of B-cell differentiation. The B-cell receptor on a naive B-cell binds to the antigen, initiating the activation process. Activated B-cells undergo rapid proliferation within specialised structures known as germinal centres, which are found within secondary lymphoid organs. The rapidly proliferating antigen-specific B-cells are known as centroblasts. Some of these cells become centrocytes in the light zone, before differentiating into either plasma cells or memory B-cells. Plasma cells secrete antibodies that neutralise or help eliminate the antigen. Memory B-cells remain in the secondary lymphoid organs or circulate throughout the lymphatic and circulatory systems, waiting for future encounters with the same antigen. *Source:* Created with BioRender.com.

cells undergo *somatic hypermutation* which introduces random mutations in the genes encoding the variable regions. B-cells with higher-affinity B-cell receptors due to somatic hypermutation receive stronger survival signals, leading to their preferential expansion. Some activated B-cells differentiate into plasma cells. These plasma cells secrete large quantities of antibodies specific to the encountered antigen.

Memory Functions

Memory T-cells

Within the expanding population of activated T-cells, a small subset differentiates into memory T-cells, a process influenced by factors such as the strength and duration of the antigen stimulation, the presence of co-stimulatory signals and cytokine signalling. Differentiated memory T-cells receive distinct signals that programme them for long-term survival, including the expression of specific transcription factors, such as T-bet for CD8+ memory T-cells and TCF1 for CD4+ memory T-cells. Certain cytokines, such as IL-7 and IL-15, promote the survival and maintenance of memory T-cells. Memory T-cells migrate to various tissues, including lymphoid organs, peripheral tissues and sites of previous infection, where they provide local immune surveillance and rapid response to re-infection.

Memory B-cells

A subset of activated B-cells differentiates into long-lived memory B-cells. Like memory T-cells, memory B-cells persist in the body and provide immunological memory, responding quickly upon re-exposure to the same antigen, leading to a faster and more robust secondary immune response. The antibody molecules produced in the primary response are usually IgM. However, in the secondary response, antibody molecules are often produced from rearranged genes in which the information encoding the variable portion of the heavy chain is recombined with a different constant gene segment. The specificity of the antibody remains the same, but the portion of the molecule which is responsible for its effector function may be changed. This process is known as *class switching*. Memory B-cells that have undergone class switching are referred to as 'switched' *memory B-cells*. The choice of which class of immunoglobulin to switch to depends on the nature of the pathogen and the cytokine signals received. IgG antibodies are the most common in blood and have various roles including the neutralisation of pathogens and complement activation. IgA antibodies are often found in mucosal areas, such as the gut or lungs, and are involved in neutralising pathogens at these sites. IgE antibodies are involved in responses to parasites and are also responsible for allergic reactions (see later).

Atypical memory B-cells are a distinct subset of memory B-cells described in chronic infectious diseases, for example, in malaria. They exhibit an altered functionality compared to conventional memory B-cells and have a reduced capacity to differentiate into antibody-secreting cells. They also have a propensity to produce autoantibodies and pro-inflammatory cytokines and have also been found in some autoimmune diseases such as systemic lupus erythematosus (SLE) and rheumatoid arthritis (RA).

Unlike conventional memory B-cells, which are generated after exposure to specific antigens, *innate IgM memory B-cells* do not require T-cell help for activation. These B-cells are characterised by their high expression of surface IgM antibodies and their lack of somatic hypermutation. Innate IgM memory B-cells can produce low affinity, polyreactive IgM antibodies that can bind to a wide range of pathogens as well as to some self-antigens.

Long-lived plasma cells also play a crucial role in maintaining immune memory. Their primary function is the sustained production and secretion of antibodies specific to antigens that the immune system has previously encountered. These cells reside mainly in the bone marrow, where they can survive for years or even decades without re-exposure to the antigen. By continuously producing antibodies, long-lived plasma cells provide long-term protection against recurring infections by the same pathogen.

Autoimmunity

As described in the foregoing sections, the adaptive immune system can normally discriminate between self and foreign antigens. This is crucial to preventing the immune system from attacking the body's own tissues. However, under certain circumstances, the mechanisms that prevent self-reactivity break down. This concept is known as *autoimmunity*. The earliest evidence for the existence of autoimmune disease was the identification of *autoantibodies* to red blood cells in a haematological complication of syphilis. It is now recognised that almost every organ and system of the body can be affected by autoimmune disease.

Immunological Tolerance

Immunological tolerance prevents the development of harmful immune responses against self-antigens. It involves the suppression or regulation of immune reactions to self-components, ensuring that the immune system targets foreign invaders while sparing healthy tissues. There are two major forms of immunological tolerance:

Central Tolerance

Central tolerance occurs during the maturation of immune cells in the thymus and bone marrow. In the thymus, developing T-cells undergo a process called *negative selection*, in which T-cells with receptors that recognise self-antigens are eliminated. Similarly, in the bone marrow, developing B-cells undergo negative selection to remove self-reactive B-cells. These processes ensure that potentially self-reactive lymphocytes are eliminated or rendered functionally tolerant before they enter peripheral tissues.

Peripheral Tolerance

Peripheral tolerance operates outside the primary lymphoid organs and provides additional layers of tolerance control. Mechanisms include:

- *Anergy:* This occurs when B- or T-cells encounter an antigen without the necessary co-stimulation, causing them to become unresponsive or anergic.
- *Immune privilege sites* in the body, such as the brain, eye and testes, limit the ability of the immune system to respond there.
- *Regulatory T-cells*: As described earlier, Tregs are a subset of CD4+ T-cells characterised by the expression of the transcription factor FOXP3. They have a crucial role in maintaining immune tolerance by suppressing the activation and proliferation of autoreactive T-cells.
- *Immune checkpoints* are interactions involving ligand receptor pairs, usually on two different cells, that prevent excessive immune activation. They have emerged as critical drug targets in some diseases, for example, cancer.

In some situations, autoimmunity can be beneficial. Thus, so-called *physiological autoimmunity* can help clear degraded self-antigens, contributing to the maintenance of homeostasis. Moreover, autoantibodies, for example antinuclear antibodies and rheumatoid factor, can be detected in healthy individuals. In some cases, autoantibodies can be both detrimental and helpful at the same time. For example, in malaria (see also Case Study CS8), autoantibodies contribute to the disease process, causing complications such as cerebral malaria, anaemia and acute kidney injury, but they can also bind to malarial antigens and form immune complexes that activate complement pathways, leading to parasite clearance. The next section describes the situation in which autoimmunity is harmful to the host, in other words, '*pathological autoimmunity*'.

The Spectrum of Autoimmune Diseases

Autoimmune diseases can be classified as either *organ-specific* in which the autoantigen is localised in one organ only (e.g. thyroid peroxidase and thyroglobulin in Hashimoto thyroiditis) or *systemic* in which the autoantigen is widespread. An example of a systemic autoimmune disease is SLE, which is characterised by the involvement of multiple organ systems and a wide range of clinical manifestations (Fig. 6.9). Some of the more common autoimmune diseases together with their target antigens are shown in Table 6.2. There can be an overlap between different autoimmune diseases. For example, patients with SLE may have the clinical features of scleroderma and RA and patients with autoimmune thyroid disease often have gastric autoimmune disease. Such patients are said to have '*overlap syndrome*'. However, there is usually little overlap between organ-specific and systemic autoimmune diseases.

Different autoimmune processes may occur in the same tissue, leading to different clinical outcomes. For example, Hashimoto thyroiditis involves the destruction of thyroid cells resulting in hypothyroidism, whereas Graves disease involves the opposite effect of hyperthyroidism due to autoantibodies triggering the thyroid-stimulating hormone (TSH) receptor. Conversely, different autoimmune processes affecting the same tissues may have similar clinical effects. Thus, Goodpasture syndrome (anti-GBM disease; Case Study CS32) and SLE both cause glomerulonephritis, but through type II and III hypersensitivity mechanisms, respectively (see later).

(A)

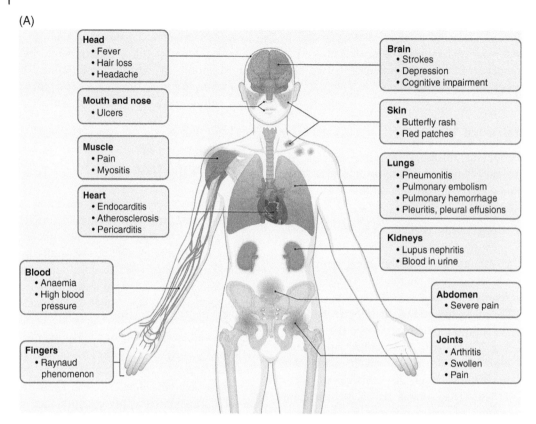

Figure 6.9 Systemic lupus erythematosus (SLE). (A) SLE is an autoimmune disease in which the immune system attacks its own tissues, causing widespread inflammation and tissue damage in affected organs, including the joints, skin, brain, lungs, kidneys and blood vessels. Slightly modified with acknowledgment to Wendy Jiang. *Source:* Created by Biorender.com. https://medlineplus.gov/genetics/condition/systemic-lupus-erythematosus/ https://medlineplus.gov/ency/article/000435.htm. (B) Immunopathogenesis of SLE. In genetically predisposed individuals with a propensity to loss of self-tolerance, certain triggers, e.g. infection, disrupt the normal clearance of nucleic acids and increase the formation of NETs. Chromatin fragments can bind directly to the B-cell receptor of autoreactive B-cells, which produce autoantibodies. These autoantibodies can bind to the antigen, forming immune complexes. The immune complexes can engage FcγRIIB (CD32b), an inhibitory Fc receptor on B-cells. In SLE, the balance between this inhibition and activation by the B-cell receptors is tipped towards activation, leading to the production of more autoantibodies. The excessive production of the B-cell activating factor (BAFF) is characteristic of SLE and contributes to the survival and activation of autoreactive B-cells (inset), thereby exacerbating the autoimmune response and disease severity. The autoantibodies form more immune complexes with self-antigens, perpetuating the cycle of inflammation and tissue damage. The immune complexes can be internalised in plasmacytoid dendritic cells, through Fc receptors such as FcγRIIa (CD32a). Once internalised, the chromatin components engage endosomal TLR9 (not shown), which recognises unmethylated CpG DNA motifs. This triggers a signalling cascade that culminates in the robust production of Type I interferons, such as IFN-α. These interferons then act in both autocrine and paracrine fashions to further activate plasmacytoid dendritic cells and conventional dendritic cells. Conventional dendritic cells that have also taken up self-antigens, present them to T-cells and produce cytokines that promote Th17 responses. T follicular helper (Tfh) cells play a critical role in promoting the activation and differentiation of autoreactive B-cells. *Source:* Created with BioRender.com.

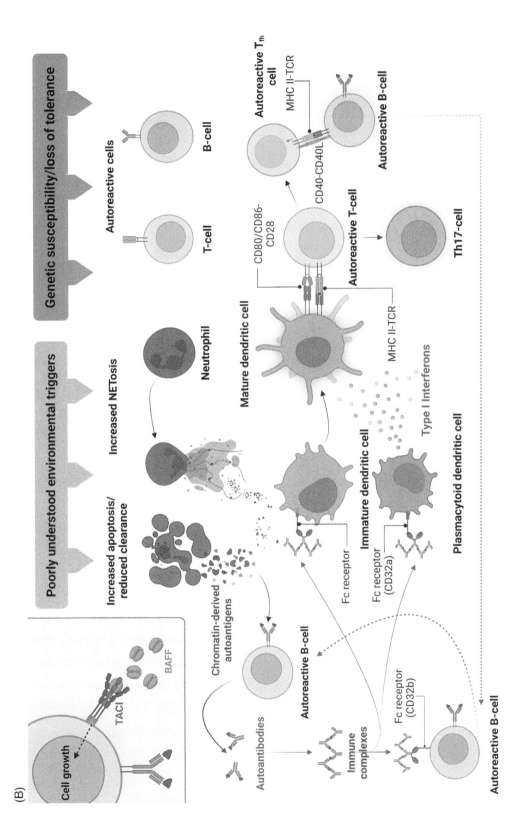

Figure 6.9 (Continued)

Table 6.2 Autoimmune diseases, target autoantigens and MHC associations.

Autoimmune disease*	Antigen	MHC susceptibility
Hashimoto thyroiditis	Thyroid peroxidase, thyroglobulin	HLA-DR3, HLA-DR5
Graves disease	Thyroid-stimulating hormone receptor	HLA-B8, HLA-DR3
Type 1 diabetes (CS71)	Insulin, glutamic acid decarboxylase, islet antigens	HLA-DR3, HLA-DR4
Coeliac disease (CS20)	Tissue transglutaminase, gluten proteins	HLA-DQ2, HLA-DQ8
Pernicious anaemia (CS23)	Intrinsic factor, gastric parietal cell antigens	HLA-DR4, HLA-DR2
Vitiligo	Melanocyte antigens	HLA-A2, HLA-DR4
Primary biliary cholangitis	Mitochondrial antigens (PDC-E2, BCOADC-E2)	HLA-DR3, HLA-DRB1
Autoimmune hepatitis	Liver-specific antigens	HLA-DR3, HLA-DR4
Rheumatoid arthritis (CS29)	Rheumatoid Factor, citrullinated proteins	HLA-DRB1
Systemic lupus erythematosus	Double-stranded DNA, Sm, Ro (SS-A), La (SS-B)	HLA-DR2, HLA-DR3, HLA-DRB1
Multiple sclerosis (CS25)	Myelin basic protein, myelin oligodendrocyte glycoprotein	HLA-DR2, HLA-DR4
Sjögren syndrome	Ro (SS-A), La (SS-B)	HLA-DR3, HLA-DR2
Psoriasis	Keratin, epidermal antigens	HLA-Cw6
Myasthenia gravis	Acetylcholine receptor	HLA-DR3, HLA-DRB1
Goodpasture syndrome (CS32)	Collagen IV	HLA-DR2, HLA-DRB1
Polymyositis	Jo-1 (Histidyl-tRNA Synthetase)	HLA-DR3, HLA-DRB1

* Associated Case Study indicated in brackets.

Mechanisms of Tissue Injury

Autoantibodies

Autoantibodies are major contributors to tissue damage in autoimmune diseases. They can bring about the cytotoxic destruction of cells through cell surface binding and lysis. Complement activation and/or ADCC are the main pathways of cell destruction. NK cells and eosinophils mediate ADCC by releasing substances that cause cell damage. Macrophages can mediate ADCC by phagocytosis. Another pathogenic mechanism involves immune complex-mediated damage, as seen, for example, in SLE. Autoantibodies can also interact with cell surface receptors, either activating them (as in Graves disease) or blocking them (e.g. anti-acetylcholine receptor antibodies in myasthenia gravis).

Autoreactive T-cells

Unlike autoantibodies, autoreactive cytotoxic T-cells usually act directly on the target tissue, recognising their target cells through the binding of their TCR to the combination of MHC class I and autoantigen-derived peptides on the target-cell surface. Once bound, cytotoxic T-cells kill target cells either by the secretion of cytotoxic granules containing perforin and granzyme B, which cause cell

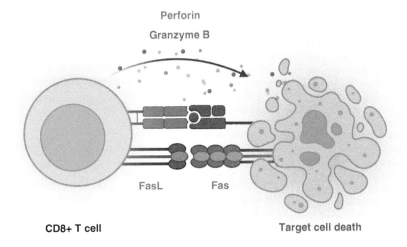

Figure 6.10 Major mechanisms of killing by cytotoxic T-cells. CD8+ cytotoxic T-cells (CTLs), play a crucial role in eliminating target cells, such as virus-infected cells or tumour cells. Once their TCRs have recognised foreign peptide in conjunction with MHC class I, they can deploy multiple mechanisms to induce cell death. One of the key pathways involves the engagement of the Fas-Fas ligand (FasL) interaction (see also Chapter 3). CD8+ T-cells express FasL on their surface, and when they recognise a target cell expressing Fas receptors, they trigger a cascade of intracellular events that result in the activation of caspases, ultimately leading to apoptotic death of the target cell. Additionally, CD8+ T-cells can directly release cytotoxic granules containing perforin and granzyme B. Perforin forms pores in the target cell membrane, allowing granzyme B to enter the cell. Granzyme B then activates caspases, leading to apoptosis. *Source:* Created with BioRender.com.

membrane disintegration and induce apoptosis, or by activation of the Fas-Fas ligand pathway which induces apoptosis of the target cell (Fig. 6.10). This process is particularly relevant in organ-specific autoimmune diseases, where the destruction of specific cells or tissues contributes to disease pathology. For example, in type 1 diabetes, autoreactive cytotoxic T-cells target and destroy pancreatic β-cells (Chapter 13).

In addition to direct cell killing, T-cells also contribute to tissue damage through the release of pro-inflammatory cytokines. Upon activation, T-cells can produce cytokines, such as TNF-α and IFN-γ, which promote inflammation by recruiting and activating other immune cells. Chronic production of pro-inflammatory cytokines by autoreactive T-cells leads to sustained inflammation and tissue destruction in autoimmune diseases such as RA (Case Study CS29) and multiple sclerosis (Case Study CS25).

Mechanisms Leading to Loss of Tolerance

Understanding the mechanisms that lead to loss of tolerance is essential for unravelling the complex pathogenesis of autoimmune diseases and, in turn, developing more effective therapies. Some of the mechanisms that lead to the loss of tolerance are discussed below:

T-cell Bypass

T-cell bypass occurs when B-cells are activated independently of T-cell help. This can occur when *superantigens* are present. Superantigens are unique microbial proteins produced by certain infectious agents, including bacteria and viruses. Unlike conventional antigens that are presented by specific MHC molecules, superantigens can bind simultaneously to the β-subunit of the TCR and the MHC class II molecule outside the peptide-binding groove leading to the activation of a broader repertoire of T-cells and consequently polyclonal B-cell activation. This can result in the production of a diverse array of antibodies, some of which may be self-reactive.

T-cell–B-cell Discordance

In certain instances, the antigen recognised by B-cells can be different to the antigen recognised by the T-cell that provides help to the B-cell. This can occur when B-cells endocytose antigens and present fragments of them on their MHC class II molecules to T-cells. If the co-endocytosed antigens are different from the antigen recognised by the B-cell receptor, the B-cell can still receive help from T-cells specific to the co-endocytosed antigens. This phenomenon is referred to as T-cell–B-cell discordance. For example, in coeliac disease, the B-cells recognising tissue transglutaminase (tTG) are helped by T-cells that recognise gliadin. This leads to the production of autoantibodies against tTG, which are a hallmark of coeliac disease (Case Study CS20).

Molecular Mimicry

Molecular mimicry occurs when an exogenous antigen shares structural similarities with host antigens. Antibodies produced against the exogenous antigen then also bind to host antigens, leading to autoimmunity. For example, in rheumatic fever (Case Study CS37), which is triggered by *Streptococcus pyogenes* infection, molecular mimicry between the bacterial M protein and human lysoganglioside leads to the development of cardiac-reactive T-cells.

Epitope Spreading

Epitope spreading is the phenomenon in which the immune response broadens its target from initially recognising and responding to dominant epitopes, to include subdominant and cryptic epitopes. One well-studied example of epitope spreading is observed in experimental autoimmune encephalomyelitis (EAE), an animal model for multiple sclerosis. In EAE, the immune response is initially triggered by myelin antigens, such as myelin basic protein or proteolipid protein, leading to the activation of autoreactive T-cells and the production of pro-inflammatory cytokines. As the disease progresses, there is diversification of the immune response, and additional myelin antigens, including myelin oligodendrocyte glycoprotein, are targeted by autoreactive T-cells. This expansion in epitope recognition occurs through several mechanisms, one of which involves the processing and presentation of new antigens by antigen-presenting cells; as inflammation and tissue damage progress, the antigen-presenting cells encounter a wider range of self-antigens, leading to the activation of T-cells specific to these newly presented epitopes.

Factors Contributing to Susceptibility to Autoimmune Disease

The aetiology of autoimmune diseases is complex and involves the interplay between genetic and environmental factors.

Genetics

Specific alleles within the MHC region are associated with either an increased or decreased susceptibility to autoimmune diseases (Table 6.2). As discussed earlier, the MHC region is highly polymorphic, meaning it exhibits extensive genetic variation between individuals. This genetic diversity can affect

- The repertoire of self-antigens presented by MHC molecules.
- The binding affinity and specificity of MHC-peptide complexes to TCRs. Certain MHC

alleles may present self-antigens in a way that enhances T-cell recognition, increasing the likelihood of the activation of autoreactive T-cells.

- The efficiency of thymic selection, potentially leading to the escape of autoreactive T-cells that can contribute to autoimmune diseases.

In addition, there are rare monogenic autoimmune diseases in which a single gene defect can lead to the loss of tolerance. For example, *autoimmune polyendocrinopathy syndrome type 1* (APS1) is caused by a mutation in the autoimmune regulator (AIRE) gene, leading to impaired negative selection in the thymus and the presentation of self-antigens. Similarly, *autoimmune lymphoproliferative syndrome* (ALPS) is characterised by the accumulation of a polyclonal population of double-negative T-cells (meaning they express neither CD4 nor CD8). This condition results from mutations in either the Fas or FasL gene (Chapter 3). The Fas-FasL pathway is necessary for the deletion of autoreactive T-cells.

Environmental Factors

Various environmental factors are implicated as aetiological factors in autoimmunity. They include;

Diet

Certain dietary factors contribute to autoimmunity. For example, deficiencies in vitamin D have been linked to an increased risk of multiple sclerosis and RA. On the other hand, diets rich in fruits, vegetables and omega-3 fatty acids may have anti-inflammatory effects that could reduce the risk of autoimmune diseases.

Microbiota

The gut microbiota plays a critical role in the development and function of the immune system. Imbalances in the gut microbiota, often referred to as *dysbiosis*, are associated with a variety of autoimmune diseases, including type 1 diabetes, RA and inflammatory bowel disease.

Infectious Agents

Certain infections can trigger autoimmune responses, for example, through molecular mimicry, as described earlier. *Streptococcus pyogenes* which can cause rheumatic fever is the classical example. Other bacteria implicated in autoimmunity include *Borrelia burgdorferi* which causes Lyme disease, and which can, in some cases, trigger an autoimmune response known as Lyme arthritis. *Campylobacter jejuni* is a common cause of food poisoning but is also linked to the development of Guillain-Barré syndrome, an autoimmune condition affecting the peripheral nerves. Viruses implicated in autoimmunity include the Epstein-Barr virus (SLE, RA and multiple sclerosis), and the Human immunodeficiency virus (e.g. Sjögren syndrome).

Xenobiotics

These are substances foreign to the body. Examples include tobacco smoke, which is associated with an increased risk of RA and SLE, and certain foodstuffs, for example, gluten which can trigger coeliac disease. Certain drugs, for example, hydralazine (a medication used to treat high blood pressure) and procainamide (used for heart arrhythmias) can induce a condition known as *drug-induced lupus erythematosus*. This is an autoimmune condition with similarities to SLE, but which usually resolves once the medication is discontinued. Another example is the antibiotic penicillin, which can cause a variety of immune reactions, ranging from mild rash to severe anaphylaxis (see later). Certain chemicals and pollutants in the environment can trigger autoimmune reactions. For example, exposure to silica dust is associated with an increased risk of developing RA, SLE and systemic sclerosis. Likewise, exposure to heavy metals, such as mercury and lead, has been linked to autoimmunity. For example, studies have suggested a potential link between mercury exposure and the development or worsening of SLE.

Biological Sex and Autoimmunity

Autoimmune diseases disproportionately affect women compared to men, with about 78% of those affected being female. The reasons for this disparity are complex and may involve both biological and environmental factors. Hormonal differences are likely to play a significant role. For instance, oestrogen can stimulate immune responses, potentially increasing the risk of autoimmunity. Genetically, women have two X chromosomes, and recent research suggests that some genes on the X chromosome with immune functions might be involved in this increased susceptibility.

Pregnancy

The immunological changes that occur during pregnancy are believed to contribute to variations in the course of autoimmune diseases. Some of the mechanisms underlying these variations are outlined below:

- *Immunological tolerance:* Pregnancy is a unique immunological state in which the maternal immune system needs to tolerate the semi-allogeneic foetus (half of the foetal genes are foreign to the mother, being of paternal origin). To facilitate this, there is a shift from a Th1 (pro-inflammatory) immune response to a Th2 (anti-inflammatory) dominant response. Many autoimmune diseases, like RA, are Th1-driven; hence, the shift to a Th2 state can lead to symptom improvement. However, diseases, such as SLE, which have a more pronounced Th2 component, may worsen.
- *Regulatory T-cells (Tregs):* Tregs play a crucial role in maintaining self-tolerance and preventing autoimmune reactions. Their numbers and function usually increase during pregnancy, promoting an immunosuppressive environment, which can modulate autoimmune disease activity.
- *Hormonal changes:* High levels of progesterone and human chorionic gonadotropin (hCG) during pregnancy have immunomodulatory effects. For example, progesterone promotes Th2 responses and can enhance the function of Tregs.
- *Cortisol levels:* The body naturally increases cortisol production during pregnancy. Cortisol has anti-inflammatory properties that can reduce the severity of autoimmune diseases.

After childbirth, the immunological environment rapidly shifts back to its pre-pregnancy state.

Hypersensitivity

The term hypersensitivity is used to describe situations in which the immune system reacts inappropriately or excessively to antigens, which otherwise appear to be relatively harmless. Hypersensitivity reactions can be classified into four different types based on their underlying immunological mechanisms and clinical manifestations (Fig. 6.11). Hypersensitive reactions may also be directed against 'self-antigens' in the context of autoimmune diseases. Multiple factors determine whether a hypersensitive, rather than a normal, immunological response occurs, including the genetic makeup of the individual, the physical and chemical properties of the antigen and the nature of the exposure. A particular disease may involve more than one hypersensitivity mechanism.

Type I Hypersensitivity

Type I hypersensitivity reactions (often referred to as '*allergic*' reactions) appear within minutes of exposure to the antigen (in this case, often referred to as the *allergen*), Individuals predisposed to this group of conditions are referred to being as '*atopic*', and an atopic tendency can often be traced back through families. Atopic patients are defined

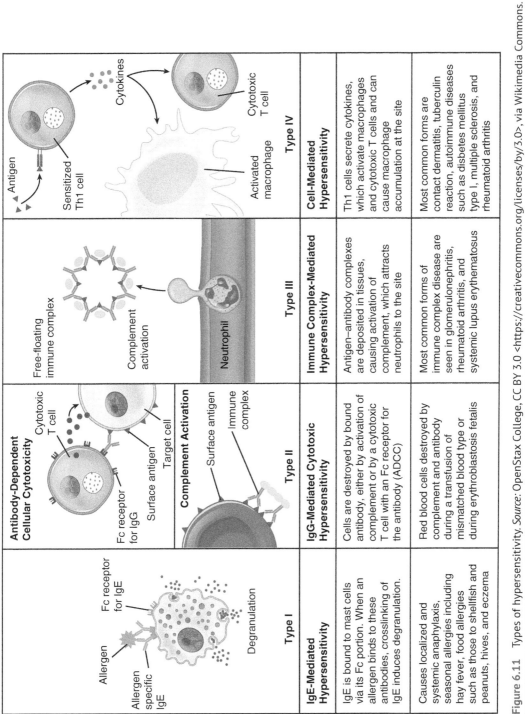

Type I	Type II	Type III	Type IV
IgE-Mediated Hypersensitivity	**IgG-Mediated Cytotoxic Hypersensitivity**	**Immune Complex-Mediated Hypersensitivity**	**Cell-Mediated Hypersensitivity**
IgE is bound to mast cells via its Fc portion. When an allergen binds to these antibodies, crosslinking of IgE induces degranulation.	Cells are destroyed by bound antibody, either by activation of complement or by a cytotoxic T cell with an Fc receptor for the antibody (ADCC)	Antigen–antibody complexes are deposited in tissues, causing activation of complement, which attracts neutrophils to the site	Th1 cells secrete cytokines, which activate macrophages and cytotoxic T cells and can cause macrophage accumulation at the site
Causes localized and systemic anaphylaxis, seasonal allergies including hay fever, food allergies such as those to shellfish and peanuts, hives, and eczema	Red blood cells destroyed by complement and antibody during a transfusion of mismatched blood type or during erythroblastosis fetalis	Most common forms of immune complex disease are seen in glomerulonephritis, rheumatoid arthritis, and systemic lupus erythematosus	Most common forms are contact dermatitis, tuberculin reaction, autoimmune diseases such as diabetes mellitus type I, multiple sclerosis, and rheumatoid arthritis

Figure 6.11 Types of hypersensitivity. *Source:* OpenStax College, CC BY 3.0 <https://creativecommons.org/licenses/by/3.0>, via Wikimedia Commons. https://commons.wikimedia.org/wiki/File:2228_Immune_Hypersensitivity_new.jpg.

as those suffering from a combination of three clinical conditions: allergic rhinitis, eczema and asthma.

Type I hypersensitivity involves the interaction between allergen, allergen-specific IgE and tissue mast cells. IgE is normally present in very low concentrations in the serum, but levels are increased in patients with parasitic infections, in most atopic patients, as well as in many apparently healthy individuals. One of the key physiological roles of IgE is the eradication of parasitic infections (Fig. 6.12). Notably, Type I hypersensitivity appears to be a particular problem in areas of the world where such infections are no longer endemic. To mount a Type I response, an individual's immune system must previously have encountered the antigen/allergen and their B-cells have responded by producing antigen/allergen-specific IgE. Mast cells, found in the mucosae of the airways and gut, as well as in skin and connective tissues, bind this IgE via receptors for the Fc portion of IgE. On re-exposure to allergen, the IgE molecules on the surface of the mast cells bind allergen via their available antigen-binding (Fab) portions and become cross-linked. This initiates a series of intracellular signals resulting in the release of mast cell cytosolic granular contents. These granules contain pre-formed mediators such as histamine, heparin, lysosomal enzymes and various proteases. In addition to this immediate response, allergen exposure activates the *de-novo* synthesis of leukotrienes, prostaglandin, thromboxanes and chemotactic factors. This second group of mediators, which are released later (usually around 4–6 hours after allergen exposure) have important clinical consequences in exacerbating disease. The physiological effects of early and late phase mediators include allergic rhinitis, asthma, eczema and urticaria and are summarised in Table 6.3. Allergic rhinitis is characterised by facial flushing, itching of the nose, mouth and eyes and watery nasal discharge (rhinorrhoea), following exposure to allergens such as pollens (hay fever) or dust mites (perennial rhinitis).

These symptoms are caused by vasodilatation and transudation of vascular fluid due to local histamine release. Allergens which penetrate further into the respiratory tree and lodge in the bronchi can cause asthma, with narrowing of the airways due to constriction of smooth muscle and the accumulation of fluid (Case Study CS30).

The most severe form of Type I hypersensitivity is an anaphylactic reaction and is a life-threatening condition, generalised vasodilatation causing a fall in blood pressure; the patient often collapses with loss of consciousness. There may be other features such as skin rash, abdominal pain, vomiting, diarrhoea and breathing difficulty due to bronchoconstriction and oedema. These clinical features are the result of the widespread activation of mast cells and circulating basophils with systemic release of their mediators.

Type II Hypersensitivity

Type II hypersensitivity reactions are also mediated by antibody, but in contrast to Type I reactions the antibodies involved are either of the IgG or IgM class. A feature of Type II reactions is that the antibody response is directed against antigens expressed on the cell surface and therefore the damage that results from these responses tends to be limited to a particular organ or cell type. The pathogenic antibodies in these reactions bind to cells via their Fab portion. The ensuing cell damage is a consequence of:

- Activation of the complement cascade at the cell surface leading to the production of the 'anaphylatoxins', C3a and C5a (Chapter 5), with subsequent recruitment of inflammatory cells. Ultimately, the assembly of the terminal complement pathway components C5–C9 into the membrane attack complex (MAC) causes pore formation in the cell wall (Chapter 5). Cellular damage is enhanced

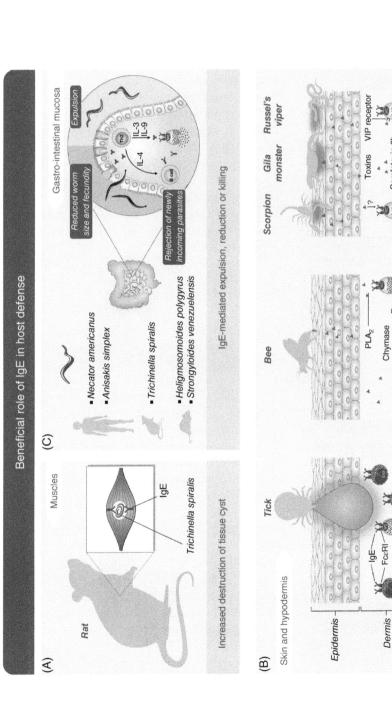

Figure 6.12 Evolutionary advantage of IgE-dependent immune responses. (A) In the case of *Trichinella spiralis*, the presence of parasite-specific IgE around muscle cysts contributes to their destruction. (B) Repeated exposure to ticks can lead to acquired resistance characterised by strong infiltration of IgE-bearing basophils and mast cells associated, which is associated with shorter tick feeding times and reduced engorgement weight. Another example is the protection provided by IgE-dependent release of powerful proteases such as chymase, tryptase and carboxypeptidase A from mast cells, which degrade animal-derived venoms (e.g. phospholipase A2 (PLA2) from bee venom, sarafotoxin from snakes and helodermin from the Gila monster). (C) IgE-dependent immune responses play a significant role in the expulsion, reduction or killing of parasites. During a secondary immune response to these parasites, the activation of Th2-driven responses leads to the production of specific IgE antibodies that bind to parasite antigens and activate mast cells through their FcεRI, triggering the release of various mediators, such as histamine, proteases and cytokines. *Source:* https://onlinelibrary.wiley.com/doi/10.1111/all.14570/John Wiley & Sons.

Table 6.3 The physiological effects of early and late phase mediators in Type 1 hypersensitivity reactions.

Mediators	Effect
Preformed (early)	
Histamine	Vasodilatation, increased vascular permeability, bronchoconstriction
Heparin	Anticoagulation
Lysosomal enzymes	Proteolysis
Newly synthesised (late)	
Leukotrienes (LTC4, LTD4, LTB4)	Vasodilatation, bronchoconstriction, chemotaxis
Prostaglandins, thromboxanes	Vasodilatation, platelet activation, bronchoconstriction
Platelet-activating factor	Platelet activation

via the deposition of C3b which allows the binding of phagocytic cells.

- The direct interaction of the Fc portion of the bound antibody with Fc receptor-bearing cells (e.g. NK cells and phagocytes) resulting in ADCC/ADCP.

Complement activation and ADCC are part of the normal armamentarium for fighting infection. A normal response causes damage to cells bearing foreign (e.g. microbial or tumour) antigens. However, these responses may be considered abnormal or hypersensitive in two situations: (i) when they are directed against an individual's own cellular antigens or (ii) when the scale of the response to a foreign antigen is such that it causes damage to the individual's own cells disproportionately to the potential hazard of the antigen(s) concerned.

Typical examples of Type ll hypersensitivity include transfusion reactions, autoimmune haemolytic anaemia, hyperacute graft rejection, Goodpasture syndrome (anti-GBM disease; Case Study CS32) and Graves disease. Transfusion reactions and hyperacute graft rejections involve the recognition of truly foreign antigens and ought to be preventable. In autoimmune haemolytic anaemia, the autoantibodies are directed against self-antigens on red blood cells (Chapter 7). Goodpasture syndrome is characterised by autoantibodies to glomerular basement membrane which can cause glomerulonephritis, but the autoantibodies also react with pulmonary basement membranes, causing severe pulmonary haemorrhage. Graves disease illustrates how antibody recognition of a cell-surface antigen may result in modulation of the target cell's function. The autoantibodies in this case are directed against the TSH receptor. The result is that the antibody mimics the effect of TSH and causes the thyroid gland to secrete an excess of thyroxine, causing hyperthyroidism.

Type lll Hypersensitivity

Type lll reactions are also antibody mediated but in contrast to Type II reactions, the antigenic targets of Type III reactions are soluble and not cell membrane bound. The combination of soluble antigen and specific antibody, of IgG or IgM class, results in the formation of immune complexes. These immune complexes circulate in the bloodstream and, as a consequence, the damage caused is usually not limited to one particular cell type or organ but may occur at remote sites throughout the body.

Immune complexes are formed during normal antibody responses, as a means of assisting antigen disposal, but normally they are cleared quickly by the monocyte/macrophage system, particularly by Kupffer cells in the liver. When immune complexes persist, either in the circulation or as tissue deposits, they activate inflammatory pathways, and the response becomes hypersensitive. Antigens that cause immune complex formation may be either exogenous (infectious or environmental agents) or endogenous (self-antigens in autoimmune responses). Two classic examples of Type lll reactions demonstrate the effector mechanisms involved.

The *Arthus reaction* occurs in animals that have been repeatedly immunised with antigen, resulting in high levels of antigen-specific IgG. An intradermal injection of antigen causes the rapid local accumulation of immune complexes and a reaction, which peaks approximately 6–24 hours later. The immune complexes cause local complement activation, with recruitment of neutrophils. The release of lysosomal enzymes causes vascular damage resulting in oedema and haemorrhage, which is seen as raised red areas on the skin.

Serum sickness is a condition that was seen in the pre-antibiotic era when patients were treated with large doses of antibodies made in other animal species, e.g. horses. The 'foreign' immunoglobulin protein is antigenic and stimulates the production of human antibodies and the formation of immune complexes which cause urticaria (commonly known as 'hives', a skin condition characterised by the sudden appearance of itchy, raised and often red welts or wheals on the skin), arthralgia (joint pain) and glomerulonephritis. Immunofluorescence tests can demonstrate the deposition of immune complexes and complement components within glomeruli and small blood vessels. Symptoms persist for as long as immune complexes are present in the circulation.

Several factors influence the persistence of immune complexes. The size of immune complexes is an important determinant; larger immune complexes bind more avidly to Fc receptors and are more easily phagocytosed, while smaller complexes are more likely to get trapped in the vasculature. Low-affinity antibodies favour the generation of small/intermediate immune complexes that are difficult to clear. Deficiencies in certain components of the complement cascade, for example in some patients with SLE, can lead to the persistence

of immune complexes. Other physicochemical properties of the immune complex, including electrostatic charge and degree of glycosylation, influence their ultimate tissue destination; positively charged antigens have an affinity for negatively charged basement membranes, for example in glomeruli. Haemodynamic factors such as blood pressure and turbulence also affect immune complex deposition; high blood pressure and increased filtration rate contribute to immune complex accumulation in glomeruli. Turbulence occurs at vessel bifurcation points and increases immune complex deposition at these sites. Chronic antigen exposure, for example in chronic infections (e.g. Hepatitis B) or autoimmune conditions (e.g. SLE), allows continuous immune complex formation.

Classical examples of human disease involving Type lll hypersensitivity include immune complex-mediated glomerulonephritis (Case Study CS31), SLE and extrinsic allergic alveolitis. In immune complex-mediated glomerulonephritis, the nature of the glomerular damage is greatly influenced by the type of immune complex involved and its rate of deposition. With rapidly deposited immune complexes, a proliferative response is likely (e.g. post-streptococcal glomerulonephritis), whereas with slower deposition, membranous glomerulonephritis is more likely. In extrinsic allergic alveolitis, the patient has pre-formed IgG antibodies to inhaled allergen and develops an Arthus-type reaction in the alveoli upon exposure. The clinical features of acute alveolitis usually occur about 6 hours after exposure, corresponding to immune complex formation and the recruitment of the damaging effector mechanisms, including complement and phagocytes.

Type lV Hypersensitivity

Type IV hypersensitivity reactions are cell mediated and therefore differ from the other types in that they cannot be transmitted from animal to animal by injection of serum. Type IV responses are also referred to as 'delayed-type hypersensitivity' as the reactions occur 12 hours or more following exposure to antigen. The common feature is the involvement of T lymphocytes (particularly the CD4+T-helper subset), and the requirement for antigen-presenting cells. Examples of Type IV hypersensitivity reactions are described below.

Contact Hypersensitivity

This phenomenon is caused by *haptens* – low molecular weight antigens, which alone are incapable of eliciting an immune response. However, when they bind to larger proteins (carriers), they can become immunogenic. A common example is hypersensitivity to nickel which is a constituent of costume jewellery, watches and buttons. The first step is the *sensitisation phase*. This is when nickel ions penetrate the skin and bind to TLR4 on the surface of specialised epidermal dendritic cells, known as Langerhans cells. Once activated, the Langerhans cells migrate to the local lymph node and present nickel to naive T-cells via MHC II. This interaction results in the proliferation of antigen-specific T-cells and memory T-cells, which migrate to the skin. The *elicitation phase* occurs upon re-exposure; the re-entering nickel ion is taken up by the dendritic cell and again presented to antigen-specific T-cells via MHC II. This stimulates the activation of skin-resident memory T-cells and recruits memory T-cells from the circulation, resulting in inflammation and cellular damage, which manifests clinically as allergic nickel dermatitis.

Tuberculin Hypersensitivity

In contrast to contact hypersensitivity, tuberculin hypersensitivity is a phenomenon chiefly involving the dermis. This form of immune reaction is most famously used in the Mantoux test (also known as the tuberculin skin test), a screening tool for tuberculosis. In this test, a small amount of *tuberculin* – a protein purified from *Mycobacterium tuberculosis* – is injected into the skin. If a person has been previously exposed to this bacterium, their immune system will have developed a sensitivity to this protein. Within 48–72 hours, an immune response will occur at the injection site, resulting in a hard, raised red bump. When tuberculin is injected intradermally, it is recognised by the memory T-cells (specifically, resident memory T-cells) present in the skin, which were primed during the initial exposure to the bacterium. The activated memory T-cells release a plethora of cytokines, most notably IFN-γ and TNF-α. These cytokines attract more immune cells to the site of injection. Neutrophils, monocytes and additional lymphocytes are drawn to the site; the hard red bump being caused by the accumulation of these cells.

Granulomatous Hypersensitivity

If antigen persists, the initial tuberculin response may develop into a granulomatous hypersensitivity which is the most severe, and clinically most important, form of Type IV hypersensitivity. Granulomas are collections of macrophages, some of which coalesce to form giant cells, surrounded by a cuff of small lymphocytes. The macrophages often have the appearance of epithelial cells and are known as 'epithelioid' cells. Granulomas are formed when the immune system fails to remove foreign antigens, which are then allowed to persist, usually within macrophages. Granuloma formation is not confined to skin reactions but can occur in many organs.

Studies of granulomatous conditions in experimental situations have focused attention on the factors involved in the initiation and persistence of the lesions. Antigen presentation is associated with production of IL-1 by antigen-presenting cells, causing the activation of T-cells, which in turn produce IL-2 and express the IL-2 receptor on their surface, which contributes to their proliferation. High levels of IL-1 have been demonstrated in early experimental lesions, suggesting that macrophage production of this cytokine is an important early event in the recruitment of cells for granuloma formation.

Human diseases associated with significant granuloma formation include tuberculosis, leprosy, schistosomiasis, leishmaniasis and sarcoidosis (Case Study CS27). For example, in pulmonary tuberculosis, much of the lung damage is due to granuloma formation. Reactivation of tuberculosis, which is often seen in later life, is associated with an age-related reduction in memory T-cell function, allowing renewed mycobacterial growth.

Leprosy is an excellent example of how the immune status of an individual determines the clinical manifestation of disease. There are two major forms of leprosy, tuberculoid and lepromatous leprosy. *Tuberculoid* disease is largely asymptomatic and characterised by hypopigmented areas of the skin which appear histologically as typical granulomatous lesions with few detectable mycobacteria. These patients have potent Type IV hypersensitivity responses to *M. leprae*, the CD4+ T-cell population being predominantly Th1 cells, resulting in more effective clearance of organisms. In *lepromatous* leprosy, there are widespread skin lesions containing numerous bacilli but few lymphocytes. These patients have a predominantly Th2 response to the organism and little granuloma formation. Consequently, *M. leprae* can proliferate and disseminate more freely in lepromatous leprosy and the patient suffers the systemic effects of the infection. Host response in the form of hypersensitivity may therefore be advantageous under certain circumstances.

Immunodeficiency

Immunodeficiency disorders can be either *primary* (in which there is often a genetic cause) or *secondary* (when another illness or treatment causes the immunodeficiency). Both types reduce the immunological response, rendering the individual more susceptible to a wide range of infections. Primary immunodeficiency (PID) was thought of as a group of very rare diseases that mainly affected children. However, the advent of next-generation sequencing has led to an explosion in the recognition and characterisation of these disorders, particularly those with autosomal recessive inheritance. There are now more than 450 PID disorders defined. PID is largely classified on the basis of the component(s) of the immune system affected, for example, whether they primarily affect antibody production, cell-mediated immunity, neutrophil function or complement activity. Secondary immunodeficiency (SID) was mainly recognised as a consequence of systemic ill health, most notably HIV infection, malnutrition and immunosuppressive therapies. However, recently, there has been a huge increase in SID occurring as a consequence of the many new biological therapies that have been successfully introduced to treat a range of autoimmune, haematological and rheumatological disorders.

In the following sections, we describe the most well-defined PID.

Deficiency of Antibody-Mediated Immunity

X-linked Agammaglobulinaemia

X-linked agammaglobulinaemia (XLA) is the prototypic disorder of humoral immunity. It is caused by a deficiency of the enzyme, Bruton tyrosine kinase, leading to the arrest of B-cell development at the pre-B stage. Males with this disease do not have mature immunoglobulin-bearing B lymphocytes in peripheral blood, do not have plasma cells or secondary lymphoid follicle formation in lymphoid tissues and have severe hypogammaglobulinaemia (reduced levels of immunoglobulin in blood). No other compartment of immune function is affected; cell-mediated immunity is intact. Therefore, individuals with XLA serve as a model for discerning the role of antibody in host defence.

The BTK gene, which encodes Bruton tyrosine kinase, is on the X chromosome. BTK deficiency is caused by mutations in the BTK gene. In normal females, one of the two X chromosomes in each cell is randomly inactivated. This also happens in all the cells of female XLA carriers. However, in the case of female XLA carriers, only the B-cells that inactivate the abnormal X chromosome survive to maturity. This is because BTK is required for B-cell development. Thus, all the B-cells of female carriers display inactivation of the X chromosome with the defective BTK gene. This is so-called *skewed X-inactivation.*

Boys with XLA are usually protected from infection by trans-placentally acquired maternal IgG for the first 3–4 months of life. Thereafter, chronic or recurrent bacterial and viral infections are the predominant clinical manifestation of XLA. Otitis media, pneumonia, diarrhoea and sinusitis occur most often, usually in combination with each other. Infections with heavily encapsulated bacteria such as *Streptococcus pneumoniae* and *Haemophilus influenzae* are common because the host defence against these organisms relies upon the ability of antibody to bind to the polysaccharide capsule with subsequent complement activation and phagocytosis. Infections may begin on mucosal surfaces but are not limited to them because there is an absence of serum immunoglobulin. As a result, localised bacterial infections may spread through the blood to other tissues such as the meninges, joints and bones. Patients with XLA generally recover uneventfully from viral infections, but not without exception. Enteroviruses which usually cause a self-limiting mild gastroenteritis in normal children can cause chronic disseminated infections in XLA patients.

The mainstay of treatment is immunoglobulin replacement therapy (IGRT) which requires regular infusions of replacement IgG for the rest of the patient's life.

Isolated IgA Deficiency

Isolated IgA deficiency provides an important contrast to XLA. In this disorder, individuals have a complete, or almost complete, lack of serum and mucosal IgA but have normal levels of all other immunoglobulin classes and have normal cell-mediated immune functions. IgA deficiency is common, affecting 1:600 of the population in the United Kingdom and Ireland. The vast majority of these people are very well and require no specific immunological treatment. However, some have other associated immunological defects and it is important to identify those people who require specific interventions. Many of the clinical features of this condition can be explained by the unique biological properties of IgA. It is the predominant immunoglobulin class on mucosal surfaces, although IgA comprises only 15% of immunoglobulin in serum. IgA is secreted onto mucosal surfaces as a macromolecular complex consisting of two IgA molecules joined to a J chain and a secretory component. Most patients with IgA deficiency lack both serum and secretory IgA, but there are rare cases in which there is a deficiency of secretory, but not of serum, IgA.

Unlike the major serum immunoglobulin classes, IgG and IgM, IgA is largely silent as a mediator of inflammatory responses. It does not activate complement or promote opsonisation, but functions in antimicrobial defence by inhibiting microbial adherence and by neutralising viruses and toxins. IgA also has an important role in antigen clearance, preventing soluble antigens from penetrating the mucosa and entering the systemic circulation.

Patients with IgA deficiency may have an increased susceptibility to infection. Because IgA is the predominant immunoglobulin on mucosal surfaces, most such infections in IgA-deficient patients are confined to the mucosal surfaces of the respiratory and gastrointestinal tracts. Systemic infections, including bacterial sepsis, are no more common than among the general population. IgA-deficient patients also have an increased susceptibility to autoimmune diseases, but the underlying mechanisms are poorly understood. Individuals lacking IgA are also at risk of certain other complications which include:

- *False-negative coeliac disease*: Typical IgA-based coeliac disease screening tests may give false-negative results in patients with low or absent IgA antibodies. Immunology laboratories will often test for IgG anti-endomysial antibodies (EMA-IgG) instead. Endomysial antibodies target the endomysium, a connective tissue layer that surrounds individual muscle fibres.
- *False-positive pregnancy test*: Pregnancy tests detect the presence of hCG, a hormone produced during pregnancy. In some cases of isolated IgA deficiency, IgG antibodies against hCG may be present causing a false-positive result in a pregnancy test.

Hyper-IgM Syndrome

Hyper-IgM (HIGM) syndrome is characterised by raised levels of IgM and decreased levels, or absence, of other immunoglobulins. The most common form of HIGM is Type 1 (HIGM1). HIGM1 is X-linked and caused by mutations in the CD40LG gene, which encodes the CD40 ligand. The interaction between CD40 ligand on T-cells and its receptor, CD40, on B-cells is crucial for B-cell maturation and differentiation, including class switching (in which B-cells change the type of antibody they produce from IgM to IgG, IgA or IgE, see earlier text). Individuals with HIGM1 have an increased susceptibility to recurrent infections, particularly with opportunistic organisms such as *Cryptosporidium* and *Pneumocystis jirovecii*. They are also more prone to autoimmune disorders, ascending cholangitis and some forms of cancer, especially of the liver. The mainstay of treatment for HIGM is early identification in infancy and stem cell transplantation, which often results in complete correction of the immunological defect.

Common Variable Immunodeficiency

Common variable immunodeficiency (CVID) is a heterogeneous group of disorders characterised by impaired B-cell differentiation and lower levels of most, or all, of the immunoglobulin types. Mutations in several genes are implicated, including ICOS, CD19, CD81 and TNFRSF13B. However, it is worth noting that CVID is complex and most likely involves the interaction of multiple genetic and environmental factors. Many cases are sporadic, meaning they occur in individuals with no family history of the disorder. However, some cases are familial. Individuals with CVID are prone to bacterial infections, particularly affecting the lungs, sinuses and ears. They are also prone to inflammatory disease of the lungs and gastrointestinal tract, as well as granulomatous diseases. CVID also increases the risk of some autoimmune disorders (e.g. RA, SLE) and certain types of cancer (e.g. non-Hodgkin lymphoma, gastric cancer). The mainstay of management for patients is long-term IGRT. However, in common with all other PIDs, careful long-term follow-up is required to monitor for the occurrence of associated inflammatory, autoimmune and malignant disorders.

Deficiency of Cell-Mediated Immunity

DiGeorge Syndrome (CATCH22 Syndrome)

DiGeorge syndrome is caused by a deletion of the chromosomal region 22q11.2 (Chapter 10). The acronym CATCH 22 refers to the typical clinical features of

- **Ca**rdiac defects
- **T**hymic hypoplasia/aplasia
- **C**left palate
- **H**ypocalcaemia.

Because T-cells develop in the thymus, thymic hypoplasia leads to T-cell deficiency. All T-cell sub-populations are affected equally, so that there is a deficiency of both CD4+ T-helper cells and CD8+ cytotoxic T-cells. Affected infants have low levels of lymphocytes in the blood circulation (most peripheral blood lymphocytes are T-cells) and are depleted of T-cells in the paracortical regions of lymph nodes and spleen. The deficiency of cell-mediated immunity results in increased susceptibility to infections, particularly with intracellular pathogens. DiGeorge syndrome patients are also highly susceptible to viral infections because they lack cytotoxic T-cells to kill virus-infected host cells. As a result, relatively common viruses such as Varicella-zoster virus (chickenpox) and rotaviruses (gastroenteritis) may cause fatal infections in affected children. Most patients with DiGeorge syndrome have sufficient residual T-cells to allow the normal or near normal development of B-cells. Although thymic transplants are considered in severe cases, the majority of DiGeorge syndrome children do not require long-term specific immunological treatment, but may, in the future, require the services of cardiologists, plastic surgeons or endocrinologists, as well as educational support.

Severe Combined Immunodeficiency

Severe combined immunodeficiency (SCID) describes a group of disorders characterised by severe functional abnormalities of both B- and T-lymphocytes. Babies with SCID appear healthy at birth but are highly susceptible to severe recurrent infections, including pneumonia and meningitis caused by bacterial, fungal or viral pathogens. They may also manifest features of maternofoetal graft versus host disease (GvHD), such as jaundice or desquamating skin rash, caused by the transfer of maternal T-cells into the infant's circulation. Children with SCID fail to grow or gain weight (failure to thrive). Without treatment, children with SCID usually die within the first year of life due to severe, recurrent infections. Some forms of SCID are X-linked, while others are autosomal recessive. In cultures in which marriages between first

cousins are common, there is an elevated risk of SCID in the offspring. Because of its rarity, the diagnosis of SCID is often delayed. However, many countries have now introduced SCID testing into their newborn screening programmes with demonstrable improvements in survival.

The most common type is *X-linked SCID* and is caused by mutations in the IL2RG gene. The IL2RG gene encodes the interleukin-2 receptor gamma (IL-2Rγ) chain (also known as CD132), which is a component of several receptor complexes, including those that bind IL-2, IL-4, IL-7, IL-9, IL-15 and IL-21; as might be anticipated, the IL-2Rγ chain is crucial for lymphocyte development and function.

Other types of SCID include:

- *ADA-deficient SCID* is caused by mutations in the ADA gene, which leads to a deficiency of the adenosine deaminase enzyme. ADA is essential for the breakdown of toxic metabolites in cells, and without it, toxic compounds accumulate, causing severe damage to lymphocytes, particularly T-cells.
- *IL7R-deficient SCID* is caused by mutations in the IL7R gene, resulting in IL-7 receptor deficiency, leading to impaired T-cell development and function.
- *JAK3-deficient SCID*. Mutations in the JAK3 gene disrupt the function of the JAK3 protein, which is involved in the signalling response to cytokines which is important for T-cell development.
- *RAG1/RAG2 SCID* is caused by mutations in the RAG1 or RAG2 genes, which are essential for V(D)J recombination, described above.
- *IL2RA (CD25) deficiency* is caused by mutations in the IL2RA gene, which encodes the alpha chain of the IL-2 receptor.
- *ZAP70 deficiency*: Mutations in the ZAP70 gene lead to a deficiency of the ZAP70 protein, which is critical for T-cell receptor signalling.

Ataxia Telangiectasia

Ataxia telangiectasia (AT) is an autosomal recessive disorder with a range of clinical manifestations, primarily characterised by progressive neurodegeneration, telangiectasias (small, dilated blood vessels), immunodeficiency and a heightened risk of developing malignancies, especially lymphomas and leukaemias. The causative gene, ataxia telangiectasia mutated (ATM), is centrally involved in DNA repair, particularly in response to DNA double-strand breaks, and also in cell cycle control. When DNA damage is detected, the ATM protein becomes active and coordinates the response, which includes stopping the cell cycle to allow for repair, activating repair proteins, and if the damage is too severe, initiating apoptosis to prevent the propagation of damaged DNA. A major downstream effector of ATM is p53 (Chapter 11). In individuals with AT, mutations in the ATM gene result in a non-functional or absent ATM protein, leading to unchecked DNA damage and genomic instability. Sadly, this is a progressive condition and survival beyond the second decade of life is rare. At present, there is no definitive treatment.

X-linked Lymphoproliferative Disease

X-linked lymphoproliferative (XLP) disease is a rare genetic disorder characterised by an extreme vulnerability to Epstein-Barr virus (EBV) infection. While most people experience only a mild illness when infected with EBV, those with XLP can develop severe, life-threatening infectious mononucleosis, dysgammaglobulinaemia (abnormal levels of immunoglobulins), lymphoma and aplastic anaemia. The most common type of XLP, XLP1, is caused by mutations in the SH2D1A gene. This gene encodes the protein known as **s**ignalling **l**ymphocyte **a**ctivation **m**olecule (SLAM) **a**ssociated **p**rotein (SAP). SAP binds to and enhances signal transmission through the SLAM family receptors, which are expressed on the surface of various immune cells. Loss of SAP function impairs the ability of T-cells and NK-cells to recognise and destroy EBV-infected cells. There is considerable heterogeneity in

the clinical severity of this condition; however, stem cell transplant is a potential treatment for a significant number of patients.

Wiskott-Aldrich Syndrome

Wiskott-Aldrich syndrome (WAS) is a rare X-linked recessive disorder characterised by the triad of eczema, thrombocytopenia (low platelet count) and immunodeficiency. The underlying cause of WAS is mutation of the WAS gene which encodes the Wiskott-Aldrich syndrome protein (WASP). Mutations in the WAS gene result in a defective or absent WASP, which alters actin cytoskeleton dynamics. The main features of WAS include

- *Spontaneous nose bleeds*, caused by abnormal platelet function and thrombocytopenia; individuals with WAS are prone to bleeding episodes, including spontaneous nose bleeds (epistaxis). The small and less adhesive platelets (microthrombocytes) in WAS contribute to the bleeding tendency.
- *Bloody diarrhoea*: Gastrointestinal manifestations, such as bloody diarrhoea, can occur as a consequence of an increased susceptibility to infections.
- *Early onset eczema*, characterised by itchy and inflamed skin, typically developing within the first month of life.
- *Autoimmune disorders* are found in the majority of children with WAS. The underlying mechanism is not clear but is likely related to the dysregulation of immune responses.
- *Altered immunoglobulin levels*: IgM levels are reduced, while IgA and IgE levels are elevated. IgG levels can be normal, reduced or elevated.
- *Increased risk of cancers,* primarily lymphoma and leukaemia. The underlying reasons for this increased susceptibility are not fully understood.

Early diagnosis in infancy followed by stem cell transplant can be an effective treatment.

Deficiency of Phagocytic Cell Function

Leucocyte Adherence Deficiency

Leucocyte adherence deficiency is typically caused by mutations in the ITGB2 gene (which encodes the integrin beta-2 subunit) or SLC35C1 (which encodes solute carrier family 35 member C1). These mutations lead to dysfunctional or absent adhesion molecules, resulting in impaired chemotaxis and phagocytosis. Affected individuals have an increase in the number of circulating blood leucocytes because the cells cannot adhere to vascular endothelium and enter tissues. Because they have difficulty mobilising those leucocytes to the sites of infection, affected individuals are unusually susceptible to bacterial infections, especially those that begin along body surfaces such as the skin, gingivae, perirectal area and the lungs.

Chronic Granulomatous Disease

Chronic granulomatous disease (CGD) results from a different abnormality in phagocyte function – the inability to kill ingested microorganisms. Most cases of CGD are transmitted as an 'X-linked trait', meaning they result from mutations on the X chromosome. The affected gene encodes the gp91 protein, also known as p91-PHOX (where 91 represents the weight of the protein in kilodaltons, and 'gp' stands for glycoprotein). Mutations in this gene lead to the absence or dysfunction of the gp91 protein, which is a critical component of the NADPH oxidase complex. In addition to X-linked CGD, the condition can also be transmitted in an autosomal recessive fashion. Autosomal recessive forms of CGD result from mutations in other genes, notably, CYBA, NCF1, NCF2 and NCF4, which affect other PHOX proteins involved in the NADPH oxidase complex.

NADPH oxidase is the enzyme responsible for generating ROS, which has a crucial role in killing and destroying ingested microorganisms.

Thus, affected individuals have phagocytes with normal mobility and normal phagocytosis, meaning microorganisms are effectively trapped within phagocytic cells. However, because the microorganisms cannot be killed, granulomas and abscesses form within various tissues including the skin, lymph nodes, spleen and liver. Early diagnosis and stem cell transplant can be curative.

Chédiak-Higashi Syndrome

Chédiak-Higashi syndrome (CHS) is a phagocytic deficiency disease caused by mutations in the CHS1/LYST gene, leading to abnormal intracellular protein trafficking and dysfunctional lysosomes. This results in impaired phagocytosis, making individuals susceptible to recurrent infections, neurodegeneration, peripheral neuropathy, prolonged bleeding and an increased risk of lymphoma. The presence of giant lysosomes and inclusion bodies in various cell types, including phagocytes, underscores the systemic impact of the disorder.

Disorders of the Complement System

Patients with a genetically determined deficiency of complement components are generally more susceptible to bacterial infections. The type of bacteria that most often cause these infections depends upon the role of the specific complement component that is missing. For example, deficiencies of the early components (C1, C4, C2 and C3) interfere with the opsonic action of complement, and affected individuals have increased susceptibility to infection from heavily encapsulated bacteria (e.g. *S. pneumoniae and H. influenzae*) for which opsonisation is the primary host defence. Deficiencies in the early complement components also impair the ability of the complement system to remove immune complexes. As a result, the excessive accumulation of immune complexes in various tissues can lead to the development of immune complex-mediated disorders, such as SLE. Individuals with deficiencies of the terminal components (C5-9, forming the MAC, see Chapter 5) have an increased susceptibility to microorganisms against which the bacteriolytic actions of complement are most important. These include Gram-negative bacteria, particularly *Neisseria* spp., which cause meningococcal disease.

Complement regulatory protein deficiencies can also cause disease. For example, *C1 esterase inhibitor deficiency*, causing hereditary angioedema (HAE). HAE is characterised by recurrent episodes of swelling in the skin, gastrointestinal tract and upper airways. These episodes are caused by uncontrolled activation of the complement system and the release of inflammatory mediators, particularly bradykinin (Chapter 5).

Transplantation Immunology

Transplantation is the relocation of cells, tissues or organs. An *autologous transplant* (also known as an *autograft*) involves relocation within the same individual (e.g. the use of skin tissue from one part of the body to repair damage in another; or the removal, modification and restoration of bone marrow to the same individual). In an *isograft*, relocation is from a donor to a genetically identical (*syngeneic*) recipient. Unfortunately, few individuals have identical twins and therefore in clinical practice, most transplants are *allografts* between genetically different (*allogeneic*) donors and recipients. To the recipient (or host), the graft is 'foreign' and the immune system of the host is stimulated to attack and destroy the graft (*graft rejection*). The most significant problem in improving the success rate of transplantation is that of minimising and controlling graft rejection. Unfortunately, many transplants that are clinically indicated are not possible because of the shortage of suitable donor organs. This has stimulated research into transplantation between animals of different species (*xenografts*), particularly grafts from pigs to humans. The genetic (*xenogeneic*), immunological and more general physiological differences between donor and recipient are complex and need to be more clearly understood before the significant problems of xenograft rejection can be resolved.

Types of Rejection

Allograft rejection most commonly occurs several days to weeks after transplantation. This is referred to as *acute rejection*. *Chronic rejection* takes place over a much longer time scale (months or years). A much more rapid immune response is sometimes encountered. This is *hyperacute rejection* and takes place within 24 hours of transplantation and is caused by the presence in the host, of pre-formed anti-graft antibodies. On encountering graft antigens, these antibodies activate complement and stimulate an acute inflammatory response against the graft. The presence of pre-formed antibodies implies the presence of memory B-cells, formed by a previous exposure to antigens present on the graft. There are three ways in which the host might have previously been exposed to antigens from another individual.

- Through repeated blood transfusions leading to exposure to surface antigens on donor white blood cells which are common to graft antigens.
- In females, through exposure during pregnancy to foetal antigens, encoded by paternally inherited genes, which are common to antigens on the graft.
- Through a previous graft, which has antigens in common with the newly transplanted organ.

Fortunately, hyperacute rejection is normally avoided through careful blood grouping and monitoring for the presence of antigraft antibodies in the host prior to transplantation.

Characteristics of Acute Allograft Rejection

Early studies of acute allograft rejection in the 1940s revealed key aspects of the mechanism of rejection. Studies of skin grafts in mice showed that grafts between mice of the same strain (syngeneic) were successful, whereas grafts between different strains resulted in rejection within 10–14 days. If a second graft from the same donor strain was given to the same recipient mouse, rejection was quicker (within 5–6 days). This demonstrated a 'memory' component to the acute rejection response. The initial 'first set' response was shown to result from the activation of naïve T lymphocytes in response to graft antigens. On secondary exposure to the same graft antigens, memory T-cells formed in the primary response are activated in the more rapid, 'second set'.

Studies focused on determining the genetic basis of allograft rejection used *congenic* mice (mice that are genetically identical except for a single gene locus). The results of skin grafts between different donor and recipient strains led to the discovery that genetic variation in the MHC genes is mostly responsible for tissue rejection.

The Role of the Major Histocompatibility Complex

The high level of polymorphism at the MHC locus confers a significant evolutionary advantage in protecting the population against infection. Unfortunately, the immune system did not evolve to allow transplantation. HLA genes are incredibly diverse in the human population, with HLA-B being the most polymorphic among class I genes and HLA-DR being the most polymorphic among class II genes. This high degree of polymorphism means that each person has a unique set of HLA molecules, except for identical twins. This uniqueness contributes to the difficulty in finding compatible organ donors, as even a single mismatch can trigger an immune response leading to organ rejection. HLA typing, sometimes referred to as 'tissue typing', is therefore crucial in matching organ donors and recipients. Although all HLA loci can contribute to organ acceptance or rejection, the clinical focus has traditionally been on HLA-A, -B and -DR due to their high levels of polymorphisms. Matching these loci as closely as possible reduces the likelihood of an immune response and subsequent organ rejection. It is important to remember, however, that even with a complete HLA match, other factors, such as non-HLA antibodies, can contribute to rejection. Therefore, immunosuppressive therapy is generally necessary even with well-matched donors and recipients.

Presentation of Graft Antigens

In transplant rejection, the immune response against the transplanted organ involves both CD4+ and CD8+ T-cells. Local inflammation attracts recipient dendritic cells to the transplanted tissues through inflammatory and danger signals. Dendritic cells engulf damaged cells and present alloantigens on their surfaces. They are professional antigen-presenting cells and express both MHC class I and II, allowing them to activate both CD8+ and CD4+ T-cells.

The DCs migrate to secondary lymphoid tissues where they present alloantigens to recipient T-cells. Alloantigen recognition can occur through three pathways; direct, semi-direct and indirect (Fig. 6.13). In the direct pathway, recipient T-cells are activated by MHC molecules from the donor that are seen as foreign. The semi-direct pathway involves recipient dendritic cells and other antigen-presenting cells that present foreign MHC molecules they have taken up from donor cells through phagocytosis. For both these pathways, the specific peptide source is not the key factor. However, in the indirect pathway, foreign proteins get broken down by recipient antigen-presenting cells, and the resulting peptides from this process are then presented via the recipient's own

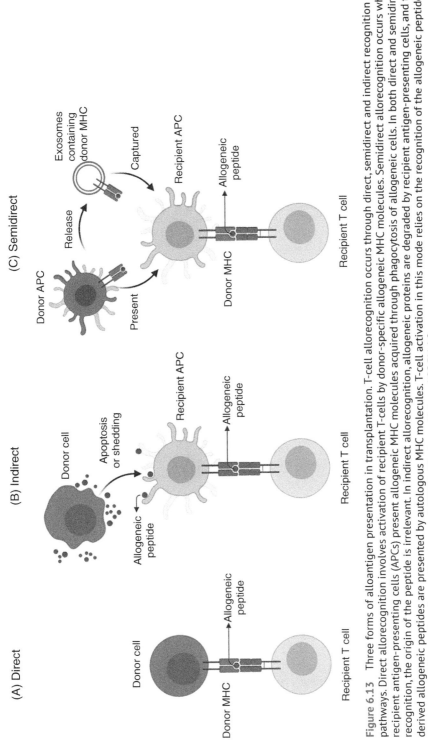

Figure 6.13 Three forms of alloantigen presentation in transplantation. T-cell allorecognition occurs through direct, semidirect and indirect recognition pathways. Direct allorecognition involves activation of recipient T-cells by donor-specific allogeneic MHC molecules. Semidirect allorecognition occurs when recipient antigen-presenting cells (APCs) present allogeneic MHC molecules acquired through phagocytosis of allogeneic cells. In both direct and semidirect recognition, the origin of the peptide is irrelevant. In indirect allorecognition, allogeneic proteins are degraded by recipient antigen-presenting cells, and the derived allogeneic peptides are presented by autologous MHC molecules. T-cell activation in this mode relies on the recognition of the allogeneic peptide-autologous MHC complex. *Source:* Demkes *et al.* (2021) / Springer Nature / CC BY 4.0.

MHC molecules. In this scenario, T-cell activation hinges on the recognition of the combination of the foreign peptide and the recipient's MHC molecule.

The number of T-cells that can recognise alloantigens is typically much higher compared to those that recognise microbial antigens. Thus, the vigorous T-cell response observed in transplant rejection, driven by the recognition of alloantigens, plays a major role in the demise of the transplanted tissues.

Controlling Rejection

Allograft survival depends on the effective control of immunological rejection. As syngeneic grafts are not rejected, there are clear advantages in trying to ensure a close match of HLA alleles between donor and recipient. Controlling transplant rejection involves a multifaceted approach aimed at modulating the recipient's immune response to the transplanted organ or tissue. Several strategies are employed to achieve immunosuppression and promote long-term graft acceptance. They include:

Immunosuppressive Medications

Immunosuppressive drugs are a cornerstone of transplant rejection management. Commonly used immunosuppressive drugs include calcineurin inhibitors (such as cyclosporine and tacrolimus), corticosteroids and antimetabolites (such as mycophenolate mofetil). These medications help prevent or reduce the activation of T-cells and the production of pro-inflammatory cytokines, thereby suppressing the immune response against the graft.

Induction Therapy

Induction therapy involves the administration of more potent immunosuppressive medications during the early post-transplant period to provide intense immunosuppression. This can include antibodies that specifically target and deplete certain types of immune cells, such as anti-thymocyte globulin (ATG) or anti-CD25 antibodies. Induction therapy aims to minimise the risk of acute rejection episodes and improve the initial success of the transplant.

Maintenance Immunosuppression

Following induction therapy, maintenance immunosuppression provides ongoing immune suppression to prevent rejection in the long-term. This typically involves a combination of different immunosuppressive medications tailored to the individual patient's needs. The dosage and combination of drugs may be adjusted over time to balance the need for immune suppression with the risk of side effects.

Haemopoietic Stem Cell Transplantation

Haemopoietic stem cell transplantation (HSCT) involves the infusion of healthy hematopoietic stem cells into the patient. The donor stem cells can be obtained from bone marrow, peripheral blood or umbilical cord blood. Before the transplantation, the patient typically undergoes *conditioning therapy*, which consists of high-dose chemotherapy and/or radiation designed to kill any abnormal cells, while also destroying the patient's own bone marrow to 'make room' for the donor stem cells to engraft. After conditioning, the donor stem cells are infused into the patient's blood whereupon they migrate to the bone marrow. HSCT is accompanied by a high risk of significant complications, including graft failure, graft-versus-host disease (see Case Study CS33), infections and long-term effects, such as infertility. There are two main types of HSCT:

- Autologous HSCT in which the patient's own stem cells are collected and then returned to them after the conditioning therapy. This type of transplant is commonly used in the treatment of diseases which do not affect the patient's own stem cells, for example in certain types of cancer.

- Allogeneic HSCT involves the use of stem cells from a matched donor, which may be a related or unrelated individual. Allogeneic transplants are used to treat diseases in which the patient's own stem cells are abnormal, for example in leukaemias, aplastic anaemia (failure of the bone marrow to produce adequate numbers of new blood cells) and some inherited immune system disorders.

Summary

The innate immune system is the body's rapid and non-specific defence against various pathogens and encompasses a diverse array of physical barriers, immune cells and soluble factors. Adaptive immunity is a vital defence mechanism in higher organisms, consisting of cell-mediated (T-cell) and antibody-mediated (B-cell) components. It exhibits specificity, diversity, memory and the ability to differentiate self from non-self. B- and T-cells generate a wide range of antigen receptors through gene rearrangement, enabling recognition of diverse antigens. The adaptive immune system also generates memory cells that mount faster and more effective responses upon antigen re-encounter. This complex process of adaptive immunity ensures a tailored immune response and long-term protection against pathogens. Immune disorders include autoimmunity which involves the breakdown of mechanisms that prevent the immune system from attacking the body's own tissues, hypersensitivity, when there are excessive or inappropriate immune responses, and immunodeficiency in which there is an impaired or weakened immune response. Transplant immunology is a specialised field within immunology that focuses on understanding and managing the immune responses involved in organ transplantation.

Key Points

1) The innate immune system provides immediate, non-specific, defence against a wide range of pathogens through physical and chemical barriers, such as the skin, mucous membranes and acidic environments.

2) Innate immune cells, including neutrophils, macrophages, eosinophils, basophils, mast cells, dendritic cells, natural killer cells and gamma delta T-cells, work together with soluble molecules to recognise and eliminate pathogens.

3) Pattern-recognition receptors (PRRs), such as Toll-like receptors (TLRs) and cytosolic PRRs, such as NOD-like receptors (NLRs), play a crucial role in detecting and responding to microbial patterns and damaged cells, initiating signalling pathways that lead to the production of pro-inflammatory cytokines and the activation of immune responses.

4) Adaptive immunity is characterised by specificity, memory and the ability to discriminate between self- and non-self-antigens.

5) The adaptive immune system generates a diverse repertoire of B- and T-cells through gene rearrangement, resulting in unique antigen receptor specificities for each individual.

6) Upon encountering an antigen, adaptive immune cells undergo clonal selection and expansion, leading to the production of effector cells and memory cells, the latter provides a rapid and enhanced immune response upon re-exposure to the same antigen.

7) Autoimmunity occurs when the mechanisms that prevent self-reactivity in the

immune system break down, leading to the immune system attacking the body's own tissues and organs.

8) Immunological tolerance, both central and peripheral, plays a crucial role in preventing potentially harmful autoimmune responses.

9) Hypersensitivity reactions involve excessive or inappropriate immune responses that can cause damage to tissues or organs. They are classified into four main types (Types I–IV) based on the mechanisms and immune components involved in the response.

10) Immunodeficiency disorders result in reduced or absent immune responses, making individuals more susceptible to infections or malignancies. They can be categorised as either primary (genetic or congenital) or secondary (acquired).

11) Histocompatibility is important in preventing transplant rejection. Mismatch between host and recipient can have devastating consequences.

12) Graft rejection can be hyperacute, acute or chronic and is mediated primarily by alloreactive T-cells.

Further Reading

Blausen.com staff (2014) Medical gallery of Blausen Medical 2014. *WikiJournal of Medicine* **1**(2). https://doi.org/10.15347/wjm/2014.010. ISSN 2002-4436.

Demkes J., Rijken S., Szymanski M.K. *et al.* (2021) Requirements for proper immunosuppressive regimens to limit translational failure of cardiac cell therapy in preclinical large animal model. *Journal of Cardiovascular Translational Research* **14**: 88.

Murphy K., Weaver C. & Berg L. (2022) *Janeway's Immunobiology* (10th edn). Norton & Company.

Perera S. (2022) *At the Edge of Mysteries, The Discovery of the Immune System.* Hero Press, London.

Part 5

Disorders of Blood and Blood Vessels

7

Anaemia

Introduction

Anaemia is defined as a reduction in blood haemoglobin concentration. This is reflected in a reduction in the oxygen-carrying capacity of the blood. Anaemia can be a consequence of many disorders that underlie defective erythrocyte production or excessive loss of erythrocytes from the circulation. Therefore, whenever a patient is found to be anaemic, it is essential to investigate the underlying cause. This chapter considers the fundamental causes of anaemia and its classification.

Blood Formation

Red Cell Formation

Haemopoiesis (blood cell production) takes place in the bone marrow. All blood cells (erythrocytes, leucocytes and platelets) are derived from a single clone of primitive cells – the *pluripotent stem cell*. Stem cells have the ability to divide and differentiate under the influence of specific growth factors, released by stromal cells, including macrophages and fibroblasts, in the bone marrow. Initially, the differentiated stem cells form *myeloid* and *lymphoid* stem cells (Fig. 7.1). Myeloid stem cells give rise to the progenitors of three principal cell lines:

- The myeloid–monocyte series (producing granulocytes and monocytes).
- The megakaryocyte series (producing platelets).
- Erythroid progenitor cells (producing erythrocytes – the red blood cells).

Erythropoiesis produces mature red blood cells. Erythropoiesis takes place in several stages as erythroid precursor cells develop from relatively undifferentiated pro-normoblasts to late

The Biology of Disease, Third Edition. Edited by Paul G. Murray, Simon J. Dunmore, and Shantha Perera.
© 2024 John Wiley & Sons Ltd. Published 2024 by John Wiley & Sons Ltd.
Companion website: www.wiley.com/go/murray/biologyofdisease3e

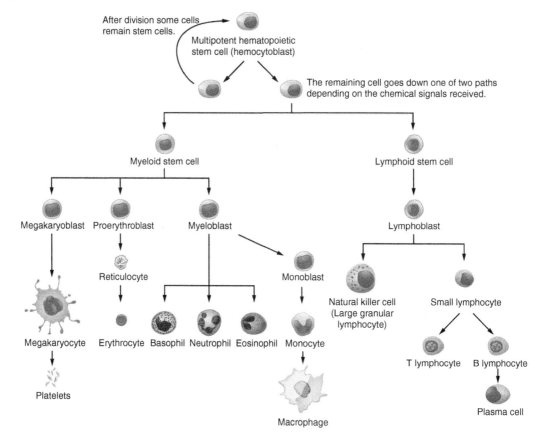

After division some cells remain stem cells.

Multipotent hematopoietic stem cell (hemocytoblast)

The remaining cell goes down one of two paths depending on the chemical signals received.

Myeloid stem cell

Lymphoid stem cell

Megakaryoblast   Proerythroblast   Myeloblast

Lymphoblast

Reticulocyte

Monoblast

Natural killer cell (Large granular lymphocyte)

Small lymphocyte

Megakaryocyte   Erythrocyte   Basophil   Neutrophil   Eosinophil   Monocyte

T lymphocyte   B lymphocyte

Platelets

Macrophage

Plasma cell

Figure 7.1 Haematopoiesis is the process through which blood cells are formed in the body. It begins with multipotent haematopoietic stem cells, also known as haemocytoblasts. These stem cells have the potential to differentiate into the two main precursors, myeloid stem cells and lymphoid stem cells. Myeloid stem cells give rise to granulocytes (including basophils, neutrophils, and eosinophils), platelets, red cells and monocytes/macrophages. Lymphoid stem cells give rise to natural killer cells, T-lymphocytes and B-lymphocytes. It should be noted that in addition to monocytes, various other cells possess phagocytic capabilities (not shown). These include osteoclasts (bone), microglia (central nervous system), Langerhans cells (skin), and Kupffer cells (liver). Some of these cells, like osteoclasts, may derive from monocyte precursors, while others, such as Kupffer cells and Langerhans cells, likely originate during embryogenesis. *Source:* Rice University / licensed under CC BY 4.0.

normoblasts. The normoblasts lose their nucleus before being released into the circulation as *reticulocytes* (young erythrocytes). Normal erythropoiesis is dependent on the availability of nutrients, including folate and vitamin B12.

Haemoglobin Synthesis

Haemoglobin is the oxygen-carrying component of erythrocytes. A typical red blood cell contains over 600 million haemoglobin molecules, each of which is a tetramer of four *haem* groups and four globin chains. The assembly of haem takes place in the mitochondria of erythrocyte precursor cells in red bone marrow. Four haem molecules each combine with α- or β-globin chains synthesised in ribosomes to form haemoglobin. Normal adult haemoglobin contains two types of globin chains – two α-chains (each 141 amino acids) and two β-chains (each 146 amino acids).

Several factors can interfere with the synthesis of haemoglobin, including lack of essential nutrients (as in iron deficiency), inherited structural abnormalities (usually in the globin chains) or defects in the efficient assembly of haemoglobin molecules.

Causes of Anaemia

A preliminary diagnosis of anaemia is based on establishing a reduction in total blood haemoglobin concentration, though this gives no indication of the cause. The reference ranges for haemoglobin concentrations differ for males and females – it is usual for adult females to have lower haemoglobin levels than their male counterparts; this is mainly because of the physiological effects of menstruation and the consequent monthly loss of blood. Haemoglobin levels are also influenced by age. In general, newborns tend to have high haemoglobin levels, which decrease over the first few months of life. After infancy, haemoglobin levels gradually increase again, reaching adult levels during adolescence. In adults, haemoglobin levels remain relatively stable, but may decrease slightly in the elderly, especially in those aged over 65 years.

A low haemoglobin concentration does not always signify anaemia: an increase in plasma volume causes an apparent anaemia (*pseudoanaemia*) though the total body haemoglobin content is normal. This is seen, for example, in pregnancy (during which a mild reduction in blood haemoglobin concentration is common) and can be difficult to differentiate from genuine anaemia of pregnancy.

Anaemia is characterised by a reduction in the concentration of haemoglobin in the blood or in numbers of circulating erythrocytes. Usually, if one of these is reduced then the other is also reduced, because haemoglobin is contained within erythrocytes. However, there are circumstances in which total blood haemoglobin concentration may be mildly reduced, but with a normal red blood cell count. This occurs in the early stages of *iron deficiency*

anaemia and in some *thalassaemia syndromes*. In such cases, despite the normal red cell numbers, individual erythrocytes are reduced in size and contain less haemoglobin. As the condition progresses and anaemia becomes more severe, the red cell count falls. Most other types of anaemia are characterised by reductions in both total haemoglobin concentration and red cell count, with erythrocytes of normal or increased cell size.

Defective Haemoglobin Synthesis

Defective haemoglobin synthesis falls into two groups: – defects of haem synthesis and defects of globin synthesis. Defects in haem synthesis can be due to lack of available iron for synthesis or a failure of the enzyme-mediated assembly of porphyrin rings. Defective globin chain synthesis may be *qualitative* (resulting in structurally abnormal haemoglobins) or *quantitative* (resulting in reduced or delayed synthesis of one or other globin chains). Defects in haemoglobin synthesis generally result in red blood cells which are smaller than normal (*microcytic*) and contain reduced amounts of haemoglobin (*hypochromic*).

Iron Deficiency

Iron deficiency, the cause of the commonest nutritional anaemia, results in diminished haem synthesis, and hence reduced haemoglobin within erythrocytes. Iron deficiency anaemia also occurs when the rate of erythropoiesis increases to compensate for blood loss or the demands of pregnancy. This is because dietary sources of iron may be adequate to support the normal rate of erythropoiesis, but are insufficient to maintain a higher rate. Chronic blood loss is an important cause of iron deficiency anaemia. During haemorrhage, iron is irretrievably lost from the body and dietary sources of iron are insufficient to compensate for the loss. Malabsorption of iron is another cause of iron deficiency.

Iron deficiency is characterised by low serum iron, low serum ferritin (reflecting depleted iron stores) and a raised concentration of serum transferrin, indicating a compensatory increase in iron transport capacity. Causes of apparent iron deficiency (blood count results which mimic those of iron deficiency with low serum iron) include defects in plasma iron transport in chronic inflammatory diseases such as rheumatoid arthritis; laboratory results differ from those in iron deficiency in that serum ferritin concentrations are often normal or high and serum transferrin is usually low. This type of anaemia is normally unresponsive to iron therapy.

Porphyrias

The porphyrias are a group of disorders characterised by an enzyme deficiency in the haem biosynthetic pathway. They are classified into *erythropoietic* or *hepatic* types depending on the nature of the enzyme deficiency. The most common acute porphyria is known as *acute intermittent porphyria* (AIP), an autosomal dominant disorder affecting an enzyme called porphobilinogen (PBG) deaminase. Clinical effects of the porphyrias include photosensitivity, dermatitis and neurological disorders resulting from accumulation of porphyrins in the tissues.

Sideroblastic Anaemia

Sideroblastic anaemia is a rare type of anaemia in which there is an abnormal accumulation of iron in erythrocyte precursors. A characteristic morphological feature is 'ring sideroblasts', which are erythroblasts containing a ring of iron surrounding their nuclei (Fig. 7.2).

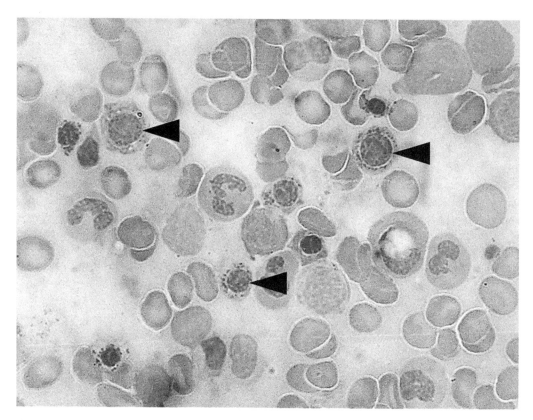

Figure 7.2 Ring sideroblasts: Image shows the characteristic abnormal deposition of iron (turquoise colour, stained with Perl's method) in the mitochondria of red cell precursors forming a ring around the nucleus (black arrowheads). *Source:* Paulo Henrique Orlandi Mourao / Wikimedia Commons / CC BY-SA 3.0.

Sideroblastic anaemia can be either congenital or acquired. *Congenital sideroblastic anaemia* is typically an inherited condition resulting from mutations in genes responsible for haem synthesis, for example, those encoding aminolevulinic acid synthase (ALAS) or ferrochelatase (FECH). *Acquired sideroblastic anaemia* can occur following administration of drugs such as isoniazid or chloramphenicol, as well as heavy metals, like lead or zinc. Chronic alcoholism can also cause acquired sideroblastic anaemia.

Inherited Defects in Globin Chain Synthesis

Thalassaemia Syndromes

The α-globin and β-globin chains are required in sufficient quantities to enable the synthesis of haemoglobin molecules with two of each type of chain. The *thalassaemias* are quantitative globin chain defects associated with an imbalance in the rate of synthesis of α-globin and β-globin chains, usually owing to decreased synthesis of one or other chain. The effects are variable, depending on the degree to which globin chain synthesis is suppressed.

Two closely linked genes (HBA1 and HBA2) on the short arm of chromosome 16 encode α-globins, so the synthesis of these chains is controlled by four genes in each cell. *α-thalassaemia* is caused by deletion or alteration of one or more of these genes, the severity of the clinical outcome varying with the number of genes affected. Severe forms of α-thalassaemia, with deletion of all four genes, are incompatible with life, and pregnancy with an affected foetus usually fails to reach full term. Less severe forms of α-thalassaemia (deletion of three of the four genes) are associated with variable survival and quality of life. Typically, there is moderate anaemia, and the formation of unstable β-chain tetramers (referred to as haemoglobin H) may be a feature. α-thalassaemia traits (deletion of one or two genes) are of little clinical significance to affected individuals, but

genetic counselling is essential if affected partners are in a child-bearing relationship.

β-globin chains are encoded from the HBB gene on the short arm of chromosome 11. Severe *β-thalassaemia* is associated with profound anaemia from the first year of life and dependence on regular transfusions. Iron overload resulting from transfusions becomes a problem, though the risk can be reduced by administering iron chelating agents. β-thalassaemia minor is much less severe, with a blood picture superficially similar to that of iron deficiency.

Haemoglobinopathies

The correct sequence of amino acids in the four globin chains is essential for an efficient function of the haemoglobin molecule. Structural globin chain defects *(haemoglobinopathies)* are usually due to base substitutions (point mutations) in the DNA which encodes the polypeptide globin chains, resulting in single amino acid substitutions. Such changes have variable outcomes, depending on the nature and site of the amino acid substitution, ranging from no clinical effect to severe consequences such as altered oxygen affinity, or destabilisation of the haemoglobin molecule resulting in reduced erythrocyte lifespan.

One of the most important of the single amino acid substitution defects is the replacement of glutamic acid by valine at position 6 in the β-globin molecule to produce *sickle haemoglobin* or haemoglobin S. When inherited as the homozygous condition (known as *sickle cell disease*), under conditions of low oxygen tension, haemoglobin S precipitates as long filaments, distorting the shape of the erythrocytes (Fig. 7.3A). The resulting sickle cells lodge in small blood vessels causing painful infarctions and reduced red cell lifespan. The terms 'sickle cell crisis' or 'sickling crisis' encompass various distinct acute conditions that occur in individuals with sickle cell disease. These episodes lead to anaemia and can manifest in multiple forms, including vaso-occlusive crisis, aplastic crisis, splenic

(A)

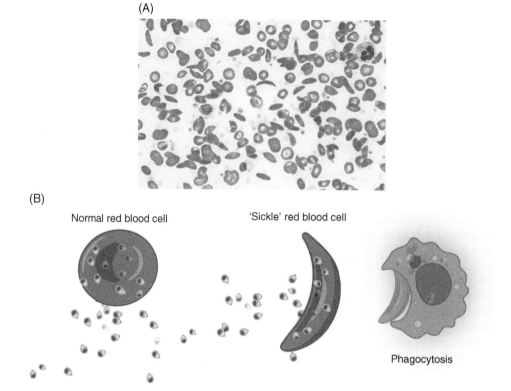

(B)

Figure 7.3 (A) Sickle cell anaemia: Peripheral blood smear showing 'sickle'-shaped red blood cells. *Source:* Ed Uthman. / Wikimedia Commons / CC BY 2.0. (B) Possible advantage of sickle cell trait: Normal red blood cells provide a fertile ground for the malarial parasite to replicate, resulting in a significant burden of infection and associated symptoms. In contrast, red blood cells in individuals with sickle cell trait can undergo sickling under low-oxygen conditions or other stresses. This sickling can inhibit the parasite's life cycle or can flag the cells for faster removal from the circulation by phagocytosis. *Source:* Created with BioRender.com.

sequestration crisis, haemolytic crisis and others. Typically, a sickle cell crisis lasts between 5 and 7 days. Although infection, dehydration and acidosis can trigger sickle cell crisis by promoting the sickling of red blood cells, in many cases, there is no identifiable underlying cause.

Another of the haemoglobinopathies, haemoglobin C disease, is also characterised by reduced red cell lifespan and is due to the replacement of glutamic acid by lysine at the same β-chain locus as haemoglobin S.

When either condition is inherited as the heterozygous 'trait' (one gene coding for normal haemoglobin and the other coding for haemoglobin S or C), some normal adult haemoglobin is produced alongside the defective haemoglobin and subjects are generally symptom-free. In each case, the heterozygous condition confers some protection against malaria due to the protozoan parasite *Plasmodium falciparum*, which helps to explain their natural geographical distribution (Fig. 7.3B; see also Case Study CS8).

The precise cause of anaemia in the different haemoglobinopathies and thalassaemia syndromes is often complex. Increased rates of *haemolysis* leading to reduction in erythrocyte lifespan and ineffective erythropoiesis (IE) (see later) both contribute. Combinations of inherited structural haemoglobin defects and thalassaemia minor can produce complex and very severe forms of anaemia.

Diminished Erythrocyte Production

In this group of anaemia, haemoglobin synthesis is normal, but erythropoiesis is reduced. Defects in erythropoiesis may be due to nutrient deficiency, bone marrow *hypoplasia* (underactivity), infiltration of bone marrow with malignant cells or various chronic disease states.

Nutrient deficiency

Certain vitamins and other nutrients are essential for normal blood formation. Of particular importance in erythropoiesis are the B group vitamins, vitamin B12 and folate, which are essential for DNA synthesis. Lack of either of these vitamins is associated with disordered maturation of erythrocyte precursors in the bone marrow. In prolonged vitamin B12 or folate deficiency, erythrocyte precursor cells display a characteristic morphological alteration in which nuclear maturation lags behind that of the cytoplasm (*megaloblastic change*). These abnormal precursor cells are known as *megaloblasts*. The mature erythrocytes resulting from megaloblastic erythropoiesis are much larger than normal and are known as *macrocytes*, though they are significantly reduced in number. Deficiencies of vitamin B12 and folate have three potential causes – dietary insufficiency, malabsorption and metabolic interference:

Vitamin B12
Dietary deficiency of vitamin B12 is rare, but since the vitamin is present only in foods of animal origin, strict vegetarians are at risk. Absorption of vitamin B12 from the intestinal tract requires the presence of a glycoprotein, *intrinsic factor* (IF), secreted by parietal cells in the fundus and body of the stomach. IF binds to vitamin B12 and the complex is absorbed in the distal ileum, so lack of IF results in vitamin B12 malabsorption. Vitamin B12 malabsorption is a feature of *pernicious anaemia*, an autoimmune condition characterised by the presence of autoantibodies to parietal cells and/or IF (see also Case Study CS23). Disturbance of vitamin B12 metabolism is rare but prolonged exposure to nitrous oxide anaesthesia has been shown to cause megaloblastic change.

Folate
Folate deficiency can arise due to dietary lack, especially during periods of heightened requirements, such as in pregnancy, or in conditions characterised by rapid cell turnover, such as psoriasis, neoplastic disorders and haemolytic disorders. Malabsorption is a hallmark of certain digestive tract disorders, notably coeliac disease, and can lead to folate deficiency. Chronic alcohol consumption also contributes to folate deficiency. Alcohol interferes with folate absorption in the small intestine and can also impair the conversion of folate to its active form, tetrahydrofolate (THF), leading to reduced folate availability for essential cellular processes. This deficiency is compounded by the fact that alcohol can lead to poor dietary choices and inadequate nutrient intake, further exacerbating the risk of folate deficiency. Certain drugs also cause folate deficiency. For example, methotrexate, which is primarily used to treat people with leukaemia or autoimmune diseases, competes with folate for the enzyme folic acid reductase, resulting in reduced purine synthesis and causing megaloblastic changes in the bone marrow.

Bone Marrow Hypoplasia

Reduced erythropoiesis can be caused by bone marrow damage, leading to hypoplastic or aplastic anaemia (anaemia due to reduced haemopoietic activity). There are several potential causes, though many cases are of unknown aetiology. Some are *iatrogenic* (physician induced), caused by exposure to cytotoxic drugs, or other chemicals and radiotherapy. Thyroid, liver or renal disease, or viral infections, may also result in bone marrow hypoplasia.

Ineffective Erythropoiesis

IE is a failure of erythrocyte precursor cells to reach maturity, resulting in their destruction in the bone marrow. Some degree of IE occurs in normal haemopoiesis affecting up to one in eight erythrocyte precursors. An increase in IE contributes to many forms of anaemia, including those seen in the thalassaemia syndromes, megaloblastic anaemia and some leukaemias.

Chronic Disease States

Inflammatory or neoplastic disease may cause diminished red cell production. In certain chronic inflammatory disorders (e.g. rheumatoid arthritis), anaemia may be due to a defect in iron utilisation. Chronic renal failure is also associated with impaired erythropoiesis because the kidney is a major source of the hormone erythropoietin, which stimulates the later stages of red cell production.

Malignant Blood Disorders

Leukaemias and other malignant blood diseases are an important cause of anaemia due to reduced erythropoiesis. Leukaemias are categorised as *acute* or *chronic* depending on the degree of maturation of the malignant cells and are further subdivided according to the particular haemopoietic cell line involved. Characteristically, in these malignancies, the bone marrow contains large numbers of immature leukaemic cells which displace normal haemopoietic cells, and there may also be some megaloblastic changes in erythropoiesis. Most leukaemias are characterised by anaemia (often severe in the acute forms of leukaemia), *thrombocytopenia* (reduction in blood platelet count) and *neutropenia* (reduction in circulating neutrophils), increasing susceptibility to spontaneous bleeding and severe bacterial infection.

Excessive Erythrocyte Destruction (Haemolytic Disorders)

The normal red cell has a mean lifespan in the circulation of 110–120 days. Reduced erythrocyte lifespan may result in anaemia, though this is not always the case because the bone marrow is usually able to increase the production of erythrocytes to compensate for the premature loss, stimulated by elevated levels of erythropoietin Despite this, if red cell lifespan is very short, or the defect is chronic, premature cell death may not always be fully compensated for by an increase in the rate of erythropoiesis.

Haemolysis releases a significant amount of haemoglobin into the bloodstream. This haemoglobin is broken down, releasing iron which accumulates in tissues as haemosiderin. *Haemosiderosis* is the excessive deposition of iron in tissues, and this occurs particularly in the liver, spleen and bone marrow. Over time, the iron overload can lead to tissue damage. Haemolysis also results in the release of a large amount of bilirubin. Bilirubin is normally processed by the liver and excreted in bile. However, in haemolytic anaemia, the increased breakdown of red blood cells can overwhelm the liver's ability to process bilirubin, leading to its accumulation in the blood. Elevated levels of bilirubin cause jaundice.

There are two major groups of haemolytic disorders – those due to *intrinsic* and those due to *extrinsic* erythrocyte defects.

Intrinsic Erythrocyte Defects

Intrinsic erythrocyte defects leading to haemolytic disorders are usually congenital and can be subdivided into the following:

Alterations in haemoglobin structure which can decrease erythrocyte lifespan due to instability of the haemoglobin molecule itself or its defective interaction with other cell structures. The notable example here is sickle cell anaemia which is described above.

Deficiencies or structural defects of intracellular enzymes, such as *pyruvate kinase* (PK) or *glucose-6-phosphate dehydrogenase* (G6PD), affect the ability of erythrocytes to metabolise nutrients and provide for their energy needs. In these cases, the red cells may not survive for as long as normal in the circulation.

Erythrocyte membrane defects are usually associated with abnormalities of the cytoskeleton and become clinically significant if they alter the biconcave disc shape and reduce erythrocyte deformability, reducing their ability to enter small blood vessels. The affected cells become trapped in the splenic sinusoids and are prematurely removed from the circulation.

Hereditary spherocytosis (HS) is the most common of the red cell membrane disorders and is caused by mutations in SPTA1, SPTB, ANK1, EPB42 or SLC4A1. These mutations lead to an abnormal red cell membrane skeleton deficient in α- or β-Spectrin, Ankyrin, Protein 4.2 or Band 3 proteins, respectively. These proteins are instrumental in constructing the scaffold and vertical connections of the red cell membrane skeleton with the lipid membrane. Individuals affected by autosomal dominant forms of HS arising from mutations in SLC4A1, SPTB or ANK1 genes, or autosomal recessive HS caused by EPB42 mutations, exhibit a spectrum of clinical presentation, ranging from *compensated haemolysis* to a moderately severe form of anaemia. In compensated haemolysis, the rate of erythropoiesis increases to compensate for red cell loss. However, if the rate of production fails to keep pace with the rate of destruction, anaemia results. Thus, there is normally a substantial reserve capacity in the bone marrow which can compensate for reduced red blood cell lifespan, though periodically the bone marrow may fail to compensate fully, particularly if the patient has an infection or health is compromised in some other way. Patients may experience occasional periods of anaemia *(haemolytic crisis)* interspersed with months or years of good health. *Hereditary elliptocytosis* (HE) is another inherited defect in the red cell membrane. Although HE is also caused by mutations in the SPTA1, SPTB or EPB41 genes, these mutations change the functional properties of spectrin tetramers rather than causing a deficiency (as observed in HS), contributing, in turn, to the distinctive elliptical shape of the red blood cells.

Paroxysmal nocturnal haemoglobinuria (PNH) is a rare and life-threatening blood disorder characterised by the excessive destruction of red blood cells by complement, owing to a deficiency in complement inhibitors. The disease is also associated with a higher incidence of thrombosis. PNH is distinctive as it is the only acquired haemolytic anaemia resulting from an intrinsic defect in the cell membrane, specifically a deficiency of glycophosphatidyl inositol (GPI). This leads to the absence of protective exterior surface proteins that typically attach via a GPI anchor. It can manifest as a standalone condition (primary PNH) or develop alongside other bone marrow disorders like aplastic anaemia (secondary PNH). Though not all individuals exhibit the characteristic red urine that initially gave the disease its name, common symptoms include anaemia-related fatigue, shortness of breath and palpitations.

Extrinsic Erythrocyte Defects

Extrinsic defects are usually acquired and affect erythrocytes which are not intrinsically abnormal in any way. Examples include infection with malarial parasites, other systemic infections, severe trauma or burns and heart valve defects. Some non-steroidal anti-inflammatory drugs can also cause haemolytic anaemia, triggering oxidative stress which damages red blood cells.

The acquisition of autoantibodies to antigens on the red cell surface (such as those of the rhesus system) is a cause of *autoimmune haemolytic anaemia*, which can be triggered by exposure to some drugs. Examples include, chemotherapy drugs, such as mitomycin C, immune checkpoint inhibitors, such as pembrolizumab and nivolumab, and antibiotics like penicillins or cephalosporins.

Nutritional Aspects of Haemolytic Disorders

In most haemolytic disorders, haemoglobin is broken down extravascularly, usually in the spleen, and the iron and globin chains are recycled for further use within the body. For this reason, there is not usually any iron deficiency in most haemolytic disorders. However, the increased rate of erythropoiesis which is necessary to compensate for increased red cell destruction can deplete folate and cause megaloblastic changes. For this reason, folic acid is often administered in chronic haemolytic disorders to avoid any megaloblastic changes which could arise due to folate depletion.

Blood Loss

Haemorrhage results in erythrocytes leaving the circulation prematurely. The effects on erythropoiesis vary depending on whether the blood loss is *acute* or *chronic*. Acute haemorrhage (i.e. sudden and of short duration) of more than 1 L leads to a fall in total blood haemoglobin over the first 24 hours. Normally, the bone marrow makes up the deficit in red cell numbers within a few weeks, as long as bleeding does not continue. If anaemia results from acute blood loss, it is temporary and not associated with any obvious morphological changes to the erythrocytes, other than a reduction in the total red cell count, and is described as *normochromic* and *normocytic* (see below). In the early stages of prolonged (or chronic) blood loss, similar compensatory mechanisms are initiated. However, a continuous supply of iron is needed for the synthesis of haemoglobin and there is often insufficient iron available from dietary sources to compensate for that which continues to be lost. Thus, in chronic blood loss of a degree sufficient to cause anaemia, the eventual outcome is invariably iron deficiency. Lack of iron leads to impaired haemoglobin synthesis and this, added to the requirement for increased erythropoiesis to replace lost red blood cells, results in *microcytic hypochromic anaemia* (see below). Chronic blood loss, including that due to menstruation, is an important cause of iron deficiency anaemia. However, as mentioned earlier, iron deficiency may also be caused by malabsorption or a poor diet.

Classification of Anaemia

When a patient is first recognised to be anaemic, the underlying cause is not always immediately apparent, so laboratory data are used as a basis for further investigations. Anaemia is initially classified based on erythrocyte size and morphology. The average size of the erythrocytes, measured as *mean cell volume* (MCV) is most useful in defining the type of anaemia, which is provisionally classified as *normocytic* (normal MCV), *microcytic* (reduced MCV) or *macrocytic* (high MCV). *Mean cell haemoglobin (MCH)* (the amount of haemoglobin in the average erythrocyte) and the microscopic morphology of erythrocytes on stained blood films are important. If staining intensity is reduced, the red cells are described as *hypochromic*, which usually correlates with reduced MCH. Measurement of total haemoglobin concentration, erythrocyte count (RBC), MCV and MCH enable different types of anaemia to be described as *microcytic hypochromic*, *macrocytic normochromic* and *normocytic normochromic* (Table 7.1).

Classification of anaemia based on cell size gives little information on the specific *aetiology* (underlying cause) in individual cases. However, it does indicate some possible causes and helps to exclude others, so that further investigations to identify the precise cause(s) of the anaemia can be initiated. As with other disorders, the aetiology of anaemia can be multifactorial: Table 7.2 lists possible causes of microcytic, normocytic and macrocytic anaemias.

Table 7.1 Laboratory classification of anaemia.

Red cell characteristics	Hb	RBC	MCV	MCH
Microcytic hypochromic	Low	Normal/low	Low	Low
Normocytic normochromic	Low	Low	Normal	Normal
Macrocytic normochromic	Low	Low	High	Normal/high

Hb, haemoglobin concentration; MCH, mean red cell haemoglobin content; MCV, mean red cell volume; RBC, red cell count.

Table 7.2 Classification of anaemia on the basis of cell size.

Red cell size	Red cell features	Possible underlying causes
Microcytic anaemia		
Reduced mean cell volume (MCV <80 fL approx.)	Microcytic and hypochromic or microcytic without hypochromic red blood cells	Iron deficiency anaemia Anaemia of chronic inflammatory disease (e.g. rheumatoid arthritis) Thalassaemia
Macrocytic anaemia		
Increased mean cell volume (MCV >92 fL approx.)	Megaloblastic (macrocytosis due to megaloblastic erythropoiesis)	Vitamin B12 or folate deficiency or acute leukaemia with megaloblastic changes in the bone marrow
	Non-megaloblastic (macrocytosis without megaloblastic change)	Some cases of liver disease Alcoholism (but note that folate deficiency may have a role) Some haemolytic anaemias
Normocytic anaemia		
Normal mean cell volume	Usually normochromic, normocytic anaemia	Acute blood loss Some haemolytic anaemias Some anaemias of chronic disorder (e.g. chronic renal failure) Hypoplastic or aplastic anaemias, malignancy, pregnancy

Summary

Anaemia is characterised by a reduction in blood haemoglobin concentration, resulting in a decreased oxygen-carrying capacity. Anaemia can be caused by different underlying disorders that include nutritional deficiencies, bone marrow disorders, haemolytic disorders and chronic diseases. Understanding the specific aetiology of anaemia is crucial for effective management and treatment. Classifying anaemia, for example, based on erythrocyte size and morphology, can help guide further investigations and tailor interventions accordingly.

> **Key Points**
>
> 1) The four basic causes of anaemia are defective haemoglobin synthesis, diminished erythrocyte production excessive erythrocyte destruction and blood loss.
> 2) Iron deficiency anaemia can arise from poor diet or malabsorption; chronic blood loss is also an important cause.
> 3) The haemoglobinopathies and thalassaemia syndromes are inherited conditions affecting haemoglobin synthesis and structure.
> 4) Vitamin B12 and folate deficiencies cause megaloblastic changes in erythropoiesis.
> 5) Erythropoiesis is reduced in aplastic anaemia and in cases in which the bone marrow is infiltrated by leukaemic or other malignant cells.
> 6) Haemolytic disorders may be inherited or acquired.

Further Reading

Benz Jr. E., Berliner N. & Schiffman F. (eds) (2017) *Anaemia: Pathophysiology, Diagnosis, and Management.* Cambridge University Press, Cambridge. https://doi.org/10.1017/9781108586900.

Chaparro C.M. & Suchdev P.S. (2019) Anaemia epidemiology, pathophysiology, and etiology in low- and middle-income countries. *Annals of the New York Academy of Sciences* **1450**:15–31. https://doi.org/10.1111/nyas.14092. 31008520; PMC6697587.

8

Disorders of Haemostasis

Introduction

The normal haemostatic (or 'blood coagulation') system helps to ensure that blood is confined to, and flows freely within, the circulatory system. The process involves platelets, vascular endothelium and the blood coagulation cascade. Haemostatic mechanisms thus contribute to the maintenance of efficient blood flow and to the arrest of bleeding from injured blood vessels. This chapter considers basic haemostatic processes and discusses some of the disorders that result from defects in haemostasis.

Overview of Haemostasis

Complex processes have evolved to prevent excessive blood loss after injury and to maintain circulatory integrity. These haemostatic processes depend upon interactions between

- The blood vessel wall, in which damage to the endothelium initiates clotting processes.
- Circulating blood platelets, which are activated by vessel wall injury or other tissue damage.
- The blood coagulation and fibrinolytic mechanisms, and their respective inhibitors.

Dysregulated haemostasis can lead to either *bleeding* or *thrombosis* (formation of a clot within the vascular system). Expressed very simply, most bleeding disorders are due to underactive haemostasis, and some thrombotic disorders are due, at least in part, to overactive haemostatic mechanisms.

Arrest of Bleeding

The physiological process which arrests bleeding following damage to a blood vessel (Fig. 8.1) is initiated by *vasoconstriction* (narrowing of blood vessels) which reduces blood

The Biology of Disease, Third Edition. Edited by Paul G. Murray, Simon J. Dunmore, and Shantha Perera.
© 2024 John Wiley & Sons Ltd. Published 2024 by John Wiley & Sons Ltd.
Companion website: www.wiley.com/go/murray/biologyofdisease3e

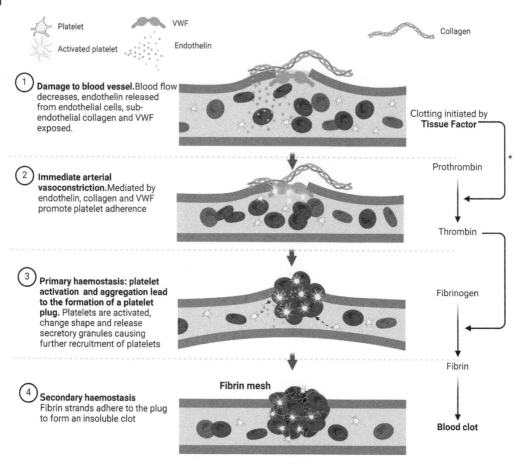

Figure 8.1 Primary and secondary haemostasis. The physiological process that stops bleeding after blood vessel damage involves several steps. (1) Initial damage causes the release of endothelin and activates a reflex neurogenic mechanism, which (2) leads to immediate narrowing of the blood vessels to reduce blood flow and limit blood loss. When the endothelial surfaces are disrupted, the underlying subendothelial collagen fibres are exposed. Platelets adhere to these collagen fibres through receptors on their surface (GPVI, see Fig. 8.3) and to von Willebrand factor (VWF, sometimes written as vWF) present in the exposed subendothelial matrix. (3) This leads to platelet activation, resulting in a change in their shape and the release of secretory granules and the formation of a platelet plug. This plug seals the damaged vessel. Concurrently, the coagulation cascade is initiated by Tissue Factor, exposed upon vessel injury, ultimately leading to the generation of fibrin. * denotes multiple steps as shown in Fig. 8.4. (4) The fibrin meshwork entangles red blood cells, platelets and other blood components, forming a stable blood clot or *thrombus* at the injury site. The clot provides mechanical support, prevents further bleeding and aids in wound healing. After it is formed, the clot undergoes stabilisation and contraction. *Source:* Created with BioRender.com.

flow and limits blood loss. Vasoconstriction is mediated by a reflex neurogenic mechanism and is further promoted by the release of endothelin by endothelial cells. Disruption to the endothelial surfaces exposes the underlying subendothelial collagen fibres and von Willebrand factor (VWF, which may also be written as vWF), to which platelets adhere, an effect mediated by receptors on platelets (Fig. 8.3). The adhered platelets undergo activation, leading to changes in their shape, the release of their granules and exposure of Integrin αIIbβ3 receptors on their surface. Platelet activation is facilitated by various signalling molecules, including adenosine diphosphate (ADP), thromboxane A2 (TXA$_2$) and thrombin. Within

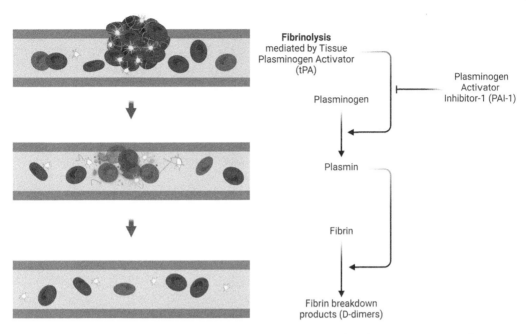

Figure 8.2 Fibrinolysis dissolves blood clots to restore normal blood flow. Plasminogen, an inactive precursor protein, is converted to its active form, plasmin, a process triggered by tissue plasminogen activator (tPA) released by endothelial cells (or which can be administered therapeutically). Plasmin cleaves fibrin, releasing fibrin degradation products (e.g. D-dimers). Plasminogen activator inhibitor-1 (PAI-1), also released by endothelial cells, inhibits tPA activity. *Source:* Created with BioRender.com.

several minutes, the secreted products recruit additional platelets that aggregate to form a platelet plug, which seals the damaged vessel. Simultaneously with platelet activation, the coagulation cascade is initiated. This involves a series of enzymatic reactions that culminate in the conversion of the inactive plasma protein, fibrinogen into fibrin. The coagulation cascade is triggered by Tissue Factor (TF), which is exposed upon vessel injury. The fibrin meshwork entangles red blood cells, platelets and other blood components, forming a stable blood clot or *thrombus* at the site of injury. The clot provides mechanical support, prevents further bleeding and facilitates wound healing. After the clot is formed, it undergoes stabilisation and contraction. Platelets play a role in this process by contracting, squeezing the clot and making it more compact. Because the process is so potent, control mechanisms are required to prevent its inappropriate activity. The control mechanisms include

- Blood flow, which washes away activated coagulation factors.
- The production, by vascular endothelium, of prostacyclin, PGI_2, which inhibits platelet activity.
- Activation of *fibrinolysis,* which dissolves clots (Fig. 8.2).
- The presence of inhibitors of plasma coagulation factors.

Platelet Structure and Physiology

Platelets have a fundamental role in haemostasis. They are cytoplasmic fragments derived from megakaryocytes and can survive in the circulation for up to 10 days. Platelets have a membrane-associated outer zone which is involved in platelet adhesion and aggregation and which interacts with the blood coagulation cascade. There are several important receptors on the platelet surface (Fig. 8.3):

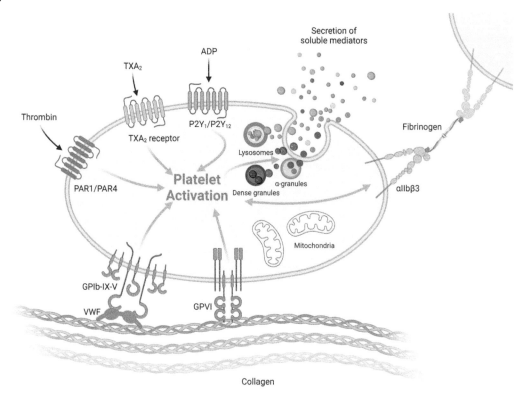

Figure 8.3 Platelet activation. Platelets express several important receptors: The Glycoprotein Ib/IX/V (GPIb/IX/V) complex binds to von Willebrand factor (VWF), facilitating platelet tethering and initial adhesion to the damaged vessel wall. Glycoprotein VI (GPVI) interacts with exposed collagen fibres in the subendothelial matrix, promoting platelet adhesion and activation. Integrin αIIbβ3 binds to fibrinogen, establishing a connection between adjacent platelets. Thrombin activates platelets through two protease-activated receptors, PAR-1 and PAR-4. P2Y1 and P2Y12 receptors are activated by adenosine diphosphate (ADP) released by activated platelets and other cells at the injury site. Activation of these receptors enhances platelet aggregation and amplifies the platelet response. TXA$_2$ receptor mediates the effects of TXA$_2$, a potent platelet agonist. Once platelets are activated, they release the contents of their granules. There are three types of platelet granules: α-granules are abundant and contain adhesive proteins like VWF, fibrinogen, fibronectin, vitronectin and thrombospondin. These proteins are crucial for platelet adhesion and the formation of a stable blood clot. α-granules also release platelet factor 4 (PF4), interleukin-8 (IL-8), platelet-derived growth factor (PDGF), transforming growth factor-β (TGF-β) and vascular endothelial growth factor (VEGF). Dense granules contain ADP, ATP, serotonin, calcium, pyrophosphate and polyphosphate. Platelet lysosomes contain acid hydrolases such as cathepsins, hexosaminidase, β-galactosidase, arylsulfatase, β-glucuronidase and acid phosphatase. *Source:* Created with BioRender.com.

- Glycoprotein Ib/IX/V (GPIb/IX/V) complex, which binds VWF and is essential for platelet tethering and initial adhesion to the damaged vessel wall.
- Glycoprotein VI (GPVI), which binds to exposed collagen fibres in the sub-endothelial matrix, facilitating platelet adhesion and activation.
- Integrin αIIbβ3 (also referred to as GPIIb/IIIa), which binds to fibrinogen and forms a bridge between adjacent platelets.
- Protease-activated receptors (PARs), PAR-1 and PAR-4, which are bound by thrombin, leading to platelet activation
- P2Y1 and P2Y12 receptors, which are activated by ADP released by activated platelets and

other cells at the site of injury. Activation of these receptors enhances platelet aggregation and further amplifies the platelet response. For example, stimulation of P2Y1 and P2Y12 receptors leads to the activation of integrin αIIbβ3 (GPIIb/IIIa) on the platelet surface.

- TXA$_2$ receptor, also known as the TP receptor, mediates the effects of TXA$_2$, a potent platelet agonist. Activation of the TP receptor promotes platelet activation, aggregation and vasoconstriction.

Once activated, platelets release their granule contents, further potentiating platelet aggregation. There are three types of platelet granules:

α-granules are the most abundant secretory organelles in platelets, with approximately 50–80 per platelet. They contain a wide range of adhesive proteins that play crucial roles in primary haemostasis. These adhesive proteins include VWF, fibrinogen, fibronectin, vitronectin and thrombospondin. These proteins are essential for platelet adhesion and contribute to the formation of a stable thrombus during clotting. In addition to adhesive proteins, α-granules also contain numerous mediators with diverse functions in coagulation, wound repair, inflammation and angiogenesis. Some of the important mediators are platelet factor 4 (PF4), interleukin-8 (IL-8), platelet-derived growth factor (PDGF), transforming growth factor-β (TGF-β) and vascular endothelial growth factor (VEGF).

Dense granules are the second major secretory compartment in platelets, with three to eight dense granules per platelet. They mainly contain small molecules such as ADP, ATP, serotonin, calcium, pyrophosphate and polyphosphate. They have a crucial role in primary haemostasis by acting as a feedback mechanism to stimulate the platelet P2Y12 receptor through the release of ADP. Dense granules also contain various membrane proteins, including lysosomal membrane proteins, CD63 and LAMP1/2, as well as non-lysosomal proteins such as P-selectin, GPIb and

αIIbβ3 integrin. Upon platelet activation, these membrane proteins are translocated to the cell surface, contributing to platelet activation.

Platelet lysosomes contain various acid hydrolases, including cathepsins, hexosaminidase, β-galactosidase, arylsulfatase, β-glucuronidase and acid phosphatase. The secretion of their lysosomal contents contributes to the functions of platelets in receptor cleavage, fibrinolysis, degradation of extracellular matrix and vasculature remodelling.

The Blood Coagulation Cascade and Its Control

The blood coagulation cascade (Fig. 8.4) results in the production of fibrin, which is incorporated into the haemostatic plug. The blood coagulation cascade exhibits the typical features of other biological cascades, i.e. amplification, specificity and control. This potent mechanism is balanced by several naturally occurring inhibitors which act to inhibit fibrin clot formation (notably the *serine protease inhibitors*) and the fibrinolytic system which digests clots once they are formed.

It is convenient to depict two separate, somewhat distinct, coagulation mechanisms with a common final pathway. The *intrinsic* coagulation pathway refers to the process of clot formation that occurs when blood is exposed to an artificial surface such as glass, as seen in laboratory tests like the activated partial thromboplastin time (aPTT). It involves factors present in the blood itself, and the activation of this pathway occurs when factors in contact with the surface initiate a series of reactions leading to clot formation. The *extrinsic* coagulation pathway, on the other hand, is triggered by molecules present outside of the blood, often involving the release of TF from damaged tissues. This pathway plays a significant role in the initial stages of coagulation and is assessed using laboratory tests like the prothrombin time (PT).

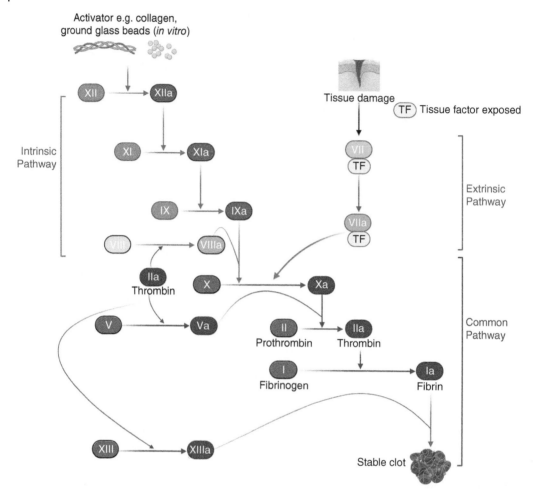

Figure 8.4 The coagulation cascade. Upon vessel damage, the *extrinsic pathway* is activated, in which Factor VII interacts with the now exposed Tissue Factor (TF), forming the TF-VIIa complex. TF-VIIa activates Factor X to form Factor Xa. The *intrinsic pathway* involves the formation of a primary complex, e.g. on collagen which activates Factor XII, and in turn, Factor XI and Factor IX, leading to the activation of Factor X to form Factor Xa. The intrinsic pathway has a minor role in initiating clot formation and is more important for inflammation and innate immunity. The common pathway is the convergence point of the intrinsic and extrinsic pathways. Here, a complex, consisting of Factor Xa and Factor Va, activates prothrombin to form thrombin. Thrombin converts fibrinogen to fibrin. Thrombin also activates Factor V and Factor VIII and thus creates a positive feedback loop which generates more thrombin. Thrombin also stabilises the fibrin network by activating Factor XIII. Thrombin is not solely generated through these pathways but is also produced by activated platelets during the initiation of the platelet plug. Negative regulators of these pathways include thrombin itself. *Source:* Created with BioRender.com.

Natural Anti-coagulants

Natural anticoagulant proteins prevent excessive clot formation. They act at different points in the coagulation cascade and include

- Antithrombin III, which inhibits several coagulation factors, including thrombin and Factor Xa.

- Protein C, in combination with its co-factor, protein S, inhibits Factors Va and VIIIa.
- Tissue factor pathway inhibitor (TFPI), a high-affinity inhibitor of TF, Factor VIIa and Factor Xa.
- Thrombin also acts as a negative feedback by activating plasminogen to form plasmin and stimulating the production of

antithrombin III. Plasmin acts directly on the fibrin mesh to break it down.

Prostaglandin Metabolism in Platelets and Blood Vessels

We have seen that platelet aggregation is stimulated by exposure to a range of substances including collagen and thrombin, the latter being formed in the coagulation cascade. Thrombin binds to receptors on the vascular endothelium and on platelet membranes. This causes the release of arachidonic acid which is metabolised by cyclooxygenase to prostaglandin G_2 (PGG$_2$). PGG$_2$ is further metabolised to TXA$_2$ in platelets and to PGI$_2$ (prostacyclin) in the vascular endothelium (see also Chapter 5). TXA$_2$ and PGI$_2$ have opposing effects on platelet adenylate cyclase: PGI$_2$ stimulates activity of this enzyme causing an increase in cyclic adenosine monophosphate (c-AMP), thus lowering free calcium ion concentrations and reducing platelet adhesion and aggregation. PGI$_2$ is also a vasodilator. In contrast, TXA$_2$ is a vasoconstrictor and a potent inducer of platelet aggregation.

Role of the Vascular Endothelium in Haemostasis

In the presence of small amounts of thrombin, healthy vascular endothelium opposes coagulation by producing PGI$_2$, which inhibits platelet aggregation. Other substances opposing clot formation include tissue plasminogen activator (tPA) which stimulates fibrinolysis, and the coagulation inhibitors, antithrombin III and thrombomodulin.

Conversely, endothelial cells synthesise VWF a procoagulant protein which, as we have seen, also promotes platelet adhesion and aggregation. Thus, healthy vascular endothelium contributes to both the promotion and the inhibition of blood coagulation. Damage to vascular endothelium activates platelets and blood coagulation mechanisms. This may be desirable (to arrest bleeding) or undesirable, for example, when damage to the intima of the blood vessel during atherosclerosis promotes the formation of an unwanted thrombus.

Haemostatic Disorders

Haemostatic disorders can be divided into two types:

- Disorders of underactivity, in other words, bleeding disorders, such as *haemophilia A, Christmas disease* or *thrombocytopenic purpura.*
- Disorders of overactivity, the so-called thrombotic disorders such as *thrombotic stroke, coronary heart disease* and *deep vein thrombosis.*

Bleeding Disorders

Abnormal bleeding may arise from deficiencies in platelet numbers or function, blood coagulation factor disorders (either congenital or acquired) or defects in the vascular wall. Of these, platelet defects are the most common, followed by coagulation factor disorders. Bleeding due primarily to vascular disorders is rare. Increased fibrinolytic activity or raised coagulation inhibitors are also implicated in excessive bleeding.

Platelet Disorders

Platelet disorders are usually acquired. *Thrombocytopenia*, the term that describes reduced platelet numbers, usually presents with superficial bleeding manifest as *petechiae* (small haemorrhages in the skin, 1–2 mm in diameter) or *purpura* (slightly larger haemorrhages, >3 mm) and may be due to either a failure of platelet production or shortened platelet survival. *Immune thrombocytopenic purpura* (ITP) is autoimmune in origin and characterised by the presence of circulating immunoglobulin G (IgG) antibodies directed against platelet proteins, usually against platelet membrane glycoproteins. ITP is usually chronic; however, acute forms can arise in children following an infection. Drugs can also induce an autoimmune response to platelets, and some drugs depress megakaryocyte production. Other conditions that can cause thrombocytopenia include bone marrow disorders, such as aplastic anaemia, leukaemia and HIV infection. In cases of *hypersplenism* (spleen is more active, but may not always be increased in size) or *splenomegaly* (spleen is increased in size, and may or not be more active), the spleen may trap and sequester a significant number of platelets, leading to a decrease in circulating platelet count. Fluid replacement, for example when there is blood loss during surgery can result in *dilutional thrombocytopenia*, in which the overall platelet count decreases due to an increase in blood volume without a proportional increase in platelet numbers.

Defects in platelet function rarely cause major spontaneous bleeding, though they may cause blood loss after dental extraction or surgery. Ingestion of aspirin or other non-steroidal anti-inflammatory agents blocks the activity of platelet cyclooxygenase, causing a characteristic transient defect in platelet function. However, this is rarely sufficient to cause bleeding alone, but may exacerbate bleeding from other lesions, e.g. ulceration or malignancy of the gastrointestinal tract. Rare inherited disorders of platelet function include Glanzmann thrombasthenia and Bernard-Soulier syndrome. Mutations in the ITGA2B or ITGB3 gene cause Glanzmann thrombasthenia. These genes encode the two subunits of integrin αIIbβ3. Bernard-Soulier syndrome is caused by mutations in one of three genes – GP1BA, GP1BB or GP9. The proteins encoded by these genes are components of the GPIb/IX/V complex.

Inherited Coagulation Disorders

The best known of the inherited bleeding disorders are *haemophilia A* and *haemophilia B* (Christmas disease), caused by deficiencies in Factor VIII and Factor IX, respectively. The inheritance of both disorders is X-linked. The estimated incidence of haemophilia A is approximately 1 in 5000 to 1 in 10 000 male births, whereas that of haemophilia B is approximately 1 in 25 000 to 1 in 30 000 male births. Both are characterised by painful spontaneous bleeding into joints and muscles, depending on the severity of the deficiency. The concentration of the deficient coagulation factor is typically in the region of 0–5% of normal, corresponding to a range of clinical severity from severe to mild haemophilia. Treatment of active bleeding is by administration of the deficient coagulation factor (Factor VIII or Factor IX), while some patients with severe deficiency may require coagulation factor replacement on a prophylactic basis.

Von Willebrand disease (VWD; see also Case Study CS48) is an inherited autosomal dominant bleeding disorder due to a deficiency of VWF, resulting in reduced Factor VIII coagulant activity and abnormal platelet function. Again, severity varies, and in most cases is mild, with patients experiencing episodic bleeding, particularly after a haemostatic challenge, such as minor surgery or dental extraction. In contrast to haemophilia, bleeding in VWD tends to be superficial, typically from mucous membranes. However, a small number of patients with VWD experience spontaneous bleeding similar to that observed in haemophilia A, requiring treatment with Factor VIII.

Acquired Coagulation Disorders

Acquired coagulation factor disorders are conditions in which there is a temporary or long-term deficiency or dysfunction of specific coagulation factors. They develop in individuals who do not have an inherited predisposition to bleeding disorders but arise instead as a consequence of various underlying medical conditions or external stimuli. A common acquired coagulation factor disorder is liver disease-associated coagulopathy. The liver produces many coagulation factors, so that in liver disease, such as cirrhosis or hepatitis, the synthesis of clotting factors can be impaired. In addition, because of the malabsorption associated with *cholestatic* ('obstructive') liver disease there is often also a reduction in several blood coagulation factors, particularly Factors II, VII, IX and X which depend upon absorbed vitamin K for their synthesis.

Management of acquired coagulopathy in liver disease involves addressing the underlying liver condition and may include supportive measures such as vitamin K supplementation, blood product transfusions or specific clotting factor replacement therapies. Another acquired coagulation disorder is *disseminated intravascular coagulation* (DIC), characterised by systemic activation of coagulation and widespread deposition of *microthrombi* with a consequent depletion in platelets and coagulation factors, and in turn a high risk of spontaneous bleeding. DIC is often triggered by underlying medical conditions such as infection, malignancy or complications during pregnancy.

Thrombotic Disorders

Thrombosis

During the 19th century, Virchow described three factors contributing to thrombosis, which later became known as *Virchow's triad*. They are as follows:

- Damage to the inner surface of the vessel wall.
- Disturbed blood flow.
- Hypercoagulability.

One or more of the components of Virchow's triad contribute to most arterial and venous thromboembolic disorders. However, certain elements predominate in venous and arterial thrombosis. Thus, in general, venous thromboembolism is triggered by changes in blood coagulation factors or their inhibitors, and/or reduced blood flow. In contrast, arterial thrombosis is usually associated with damage to the vascular endothelium and/or increased platelet activity. However, these are not mutually exclusive: for example, increased blood fibrinogen levels are linked to an increased risk of coronary artery occlusion, and platelets are certainly implicated in venous thrombosis.

Venous Thromboembolism

An *embolus* is defined as a thrombus (or other forms of solid material, such as fat, or an air

bubble) carried in the circulatory system (see also Chapter 9). *Thromboembolism* is the occlusion of a blood vessel by a thrombus which originated in a vessel elsewhere and which has been carried in the circulatory system from its original site. Venous thrombi form in the deep veins of the legs, typically after surgical operations when there is prolonged bed rest accompanied by reduced blood flow and a post-operative hyper-coagulable state. The fibrinolytic system is activated and gradually dissolves any clots, but fragments may break off to form emboli. These can pass through the vena cava and the right chambers of the heart and lodge in the pulmonary circulation resulting in *pulmonary embolism* (PE). PE results in reduced perfusion of the lungs, varying in severity depending on the site and size of the occluded vessel(s). Acute chest pain and haemoptysis are among the typical clinical features; however, small pulmonary emboli may produce few signs and symptoms, whereas large emboli can result in sudden death.

Unexplained or recurrent venous thrombosis may be due to other causes. *Thrombophilia* is the term used to describe an increased tendency to develop thrombosis. Thrombophilia can be caused by

- Inherited deficiencies of anticoagulant proteins, such as protein C, protein S and antithrombin III.
- The presence of a mutation in Factor V, known as Factor V Leiden. This mutation impairs proper binding of protein C, which normally inhibits the pro-clotting activity of Factor V, leading to a hypercoagulable state. Factor V Leiden is the most common hereditary hypercoagulability disorder amongst ethnic Europeans.
- Anti-phospholipid syndrome, an autoimmune disorder characterised by the presence of anti-phospholipid antibodies. Anti-phospholipid syndrome provokes thrombosis in both arteries and veins and causes pregnancy-related complications that include an increased risk of miscarriage, stillbirth,

pre-term delivery and severe pre-eclampsia. Diagnosis requires one clinical event (i.e. thrombosis or a pregnancy complication) and two positive autoantibody blood test results, at least 3 months apart. The autoantibodies are directed against phospholipids and include anti-cardiolipin antibodies, anti-beta-2 glycoprotein I antibodies and anti-phosphatidylserine/prothrombin antibodies. Patients with this syndrome have the so-called lupus anticoagulant (LA), autoantibodies that interfere with phospholipid-dependent coagulation tests, prolonging clotting times. The name is a misnomer as they are not true anti-coagulants, but rather inhibitors of the coagulation process in laboratory tests. They are also not specific to lupus. The detection of LA is one of the criteria for diagnosing anti-phospholipid syndrome, along with the presence of anti-phospholipid antibodies and a history of thrombosis or pregnancy complications.

- Elevated homocysteine levels are linked to an increased risk of thrombosis through various mechanisms, including increased TF expression, attenuated anticoagulant processes, enhanced platelet reactivity, increased thrombin generation, augmented Factor V activity, impaired fibrinolytic potential and vascular injury/endothelial dysfunction.
- Prothrombin gene mutations, which increase the risk of thrombophilia.
- Other conditions, such as cancer (particularly malignancies involving the pancreas, stomach, lung or ovary), pregnancy, hormone replacement therapy and obesity.

Arterial Thrombosis

Atherosclerosis, which is discussed in detail in Chapter 9, is a significant risk factor for arterial thrombosis. Atheromatous lesions, characterised by the accumulation of fibrofatty plaques within the arterial walls, can activate the haemostatic system leading to the formation of thrombi within the affected vessels. This can result in the obstruction or occlusion of blood flow, depriving the onward tissues of

their necessary oxygen and nutrients. When an atherosclerotic plaque ruptures or erosion occurs, it exposes the underlying components, such as collagen and TF, to circulating blood. This triggers platelet activation and the coagulation cascade, leading to the formation of a thrombus. If the thrombus develops in the coronary arteries, which supply blood to the myocardium, it can lead to myocardial ischaemia. Reduced blood flow and oxygen supply to the heart muscle can result in symptoms such as chest pain (angina) and potentially progress to a more severe condition known as myocardial infarction or in other words, heart attack (Case Study CS40). Myocardial infarction usually occurs when there is complete occlusion of a coronary artery resulting in a prolonged lack of blood flow to the affected area of the heart, leading to irreversible damage to the myocardium.

Clinical Considerations

Warfarin and heparin have been the mainstay of anti-coagulant therapy for many years. *Warfarin* inhibits vitamin K epoxide reductase, which is responsible for the recycling of vitamin K. Without sufficient active vitamin K, clotting factors II, VII, IX, and X have decreased activity. Protein C and protein S are also inhibited, but to a lesser degree. *Heparin* binds to the enzyme inhibitor anti-thrombin III (AT), causing a conformational change that results in its activation through an increase in the flexibility of its reactive site loop. Activated AT then inactivates thrombin, Factor Xa and other proteases.

Unlike traditional anticoagulants, *direct oral anticoagulants* (DOACs) act directly on key factors involved in blood clot formation, offering a more targeted and predictable mode of action. DOACs include factor Xa inhibitors (e.g. apixaban, rivaroxaban and edoxaban) and direct thrombin inhibitors (e.g. dabigatran).

Monitoring the effectiveness of anti-coagulant drugs is important. This is done by measuring the PT and the aPTT. PT is the time it takes for a clot to form in a sample of blood after the addition of specific clotting factors, including TF. It primarily assesses the functionality of the extrinsic pathway of coagulation and is used to monitor the effectiveness of warfarin (which primarily affects the extrinsic pathway). Results are reported as the International Normalised Ratio (INR), which standardises the PT results across different laboratories. The aPTT test measures the time it takes for a clot to form in a sample of blood after the addition of certain clotting factors and an activator substance called partial thromboplastin. aPTT assesses the functionality of the intrinsic pathway and is used to monitor the effectiveness of heparin (which primarily affects the intrinsic pathway).

Summary

Haemostasis is a complex process that involves interactions between the blood vessel wall, platelets and the coagulation and fibrinolytic systems. It functions to prevent excessive bleeding and maintain circulatory integrity. Dysregulated haemostasis can result in bleeding disorders or thrombotic disorders. Bleeding disorders can be caused by platelet defects, inherited coagulation factor deficiencies or defects in the vascular wall. Thrombotic disorders, on the other hand, involve the formation of abnormal blood clots and can be triggered by factors such as damage to the vessel wall, disturbed blood flow and hypercoagulability. Understanding the mechanisms and regulation of haemostasis is crucial for the diagnosis and management of bleeding and thrombotic disorders.

Key Points

1) Activation of haemostatic mechanisms involving blood platelets and the coagulation cascade follows damage to the vascular endothelium. These haemostatic mechanisms prevent excessive blood loss.
2) Deficiencies in haemostatic responses cause bleeding disorders.
3) Overactivity of haemostatic mechanisms increases the risk of arterial and/or venous thromboembolism.
4) Reduced blood flow and venous stasis, together with raised coagulation factor concentrations (particularly postoperatively), or reduced coagulation inhibitors, predispose to venous thromboembolism.
5) In arteries, where endothelial injury is associated with atheromatous deposits, haemostatic activation exacerbates the damage, often resulting in thrombus formation.
6) Therapeutic measures to reduce platelet reactivity reduce the risk of arterial thrombosis. Oral anticoagulant drugs reduce plasma concentrations of vitamin K-dependent coagulation factors thereby reducing the likelihood of further clot formation, particularly after deep vein thrombosis or pulmonary embolism.

Further Reading

Hoffman R., Benz E., Silberstein L., *et al.* (2022) *Hematology, Basic Principles and Practice* (8th edn). 9780323733885; eBook ISBN: 9780323733892.

Key N.S., Makris M. & Lillicrap D. (2017) *Practical Hemostasis and Thrombosis* (3rd edn). 978-1-118-34471-2.

White G.C., Roberts H.R. & Key N.S. (2020) The biology of haemostasis and thrombosis. In: Firth J., Conlon C. & Cox T. (eds) *The Oxford Textbook of Medicine* (6th edn).

9

Vascular Disorders

Introduction

Cardiovascular disorders encompass a wide range of conditions affecting the heart and blood vessels. Prominent among these is atherosclerosis, an inflammatory disease that often clinically manifests as coronary heart disease (CHD), a major cause of disability and death. As well as atherosclerosis, this chapter considers other forms of vascular injury such as aneurysms and vasculitis and should be read alongside the companion chapter on *Disorders of Haemostasis* (Chapter 8). For completeness, we also include here a brief section on the cardiomyopathies. Other important cardiovascular disorders, including myocardial infarction, heart valve disease and peripheral artery disease, are considered in detail within individual case studies.

Atherosclerosis

Arteriosclerosis is a general term used to describe a group of disorders characterised by the thickening, hardening and loss of elasticity of arterial walls. This term is sometimes used interchangeably with *atherosclerosis*, which is a specific type of arteriosclerosis that involves the build-up of plaque in the arteries. These terms should be distinguished from *arteriolosclerosis* which refers to thickening and hardening of small arteries, called arterioles, and which is associated with hypertension, type 2 diabetes and chronic kidney disease. This section is concerned with atherosclerosis.

Atherosclerosis is characterised by a progressive accumulation within the arterial wall, of cholesterol-containing particles, cell fragments and support matrix materials. These lesions

The Biology of Disease, Third Edition. Edited by Paul G. Murray, Simon J. Dunmore, and Shantha Perera.
© 2024 John Wiley & Sons Ltd. Published 2024 by John Wiley & Sons Ltd.
Companion website: www.wiley.com/go/murray/biologyofdisease3e

occur principally in the large- and medium-sized elastic and muscular arteries and can lead to a reduction in the blood supply to the heart, brain or extremities, resulting in *ischaemia* (reduced oxygen supply) in the tissues supplied by the affected arteries. Clinical examples of the effects of ischaemia are *claudication* (pains in the calves and legs when exercising, relieved by rest; Case Study CS38) and *angina* (chest pain). There are two different clinical forms of angina, stable and unstable angina, which are different in terms of the pattern and severity of the chest pain. *Stable angina* typically follows a predictable pattern; chest pain is usually moderate, short-lived (minutes), and consistent in intensity occurring during physical exertion and relieved by rest or medication. Stable angina tends to remain stable over time, with a consistent pattern of symptoms. The frequency and intensity of episodes may increase gradually over several months or years. *Unstable angina*, on the other hand, does not follow a predictable pattern and can occur at rest or with minimal exertion. The pain may occur suddenly or become more frequent, severe or prolonged. Stable angina is a symptom of underlying coronary artery disease, but does not usually indicate an immediate risk of heart attack. In contrast, unstable angina is considered a medical emergency as it indicates an increased risk of a heart attack (Case Study CS40) or other serious cardiovascular events.

A life-threatening complication of atherosclerosis is rupture of the atherosclerotic lesion and if this occurs in the coronary arteries that supply the myocardium, the resulting ischaemia may cause death of an area of myocardial tissue (*myocardial infarction*) producing the clinical effects of a 'heart attack'.

Aetiology and Pathogenesis

The normal human artery (Fig. 9.1) is composed of: the *tunica intima,* an inner smooth, uninterrupted, monolayer of endothelial cells overlying a thin layer of connective tissue matrix scattered within which are a small number of smooth muscle cells, together with collagen and

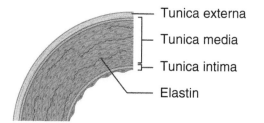

Figure 9.1   Structure of a normal human artery. The main layers of the arterial wall are the intima, media and adventitia, with the internal and external elastic lamina (shown in purple) separating these layers. *Source:* Created with BioRender.com.

glycosaminoglycans (GAG). The intima is bounded by elastic tissue (the internal elastic lamina); the *tunica media*, a thick layer of smooth muscle cells separated by small amounts of elastin, collagen and GAG; and the *tunica adventitia*, an external layer of connective tissue separated from the media by the external elastic lamina. It consists of collagen, fibroblasts and GAG.

The events that lead to atherosclerotic plaque formation are described in Fig. 9.2. Endothelial cell injury is the initiating event, causing increased vascular permeability, adhesion of leucocytes (including monocytes) and platelet activation (see Chapter 8). The release of cytokines from injured endothelial cells and the expression of adhesion molecules on endothelial cells promote the recruitment of monocytes and their migration into the subendothelial tissues. There is accumulation of lipoproteins consisting mainly of low-density lipoprotein (LDL) and its oxidized forms which are phagocytosed by monocytes that have differentiated into macrophages. These macrophages are known as 'foam' cells because of their high lipid content. The lipids form *fatty streaks*, which are the earliest visible signs of atherosclerosis. Further cytokine release attracts smooth muscle cells which migrate from the media into the intima. The smooth muscle cells proliferate and produce extracellular matrix, including collagen. Later, there is calcification and apoptosis of cells within the thickening intima with formation of a necrotic centre surrounded by a fibrous cap. An example of an advanced atheromatous plaque is shown in Fig. 9.3. The thickened atheromatous plaques

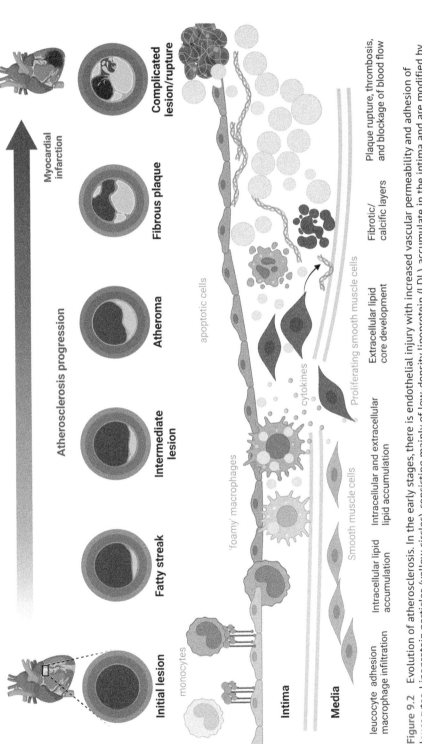

Figure 9.2 Evolution of atherosclerosis. In the early stages, there is endothelial injury with increased vascular permeability and adhesion of leucocytes. Lipoprotein particles (yellow circles), consisting mainly of low-density lipoprotein (LDL), accumulate in the intima and are modified by oxidation and glycation. Oxidative stress induces local cytokine secretion with migration of leucocytes into the intima. Monocytes differentiate into phagocytic macrophages and engulf lipoprotein particles and are described as 'foam' cells. Smooth muscle cells migrate into the intima from the media, proliferate and synthesise extracellular matrix, including collagen. This marks the transition from a fatty streak to a fibro-fatty lesion. In the later stages, calcification occurs, accompanied by cell death yielding a relatively acellular fibrous capsule surrounding a lipid-rich core, known as an atheromatous plaque. The atheromatous plaque may be eroded, or rupture, initiating thrombus formation. *Source:* Created with BioRender.com.

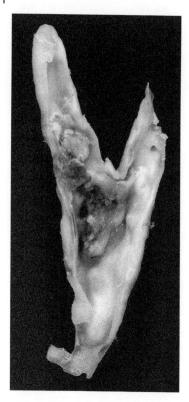

Figure 9.3 Atherosclerotic plaque. From a carotid endarterectomy specimen showing the bifurcation of the common into the internal and external carotid arteries. *Source:* Ed Uthman./Wikimedia Commons / CC BY 2.0.

can cause narrowing of the lumen, leading to complete or partial occlusion of the vessel lumen, referred to as *atherosclerotic stenosis*. Later, plaque erosion or rupture can lead to a catastrophic activation of haemostasis, resulting in thrombus formation. Matrix metalloproteases released from macrophages appear to be involved in plaque destabilisation by digesting collagen fibres. The 'shoulder' of the plaque is the most prone to rupture. Arterial flow may take the contents of the plaque and/or any thrombus to an arteriole distal to the rupture where occlusion can occur, an event known as *embolism*.

Involvement of Lipids in Atherosclerosis

Lipids are transported in the blood in a variety of particles containing greater or lesser amounts of cholesterol and triglyceride. The major particles which transport lipids are chylomicrons

(triglyceride-rich), very low-density lipoproteins (VLDL, triglyceride-rich), LDL (cholesterol-rich) and high-density lipoproteins (HDL, cholesterol-rich) (Fig. 9.4). Not all of these lipids are implicated in the development of atherosclerosis. For example, HDL confers protection against atherosclerosis, in contrast to LDL, the major particle carrying cholesterol in the blood, which is highly atherogenic. LDL can be derived from the metabolism of VLDL, through an intermediary lipoprotein called intermediate-density lipoprotein (IDL). LDL is normally cleared by blood cells and the liver by uptake on specific receptors (apolipoprotein B_{100}). However, LDL can become retained within an arterial wall. Once inside the arterial endothelium, LDL particles are oxidized by reactive oxygen species (ROS). Oxidized LDL (oxLDL) is highly pro-inflammatory, causing the release of chemical mediators, including chemoattractants that further promote the monocyte migration described above. There is then a vicious circle of more uptake of LDL which further increases the inflammation.

Risk Factors for Atherosclerosis

Non-modifiable risk factors are intrinsic characteristics or events of medical relevance that an individual cannot control or change. One key non-modifiable risk factor for atherosclerosis is *age*. With age, blood vessels, especially the endothelial cells, become progressively damaged. Moreover, advancing age often corresponds with an increased likelihood of other risk factors, such as higher cholesterol levels and hypertension (see below), which synergistically elevate the risk of atherosclerosis. *Men* are inherently more prone to atherosclerosis due to hormonal and metabolic differences; lower oestrogen levels in men diminish cardiovascular protection. Accordingly, *postmenopausal status* in women, marked by reduced oestrogen levels, increases their susceptibility. *Family history* of cardiovascular conditions underscores the genetic component; shared genes between family members can elevate LDL cholesterol levels, trigger inflammation, and amplify overall risk. Another example is homozygous defects in the

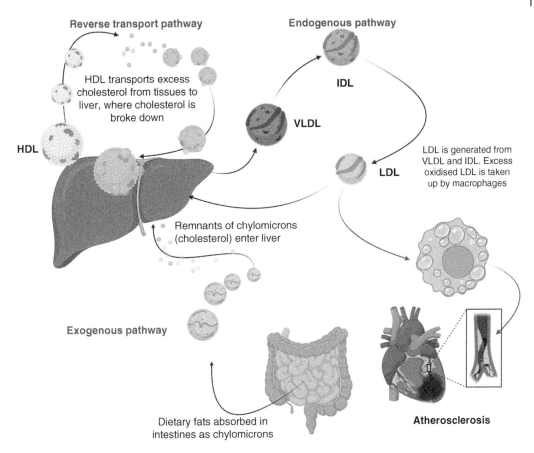

Figure 9.4 Basic pathways of lipoprotein metabolism. Lipoprotein metabolism involves several pathways that regulate the transport and distribution of lipids in the body. The three main pathways are the exogenous pathway, the reverse transport pathway and the endogenous pathway: In the *exogenous pathway*, dietary fats, including triglycerides, phospholipids and cholesterol, are absorbed by the intestines into the bloodstream as chylomicrons. Chylomicrons are large lipoprotein particles formed in the intestinal cells, and they carry dietary lipids to various tissues, particularly adipose tissue and muscle. As chylomicrons circulate, lipoprotein lipase (LPL) on the surface of blood vessel walls breaks down triglycerides within chylomicrons, releasing fatty acids for energy or storage (not shown). The remnants of chylomicrons, which contain mostly cholesterol, are taken up by the liver. In the *reverse transport pathway*, high-density lipoproteins (HDL), often referred to as 'good cholesterol', are synthesised by the liver. HDL transport cholesterol from peripheral tissues back to the liver for excretion through a process called reverse cholesterol transport. This pathway helps maintain cholesterol balance and reduces the risk of cardiovascular disease. The *endogenous pathway* primarily involves very low-density lipoproteins (VLDL) and low-density lipoproteins (LDL). In the liver, VLDL is synthesised and released into the bloodstream, carrying triglycerides and cholesterol. As VLDL particles release triglycerides to peripheral tissues, they become smaller and denser, forming intermediate-density lipoproteins (IDL). Further metabolism of IDL leads to the formation of LDL particles, which are enriched in cholesterol. LDL particles transport cholesterol to cells throughout the body, including arterial walls. Elevated levels of LDL cholesterol are associated with an increased risk of atherosclerosis and cardiovascular disease. *Source:* Created with BioRender.com.

enzymes necessary for homocysteine metabolism (cystathionine-β-synthase and methylene tetrahydrofolate reductase). These individuals develop severe atherosclerosis very early in life and may experience myocardial infarction by the age of 20. Homocysteine is toxic to endothelium, is pro-thrombotic, increases collagen production and decreases the availability of nitric oxide. Plasma homocysteine concentrations are elevated in some people who have no enzymatic defect and these individuals also have an increased risk of atherosclerosis.

Modifiable risk factors for atherosclerosis are often lifestyle factors, such as diet and exercise. One of the most significant is *hypertension*. The risk of developing atherosclerosis is 2–3 times higher in individuals with hypertension compared to those with normal blood pressure. Moreover, the risk of atherosclerosis increases progressively with higher blood pressure, even within the range considered 'normal' or 'pre-hypertensive' This means that even a small increase in blood pressure can contribute to a higher risk of atherosclerosis. Hypertension is considered to be one of the most important factors that initiates injury to the endothelium. In patients with hypertension, levels of angiotensin-II (the principal product of the renin–angiotensin system) are also often increased. Angiotensin II is a potent vasoconstrictor and can also stimulate the growth of intimal smooth muscle cells, which is a feature of atherogenesis. Hypertension is also associated with a number of additional pro-inflammatory activities which increase the concentrations of hydrogen peroxide and free radicals, reduce the formation of nitric oxide and in turn increase peripheral resistance to arterial blood flow.

Smoking is another major modifiable risk factor. Cigarette smoke contains toxins that directly damage endothelial cells. Smoking also causes vasoconstriction (as a direct result of the action of nicotine), reduces the metabolism of VLDL and increases production of IDL particles, which are important in the development of atherosclerosis because they are not taken up by the classical apolipoprotein B_{100} receptor-mediated pathway. Smoking has also been shown to reduce antioxidant protection in plasma, which may exacerbate lipoprotein oxidation. Plasma fibrinogen levels are also raised in smokers, increasing the likelihood of thrombosis.

The elevated levels of blood glucose in patients with *diabetes* (see also Chapter 13) cause increased glycation and oxidation of LDL. LDL particles are smaller and denser in diabetic individuals meaning they are more prone to both oxidation and uptake into atherosclerotic lesions. Insulin resistance is also associated with dyslipidaemia and hypertension, both of which exacerbate atherosclerosis risk.

Obesity (see also Chapter 12) is associated with multiple risk factors for atherosclerosis, including dyslipidaemia and hypertension. Adipose tissue also produces pro-inflammatory factors that promote the progression of atherosclerosis.

Infectious agents are also associated with the development of atherosclerosis, including: *Chlamydia pneumoniae*, a respiratory pathogen which has been detected within atherosclerotic plaques; *Porphyromonas gingivalis*, a pathogen usually involved in chronic periodontal disease; *Helicobacter pylori*, Cytomegalovirus (CMV), Herpes simplex viruses (HSV)-1 and -2, HIV, and others.

Metabolic Syndrome

Metabolic syndrome is a cluster of interrelated metabolic abnormalities that increase the risk of developing cardiovascular disease and type 2 diabetes. It is defined clinically by the presence of at least three of the following:

- *Abdominal obesity:* Excessive fat accumulation around the waist, with a waist circumference of >102 cm for men or >88 cm for women.
- *Elevated triglycerides:* High levels of triglycerides in the blood, defined as ≥1.7 mmol/L or being on medication for elevated triglycerides.
- *Reduced HDL cholesterol:* Low levels of HDL cholesterol, defined as <40 mg/dL (1.03 mmol/L) for men or <50 mg/dL (1.29 mmol/L) for women, or being on medication for low HDL cholesterol.
- *Hypertension:* Systolic blood pressure ≥130 mmHg, diastolic blood pressure ≥85 mmHg, or being on medication for hypertension.
- *Elevated fasting glucose:* Fasting blood glucose levels ≥100 mg/dL (5.6 mmol/L) or being on medication for elevated glucose levels (diabetes).

Aneurysms

Aneurysm is an important complication of atherosclerosis. An *aneurysm* is defined as a localised abnormal distension of a blood vessel.

The so-called *true aneurysms* involve all arterial vessel layers (i.e. intima, media, and adventitia) and are typical of aneurysms arising in atherosclerotic vessels and include ventricular aneurysms that follow transmural myocardial infarctions (myocardial infarctions that affect the full thickness of the myocardium). *False aneurysms,* also called *pseudoaneurysms,* occur when blood leaks out of a vessel but is trapped by extravascular connective tissues. False aneurysms are often caused by trauma or surgery, such as coronary angiography or arterial grafting.

Aneurysms caused by atherosclerosis commonly occur in the abdominal aorta. Most *abdominal aortic aneurysms* are asymptomatic but can rupture or result in embolism or downstream ischaemia. In contrast to abdominal aortic aneurysms, *thoracic aortic aneurysms* are most often associated with hypertension and are often symptomatic with pain in the upper back or chest and difficulty in breathing (due to compression of the aneurysm on the bronchi). *Mycotic abdominal aortic aneurysm* is a rare type of aneurysm that occurs when a blood vessel is infected with bacteria or fungi which weakens the wall.

The risk of aneurysm is increased in any condition that weakens the connective tissues of the arteries, for example as a result of extracellular matrix degradation by matrix metalloproteinases in atherosclerosis, or in inherited collagen/connective disorders (e.g. Ehlers-Danlos syndrome and Marfan syndrome; see Case Study CS50).

Aortic dissection occurs commonly in the ascending aorta and is defined as a tear between the layers of the aorta (see also Case Study CS34). Blood rushes through the tear, causing the inner and middle layers of the aorta to split (dissect). Hypertension leading to medial degeneration of the artery is a major risk factor for aortic dissection. Classic symptoms are intense pain described as a tearing or ripping sensation that spreads to the neck or down the back, severe stomach pain and shortness of breath. Approximately 20% of patients with acute aortic dissection die before reaching the hospital, this rises to around about 25% at 6 hours and 50% at 24 hours.

Vasculitis

Vasculitis is inflammation of the blood vessels, often resulting in ischaemia of the tissues supplied by the involved vessels. There are two major causes of vasculitis – those mediated by immune mechanisms and those caused by direct infection of blood vessels.

Immune-mediated Vasculitis

The two major forms of immune-mediated vasculitis involve either pathogenic immune complex formation/deposition or production of anti-neutrophil cytoplasmic antibodies (ANCA).

Immune complex vasculitis occurs in autoimmune disorders and includes IgA vasculitis (Henoch Schönlein purpura), lupus vasculitis, serum sickness and cutaneous vasculitis syndromes, anti-glomerular basement membrane (GBM) vasculitis, Hepatitis C virus (HCV)-associated cryoglobulinaemic vasculitis and Hepatitis B virus (HBV)-associated vasculitis (see also Chapter 6). In this type of vasculitis, immune complexes are deposited in blood vessels with the subsequent recruitment of leucocytes, mainly neutrophils, which become activated and cause tissue damage. This is well illustrated in the case of HCV-induced cryoglobulinaemic nephropathy. HCV infection of B cells leads to the production of IgM antibodies with rheumatoid factor (RF) activity that binds HCV-IgG immune complexes. These cold-precipitable immune complexes deposit in the subendothelial space and in the mesangium of the kidney, where they activate the classical complement pathway. This leads to the formation of C3a and C5a (Chapter 5) that recruit and activate inflammatory cells leading to the tissue damage.

ANCA are a group of autoantibodies detected in a number of autoimmune disorders, but are particularly associated with systemic vasculitis. There are two types: p-ANCA (perinuclear) and c-ANCA (cytoplasmic). p-ANCA are predominantly directed against neutrophil myeloperoxidase and c-ANCA mostly react against neutrophil proteinase-3. ANCA-positive *vasculitides* (the plural form of vasculitis) include *granulomatosis with polyangiitis* (GPA;

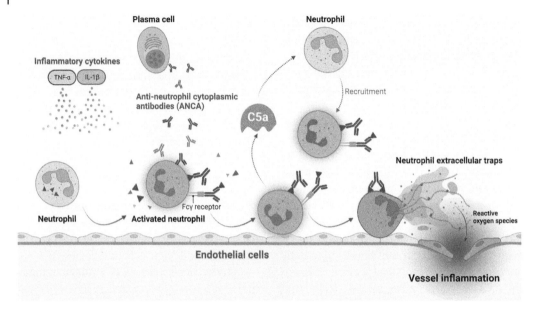

Figure 9.5 Vasculitis mediated by anti-neutrophil cytoplasmic antibodies (ANCA). ANCA-mediated vasculitides are a group of autoimmune diseases that cause inflammation and damage to blood vessels. ANCA are antibodies that target either neutrophil myeloperoxidase (MPO) or neutrophil proteinase-3 (NP3) (shown as red triangles). The immune system produces ANCA in response to an unknown trigger. Inflammatory cytokines prime the neutrophils which then express surface MPO or NP3. ANCA bind to these surface proteins and also the Fcγ receptor on neutrophils, inducing excessive activation of the neutrophils which produce cytokines, reactive oxygen species and lytic enzymes. ANCA-stimulated neutrophils release neutrophil extracellular traps (NETs) that further contribute to the local inflammatory response and endothelial damage. At the same time, more neutrophils are recruited, for example, via the complement C5a subunit (see Chapter 5). *Source:* Created with BioRender.com.

formerly known as Wegener granulomatosis, see also Case Study CS39) and *microscopic polyangiitis*. The pathogenesis of vasculitides caused by ANCA is depicted in Fig. 9.5. Other forms of immune vasculitis, including those affecting larger arteries, are outlined in Fig. 9.6.

Direct Infection

Direct infection of blood vessels is a less common cause of vasculitis and may be due to local infection, e.g. from an abscess, or due to bloodborne systemic infections, e.g. during septicaemia.

Cardiomyopathies

The cardiomyopathies are a group of diseases affecting the heart muscle that make it harder for the heart to pump blood. There are several types:

Dilated Cardiomyopathy

Dilated cardiomyopathy (DCM) is the most common type. In DCM, the ventricles become dilated and the muscle walls thin, meaning they do not contract properly. Symptoms include breathlessness on exertion or at rest, swelling of the legs, ankles and feet, fatigue, arrhythmias and dizziness or fainting. Risk factors include long-term hypertension, heart valve disease and some viral infections. Around 20–35% of DCM cases are familial, though this may be an underestimate given that not all individuals with the genetic variants responsible for DCM will develop the condition (known as *incomplete penetrance*; Chapter 10). In familial cases, DCM is usually inherited in an autosomal dominant manner, meaning that an individual needs only a mutation in one copy of a gene (inherited from either parent) to potentially develop the

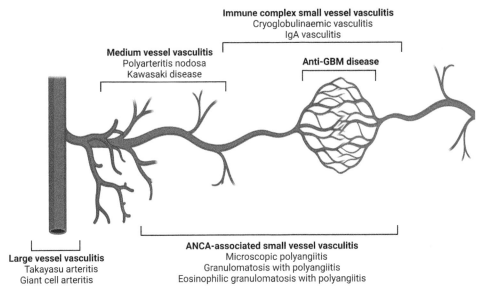

Figure 9.6 Blood vessels affected by the different forms of vasculitis. Large vessel vasculitis affects the large arteries in the body, such as the aorta and its branches. The two most common forms of large vessel vasculitis are giant cell arteritis and Takayasu arteritis. Medium vessel vasculitis affects the medium-sized arteries, such as the renal and mesenteric arteries. The commonest forms of medium vessel vasculitis are Kawasaki disease and polyarteritis nodosa. Small vessel vasculitis affects the small arteries, arterioles and capillaries. This type of vasculitis is further classified into three subtypes based on the presence or absence of specific antibodies: ANCA-associated vasculitis, immune complex small vessel vasculitis and vasculitis associated with anti-GBM antibodies. ANCA-associated vasculitis is the most common form of small vessel vasculitis and includes microscopic polyangiitis, granulomatosis with polyangiitis (formerly known as Wegener granulomatosis) and eosinophilic granulomatosis with polyangiitis (formerly known as Churg-Strauss syndrome). *Source:* Created with BioRender.com.

condition. In these cases, each child of an affected parent has a 50% chance of inheriting the mutation. The most commonly involved proteins are titin (encoded by the TTN gene), lamin A/C (LMNA gene), myosin-binding protein C (MYBPC3 gene) and myosin heavy chain 7 (MYH7 gene). Variants in the TTN gene account for up to 25% of familial cases and 18% of sporadic cases.

Restrictive Cardiomyopathy

Restrictive cardiomyopathy is less common and can occur at any age, but often affects older adults. In this form of cardiomyopathy, the heart muscle in the ventricles becomes rigid, reducing the amount of blood they can hold and leading to reduced blood flow from the heart. Causes include amyloidosis, haemochromatosis and sarcoidosis.

Hypertrophic Cardiomyopathy

Hypertrophic cardiomyopathy (HCM) occurs when the walls of the ventricles (usually the left ventricle) thicken. Despite the thickened walls, the ventricle size often remains normal. However, the thickening may block blood flow out of the ventricle – if blockage is severe, this can cause shortness of breath, chest pain, fainting, and in some cases, sudden cardiac death, which can sometimes be seen in young athletes who die during vigorous physical activity. Many people with HCM, however, have few, if any, symptoms, and can live a normal life.

Arrhythmogenic Right Ventricular Cardiomyopathy

This is a rare type of cardiomyopathy that involves the replacement of myocardial tissue by fibro-fatty tissue, causing the right ventricles to dilate and contract poorly. Over time, the ability of the right ventricles to pump blood deteriorates.

Summary

Atherosclerosis develops over a period of years and can cause gradual arterial occlusion which, when this affects the myocardial blood supply, can present as angina. Alternatively, it may present with more sudden, and potentially catastrophic, myocardial infarction. Endothelial injury, hyperlipidaemia, activation of the blood coagulation cascade and activation of platelets all contribute to its pathogenesis. The most effective strategy to reduce the incidence of CHD is to prevent atherosclerosis. Risk factors for atherosclerosis should be identified in individual patients and treatment plans tailored to reduce risk factors, principally by modification of lifestyle. Lifestyle changes should therefore focus on the following: cessation of smoking, identification and treatment of individuals with hypertension, reduction of the saturated fat content in the diet, identification and treatment of diabetes and modification of haemostatic mechanisms in patients with a high risk of thrombosis. Atherosclerosis and hypertension are major risk factors for different forms of aneurysm. Vasculitis is inflammation of blood vessels. In most cases, vasculitis is caused by an abnormal immune response that mediates the damage to blood vessels.

Key Points

1) Coronary heart disease is a disorder that causes substantial global morbidity and mortality. The major pathological process associated with coronary heart disease is atherosclerosis. Atherosclerosis is also a major risk factor for the development of some aneurysms.
2) Atherosclerotic plaques develop slowly and are characterised by a gradual accumulation of lipids, connective tissues, macrophages and smooth muscle cells within the arterial wall.
3) Endothelial injury is implicated in the development of atherosclerosis. Free radicals are thought to modify lipoproteins, enhancing their uptake into atherosclerotic lesions.
4) Activation of both the coagulation cascades, and of platelets, contributes to thrombosis.
5) Complications of atherosclerosis include necrosis and fissuring of the fibrous plaque, which may activate haemostatic mechanisms leading to thrombosis. Embolism occurs when a fragment of atheroma or thrombus breaks away and lodges elsewhere.
6) Strategies to prevent atherosclerosis include dietary modification, cessation of smoking and effective treatment of hypertension and diabetes.
7) The vasculitides are a diverse group of entities most of which are the result of immune complex disease or the activation of neutrophils by autoantibodies.

Further Reading

Camm A.J., Lüscher T.F., Maurer G. & Serruys P.W. (eds) (2018) *The ESC Textbook of Cardiovascular Medicine* (3rd edn). Oxford University Press, Oxford.

Lilly L.S. (ed) (2018) *Braunwald's Heart Disease: A Textbook of Cardiovascular Medicine* (11th edn). Elsevier, Philadelphia.

Lilly L.S. (ed) (2021) *Pathophysiology of Heart Disease: A Collaborative Project of Medical Students and Faculty* (7th edn). Wolters Kluwer, Philadelphia.

Part 6

Genetic Disorders

10

Genetic Disorders

Introduction

Over the last few decades, there have been major advances in our understanding of the pathogenesis of genetic diseases. This has been brought about in part by the increased awareness of the role of a genetic component in the pathogenesis of many diseases and partly by the increased application of molecular techniques. Together, these factors have resulted in genetics being one of the fastest-moving fields in medicine today. In broad terms, genetic disorders can be classified into single-gene disorders that usually obey Mendelian inheritance, chromosome abnormalities and complex multigenic disorders. The latter are much more common than the former two groups and are caused by, often poorly understood, interactions between multiple genes and environmental factors, each variant gene conferring a small increase in risk. Complex multigenic disorders include common diseases such as diabetes and atherosclerosis. This chapter focuses on single-gene disorders and chromosome abnormalities. A point worth emphasising here is the difference in scale between the single-gene disorders and chromosome abnormalities. Single-gene mutations (clinically, often referred to as *variants*) may involve as little as a single base substitution ranging up to tens of kilobases of DNA. In contrast, microscopically visible deletions, involve relatively

The Biology of Disease, Third Edition. Edited by Paul G. Murray, Simon J. Dunmore, and Shantha Perera.
© 2024 John Wiley & Sons Ltd. Published 2024 by John Wiley & Sons Ltd.
Companion website: www.wiley.com/go/murray/biologyofdisease3e

large amounts of DNA (the smallest chromo-somal deletion, visible by light microscopy of suitably prepared and stained chromosomes, will remove a minimum of 5 Mb of DNA). In contrast, the gain or loss of entire chromo-somes involves huge amounts of genetic mate-rial, including thousands of genes.

Types of Variants

A mutation is a change in the genetic make-up (*genotype*) of a cell. If mutations occur in the cells that form the *gametes* (the germ cells), then the mutation may be transmitted to the offspring where it will be present in every cell of an affected individual. These heritable mutations may arise as *de novo* (new) muta-tions in the germ cells of one or both parents or may be inherited by the parents from previous generations. Alternatively, mutations occur-ring in other body cells, termed *somatic muta-tions*, will not be inherited by subsequent generations but will result in genetic differ-ences between different cells of the same organism. Somatic mutations may have conse-quences for the individual, since some contrib-ute to the development of cancer.

Mutations range in size from single base changes to rearrangements or deletions involv-ing large sections of whole chromosomes or loss of gain of whole chromosomes. Mutations involving single genes are usually the result of either the substitution of a single base with another (known as a *point mutation*) or the deletion or insertion of bases. These mutations may arise within the regions of a gene which code for protein (*exons*), within the non-coding regions (*introns*) or within nearby control ele-ments, and are discussed below.

Single-gene Disorders

Mutations arising in the exons of a gene can affect the nature of the protein product (Fig. 10.1). For example, the insertion or

deletion of one or two base pairs will lead to alterations in the reading frame of the DNA. Mutations arising in this way are known as *frameshift mutations*. A point mutation may alter the code in a triplet of bases (*codon*) and lead to the replacement of one amino acid by another in the protein product. This is referred to as a *missense mutation* and if it occurs in a critical region can dramatically change the function of the encoded protein.

Point mutations and frameshift mutations can also introduce premature stop codons into a messenger RNA (mRNA) sequence, creating a *nonsense* mRNA. This can trigger *nonsense-mediated decay* (NMD), which results in the degradation of the abnormal mRNA. NMD serves as a crucial quality control mechanism in eukaryotic cells, helping to ensure that only mRNAs with intact coding sequences are trans-lated into functional proteins. In this way, NMD helps maintain the integrity of the cellular pro-teome by eliminating potentially harmful or non-functional protein products.

Mutations within gene control regions can lead to a marked reduction or total lack of tran-scription, as seen in several forms of haemo-lytic anaemia. Point mutations within introns can result in defective splicing of intervening sequences. This in turn affects the normal pro-cessing of mRNA and hence formation of the mature mRNA necessary for protein synthesis.

Patterns of Inheritance

The majority of single-gene disorders show clear patterns of Mendelian inheritance. Diseases can either be *X-linked* (i.e. the affected gene is on the X chromosome) or *autosomal* (the affected gene is on one of chromosomes 1–22). In addition, they may be *dominant*, requiring a mutation in only one copy of the gene, or *recessive*, where mutations in both cop-ies are needed for expression of the phenotype. Examples of these different patterns of inherit-ance are shown in Fig. 10.2. Some single-gene disorders do not follow Mendelian inheritance.

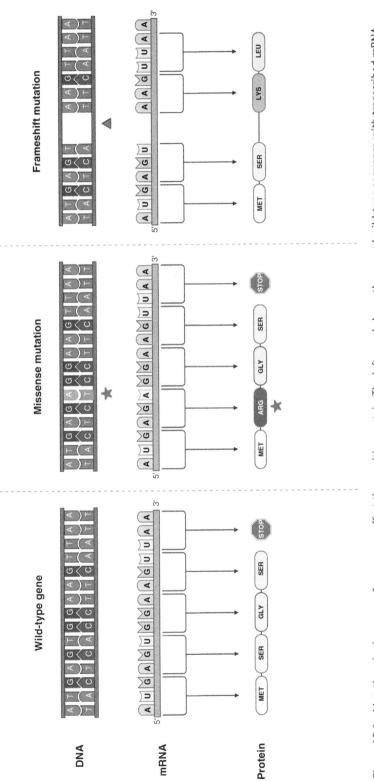

Figure 10.1 Mutations in the exons of a gene can affect the resulting protein. The left panel shows the normal wild-type sequence with transcribed mRNA and translated amino acids. Middle panel: a point mutation leads to a change in an amino acid. This has caused different amino acids to be made downstream of the deletion. Insertion of bases can have a similar effect. Additionally, point mutations and frameshift mutations may result in the creation of premature stop codons in an mRNA sequence, causing the production of a truncated protein, also known as a nonsense mRNA (not shown). *Source:* Created with BioRender.com.

They include diseases caused by trinucleotide repeat mutations and disease caused by mutations in mitochondrial genes, both of which are discussed later. In addition, rarely, autosomal dominant disorders may not obey the laws of Mendelian inheritance. For example, in some cases of *osteogenesis imperfecta*, both parents may be phenotypically normal but have more than one affected child. This can be the result of *gonadal (germline) mosaicism* in which mutations selectively occur postzygotically in germ cells of the gonads. Some cases inherited by gonadal mosaicism are often mistakenly thought to be caused by *de novo* mutations.

(A)

Autosomal dominant inheritance

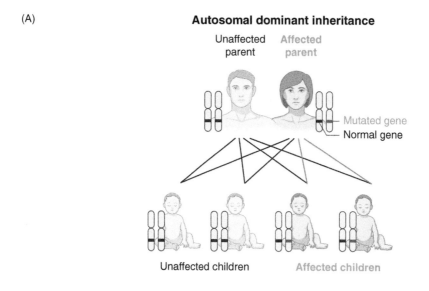

Autosomal recessive inheritance

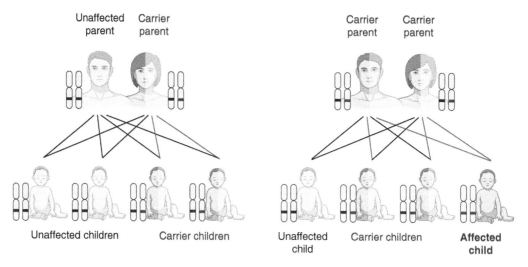

Figure 10.2   (A) Upper panel: Autosomal dominant inheritance. Lower panel: Autosomal recessive inheritance. *Source:* Created with BioRender.com. (B, overleaf) Upper panel: X-linked dominant inheritance. Lower panel: X-linked recessive inheritance. *Source:* Created with BioRender.com.

(B)

X-linked dominant inheritance

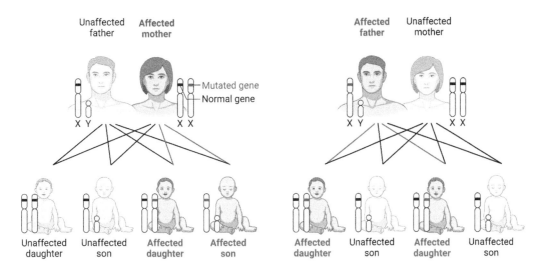

X-linked recessive inheritance

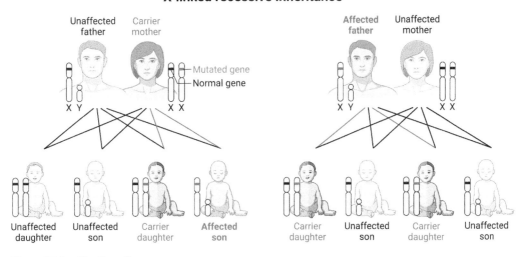

Figure 10.2 (Continued)

Molecular Basis of Mendelian Single-gene Disorders

For many classical Mendelian single-gene disorders, single base changes are a common disease-causing mechanism. For some diseases, all affected individuals carry the same pathogenic variant, whereas for other disorders, the nature of the variant may vary between affected individuals. Moreover, even if individuals share the same pathogenic variant, they may be phenotypically different, with some people, even within the same family, affected more or less severely than others. This is referred to as *variable expressivity*. *Penetrance* is a related term and is the proportion of individuals carrying the pathogenic variant who develop the disease phenotype.

An example of a disease in which all affected individuals have the same single point mutation is sickle cell anaemia (see also Chapter 7 and Case Study CS47). This autosomal recessive disease has a higher incidence among certain ethnic groups, for example, people of African descent, including African-Americans, and carries significant morbidity and mortality. It is caused by a point mutation in codon 6 of the β-globin chain that causes the substitution of L-glutamic acid by valine. As soon as haemoglobin formed from mutant β-globin becomes deoxygenated, it aggregates and forms crystalline deposits causing the red cells to assume their characteristic sickle shape. Repeated sickling damages the cell membrane and shortens red cell lifespan, leading to haemolytic anaemia, blockage of the microcirculation and infarction.

Other diseases have a more complicated pattern of point mutations. One such disease is a form of inherited colon cancer called familial adenomatous polyposis (FAP; see also Chapter 11 and Case Study CS41). In this autosomal dominant disease, affected individuals develop colorectal polyps, one or more of which usually becomes malignant by the third or fourth decade. When the gene for this disease (the APC gene) was identified, it became apparent that the majority of patients had different mutations and that most of these were either point mutations or very small insertions or deletions of a few bases. The mutations are scattered throughout the gene and over 95% of them result in the introduction of a premature stop codon which produces a truncated protein. In FAP, a mutation in only one allele of the FAP gene may be sufficient to cause disease. This is because the normal protein produced by the unaffected allele is inactivated by its interaction with the abnormal protein (referred to as a '*dominant negative effect*'). As the phenotype of patients with FAP is variable, e.g. with regard to the age of onset, it was initially believed that this would be explained by the position and nature of the mutations. On the whole, this has not been the case, although there are some exceptions. An attenuated form of the disease with fewer polyps and a delayed age of onset is caused by mutations in exon 3. The protein made from this mutated gene is very short and rapidly degraded. Only low levels of normal-sized protein (produced by the unaffected allele) are present and the subsequent effect is a milder phenotype.

Other single-gene disorders can be caused by *duplications* or *deletions*. Thus, large deletions are a major cause of *Duchenne muscular dystrophy* (DMD). This is an X-linked recessive condition seen primarily in boys which causes muscle degeneration and eventually death. 50–60% of affected boys show deletions of one or more exons of the very large dystrophin gene. These deletions occur with particularly high frequency at certain points ('hot spots') throughout the gene. The deletions cause an alteration in the reading frame of the dystrophin gene so that a premature stop codon is introduced and hence a truncated and unstable protein is produced. If muscle tissue from affected boys is examined, it can be seen that the dystrophin protein is virtually absent. In contrast, *Becker muscular dystrophy* (BMD) is a milder disease that is also caused by abnormalities in the dystrophin gene. In contrast, in BMD, deletions in the dystrophin gene occur in triplets that leave the reading frame intact. Therefore, although part of the protein is missing, it is still able to function at a lower level and the phenotype of the disease is consequently milder than that of DMD.

Single-gene disorders can affect various cellular processes, including enzymes that regulate metabolic pathways, structural proteins and receptors or genes involved in signalling pathways, among others. For example, mutations in genes encoding lysosomal enzymes can give rise to the so-called *lysosomal storage disorders*. These arise when an enzyme which normally breaks down its substrate is deficient or absent (leading to its reduced or complete loss of function) owing to mutation. This leads to the accumulation of the substrate within lysosomes. Lysosomal storage disorders include *Gaucher disease* (mutations in the GBA gene which encodes glucocerebrosidase, leading to the accumulation of glucocerebrosides), *Tay-Sachs disease* (mutations in HEXA which encodes β-hexosaminidase-A resulting in excess of GM2 gangliosides), *Niemann-Pick disease* (mutations in the SMPD1 gene encoding sphingomyelinase, leading to the accumulation of sphingomyelin) and *Hurler syndrome* (caused by mutations in the IDUA gene which encodes alpha-L-iduronidase, responsible for breaking down glycosaminoglycans).

Mutations in genes encoding structural proteins can lead to abnormalities in connective tissues, for example, *Marfan syndrome* (caused by a mutation in FBN1, one of the genes that makes fibrillin; see also Case Study CS50), *Ehlers-Danlos syndrome* (of which there are many types, commonly caused by genes that encode collagen subunits, e.g. COL5A1 and COL5A2) and *osteogenesis imperfecta* (AKA 'brittle bone' disease, caused by mutations in the COL1A1 or COL1A2 genes).

Inherited disorders that affect signal transduction include *familial hypercholesterolaemia*, which is commonly caused by mutations in the LDLR gene, which encodes the low-density lipoprotein receptor. Mutations in LDLR impair the clearance of LDL cholesterol from the bloodstream. LDL cholesterol is involved in the development of atherosclerosis (Chapter 9). In addition to LDLR mutations, mutations in the PCSK9 gene are another common cause of familial hypercholesterolaemia. PCSK9 encodes an enzyme that regulates LDL receptor degradation.

Other signalling pathway disorders include the RASopathies, caused by mutations in genes within the Ras-MAPK pathway. For example, *neurofibromatosis type 1,* which is characterised by the development of neurofibromas, is caused by mutations in the NF1 gene (see also Chapter 11). Another RASopathy is *Noonan syndrome* caused by mutations in PTPN11, SOS1, RAF1 and others and characterised by distinctive facial features, short stature and developmental delays.

Cystic fibrosis (CF) is one of the most common autosomal recessive conditions (see also Case Study CS43). It is characterised by abnormally thick mucus secretions in the lungs and by pancreatic enzyme insufficiency. The observation of elevated sweat electrolytes suggested that the causative abnormality might be a defect in the regulation of chloride ion conductance. When the *CF* gene was isolated in 1989, it was shown to encode a transmembrane protein which functions as a chloride ion channel. The gene is now referred to as the *cystic fibrosis transmembrane conductance regulator* (gene symbol is *CFTR*). Many different CFTR mutations have so far been identified, including nonsense mutations, frameshift mutations and splice site mutations. The majority of mutations occur at low frequency, but one, known as delta-F508, has been found in approximately 75% of CF cases in the United Kingdom. This mutation is a 3-base-pair (bp) deletion in exon 10 which causes the deletion of the amino acid phenylalanine from the adenosine triphosphate (ATP)-binding domain of the protein (Fig. 10.3). There are different classes of CFTR mutations (Fig. 10.4). Class I mutations result in severely reduced or absent CFTR expression. Class II mutations lead to misfolding, premature degradation and impaired protein biogenesis, reducing CFTR levels on the cell surface. Class III mutations impair the regulation of the CFTR channel, resulting in abnormal gating. Class IV mutations alter channel conductance. Class V mutations alter protein abundance by introducing promoter or splicing abnormalities. Class VI mutations destabilise the channel in post-ER compartments and/or at the plasma membrane and in Class VII mutations there is an absence of mRNA. The net effect of the loss of functional CFTR in secretory epithelial cells is reduced chloride ion and bicarbonate secretion, with increased sodium ion absorption

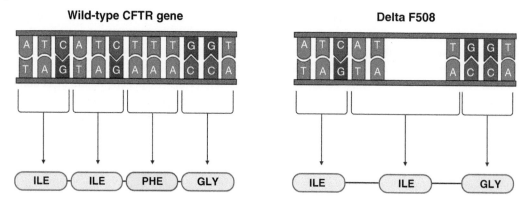

Figure 10.3 The delta-F508 mutation in the CFTR gene. This mutation is present in around 75% of cases of cystic fibrosis and is a deletion of 3 base pairs in exon 10, which results in the removal of a single amino acid, phenylalanine. 'Delta' refers to a deletion, 'F' is the symbol for phenylalanine and '508' is the number of the affected codon. *Source:* Created with BioRender.com.

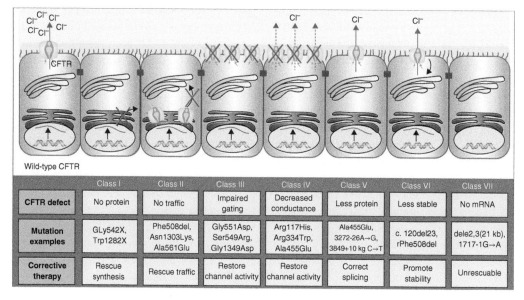

	Class I	Class II	Class III	Class IV	Class V	Class VI	Class VII
CFTR defect	No protein	No traffic	Impaired gating	Decreased conductance	Less protein	Less stable	No mRNA
Mutation examples	GLy542X, Trp1282X	Phe508del, Asn1303Lys, Ala561Glu	Gly551Asp, Ser549Arg, Gly1349Asp	Arg117His, Arg334Trp, Ala455Glu	Ala455Glu, 3272-26A→G, 3849+10 kg C→T	c. 120del23, rPhe508del	dele2,3(21 kb), 1717-1G→A
Corrective therapy	Rescue synthesis	Rescue traffic	Restore channel activity	Restore channel activity	Correct splicing	Promote stability	Unrescuable

Figure 10.4 Classes of CFTR mutations and their respective therapeutic strategies. CFTR mutations are grouped into seven functional classes, with the expectation that the same type of modulator will be applicable to all the defects in one class. Class VII mutations are not expected to be rescuable by a modulator. A therapeutic strategy for Class VII mutations could be stimulation of alternative chloride channels, gene therapy, gene editing or the addition of messenger ribonucleic acid. rPhe508del = rescued Phe508del. *Source:* Kris De Boeck 2020 /John Wiley & Sons.

and mucus secretion. In turn, there is impaired mucociliary clearance, causing recurrent infections, inflammation and lung damage, which is primarily responsible for morbidity and mortality in CF patients. New drugs targeting CFTR defects have revolutionised the treatment of CF, leading to improved chloride ion transport and reduced mucus buildup in the airways (see Case Study CS18), ultimately enhancing the respiratory function and quality of life of affected individuals.

Charcot-Marie-Tooth (CMT) syndrome is the most common inherited disorder of the peripheral nervous system affecting 1 in 2500 people. It is characterised by wasting of the distal limb muscles. CMT is subdivided into a number of types based on pathology and electrophysiology, the most common of which is type 1A. The gene for the disease has been mapped to chromosome 17 and a surprising mechanism of mutation was identified – a large duplication of DNA that includes the peripheral myelin protein-22 (PMP22) gene. The CMT1A phenotype is also seen if point mutations occur in this gene – yet another example of the same disease occurring via a number of different mechanisms. Deletion within the same region of PMP22 causes hereditary neuropathy with liability to pressure palsy (HNPP), a peripheral neuropathy.

Disease-causing 'Silent' Mutations

Silent mutations are changes in DNA that do not alter the amino acid sequence of the protein. Conventionally, silent mutations were considered to be inconsequential since they do not affect protein composition. However, silent mutations can disrupt various stages of protein production, including transcription, RNA splicing and mRNA folding. For instance, mutations in splicing enhancer motifs have been shown to interfere with intron removal and cause diseases such as CF.

Non-Mendelian Single-gene Disorders

Trinucleotide Repeat Disorders

Trinucleotide repeat disorders are typified by an increasing severity of the disease with successive generations (termed *anticipation*) and a parental bias in the transmission of the severe forms of the disorder. However, there is variation in the way in which these repeats function at the DNA or mRNA level.

The first of these diseases to be recognised was *fragile X syndrome (FXS)*, the most common inherited form of intellectual disability, affecting 1 in 7000 males. The pattern of inheritance in FXS is unusual for an X-linked dominant disease in that there are both carrier males and carrier females. In addition, a proportion of carrier females are affected and they always inherit the disease from their mothers. The gene responsible is *FMR1*. The 5-prime untranslated region of this gene contains a repeat of the trinucleotide, CGG. In normal individuals, the number of these repeats varies between 6 and 55. Carriers of the repeat have between 55 and 200 copies (the so-called *premutation* for fragile X), whereas affected individuals have over 200 copies. The clinical phenotype is only seen if there are over 200 copies. A male carrier will transmit the premutation to all his daughters. The number of CGG repeats varies only slightly during this process but remains within the premutation range so that all of his daughters are carriers of the disease and are clinically unaffected by fragile X. It is only when the premutation is transmitted by a female that there is a significant increase in the number of repeats leading to expression of the full mutation. Thus, if a carrier female transmits the mutation to her son, the number of repeats will increase significantly to the full mutation range and he may be clinically affected. The premutation can also be transmitted to a daughter with a similar increase in repeat number but in this case only about 50% of females who inherit the full mutation will have significant developmental delay. At the molecular level, expansion of the CGG repeat above 200 copies results in abnormal methylation of DNA sequences at the 5-prime end of the gene and complete shutdown of *FMR1* transcription. It should be noted here that although premutation does not lead to FXS, it is associated with two related disorders. Thus, around one-fifth of all female carriers have premature ovarian failure, known as *Fragile X-associated primary ovarian insufficiency (FXPOI)*. Moreover, around one-half of all male premutation carriers exhibit progressive neurodegeneration which begins in the sixth decade, and is known as *fragile X–associated tremor/ataxia syndrome (FXTAS)*. In both conditions, the abnormalities are caused not by the absence of FMR1 expression – the FMR1 gene is not methylated and continues to be expressed, but by the overproduction of FMR1 mRNA (which contains the expanded repeats). These abnormal mRNAs are described as 'toxic' because they bind to RNA-binding proteins and disturb their normal function.

Other diseases involving trinucleotide repeat expansion have been identified, including *myotonic dystrophy* (abbreviated to DM, which stands for *dystrophia myotonia*, the Latin name for the disorder), a progressive wasting disease of the muscles. There are two major forms: DM1 and DM2. DM1 is caused by a CTG expansion in the 3-prime untranslated region of the dystrophia myotonica protein kinase (DMPK) gene, while DM2 is caused by a CCTG expansion within intron 1 of the cellular nucleic-acid-binding protein (CNBP). DM1 can be further classified as mild DM1, classic DM1 and congenital DM1. Mild DM1 is characterised by cataracts and sustained muscle contractions (*myotonia*). Classic DM1 is characterised by muscle weakness and atrophy, myotonia, early-onset cataracts and abnormalities in cardiac conduction. Congenital DM1 is characterised by muscle weakness (*hypotonia*), difficulty breathing, intellectual disability and early death. DM2 causes similar symptoms to DM1, but is usually less severe and not associated with congenital disease. As with FXS, the greater the number of repeats, the more severe the disease, and disease increases in severity with successive generations. As with FXPOI and FXTAS, the abnormal mRNAs produced are 'toxic' because they sequester RNA-binding proteins, causing dysregulated RNA splicing which is toxic to muscle cells.

Huntington disease is one of the most serious genetic diseases affecting the CNS. An unstable CAG repeat sequence in exon 1 of the HTT gene is responsible. As with FXS and DM, there is a variation in repeat number in the normal population and an unstable expansion of the repeat number in affected individuals. There is also 'anticipation' – those with early onset disease have the highest number of repeats and a sex bias with juvenile onset disease seen with transmission from the father. The precise effect of the trinucleotide repeat remains unclear, although it is likely a gain-in-function of the abnormal Huntingtin protein, rather than a shutdown in transcription of the gene.

Diseases Caused by Mutations in Mitochondrial Genes

Mitochondrial DNA (mtDNA) encodes 13 proteins which are enzymes involved in oxidative phosphorylation. Mitochondria are normally inherited only from the mother. This is thought to be due to the fact that mitochondria in sperm are usually destroyed by the egg cell after fertilisation or are lost in the sperm tail during fertilisation. Mutations in mtDNA have deleterious effects, mostly in organs which rely on oxidative phosphorylation for normal function, for example the heart and nervous system. Thus, they can cause different diseases, including: *mitochondrial myopathy, encephalopathy, lactic acidosis and stroke-like episodes* (MELAS); *myoclonic epilepsy associated with ragged red fibres* (MERRF) and *chronic progressive external ophthalmoplegia syndrome/progressive external ophthalmoplegia* (CPEO/PEO).

Chromosome Abnormalities

Chromosome abnormalities may be broadly subdivided into two types: constitutional and acquired. *Constitutional chromosome abnormalities* are those that are either inherited from one or other parent, or arise as *de novo* mutations during gamete development (*gametogenesis*) or during early embryogenesis. The chromosome defect is found either in all cells of an affected individual or, less commonly, in a proportion of cells (*mosaicism*). Since the abnormality is present in germ cells, it can be transmitted to the next generation. *Acquired chromosome abnormalities* may arise either during foetal growth or later in cells retaining the ability to undergo cell division. Such acquired chromosomal mutations are confined to the cell in which they arise and its progeny. Non-random acquired chromosome abnormalities contribute to the development of many types of cancer (Chapter 11). Mitotic errors, resulting in the gain or loss of single chromosomes at cell division (*aneuploidy*; Chapter 11), also occur at a very low rate in all individuals. Most of these errors are lethal for the cell progeny but a few are benign and may accumulate with age. Examples are the loss of the Y chromosome which occurs in a proportion of lymphocytes in older men and the addition of an extra X chromosome which occurs in a proportion of lymphocytes in older women. Some studies have suggested a potential association between loss of the Y chromosome and an increased risk of various cancers, including leukaemia.

The Structure of Chromosomes

Chromosomes are complex macromolecular structures which are a fundamental unit of genome organisation. In order to appreciate the potential phenotypic effect of chromosome abnormalities, it is helpful to have some idea of the relationship between the size of a chromosome and the DNA base-pair (bp) sequence from which genes are transcribed. Human somatic cells each contain about 6000 million base pairs of DNA organised into 46 chromosomes (known as the *diploid* set) consisting of 22 pairs of autosomes and two sex chromosomes (XX in females and XY in males). There are around 20 000–25 000 genes in the human genome. Chromosomes are highly organised structures consisting of a single, linear, double-stranded DNA molecule which interacts with various DNA-binding proteins in a complex manner. The overall packing ratio from DNA strand to a chromosome at metaphase is about

8000:1. Chromosome 1 is the largest and contains approximately 300 megabases (i.e. 300 million base pairs) of DNA, whereas chromosome 21 contains only 50 megabases of DNA.

Types of Chromosome Abnormality

Chromosome abnormalities involve either a change in chromosome number or a change in chromosome structure.

Numerical Abnormalities

Euploidy refers to a state in which the cell has a normal, balanced set of chromosomes. In other words, there is a complete set of chromosomes that is either the same as the *haploid* number (n) or a multiple of it. In humans, euploidy typically refers to having a diploid (2n) set of chromosomes, which means having two complete sets of chromosomes, one inherited from each parent. Sometimes there is a gain of entire haploid sets of chromosomes resulting in *triploidy* (n = 69) or *tetraploidy* (n = 92).

Aneuploidy refers to an abnormal number of chromosomes – in other words, in most cases a chromosome number that is not an exact multiple of the haploid number. Aneuploidy can be caused by *non-disjunction*. This occurs when there is a failure of homologous chromosomes or sister chromatids to separate correctly during cell division, which can occur in both meiosis and mitosis. Non-disjunction leads to *monosomy* in one daughter cell and *trisomy* in the other. Thus, aneuploidy may take the form of an extra chromosome, known as trisomy (three copies present) or loss of a chromosome, referred to as monosomy (one copy present). Another cause of aneuploidy is *anaphase lag*. This is when one set of sister chromatids lag behind during anaphase and do not get incorporated into one of the daughter cells, the other pair moves normally and is incorporated into the other daughter cell. Thus, anaphase lag will cause one daughter cell to lack one paired set of chromosomes, creating a form of monosomy, while the other daughter cell receives a complete set of chromosomes and is normal. Monosomy or trisomy involving the sex chromosomes is compatible with life (the exception being monosomy Y, i.e. one Y chromosome and no X chromosome) and is usually associated with varying degrees of phenotypic abnormalities. This is because X inactivation normally ensures that only one X chromosome is actively transcribed in each cell and the Y chromosome contains a large amount of inert, highly repetitive DNA. Clinically, the best-defined sex chromosome abnormalities are Klinefelter syndrome and Turner syndrome.

Klinefelter syndrome (frequency 1:500 to 1:1000) occurs when two X chromosomes and one Y chromosome are present (47,XXY, Fig. 10.5). The clinical features include tall stature, hypogonadism (small reproductive organs) and gynaecomastia (enlargement of male breasts). Individuals are invariably infertile. Some individuals with Klinefelter syndrome may experience educational difficulties. However, many do not and lead successful and fulfilling lives. The data here must be treated with caution, as there is likely to be significant ascertainment bias which may incorrectly identify more men with intellectual difficulties (only those more severely affected will see medical attention).

Turner syndrome (frequency 1 in 10 000 females) results from complete or partial monosomy of the X chromosome and is characterised by short stature and infertility in affected females. Intelligence is usually within normal limits. In just over one half of cases, there is complete monosomy of the X chromosome (i.e. 45,X). The remainder of patients show either mosaicism or only partial monosomy, involving various structural rearrangements (e.g. deletions, rings or isochromosomes) of

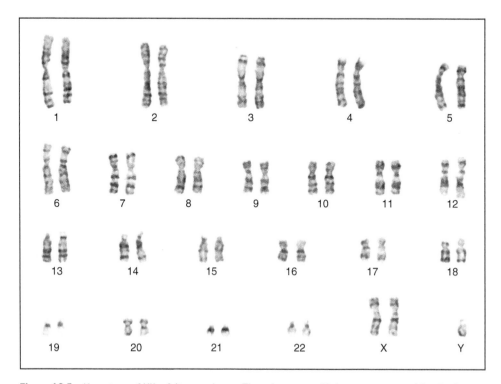

Figure 10.5 Karyotype of Klinefelter syndrome. There is an extra X chromosome, resulting in the genotype of 47,XXY, instead of the typical male genotype of 46,XY. *Source:* Nami-ja / Wikimedia Commons / Public Domain.

one X chromosome. Many live-born individuals with Turner syndrome may exhibit genetic mosaicism, meaning they have a mixture of cell lines with different chromosome compositions. This mosaicism is thought to contribute to their survival, as some cells may carry the typical 45,X chromosome arrangement, while others have a variation involving a normal cell line or a structural rearrangement of the X chromosome, sometimes without the complete 45,X pattern.

45,X/46,XY mosaicism, also known as *mixed gonadal dysgenesis*, is a chromosomal condition characterised by a mosaic pattern of sex chromosomes in affected individuals. Typically, individuals have a combination of two cell lines: one with a single X chromosome (45,X) and another with both an X and a Y chromosome (46,XY). This mosaic pattern can result in a wide spectrum of phenotypic variations, encompassing male, female or intersex presentations. The presence of the Y chromosome material in individuals with 45,X/46,XY mosaicism can influence gonadal development, potentially leading to undifferentiated or dysgenetic gonads. This condition is often associated with an increased risk of gonadal tumours.

Trisomy X, also known as *triple X syndrome*, is a chromosomal disorder in which females have an extra X chromosome, resulting in a karyotype of 47,XXX. This condition occurs relatively commonly, with a frequency of about 1 in 1000 females. However, it is often underdiagnosed, and fewer than 10% of affected individuals are aware of their condition. Common features include learning disabilities, mild dysmorphic features such as *hypertelorism* (wide-spaced eyes) and *clinodactyly* (incurved little fingers), early menopause and above average height (but note the potential for strong ascertainment bias, as discussed earlier).

In contrast, monosomy or trisomy involving autosomes generally results in the loss or gain of too much important genetic information to be compatible with live birth. However, a number of autosomal trisomies may survive to birth. Trisomy 21 (*Down syndrome*), trisomy 18 (*Edward syndrome*) and trisomy 13 (*Patau syndrome*) are the most common, in descending order of frequency.

Structural Abnormalities

Structural abnormalities (Fig. 10.6) include large deletions or insertions of genetic material, duplication of segments of a chromosome, inversions or translocations. Some of these are discussed briefly below:

Chromosome inversion occurs when a segment of a chromosome is reversed in orientation. It is caused by two breaks in a chromosome with subsequent reinsertion of a piece in the inverted order.

An *isochromosome* is an example of an unbalanced structural abnormality. It arises from a simultaneous duplication and deletion of genetic material, resulting in a chromosome containing two copies of either the long (q) arm or the short (p) arm. As a consequence, there is partial trisomy of the genes present in the isochromosome and partial monosomy of the genes in the arm that is lost.

Ring chromosomes are produced when a break occurs at both ends of a chromosome with fusion of the damaged ends. This can result in the loss of genetic material. Ring chromosomes are unstable during cell division and can form interlocking or even fused rings.

Balanced structural chromosome abnormalities represent a category of chromosomal variations in which there is a rearrangement of genetic material between chromosomes, but the overall amount of genetic material remains the same. In other words, no genetic material is gained or lost in these rearrangements. While individuals with balanced structural chromosome abnormalities typically do not exhibit noticeable health issues or developmental problems themselves, the rearrangements can have significant implications during reproduction. In the context of recurrent miscarriages (Case Study CS45), approximately 1–2% of affected individuals

Autosomal chromosomes

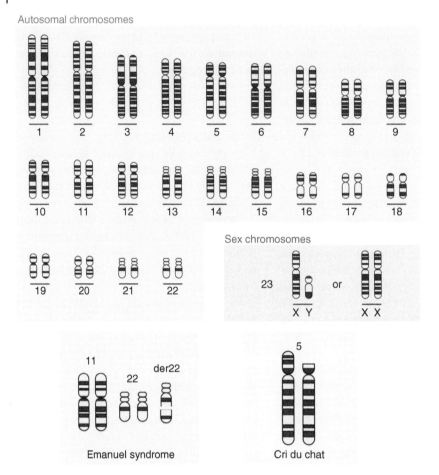

Sex chromosomes

Emanuel syndrome   Cri du chat

Figure 10.6 Examples of structural chromosome abnormalities. The upper panel shows a normal karyotype. The lower left panel shows the chromosome abnormality present in individuals with Emanuel syndrome who have an additional chromosome, which is a derivative chromosome formed by the top portion of chromosome 22 and the bottom portion of chromosome 11 (i.e. a total of 47 chromosomes). The presence of an extra chromosome in individuals with Emanuel syndrome leads to a set of characteristic clinical features that include developmental abnormalities such as a small jaw (*micrognathia*), cleft palate, heart defects, skin tags and genital defects. The lower-right panel shows a deletion involving part of one copy of chromosome 5 in an individual with Cri du chat syndrome. Loss of segments of a chromosome may remove many important genes leading to severe phenotypic abnormalities. *Source:* Created with BioRender.com.

may have a balanced form of a structural chromosome abnormality. When they pass on their genetic material to their offspring, the resulting embryos may inherit unbalanced chromosome arrangements, which can lead to developmental abnormalities, miscarriages or health issues in the offspring. Therefore, even though individuals with balanced structural chromosome abnormalities may not exhibit health problems themselves, these rearrangements can significantly impact their ability to have healthy pregnancies and offspring. Genetic counselling and thorough evaluation are essential for individuals with recurrent miscarriages to identify and understand the presence of balanced structural chromosome abnormalities and their implications for reproductive outcomes. *Robertsonian translocations* are balanced translocations that occur in acrocentric chromosomes (chromosomes 13, 14, 15, 21 and 22). When these chromosomes break at their centromeres, the two resulting long arms may fuse, resulting in a new chromosome. While individuals carrying Robertsonian translocation may have only 45 chromosomes in each of their cells, all essential genetic material is still present. Therefore, carriers are usually normal. However, their offspring may effectively inherit extra chromosomal material and express the Patau or Down phenotype (Fig. 10.7).

However, many structural abnormalities lead to chromosome imbalance as a result of the gain or loss of genetic material, usually resulting in profound physical and intellectual disability. Imbalance can be expressed as a percentage of the ratio of the length of the missing or additional segment over the total length of the haploid autosomal chromosome complement (% HAL). This crude measurement takes no account of gene content. Excess material, if less than 1% of HAL, in the form of partial trisomy may result in a viable pregnancy with frequent live birth. An excess of between 1 and 3% HAL increases the risk of a severely compromised phenotype resulting in

non-viability *in utero* and therefore abortion. Partial monosomy (i.e. a HAL deficit of 1–2%) is much less well tolerated than partial trisomy.

Structural abnormalities occur with high frequency in certain conditions associated with an increased sensitivity to chromosome breakage. These diseases include Fanconi anaemia, Bloom syndrome and ataxia telangiectasia. The type of breakage, which is not generally site-specific, is characteristic for each disease and reflects the underlying DNA repair defect in each case (see also Chapter 11).

Site-specific fragility may also occur both spontaneously and in the presence of various inducing agents. Only one site appears to be clinically important, situated towards the end of the long arm of the X chromosome at Xq27.3. Expression of this fragile site is seen in the FXS.

Uniparental Disomy

A third class of abnormality results when both chromosomes of a pair are inherited from one or other parent resulting in *uniparental disomy*. Such a finding implies that the parental origin

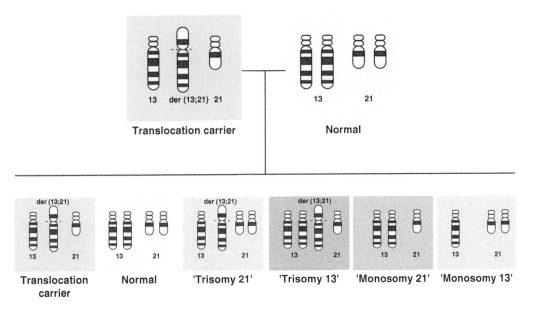

Figure 10.7 Children of a carrier of a Robertsonian translocation. Shown are the possible karyotypes of the offspring of a male carrying a Robertsonian translocation involving chromosomes 13 and 21, and a normal female. Most foetuses of Robertsonian translocation carriers (all the monosomic foetuses, and most with three copies of chromosome 13) will not survive long after birth. *Source:* Created with BioRender.com.

of a chromosome may determine gene expression on that chromosome. This phenomenon is called *imprinting*. The best example of imprinting is the relationship between the pattern of inheritance of chromosome 15 in Prader–Willi syndrome (PWS) and Angelman syndrome (AS).

PWS and AS may also result from the phenomenon of *trisomic rescue*, also known as *trisomy zygote rescue*. In this process, a zygote initially containing three copies of a particular chromosome, such as can occur when one parent has a Robertsonian translocation, loses one of the extra chromosomes. If both the retained chromosomes come from the same parent, then uniparental disomy results. If the retained chromosomes come from different parents, there are typically no phenotypic or genotypic anomalies.

PWS is characterised by infantile *hypotonia* (poor muscle tone), short stature, small hands and feet, almond-shaped eyes, hypogonadism, psychomotor retardation, hypopigmentation (pale skin) and early onset of childhood *hyperphagia* (overeating) with consequent obesity. AS is characterised by severe intellectual disability, seizures, inappropriate laughter, *ataxic gait* (unsteady posture, lurching walk), puppet-like upper limb movements, lack of speech and a large jaw. In PWS, deletions within the long arm of chromosome 15 (q11–q13) are always found on the paternally derived chromosome. Non-deletion cases can display uniparental disomy and in these cases two copies of the maternal chromosome 15 are present. These findings imply that the absence of a sequence on the paternally derived chromosome 15 at q11–q13 gives rise to the PWS phenotype. In contrast in AS, deletions are always on the maternally derived chromosome and uniparental disomy, although much less frequent than in PWS, manifests itself in the presence of two paternal copies of chromosome 15.

Chromosome Abnormality and Clinical Syndrome

Variation is the hallmark of the relationship between a chromosome abnormality and a clinical syndrome. Using the example of Down syndrome (trisomy 21), there is considerable variability in the phenotype of such individuals. Indeed no single feature is considered to be *pathognomonic* (i.e. specific) for Down syndrome. Such variation may well partly reflect the different genetic background, consisting of the other chromosomes that make up the diploid set in each individual. Alternatively, it could reflect unappreciated levels of mosaicism; some individuals with Down syndrome have mosaicism, which means they have a mixture of cells with trisomy 21 and cells with a normal chromosome complement. It should also be noted that no simple correlation exists between a particular gene locus and a particular Down syndrome phenotypic trait (e.g. brachycephaly). Rather, a highly restricted *Down syndrome critical region* (HR-DSCR) of only 34 kilobases (kb) on distal 21q22.13 is the minimal region whose duplication is shared by all Down syndrome subjects and is absent in all non-Down syndrome subjects. Another general consideration is that the influence of a gene or set of genes acts through a number of developmental pathways. Chromosome abnormalities therefore increase the probability of particular developmental anomalies, without predetermining them.

How Chromosome Abnormalities Arise

It is now possible to determine both the parental origin of chromosomal abnormalities and in some cases the mechanisms by which they arise. In general, maternal errors account for most aneuploidy states, whereas point mutations and structural rearrangements are more likely to result from paternal errors.

Maternal Errors

By the time of birth, the average human ovary contains about 2 million oöcytes in the diplotene stage of the first meiotic division. Here they remain in a resting phase until further maturation through the meiotic cycle occurs prior to ovulation. Many trisomies arise as a result of errors in segregation at the first maternal meiotic division (M1) in oöcytes. Nondisjunction is the most common segregation defect. The oöcytes formed will therefore have either an extra chromosome or one less chromosome. Fertilisation of such oöcytes with normal spermatocytes will result in either trisomic or monosomic zygotes.

Paternal Errors

De novo structural rearrangements and point mutations giving rise to a variety of inherited conditions, such as familial retinoblastoma and type 1 neurofibromatosis, show a strong bias towards a paternal origin. In contrast to oögenesis, spermatogenesis is a continuous process of production of spermatozoa from puberty onwards. Studies in mice and Drosophila, which closely resemble humans with respect to spermatogenesis, indicate that point mutations arise and accumulate in spermatogonia – premeiotic cells capable of mitotic division. In contrast, in mice and Drosophila and by extrapolation in humans, structural rearrangements may fatally disrupt meiosis. Paternal bias is accounted for because structural rearrangements may arise during postmeiotic maturation (i.e. during spermatid and spermatozoon stages).

Multigenic Disorders

Multigenic disorders are conditions influenced by multiple genes as well as environmental factors. Unlike single-gene disorders in which a mutation in a single gene is primarily responsible for the disease, or chromosome abnormalities, which have a large effect on the phenotype of an affected individual, multigenic disorders involve the interplay of numerous genes, each contributing a small effect towards the development of the disorder. The study of multigenic disorders poses significant challenges due to the complexity of the underlying genetic architecture. Genome-wide association studies (GWAS) are useful tools that can help to unravel the genetic basis of these disorders. GWAS

examine the entire genome of a large number of individuals to identify common genetic variations associated with a particular disorder. For example, they have identified gene variants which confer an increased (or decreased) risk of certain multigenic disorders, such as type 2 diabetes (e.g. variants of the TCF7L2 gene), coronary artery disease (e.g. variants of PCSK9 and APOE genes) and schizophrenia (variants of CACNA1C and DRD2 genes). *Polymorphism* is the term used to describe the presence of two or more variants of a specific DNA sequence occurring between different individuals or among populations. The most common type of polymorphism involves variation at a single nucleotide (also called a single-nucleotide polymorphism or SNP).

Potential Evolutionary Benefits of Genetic Disorders

Certain genetic disorders have been found to confer protection against infectious diseases, which might explain how some genetic diseases have become so prevalent in human populations. One of the best examples is sickle cell trait (discussed in Case Study CS8), which confers a survival advantage against malaria. Individuals with sickle cell trait have lower levels of malaria parasites in their blood. Several mechanisms have been proposed to explain the resistance to severe malaria observed in individuals with sickle cell trait, including increased phagocytosis of infected red blood cells by the spleen, premature haemolysis leading to parasite death, impaired haemoglobin digestion by the parasite and reduced expression of *P. falciparum* erythrocyte membrane protein 1 (PfEMP1) on infected red cells which results in decreased binding of infected cells to the endothelium. This protein is a major virulence factor; red cell adherence to venous endothelium is important for parasite replication.

It has also been suggested that mutations in the CFTR gene, particularly the most common delta-F508 mutation, may provide a selective advantage to individuals who are heterozygous for the mutation. Studies have shown that cells expressing the mutant form of the CFTR protein exhibit resistance to invasion by *Salmonella typhi*, the bacterium responsible for causing typhoid fever. It has also been shown that mice carrying a single copy of the mutant CFTR gene are resistant to developing diarrhoea when exposed to cholera toxin.

Summary

Genetic diseases are a significant cause of illness and death. It is estimated that 10% of all adult admissions, and up to 50% of paediatric admissions, to hospital are due to a genetic cause. Single-gene disorders, often referred to as monogenic disorders, usually follow a Mendelian pattern of inheritance. Chromosome abnormalities usually arise from structural or numerical changes in chromosomes. Monogenic disorders and chromosome abnormalities are usually absent in normal people but highly penetrant in affected individuals. Complex multigenic disorders, on the other hand, occur as a result of changes in multiple genes each with low penetrance and often also depend upon exposure to environmental factors.

Key Points
1) Mutations are heritable changes in the genetic make-up of cells.
2) If a mutation is present in the cells that form the gametes, then the mutation may be transmitted to the offspring where it will be present in every cell of an affected individual. This mutation may arise as a new mutation in the germ cells of one or both parents, or it may have been inherited by the parents from previous generations.
3) The inherited single-gene disorders are characterised by the presence of mutations involving only a single gene. These mutations may be deletions or insertions of bases or involve the substitution of one base for another (point mutation).
4) Single-gene disorders usually show clear patterns of Mendelian inheritance. They may be either X-linked (mutation is present in a gene on the X chromosome) or autosomal (mutation is present in a gene on one of chromosomes 1–22). In addition, they may be dominant, requiring a mutation in only one copy of the gene, or recessive, when a mutation in both copies of the gene is required.
5) Chromosome abnormalities may be defined as microscopically visible changes in chromosome morphology and represent mutations involving a large number of genes. The smallest chromosomal deletions detectable on light microscopy remove a minimum of several megabases of DNA.
6) Chromosome abnormalities may be either constitutional, if they are inherited from one or other parent, or arise *de novo* in the embryo.
7) Chromosome abnormalities may involve either a change in chromosome number, structure or pattern of inheritance.
8) Changes in chromosome number may involve either the gain or loss of individual chromosomes (aneuploidy) or the gain of entire haploid sets (triploidy or tetraploidy). Aneuploidy of the germ cells can give rise to either trisomic or monosomic zygotes.
9) Structural abnormalities include large deletions or insertions of genetic material, duplication of segments of a chromosome, inversions or translocations.
10) A further class of abnormality arises when both chromosomes of a pair are inherited from one or other parent. This is known as uniparental disomy in which the parental origin of a chromosome determines gene expression on that chromosome.

Further Reading

Evangelou E. (eds) (2018) *Genetic Epidemiology* 9781493978670.

Turnpenny P.D., Ellard S. & Cleaver R. (eds) (2021) *Emery's Elements of Medical Genetics and Genomics* (16th edn). 9780702079665.

Strachan T. & Andrew P. (eds) (2019) *Human Molecular Genetics* (5th edn). 9780815345893.

Part 7

Neoplasia

11

Neoplasia

CHAPTER MENU

Introduction

Cancer is a major cause of global morbidity and mortality. This chapter describes the biology of cancer, outlining the causes of cancer and the fundamental defects in cancer cells that allow them to escape the normal homeostatic controls that maintain cell number in normal tissues.

The Cellular Basis of Neoplasia

Neoplasia, literally meaning 'new growth', is a disorder characterised by the abnormal and continuous growth of cells which are no longer subject to the homoeostatic controls that maintain the appropriate number of cells in normal tissues. In most cases, these cells form a solid mass of tissue which is referred to as a *tumour* (literally meaning a 'swelling'). An exception to this is leukaemia in which the abnormal

cells arise from precursors in the bone marrow and pass into the bloodstream in the same way as normal blood cells. Neoplasms are broadly divisible into *benign* and *malignant* subgroups. *Cancer* is the term commonly used to describe the disease which results from the presence of a malignant neoplasm.

It is important to distinguish neoplasia from hyperplasia and hypertrophy, both of which also result in an increase in tissue mass (see also Chapter 3). *Hyperplasia* is an increase in cell number and can be either physiological or pathological. An example of physiological hyperplasia is the increase in breast tissue that occurs during pregnancy. An example of pathological hyperplasia is benign prostatic hyperplasia (BPH). This is a non-neoplastic enlargement of the prostate gland that occurs in many men as they age. BPH can cause urinary symptoms and lead to complications if left untreated (see also Case Study CS13). *Hypertrophy* is an increase in cell size. An example of physiological hypertrophy is the increase in skeletal muscle mass that follows exercise training. Left ventricular hypertrophy is an example of pathological hypertrophy. *Metaplasia* is the change from one mature cell type to another. For example, the transformation from columnar to squamous epithelium in the cervix is a normal physiological process that occurs in response to hormonal changes. This allows the cervix to become more resistant to trauma and infections. However, metaplasia can also be associated with an increased risk of malignancy. *Barrett oesophagus*, for example, is a condition in which the normal squamous epithelium of the oesophagus is replaced by a metaplastic columnar epithelium, usually in response to chronic acid reflux. Barrett oesophagus is a pre-malignant condition because it increases the risk of developing oesophageal adenocarcinoma. Chronic exposure to cigarette smoke or other environmental toxins can also cause metaplasia of the normal respiratory epithelium of the bronchi, replacing it with a metaplastic squamous epithelium. This process is also pre-malignant because the metaplastic cells can become *dysplastic* and progress to form lung cancer. *Dysplasia* describes a state of abnormal cell growth or development within a tissue or organ. Although dysplastic cells are not cancerous, they can represent a precancerous condition, as they have the potential to transform into cancer over time if left untreated. Dysplasia is graded by pathologists based on the extent of cellular abnormality and tissue involvement observed under the microscope. *Mild dysplasia*, also known as low-grade dysplasia, indicates minor cellular and structural changes, with most of the tissue still retaining normal features. *Moderate dysplasia* signifies more extensive abnormality with greater variance in cell shape, size and organisation. *Severe dysplasia*, or high-grade dysplasia, demonstrates the highest degree of abnormality, with cells appearing very different from normal ones, and the architecture of the tissue significantly altered. High-grade dysplasia is considered a serious condition due to its higher risk of progressing to cancer and often warrants more aggressive treatment or closer monitoring.

In the case of cervical cancer, different grades of dysplasia can help define distinct stages of pre-malignant epithelial abnormality. Thus, *cervical intraepithelial neoplasia* (CIN) is the term used to describe precancerous changes in the cells of the stratified squamous epithelium lining the ecto- (outer) cervix. CIN is classified into three stages based on the extent of abnormal cell growth: CIN1 represents mild dysplasia in which only one-third of the epithelial thickness is affected; CIN2 is moderate to marked dysplasia with two-thirds involvement and CIN3 refers to severe dysplasia/ carcinoma *in situ*, in which more than two-thirds up to the full thickness of the epithelium contains abnormal cells.

Neoplasms are believed to arise from a single *target cell* which has undergone a series of genetic changes (*mutations*) or *epigenetic alterations* (heritable changes in gene expression caused by mechanisms other than changes to the DNA sequence, e.g. DNA methylation) which allow it to escape the normal growth controls imposed upon normal cells.

A cell with a beneficial mutation or epigenetic change may continue to divide until a collection of identical cells or *clone* is formed. Cells from this clone may acquire new genetic and epigenetic changes which further enhance their growth and survival. Once a tumour is established, ongoing genetic and epigenetic alterations may give rise to multiple subclones, each with differing properties. This is referred to as *tumour heterogeneity* and is an important concept in relation to tumour progression.

Some neoplasms arise in *stem cells*. Stem cells are present in small numbers in normal tissues and have two critical functions: 1) to generate descendants which will become differentiated and perform the function of the tissue, and; 2) to renew themselves so that a stable number of stem cells remain. The evidence for a stem cell origin of cancer is most compelling in the case of leukaemia. Chronic myeloid leukaemia, for example (Case Study CS52), is characterised by the accumulation of neoplastic cells of differing myeloid lineages all of which are believed to descend from a common neoplastic myeloid stem cell.

Benign and Malignant Neoplasms

Invasion and metastasis are the hallmarks of malignancy. However, not all malignant neoplasms have the capacity to metastasise. In contrast, benign neoplasms almost never invade or metastasise. Histopathological examination of sections of a tumour by haematoxylin and eosin staining is often necessary to classify tumours as either benign or malignant. Malignant cells have higher *nuclear-to-cytoplasmic ratios* compared to benign cells, meaning the nucleus is relatively larger compared to the surrounding cytoplasm. Moreover, malignant nuclei stain more intensely with dyes like haematoxylin, referred to as *hyperchromasia*. Malignant cells are also often *pleomorphic*, meaning they vary in size and shape, whereas benign cells are more uniform in appearance. In general, malignant cells divide rapidly, whereas benign cells do not.

Classification of Neoplasms

Neoplasms are usually classified on the basis of the presumed cell or tissue of origin (Table 11.1). Thus, squamous cell carcinomas are malignant neoplasms derived from squamous epithelial cells. Since this type of epithelial cell is found in many locations within the body, squamous cell carcinoma can arise at many sites, for example, in the skin, oesophagus and cervix. Histopathology is useful in distinguishing different tumours based on their morphology. For example, squamous cell carcinomas often show keratin production, seen as 'keratin pearls' (Fig. 11.1A). In contrast, adenocarcinomas, which are derived from glandular epithelial cells, often form glandular structures (Fig. 11.1B).

Clinical Aspects of Neoplasia

Malignant neoplasms are often fatal to their host. This is primarily due to their ability to metastasise and to develop resistance to various forms of therapy. In contrast, benign neoplasms are life-threatening only in exceptional circumstances. These include situations in which the tumour is present within, or impinges upon, a vital structure. For example, benign tumours of the atrium (atrial

Table 11.1 Examples of neoplasms and their cell or tissue of origin.

Tissue or cell of origin	Benign neoplasm	Malignant neoplasm
Squamous epithelium	Squamous cell papilloma	Squamous cell carcinoma
Glandular epithelium	Adenoma	Adenocarcinoma
Cartilage	Chondroma	Chondrosarcoma
Bone	Osteoma	Osteosarcoma
Smooth muscle	Leiomyoma	Leiomyosarcoma
Striated muscle	Rhabdomyoma	Rhabdomyosarcoma
Lymphocytes	—*	Lymphoma
Haemopoietic cells	—*	Leukaemia

* Note that there are no well-defined benign counterparts of some cancers, e.g. lymphomas and leukaemias.

(A)

(B)

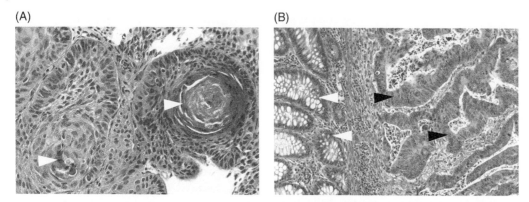

Figure 11.1 Examples of different types of cancer. Shown are micrographs of haematoxylin and eosin (H&E) stained tumours. (A) Photomicrograph of squamous cell carcinoma of the oesophagus. The tumour is moderately differentiated and keratinizing (white arrowheads). Keratinised areas are often described as 'keratin pearls'. *Source:* Michael Bonert / Wikimedia Commons / Public domain. (B) Micrograph of a colorectal adenocarcinoma showing typical glandular structures (black arrowheads). Normal colonic glands are also present (white arrowheads). *Source:* Michael Bonert / Wikimedia Commons / Public domain.

myxomas) may cause valve obstruction, cardiac insufficiency and sudden death. If the benign tumour is present within the central nervous system, then it can cause pressure damage to surrounding nervous tissue or produce serious complications as a result of associated rises in intracranial pressure.

A benign tumour may also produce a physiologically active substance in increased amounts. For example, some benign tumours of the adrenal medulla (phaeochromocytomas) secrete excessive amounts of adrenaline leading to hypertension which can precipitate myocardial infarction or cerebral haemorrhage. Likewise, islet cell adenomas of the pancreas can secrete enough insulin to produce fatal hypoglycaemia.

The production of hormones by a tumour is an example of a *paraneoplastic syndrome* – defined as signs and symptoms that occur as a consequence of the presence of a tumour (either benign or malignant), but are not due to its mass effect. Paraneoplastic syndromes may be caused by increased amounts of hormones that are normally produced by the tissue of origin (as in the examples of phaeochromocytomas and islet cell adenomas

described above), or they may be due to ectopic production of a hormone. For example, a patient with lung cancer may experience hypercalcaemia due to the abnormal production of parathyroid hormone-related protein (PTHrP) by the cancer cells. PTHrP stimulates the release of calcium from bones, leading to an increase in the level of calcium in the blood. Paraneoplastic syndromes can also be caused by an abnormal immune response. For example, in paraneoplastic cerebellar degeneration, which occurs in some people with lung, breast or ovarian cancer, autoantibodies react with the cerebellar Purkinje cell cytoplasmic antigens. The result is cerebellar defects leading to symptoms of *gait ataxia* (lack of coordination that affects a person's ability to walk, resulting in an unsteady and uncoordinated gait), *dysarthria* (speech defect characterised by slurred or slow speech that can be difficult to understand) and *tremors* (involuntary, rhythmic shaking movements).

There are many features that impact cancer prognosis. Two of the most important are grade and stage. *Grade* refers to the degree of differentiation of the tumour, in other words, how closely the tumour cells resemble their cell of origin. The tumour cells of benign neoplasms closely resemble their cell of origin and are therefore described as being *well differentiated*. Although malignant neoplasms may also be well differentiated, many are either *poorly differentiated* (cells of the tumour do not closely resemble the cell of origin) or *undifferentiated* (the cell of origin of the tumour cells cannot be determined). Undifferentiated tumours are often described as *anaplastic*. In general, undifferentiated and poorly differentiated tumours have higher growth rates and a poorer prognosis compared with well-differentiated tumours. *Stage* refers to how far the tumour has spread. The TNM system is commonly used to stage cancer and is based on the size of the tumour (T), the extent of lymph node involvement (N) and the presence of distant metastases (M).

Characteristics of Malignancy

Cancers share many biological characteristics in common. These were described by Hanan and Weinberg as the 'Hallmarks of Cancer'. We discuss each of these below.

Sustaining Proliferative Signalling

Stimulation of a normal cell into a proliferative state often depends upon an external signal in the form of a growth factor. Many growth factors bind to receptors situated on the surface of target cells. These receptors are often *receptor tyrosine kinases*. Binding of growth factor to these receptors initiates cell signalling pathways, ultimately leading to cell division. In general, tumour cells are less dependent upon normal growth factors than are normal cells. In some cases, they can bypass the normal requirements for growth factors by initiating their own constitutive growth signals either through unrestricted activation of the growth factor receptor itself or by continuous activation of the signalling pathways or of the transcription factors operating downstream of

these receptors (Fig. 11.2). Sequencing of the genomes of different cancer types has identified many of the defects responsible for this constitutive signalling. For example, nearly one-half of all melanomas contain activating mutations in the BRAF gene (see also Case Study CS68), resulting in constitutive MAP-kinase pathway signalling (Fig. 11.3). Another example is Hedgehog (Hh) signalling (Fig. 11.4). The Hh pathway is important in development but is aberrantly activated in different types of cancer, either by ligand-independent signalling driven by mutations in key pathway regulators such as the Patched (PTCH1) gene (e.g. in basal cell carcinoma and medulloblastoma) or by overexpression of Hh ligand by tumour cells (e.g. in ovarian cancer, colorectal cancer and pancreatic cancer). These and other aberrantly activated pathways controlling tumour growth have recently emerged as important targets of new cancer therapies (for examples, see the case studies associated with this chapter).

Evading Growth Suppressors

Neoplastic cells are no longer responsive to the controlling mechanisms that maintain cell number in normal tissues and continue to divide under circumstances in which normal cells cease proliferating. The result is the progressive accumulation of neoplastic cells and the formation of a tumour mass. Tumour growth rate is determined by the rate at which new tumour cells are generated through cell division, and the rate at which cells are lost from the dividing cell pool, either by cell death or by differentiation. The inactivation of tumour suppressor genes can give rise to a cell in which the effect of a growth-promoting factor goes unopposed (Table 11.2). Tumour suppressor genes usually act recessively when inactivation of both alleles is required for loss of function.

Abnormalities in the RB1 tumour suppressor gene were first detected in retinoblastoma, a neoplasm of retinal precursor cells

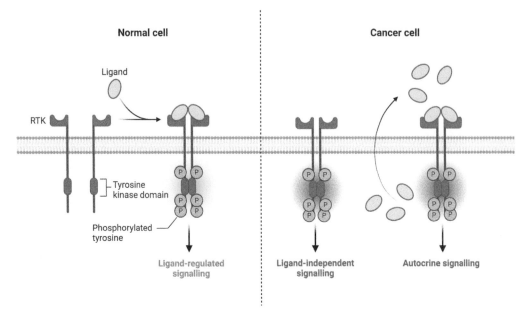

Figure 11.2 Possible mechanisms leading to reduced growth factor dependency by neoplastic cells. Left panel: Typical receptor tyrosine kinase (RTK) regulation in a normal cell. Ligation of a growth factor to the receptor on the cell membrane leads to phosphorylation of the RTK and downstream signalling. Right panel: Cancer cells may harbour mutations in RTK that enable constitutive RTK signalling in the absence of ligand, or they may secrete growth factors which stimulate autocrine signalling. *Source:* Created with BioRender.com.

(*retinoblasts*) and subsequently in other tumours, notably osteosarcomas. The retinoblastoma protein (pRb) encoded by RB1 prevents the cell from progressing through the G1 to S phase transition and is therefore a critical 'gatekeeper' of cell-cycle progression whose absence permits persistent cell proliferation. In retinoblastoma, both alleles of the *RB1* gene on chromosome 13 are inactivated, usually by deletion or point mutation. The majority of cases of retinoblastoma arise sporadically in the population. In these cases, defects in both alleles of the RB1 gene occur *somatically* (i.e. post-conception) in the same retinoblast. However, as many as one-third of retinoblastomas arise from a genetic predisposition. In these cases, a mutation is already present in one allele at conception (and consequently present in all cells of the developing retina), whereas the second occurs as a somatic event. This model explains the observed single focus of tumour formation in sporadic cases – the

tumour arising following the convergence of two rare genetic events in a single target retinoblast. Multiple tumour foci in both eyes are usually seen in heritable retinoblastoma – this is because one of the two mutations is already present in the retinal cells and only a single mutation is required to complete the triggering process.

Resisting Cell Death

Some tumour suppressor genes act by preventing apoptosis. There are two distinct pathways to apoptosis – the so-called intrinsic pathway regulated by BCL2 family members and the 'extrinsic' pathway regulated by death receptors (Chapter 3). Each pathway culminates in the activation of an otherwise latent protease (caspase 8 or 9), which initiates a cascade of proteolysis involving effector caspases responsible for the execution of apoptosis. Tumour cells commonly evolve numerous strategies

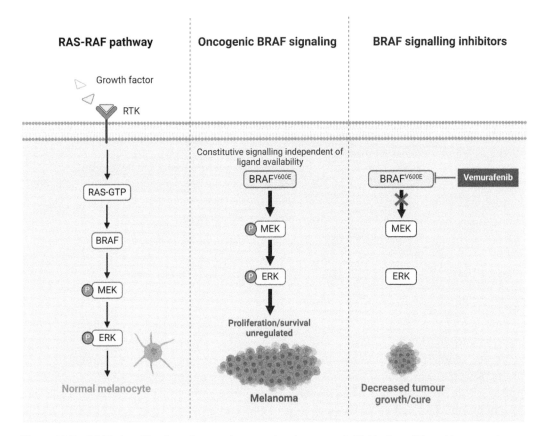

Figure 11.3 BRAF signalling in melanoma. In normal melanocytes (left), binding of ligand to a receptor tyrosine kinase activates the MAPK pathway target proteins, MEK and ERK, leading to proliferation and cell survival. In melanoma (middle), a mutation in BRAF allows constitutive activation of the pathway in the absence of ligand and receptor. The BRAF mutation shown here results in a single amino acid substitution of valine (V) to glutamic acid (E) at amino acid position 600 (V600E). This mutation accounts for most of the BRAF mutations in melanoma. BRAF inhibitors such as vemurafenib (Zelboraf®) (right) are effective in treating patients with metastatic melanoma. *Source:* Created with BioRender.com.

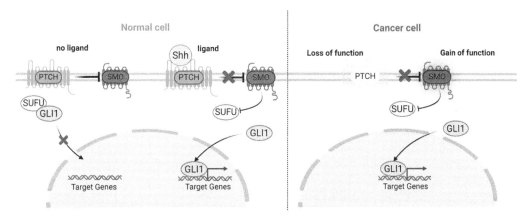

Figure 11.4 Hedgehog signalling and cancer. Left: Normal hedgehog signalling. PTCH is a receptor for the hedgehog ligand, Sonic hedgehog (Shh). In the absence of Shh, PTCH inhibits Smoothened (SMO) and there is no signalling because suppressor of fused homolog (SUFU) is bound to the GLI transcription factors. When Shh binds to PTCH, this relieves PTCH suppression of SMO. This allows activation of the GLI transcription factors. Right: Hedgehog signalling in basal cell carcinoma. The majority of mutations in sporadic basal cell carcinoma are either loss-of-function mutations in PTCH or gain-of-function mutations in SMO leading to constitutive activation of GLI transcription and tumorigenesis in the absence of Shh. *Source:* Created with BioRender.com.

Table 11.2 Examples of tumour suppressor genes.

Gene	Tumour(s)	Function of encoded protein
RB1	Retinoblastoma, osteosarcoma	Regulates G_1/S cell cycle checkpoint
TP53	Common cancers, e.g. breast, colon, lung	Regulates target genes that induce G_1/S arrest, apoptosis, senescence and DNA repair
NF1	Neurofibroma	Negative regulator of the Ras signal transduction pathway
APC	Colon cancer	Regulator of beta-catenin
BRCA1	Breast, ovarian cancer	DNA repair
BRCA2	Breast cancer	DNA repair
CDKN2A (p16)	Melanoma, pancreatic cancer	Cdk inhibitor, blocks G1 to S phase transition*

* See Chapter 3 for a description of Cdk inhibitors.

to prevent apoptosis. For example, tumours can increase the expression of anti-apoptosis regulators (e.g. Bcl-2 and Bcl-xL) and decrease the expression of pro-apoptotic factors. (e.g. Bax, Bim and Puma). Tumours can also inactivate extrinsic death pathways, for example in Hodgkin lymphoma, a type of B-cell lymphoma derived from germinal centre B cells. Normal germinal centre B cells are programmed to die by Fas (Fas Apoptotic Receptor, otherwise known as CD95 or APO-1)-induced death. However, in Hodgkin lymphoma, the tumour cells can evade this death by over-expressing c-FLIP (cellular FLICE [FADD-like IL-1β-converting enzyme]-inhibitory protein), which prevents the activation of caspases following Fas ligation (see also Chapter 3).

Various forms of stress not only activate apoptosis but can also induce *autophagy*, a process in which intracellular vesicles known as *autophagosomes* envelope intracellular organelles and then fuse with lysosomes to bring about degradation of the organelles. Knockout mice missing components of the autophagy machinery, for example the *Beclin-1* gene, exhibit increased susceptibility to cancer. Thus, like apoptosis, autophagy may also serve as a barrier to tumorigenesis. However, other studies show that in certain situations, autophagy can promote tumorigenesis, possibly by generating metabolites that can be used to support the survival of the cancer cell under nutrient-limiting conditions.

Tumour-Promoting Inflammation

The tumour microenvironment (TME) is comprised of all the non-cancerous cells and the extracellular matrix (ECM) of a tumour. Cells of the TME include fibroblasts, often referred to as cancer-associated fibroblasts, various innate and adaptive immune cells, including macrophages, and cells that comprise the blood vessels. The composition of the TME varies between cancer types and even between tumours of the same type. The TME is generally an immunosuppressive environment that is permissive for cancer growth. In contrast, normal tissue microenvironments are tumour suppressive. Thus, an important concept is that the TME develops in tandem with the evolution of the cancer cells; a process referred to as *stromagenesis*. The TME not only promotes immune evasion but also contributes to tumour cell survival, growth, invasion and metastasis, and can also mediate resistance to cancer therapies.

Enabling Replicative Immortality

Telomeres are repetitive nucleotide sequences located at the termini of chromosomes (see also Chapter 3). With successive cell divisions, telomeric sequences are lost as a result of incomplete replication. With progressive telomere shortening, cells reach their replicative limit and enter *senescence*. Thereafter, the cells will eventually undergo *crisis* characterised by gross chromosomal rearrangements and genome instability, following which almost all cells die. Rare cells that emerge from crisis are said to be *immortalized*. One way in which cells can become immortalised is by activating an enzyme known as *telomerase* which can add telomere repeat segments to the ends of telomeric DNA. Recent studies suggest that the activation of telomerase may be a late event in cancer development; the genomic instability associated with telomere deficiency contributes to the acquisition of tumour-promoting mutations early in tumorigenesis. Later, the induction of telomerase expression acts to 'fix' these mutations in immortalised cell populations.

Inducing Angiogenesis

Some tumours (e.g. renal cancer and glioblastoma) are heavily dependent for their survival upon the formation of new blood vessels, a process known as *angiogenesis*. In contrast, other tumour types are not dependent upon angiogenesis (e.g. pancreatic adenocarcinomas). Angiogenesis not only promotes the survival of tumour cells but can also facilitate metastasis. This is because the vessels produced by tumour angiogenesis are leaky, allowing the cancer cells to more easily escape into the circulation. Tumours may initially grow in close association with existing blood vessels (*host vessel co-option*) and are consequently well vascularised, or they may grow in the absence of blood vessels (*avascular initiation*). In the latter case, the tumour cells will grow until their mass outstrips their oxygen/

nutrient supply which is limited by diffusion from nearby vessels. Hypoxia in the centre of the mass induces the expression of angiogenic factors, principally vascular endothelial growth factor (VEGF) and angiopoietin 2 (Ang-2). VEGF is produced by various cell types, including endothelial cells, tumour cells and immune cells. VEGF binds to its receptors, VEGFR-1, VEGFR-2 and VEGFR-3 found on the surface of endothelial cells. The binding of VEGF activates intracellular signalling pathways that promote endothelial cell proliferation, migration and survival. Ang-2 is mainly produced by endothelial cells and acts by binding to its receptor, Tie-2, which is also expressed in endothelial cells. Ang-2 has a complex role in angiogenesis, acting as both a promoter and an inhibitor, depending on the presence or absence of VEGF. In the presence of VEGF, Ang-2 promotes angiogenesis by destabilising existing blood vessels, making them more responsive to the pro-angiogenic effects of VEGF. It does this by loosening the connections between endothelial cells and their surrounding ECM, as well as by reducing the interaction between endothelial cells and *pericytes*, which are the support cells that help maintain blood vessel stability. However, in the absence of VEGF, Ang-2 acts as an inhibitor of angiogenesis, promoting blood vessel regression by inducing endothelial cell apoptosis and preventing the recruitment of pericytes, which leads to the destabilisation and regression of the newly formed vessels.

Lymphangiogenesis is the formation of lymphatic vessels from pre-existing lymphatic vessels. Similar to angiogenesis, lymphangiogenesis is regulated by soluble factors, such as VEGF-C, VEGF-D and their receptor, VEGFR-3. Studies of different human cancers show that expression of these lymphangiogenic factors is increased in more advanced stages of disease and is associated with poor prognosis (for example, expression of VEGF-D is correlated with poor prognosis in ovarian cancer). Tumours can induce lymphangiogenesis both locally and distally in draining lymph nodes.

For example, in breast cancer, lymphangiogenesis is observed frequently in uninvolved axillary lymph nodes.

Activating Invasion and Metastasis

To spread, cancer cells must escape from their tissue of origin, invade and migrate, for example, to a vessel, in order to disseminate around the body. Examples of cancer spread are shown in Fig. 11.5. Some normal cells, for example lymphocytes, are able to migrate from one body compartment to another because they express the proteins required to do so. Lymphomas, which are tumours derived from lymphocytes, retain these homing capabilities, allowing them to spread to involve multiple lymphoid sites. Thus, the spreading of lymphomas is in part due to the fact that the lymphoma cells retain the ability to traffic within the lymphoid system. In contrast, the majority of cells in the body are not able to freely circulate, since they are firmly bound to each other and to the ECM; the tissue architecture and the

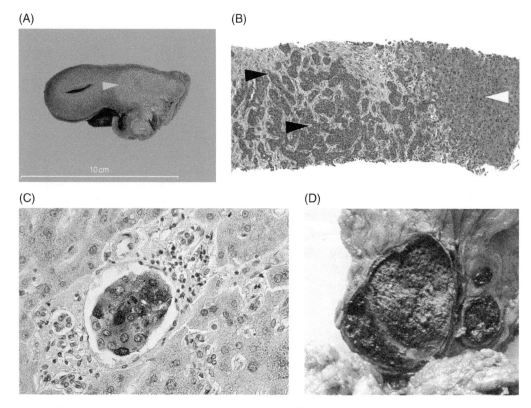

Figure 11.5 Examples of cancer invasion and metastasis: (A) Local invasion of cervical carcinoma (blue arrowhead) into the deep tissues of the myometrium. *Source:* Deilson Elgui de Oliveira / Wikimedia Commons / CC BY-SA 4.0. (B) Haematoxylin and eosin (H&E) stained section of liver with metastatic deposits of adenocarcinoma of the colon (black arrowheads). Normal liver is indicated by a white arrowhead. *Source:* Michael Bonert / Wikimedia Commons / Public domain. (C) Breast cancer metastasis to liver. A cluster of the cancer cells with their brown-staining cytoplasm is within a portal tract of the liver (monoclonal antibody b1.1, abc immunoperoxidase method, haematoxylin counterstain, ×500). *Source:* Unknown Photographer / Wikimedia Commons / Public Domain. (D) Gross appearance of lymph node metastasis from melanoma. *Source:* Department of Pathology / Wikimeida Commons / CC BY-SA 4.0.

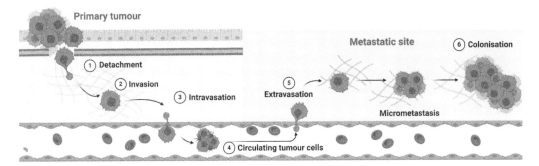

Figure 11.6 Steps in the metastasis of a typical epithelial cancer. A metastatic subclone develops within the primary tumour mass. Metastatic tumour cells detach from the primary tumour mass (1) and invade through the basement membrane and then through the underlying connective tissues (2). In order to complete the metastatic process, the cancer cells must enter blood or lymphatic vessels (3), travel as circulating tumour cells (4), leave the vessel (5) and grow within a distant tissue (6). *Source:* Created with BioRender.com.

basement membrane surrounding these cells acting as a barrier to cell movement. Cell motility is a tightly regulated process. In order to move out of a tissue, the cancer cell needs to overcome these tightly regulated processes. The key properties of the metastatic cell (Fig. 11.6) are described below:

1) *Loss of cell detachment:* Adhesion molecules are downregulated in metastatic cancers. One of the most important adhesion molecules is E-cadherin. E-cadherin is a component of adherens junctions that make tight connections between adjacent cells. As well as facilitating the breakdown of adherens junctions, loss of E-cadherin can have other effects on the cell. E-cadherin is normally tightly tethered to the actin cytoskeleton via an interaction with the β-catenin protein. Loss of E-cadherin releases β-catenin from this interaction. Normally, any free β-catenin is removed by a complex comprising various proteins, including the protein product of the tumour suppressor gene, adenomatous polyposis coli (APC; see also Case Studies CS41 and CS53); this complex targets β-catenin for degradation. However, if APC function is lost, β-catenin can translocate to the nucleus, whereupon it drives the transcription of genes that

promote cell division, including MYC and CCND1 (which encodes cyclin D1; see Chapter 3).

2) *Increased motility:* In the case of epithelial cancers, the malignant cells can 'hijack' *epithelial–mesenchymal transition (EMT)*, a normal process that allows epithelial cells to move during wound healing. EMT is regulated by a set of transcription factors, which include Snail, Slug, Twist and Zeb1/2. Included among the biological effects of these transcription factors are the loss of adherens junctions and associated conversion from an epithelial morphology to a spindle-shaped mesenchymal morphology (Fig. 11.7), with increased expression of matrix-degrading enzymes, increased motility and enhanced resistance to apoptosis.

3) *Tissue invasion:* The ability to invade requires ECM-degrading enzymes, including the matrix metalloproteinases (MMPs). MMPs are important in tissue remodelling during normal tissue repair. MMPs and other matrix-degrading enzymes can be produced by the cancer cells but may also be secreted by cells of the TME, for example by cancer-associated fibroblasts and macrophages.

4) *Extravasation:* The mechanism by which tumour cells leave the vasculature is not fully understood, but is thought to resemble the

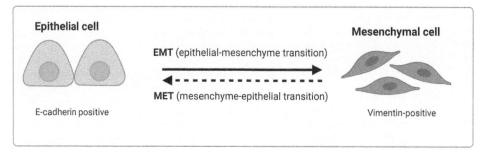

Figure 11.7 Epithelial–mesenchymal transition (EMT). EMT is a process in which epithelial cells, which are typically organised into tightly packed layers, lose E-cadherin expression and their cellular adhesion and acquire mesenchymal properties associated with the expression of vimentin. EMT is a critical step in normal wound healing, but it can also be co-opted by cancer cells to facilitate metastasis. During EMT, cancer cells become more mobile and invasive, allowing them to break away from the primary tumour and travel to other parts of the body. *Source:* Created with BioRender.com.

recruitment of leucocytes that occurs during an inflammatory response (see Chapter 5). Critical steps are the 'rolling' of tumour cells on the endothelial surface, tight adhesion of the tumour cells to the endothelial cells and eventually their trans-endothelial migration.

5) *Colonisation of the secondary site:* A rate-limiting step in the metastatic cascade is colonisation of the secondary site. Most cells that arrive at the secondary site will not survive there. One major reason for this is that the secondary site does not have the growth factors required for tumour cell survival. This is the basis of the so-called *soil and seed* hypothesis. The cancer cell – the 'seed' must find a fertile 'soil' – the tissue environment at the secondary site.

To survive, the metastatic tumour cell must associate itself with a blood vessel. All cells in the body have to maintain relatively close contact with the vascular compartment in order to allow gaseous exchange, nutrient uptake and waste removal. Metastatic cells manage this by migrating to blood vessels and then growing as a *micrometastasis* limited to a few cells deep around the vessel. This micrometastasis can remain 'dormant' for many years. In this case, *dormancy* means that the cell number stays relatively constant. However, the cells are dividing at a rate comparable to the primary

tumour and the reason why the cell number stays relatively constant is that programmed cell death occurs at a very similar rate to cell division. After a variable period of dormancy, the micrometastasis may develop its own blood supply by angiogenesis. The increased blood supply results in a decrease in apoptosis such that cell death no longer occurs at an equivalent rate to cell division, resulting in an increase in cell number and consequently allowing growth of the tumour mass.

An important question is when does metastasis occur during tumour progression, is it an early or a late event? Most studies support the so-called *late dissemination* model for metastasis in which subclones derived from the initial primary tumour acquire mutations that favour tumour spread. In the *'early dissemination'* model, multiple subclones distribute tumour cells throughout the body. After a dormant period at the secondary site, the secondary deposits evolve separately to the primary tumour. Evidence favours the late dissemination model for most cancers because of the high degree of genetic similarity observed between the primary tumour and the metastatic deposits. However, there is evidence that the early dissemination model may operate in some cancers (e.g. breast cancer).

As primary tumours evolve they become increasingly genetically unstable and this increases the likelihood of one or more sub-clones acquiring the necessary characteristics to complete the metastatic cascade described above. Re-seeding of the primary tumour by distinct clones derived from the secondary tumours has also been reported and this can contribute to further heterogeneity within the primary tumour.

The *pre-metastatic niche* refers to the microenvironment at a distant site that has been prepared or conditioned to support the survival, growth and colonisation of metastatic cancer cells. It is a concept that has emerged in the field of cancer research to help explain the process of metastasis and the establishment of secondary tumours. The pre-metastatic niche is formed before the arrival of circulating tumour cells and can be mediated by the release of factors by the primary tumour, including cytokines, growth factors and exosomes (cell-derived extracellular vesicles), which may travel to distant organs and alter the local microenvironment, making it more receptive to the incoming metastatic cells. These factors can also recruit bone marrow-derived cells, such as mesenchymal stem cells, myeloid-derived suppressor cells (MDSCs), and other immune cells, which in turn remodel the ECM at the distant site, creating a more permissive environment for the attachment and growth of metastatic cancer cells.

Deregulating Cellular Energetics

Cancer cells display markedly different metabolic activity compared to their cell of origin. One striking feature is that many cancer cells adopt *glycolysis*, normally an anaerobic process, as their means to generate energy. Moreover, they do this even in an aerobic environment suggesting there is an advantage in doing so beyond the need to survive in a hypoxic environment. This is referred to as the *Warburg effect*. Not only do many cancer cells switch to glycolysis but they also become very efficient scavengers for the available glucose in the extracellular environment, an effect they achieve through the upregulation of cell membrane glucose transporters and cytoplasmic enzymes necessary to perform glycolysis (e.g. the hexokinases). Thus, tumour cells can outcompete adjacent normal cells for the scarce glucose supply. The increased glycolysis also acidifies the local tissue environment through the release of lactic acid, a by-product of glycolysis. Normal cells are less resistant to low pH and may die, allowing further local invasion by the tumour cells. It should be noted that the increased uptake of glucose forms the basis of the widely used clinical test known as *fluorodeoxyglucose positron emission tomography* (FDG-PET). In this test, PET is used to image the uptake of FDG in the patient. FDG is very similar to naturally occurring glucose and is taken up by the cancer cell in the same way as glucose.

Despite the apparent reliance of many cancers on glycolysis, recent studies indicate that some cancers shift towards a high *oxidative phosphorylation* (OXPHOS) state that is associated with drug resistance and which represents a promising therapeutic target in certain cancer types, such as melanoma, acute myeloid leukaemia, chronic myeloid leukaemia and breast cancer. Inhibition of OXPHOS can reverse resistance to EGFR inhibition in lung adenocarcinoma, as well as to docetaxel therapy in prostate cancer, MAPK inhibition in melanoma and 5-fluorouracil treatment in colorectal cancer. Clinical trials are in progress to exploit the scientific rationale of *metabolic synthetic lethality*, in which conventional chemotherapies are combined with mitochondrial inhibitors.

Cancer-associated metabolic reprogramming can be cancer cell autonomous or may be driven by cells of the TME. Moreover, different cells within a tumour may depend upon each other's metabolic activities for their survival, a concept known as '*metabolic symbiosis*'. For example, in prostate cancer, the cancer-associated fibroblasts undergoing Warburg metabolism release lactate which is used by the cancer cells to fuel OXPHOS.

Avoiding Immune Destruction

Many cancer cells express tumour antigens which are either new cellular epitopes derived from mutated proteins (known as '*neo-epitopes*') or viral epitopes (in the case of virus-associated cancers). Cancer cells employ various strategies to evade the immune system, allowing them to grow and spread without being detected. For example, cancer cells can undergo a continuous process of genetic and phenotypic changes, which may lead to the emergence of subclones that are less immunogenic or more resistant to immune attack. This process, called *immune editing*, can result in the selection of cancer cells that are better adapted to evade the immune system. To be recognised by the immune system, cancer cells must present tumour-specific antigens on their surface using major histocompatibility complex (MHC) molecules (see Chapter 6). However, some cancer cells can downregulate MHC expression, making it difficult for immune cells to identify them. Cancer cells also secrete factors that are directly immuno-suppressive (e.g. TNF-α and IL-10) or which recruit and activate immunosuppressive cells, including regulatory T cells (Tregs), MDSC and tumour-associated macrophages (TAMs).

Immune checkpoints have emerged as important mechanisms that mediate immune evasion. Immune checkpoints are normal processes that negatively regulate immune responses and prevent autoimmunity (see Chapter 6). They are activated by ligands on tumour cells (or cells of the TME) that bind to their cognate receptors on T-cells and inhibit the function of the T-cells. Drugs, called immune checkpoint inhibitors, can block immune checkpoint interactions enabling T-cells to kill the cancer cells. One example is the programmed death ligand-1 (PD-L1) which interacts with the programmed death-1 (PD-1) receptor on T-cells. Drugs like nivolumab, which binds to PD-1, have been shown to be effective in treating people with different forms of cancer, such as melanoma (Fig. 11.8, Case Study CS68).

Genomic Instability and Mutation

DNA damage caused by exposure to chemical carcinogens or radiation is not necessarily permanent because cells have the capacity to repair DNA. *DNA repair* must take place prior to cell division to prevent the transmission of potentially harmful mutations to daughter cells. Cells can delay progression through the cell cycle to allow time for DNA repair to take place prior to cell division. Cells carrying damaged DNA also have the option of activating

Immune checkpoint inhibits T-cell activation

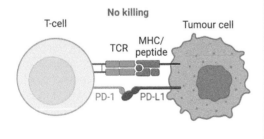

Anti-PD-1 antibodies allow T-cell killing

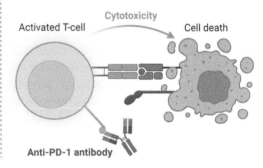

Figure 11.8 The PD-1/PD-L1 immune checkpoint in cancer. The left-hand image shows a cancer cell expressing PD-L1 which binds to PD-1 on a T-cell resulting in the inhibition of killing by the T-cell. The right-hand image shows that drugs such as nivolumab can block this interaction and allow killing of the tumour cell by the patient's own tumour-specific T-cells. *Source:* Created with BioRender.com.

apoptosis to bring about their self-destruction. The importance of DNA repair in preventing the transmission of potentially carcinogenic mutations is highlighted by the increased cancer risk observed in patients with defects in DNA repair (Table 11.3). Defects in DNA repair pathways can lead to an increased mutation rate, contributing to *genomic instability* and the accumulation of genetic changes that drive cancer initiation and progression. One of the most common defects in human cancer involves the tumour suppressor gene, TP53, which encodes the p53 protein. Abnormalities in this gene have been detected in many common cancers, including those of the colon, breast and lung. Following a range of cellular stresses, including DNA damage, TP53 is expressed and activated, resulting in either cell cycle arrest until the damage is repaired or cell death by apoptosis. In this way, potentially harmful DNA lesions are not transmitted to daughter cells and therefore do not become permanent heritable mutations. For this reason, the TP53 gene has been referred to as the 'Guardian of the Genome'.

Abnormalities in DNA repair pathways can be targeted therapeutically, for example by exploiting *synthetic lethality*, a concept in which the simultaneous inhibition of two or more genes or pathways leads to cell death (see Case Study CS64; Fig. CS64.1). The use of PARP (Poly [ADP-ribose] polymerase) inhibitors is a well-known example. PARP proteins are important in base excision repair (BER). In cancers with BRCA1 or BRCA2 gene mutations, the homologous recombination repair (HRR) pathway, which repairs DNA double-strand breaks, is compromised. When HRR is impaired, cancer cells rely on PARP-mediated BER to maintain genomic stability and survive. By inhibiting PARP, the BER pathway is also compromised, leaving cancer cells with defective HRR and BER pathways. As a result, the cancer cells cannot repair DNA damage, which leads to the accumulation of lethal levels of DNA damage and ultimately cell death.

Table 11.3   Examples of inherited defects in DNA repair associated with an increased cancer risk.

Disorder	Defect	Common neoplasms
Xeroderma pigmentosum	Defect in nucleotide excision repair pathway. Inherited mutations in at least eight genes cause xeroderma pigmentosum, e.g. XPC, ERCC2, POLH genes	Various skin cancers, including basal cell carcinoma and malignant melanoma
Bloom syndrome	Defect in BLM gene leading to high mutation rate and hyper-recombination	Various types of cancer, characterised by early onset
Fanconi anaemia	Increased susceptibility to DNA cross-linking agents. Most mutations affect FANCA, FANCC or FANCG genes	Leukaemia, particularly acute myeloid type
Ataxia telangiectasia	Increased sensitivity to radiation and spontaneous chromosome translocations, impaired ability to induce p53-mediated cell cycle arrest. Mutation in ATM gene	Leukaemias and lymphomas particularly of T-cell type, some epithelial tumours, e.g. ovaries and stomach
Hereditary non-polyposis colorectal cancer	Defect in mismatch repair genes, MLH1, MSH2, MSH6, PMS2 or EPCAM	Colorectal carcinoma, endometrial adenocarcinoma

Healthy cells with functional BRCA1 or BRCA2 genes and intact HRR pathways are not killed by PARP inhibition, as they can still rely on the HRR pathway for DNA repair. PARP inhibitors, such as olaparib, rucaparib and niraparib, are approved for the treatment of ovarian, breast and other cancers with BRCA1 or BRCA2 mutations.

Mutational signatures are patterns of specific mutations characteristic of different mutational processes. They have been identified by analysing the frequency and types of mutations present in cancer genomes and comparing them to known mutational processes. It is beyond the scope of this book to cover all the mutational signatures, but examples are listed below:

- Signature 1 is age-related and found in many cancer types. It is characterised by a high frequency of C to T transitions, particularly in non-CpG sites.
- Signatures 2 and Signature 13 are enriched for C>T and C>G substitutions and are a consequence of 'off-target' cytidine deaminase activity of the AID (activation-induced cytidine deaminase) and APOBEC (apolipoprotein B mRNA editing enzyme, catalytic polypeptide-like) family of enzymes.
- Signature 7 is found in UV-associated cancers and is enriched for C>T substitutions at sites of adjacent pyrimidines (adjacent C or T); a particularly diagnostic subset being the CC>TT dinucleotide mutation.

Chromosomal instability (CIN) is another important mechanism involved in cancer development and progression. CIN is defined as a higher than normal rate of missegregation of chromosomes (or parts of chromosomes) occurring in mitosis. CIN can result in *aneuploidy*, in other words, an abnormal chromosome number. Aneuploid cells can have fewer chromosomes (*hypoploidy*) or more chromosomes (*hyperdiploidy*). Mechanisms underpinning the development of CIN include defects in the spindle assembly checkpoint (SAC). The SAC detects the presence of misoriented or detached kinetochores during mitosis and arrests cells in metaphase until all sister chromatid pairs are properly oriented to ensure their equal separation during cell division. The SAC is a multi-protein signalling cascade that blocks activation of the anaphase-promoting complex, and thus mitotic progression. The SAC comprises several proteins located at kinetochores, including the mitotic arrest deficient (MAD) proteins (MAD1, MAD2 and MAD3), the budding uninhibited by benzimidazole (BUB) proteins (BUB1, BUB2 and BUB3/BUBR1), the monopolar spindle 1 protein (MPS1), the ROD-ZW10-Zwilch complex and the microtubule motor centromere protein E (CENPE). In some cancers, inactivation of the SAC can occur by mutation or decreased expression of these genes allowing the uneven transfer of chromosomes to daughter cells, resulting in aneuploidy. Generally, aneuploidy reduces the fitness of cells, but as cancers progress it can allow for the evolution of genetically more abnormal cells. Aneuploidy is an indicator of poor prognosis in many types of cancer. Because hyperdiploid cells make more protein this can result in *proteotoxic stress*. Drugs are in development that can induce further proteoxic stress in cancer cells resulting in their apoptosis.

Field Cancerization

Field cancerization, also known as 'field effect' or 'field defect', is a concept that describes the presence of molecular and cellular alterations in a defined field of tissue that is 'at risk' of cancer development. It suggests that the entire affected tissue or organ is predisposed to the development of multiple independent cancerous or precancerous lesions rather than just a single localised tumour. The field effect arises from the accumulation of genetic and epigenetic changes in normal cells within the tissue or organ. These alterations can be caused by various factors, including exposure to carcinogens, chronic inflammation or genetic predisposition. The affected cells may exhibit DNA mutations, changes in gene expression patterns or modifications in epigenetic marks that contribute to abnormal cellular behaviour and increased susceptibility to oncogenic transformation. An example of field cancerization is observed in certain types of oral cancer, such as squamous cell carcinoma, in which exposure to risk factors such as tobacco smoke can lead to genetic damage and epigenetic changes in the entire epithelial lining of the oral cavity, not just in a single localised lesion. Consequently, multiple areas of the oral mucosa may develop dysplasia or transform into independent cancerous tumours.

Aetiology of Neoplasia

Chemical Carcinogens

The classic example of cancer following exposure to chemicals was described in 1775 by Percival Pott who noted that chimney sweeps had a high incidence of cancer of the scrotum which he attributed to exposure to soot. Most chemical carcinogens are *electrophilic* (electron seeking) and chemically modify the DNA of exposed cells, thereby inducing mutations. Some chemical agents are not themselves carcinogenic but are converted to carcinogenic derivatives by metabolic enzymes. These are called *indirect carcinogens*. Carcinogens which

do not require chemical modification for their cancer-causing properties are known as *direct carcinogens*. Table 11.4 gives examples of chemical carcinogens.

The process of chemical carcinogenesis (Fig. 11.9) can be observed experimentally by monitoring the effect of chemical carcinogens in animals. This has led to the concepts of *initiation* and *promotion*. Initiation is the acquisition of an irreversible change (i.e. a mutation) by a cell following exposure to a carcinogen. Initiation may not be sufficient for tumour formation. However, if the cell is exposed to a second agent, known as a *promoter*, then a tumour may be formed. Application of a promoter without prior initiation will not lead to tumours. Promoters work by inducing the proliferation of the initiated clones. Some chemicals possess the capability of both initiation and promotion and are known as *complete carcinogens*.

Diet/Obesity

Chemical carcinogens present in the diet can sometimes lead to dramatic increases in the risk of certain types of cancer. For example, the

Table 11.4 Examples of chemical carcinogens.

Chemical agent	Nature of exposure	Resultant cancer
Beta-naphthylamine	Chemicals used in the rubber industry	Bladder cancer in exposed workers
Benzo(a)pyrene	Constituent of cigarette smoke*	Lung cancer
Asbestos	Inhalation of asbestos fibres	Malignant mesothelioma of pleural cavity, adenocarcinoma of the lung
Aflatoxin B_1	Produced by *Aspergillus flavus* found on groundnuts	Hepatocellular carcinoma in populations whose diet includes affected nuts
Nitrosamines	Consumption of salted fish	Nasopharyngeal cancer, gastric cancer
Ethanol	Alcohol consumption	Increased risk of liver, breast, oral, oesophageal, pancreatic cancers
Cyclophosphamide	Drug used in the treatment of cancer	Lymphomas and leukaemias

* Note there are over 50 different carcinogens present in cigarette smoke.

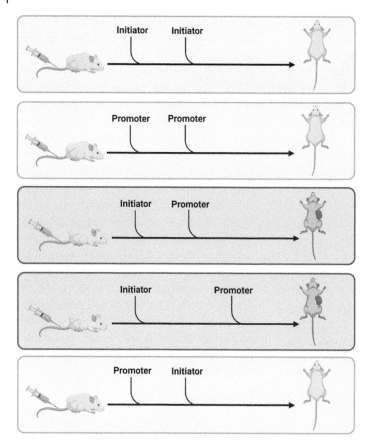

Figure 11.9 Experiments demonstrating the initiation and promotion stages of chemical-induced skin cancer in mice. From top to bottom: Application of either an initiator or a promoter alone does not result in tumour formation. Initiator followed by promoter results in tumour formation. Application of a promoter delayed for several months after initiation also results in tumour formation indicating that initiation has 'memory'. Promoter followed by an initiator does not produce tumours. *Source:* Created with BioRender.com.

consumption of nitrosamines in salted fish substantially elevates the risk of developing nasopharyngeal carcinoma among Southern Chinese populations. Another example is the consumption of aflatoxin B1 in groundnuts which leads to an elevated risk of hepatocellular carcinoma (see also Chapter 14). Apart from the presence of such potent carcinogens, a generally poor diet is also associated with an increased cancer risk. For example, it was recently estimated that around 10% of all cancer cases in the United Kingdom are linked to the deficient intake of fruit and vegetables, the consumption of red and processed meat and the deficient intake of fibre. The cancer sites

with the largest proportion of cases linked to obesity were uterus, kidney and oesophagus. Several possible mechanisms have been suggested to explain the association of obesity with an increased cancer risk (Fig. 11.10).

Radiation

The increased risk of cancer in individuals exposed to *ionising radiation* is well documented. Uranium miners, for example, have a 10-fold increased risk of lung cancer. Mortality rates from leukaemia and other cancers were increased in survivors of the Hiroshima and Nagasaki atomic explosions. Despite these

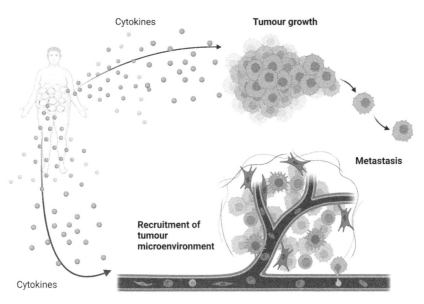

Figure 11.10 Obesity and cancer. Multiple mechanisms explain the increased cancer risk in obese individuals. There is reprogramming of adipose tissue leading to altered cytokine/adipokine secretion. Obesity also leads to alterations in systemic physiology that reflect the insulin resistance/metabolic syndrome. Local adipose tissue changes (and systemic effects) activate pro-proliferative, survival and metastatic signalling pathways either directly in the cancer cells or indirectly via the tumour microenvironment. *Source:* Created with BioRender.com.

high-profile examples, it is estimated that the most common sources of cancer-causing ionising radiation in the United Kingdom are from diagnostic tests, radiotherapy and environmental radon.

Ionising radiation is carcinogenic because it interacts with DNA and induces mutations. *Particulate radiation* (e.g. α- and β-particles) can react with DNA directly, whereas *electromagnetic radiation* (X-rays, γ-rays) is indirectly ionising by releasing energetic electrons when these rays are absorbed either directly by DNA or indirectly following an interaction with other molecules such as water; the release of electrons from water generates *free radical species,* such as the hydroxyl ion, which then react with DNA. The resulting disruption of chemical bonds leads to a variety of lesions in DNA, including base damage, intermolecular crosslinking and strand breaks.

Exposure to *ultraviolet radiation* is associated with an increased incidence of various skin cancers. In contrast to ionising radiation,

ultraviolet rays deposit energy which is insufficient to ionise molecules but is enough to excite them temporarily and make them chemically active. Ultraviolet rays can excite pyrimidine bases of DNA which then react with each other forming *pyrimidine dimers.*

Infectious Agents

The association between virus infection and the development of cancer was first described by Peyton Rous in 1911. Rous discovered the Rous sarcoma virus (RSV), a retrovirus that causes cancer in chickens (specifically, sarcomas). His work was instrumental in establishing the concept that viruses could cause cancer, a discovery for which he was awarded the Nobel Prize in Physiology or Medicine in 1966. However, it was not until 1964 that the first human tumour virus, the Epstein-Barr virus (EBV), was isolated from tumours of patients with African Burkitt lymphoma (BL). Subsequently, EBV was linked to the

Table 11.5 Examples of infectious agents associated with cancer.

Organism	Tumour association
High-risk HPV, e.g. 16,18	Cervical, oral, anal and head and neck cancers
EBV	Burkitt lymphoma, nasopharyngeal carcinoma, Hodgkin lymphoma, diffuse large B-cell lymphoma, gastric adenocarcinoma, primary CNS lymphoma and others
KSHV	Kaposi sarcoma, multicentric Castleman disease, primary effusion lymphoma
Hepatitis B virus	Hepatocellular carcinoma
Hepatitis C virus	Hepatocellular carcinoma
HTLV-1	Adult T-cell leukaemia/lymphoma
H. pylori	Gastric adenocarcinoma, MALT lymphoma
Schistosoma haematobium	Bladder cancer

development of other forms of cancer. Since the discovery of EBV, other human tumour viruses have been identified and their association with a variety of human cancers determined (Table 11.5). They include the high-risk human papillomaviruses (HPV) which are causal agents in the development of several epithelial cancers, including cervical cancer (Case Study CS6). Cervical cancer is the fourth most common cancer in women, and over 95% of cases are caused by sexually transmitted HPV. The discovery of HPV as a causative agent of cervical cancer has led to the implementation of vaccination programs; primarily targeting girls aged 9–14 before they become sexually active. Vaccinating secondary targets, such as boys and older females, is recommended where feasible and affordable. However, between 2019 and 2021, the coverage of the first dose of the HPV vaccination dropped by 25% to 15%, resulting in 3.5 million more girls missing out on HPV vaccination in

2021 compared to 2019. In December 2022, the WHO updated its recommendations to include either a one- or two-dose schedule for girls aged 9–14 years, a one- or two-dose schedule for girls and women aged 15–20 years or a two-dose schedule with a 6-month interval for women older than 21 years. This optimised HPV vaccination schedule aims to improve access to the vaccine, enabling countries to expand the number of vaccinated girls and alleviate the burden of complicated and costly follow-up required to complete the vaccination series.

The development of cancer is usually an infrequent consequence of virus infection, often occurring many years after initial infection. For example, approximately 95% of the world's adult population is infected with EBV. The majority of these individuals carries the virus asymptomatically and will not develop cancer as a consequence of EBV infection. Likewise, tumours associated with the Human T-lymphotropic virus 1 (HTLV-1) arise only infrequently in populations in which the virus is endemic.

Oncogenic viruses have evolved sophisticated strategies to induce infected cells into cell cycle and to protect them from apoptosis; all of which are designed to maximize the survival of the virus in the host and not to cause cancer. For example, the EBV gene, latent membrane protein-1 (LMP1), is a functional homologue of CD40, a cell surface receptor that regulates B-cell differentiation. The binding of the CD40 ligand (CD40L) to CD40 on normal B cells causes intracellular signalling, but when CD40L is absent this signalling is switched off. In contrast, LMP1 delivers ligand-independent constitutive intracellular signals which lead to the continued proliferation and survival of infected cells. Because LMP1 is expressed in the B cells of normal people infected with EBV, it is assumed that other cellular events are required to synergise with LMP1 in the development of EBV-associated cancers.

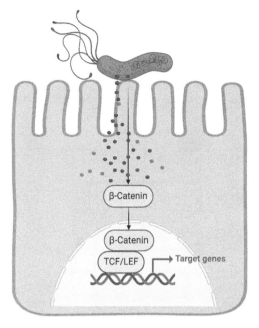

Figure 11.11 Aberrant signalling in *Helicobacter pylori*-associated gastric cancer. Some strains of *H. pylori* express a gene called cytotoxin-associated gene A (CagA), one of around 30 genes present in a DNA segment known as the cag pathogenicity island (cag PAI). Many of the cag PAI genes encode components of a syringe-like pilus structure known as a type IV secretion system (TFSS) through which CagA is delivered into the cytoplasm of gastric epithelial cells. CagA induces the activation of numerous cell signalling pathways, including beta-catenin signalling as shown here. *Source:* Created with BioRender.com.

Helicobacter pylori is a bacterium that is associated with the development of gastric adenocarcinoma. Eradication of *H. pylori* significantly decreases the risk of developing cancer in infected individuals. *H. pylori*-mediated inflammation of the gastric epithelium appears to underlie the pathogenesis of this type of gastric cancer. *H. pylori* can bind to gastric epithelial cells and insert several effector molecules into the cell which then activate intracellular signalling leading to the release of pro-inflammatory cytokines (Fig. 11.11).

Schistosomiasis is the second most common parasitic infection of humans in the world after malaria; approximately 250 million people are infected globally, mainly in tropical and subtropical regions of Africa, Asia and South America (see also Chapter 4). *Schistosoma haematobium* is now considered a causative agent in the development of squamous cell carcinoma of the urinary bladder. Several mechanisms have been suggested to explain the role of *S. haematobium* in bladder cancer. For example, fibrosis induced by schistosome eggs may induce proliferation and precancerous changes in the bladder epithelium (see also Chapter 4).

Clonorchis sinensis (see also Chapter 4) and *Opisthorchis viverrini* are parasitic flatworms, commonly known as liver flukes, that can infect humans through the consumption of raw or undercooked fish. These parasites are prevalent in certain regions of Asia, particularly in China, Korea and Vietnam. Chronic infections with *C. sinensis* and *O. viverrini* are strongly linked with the development of cholangiocarcinoma, a cancer of the bile duct epithelial cells.

Inherited Predisposition

The inheritance of a mutation in a single gene may be sufficient to greatly increase the susceptibility of an individual to one or more types of cancer. This susceptibility may be passed from one generation to the next. The classical example is retinoblastoma, which has already been discussed. Similarly, a defect in the tumour suppressor gene, neurofibromin (NF1), has been shown to be responsible for the increased risk of tumours in people with type 1 neurofibromatosis. This autosomal dominant condition is characterised by the development of multiple benign tumours derived from Schwann cells (*neurofibromas*), some of which may become malignant in later life.

Familial adenomatous polyposis (FAP; Case Study CS41) is another autosomal dominant condition in which affected people develop multiple adenomas of the colon. In adulthood,

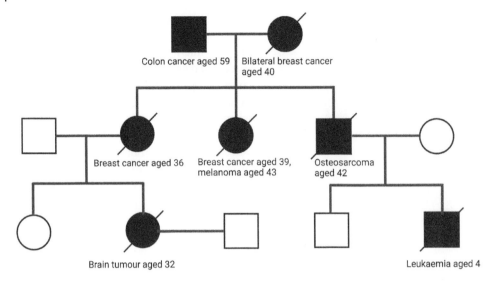

Colon cancer aged 59

Bilateral breast cancer aged 40

Breast cancer aged 36

Breast cancer aged 39, melanoma aged 43

Osteosarcoma aged 42

Brain tumour aged 32

Leukaemia aged 4

Figure 11.12 Pedigree of a family with Li–Fraumeni syndrome Squares represent males and circles females. Solid symbols represent affected individuals. A line through indicates the individual is deceased. *Source:* Created with BioRender.com.

there is a high risk of one or more adenomas transforming to carcinoma. Mutations in the APC gene account for this inherited susceptibility and are also one of the earliest detectable molecular abnormalities in sporadic colorectal cancer. Hereditary non-polyposis colorectal cancer (HNPCC or Lynch syndrome) is another, more common, form of inherited bowel cancer. Most cases of HNPCC are caused by mutations in one of the mismatch repair genes, for example, MLH1 or MSH2.

Often inherited mutations are identified as a result of familial clustering of cases. Sometimes these disorders are referred to as *cancer family syndromes*. An example is the *Li-Fraumeni cancer family syndrome*, the principal features of which are soft-tissue sarcomas in children and young adults, and the early onset of breast cancer in their mothers and other close female relatives. Leukaemia also occurs in excess. An autosomal dominant pattern of inheritance is seen and there is a high incidence of multiple primary malignancies. The Li-Fraumeni cancer family syndrome is the result of the germline transmission of TP53 mutations (Fig. 11.12).

The tendency of cancers to aggregate in families cannot always be explained by rare, high-risk, inherited mutations, but is often the result of the combined effects of multiple, common gene variants, known as *polymorphisms*, each of which is associated with a small increase in cancer risk. Alternatively, clustering of cancers within families can also be due to a shared environmental exposure or may also have occurred by chance.

Summary

Knowledge of the complex processes of tumour development, invasion, metastasis and angiogenesis is critical to an understanding of why tumours are so often lethal. The development of drugs that interfere with these processes presents exciting opportunities to reduce morbidity and mortality in cancer patients.

Key Points

1) Neoplasia, literally meaning 'new growth' is characterised by the accumulation of cells which are no longer responsive to the controlling mechanisms that maintain cell number in normal tissues.
2) Neoplasms are the tissue masses which result from the accumulation of these abnormal cells. Neoplasms may be benign or malignant. Malignant neoplasms are almost always life-threatening, whereas benign neoplasms rarely are.
3) Neoplasms are classified on the basis of the presumed cell of origin, irrespective of where they are found.
4) Neoplasms are believed to arise from a single target cell which has undergone a
series of mutations and epigenetic alterations. These changes enable the cell to escape the normal proliferative and survival constraints imposed upon normal cells.
5) Mutations may arise spontaneously, or as a result of exposure to a variety of environmental agents, including chemicals and radiant energy, or they may be inherited.
6) Local invasion of tumour cells and their ability to spread to distant sites (metastasis) are the hallmarks of malignancy.
7) Angiogenesis (the formation of new blood vessels) is an integral part of tumour progression and may be necessary for tumours to grow to a large size.

Further Reading

Hanahan D. (2022) Hallmarks of cancer: new dimensions. *Cancer Discovery* **12**: 31–46. https://doi.org/10.1158/2159-8290. CD-21-1059.

Alexandrov L.B., Kim J., Haradhvala N.J. *et al.* (2020) The repertoire of mutational signatures in human cancer. *Nature* **578**: 94–101. https://doi.org/10.1038/s41586-020-1943-3.

Niederhuber J.E., Armitage J.O., Doroshow J.H., Kastan M.B. & Tepper J.E. (eds) (2020) *Abeloff's Clinical Oncology* (6th edn). Elsevier. 978-0-323-64019-3.

DeVita V.T., Rosenberg S.A. & Lawrence T.S. (2022) *DeVita, Hellman & Rosenberg's Cancer: Principles & Practice of Oncology* (12th edn). Lippincott Williams & Wilkins (LWW). 978-1-97-518474-2.

Part 8

Obesity, Diabetes, and Environmental Diseases

12

Nutritional Disorders and Obesity

Introduction

Diseases resulting from inappropriate intake of nutrients range from micronutrient deficiencies through starvation ('protein-energy malnutrition') to obesity and related diseases.

Deficiency diseases resulting from malnutrition are now relatively uncommon in resource-rich nations. A far more significant group of disorders, which form a substantial and growing health and economic burden, are the so-called 'diseases of affluence', in particular, obesity.

Obesity is considered a disease of excessive adiposity sufficient to have significant adverse effects on health and well-being. The incidence of obesity has reached epidemic proportions in some countries. For example, in the United States, from 1999–2000 to 2017–2020, US obesity prevalence increased from 30.5% to 41.9% and the prevalence of severe obesity increased from 4.7% to 9.2%. The number of overweight or obese adults was estimated by the WHO to be >1.9 billion and >650 million, respectively, in 2016 (Fig. 12.1), meaning that 39% of the world's population were overweight and 13% (11% of men and 15% of women) were obese. The WHO estimates that the prevalence of obesity nearly tripled between 1975 and 2016 and is likely to further increase. The estimated annual medical cost of obesity in the

The Biology of Disease, Third Edition. Edited by Paul G. Murray, Simon J. Dunmore, and Shantha Perera.
© 2024 John Wiley & Sons Ltd. Published 2024 by John Wiley & Sons Ltd.
Companion website: www.wiley.com/go/murray/biologyofdisease3e

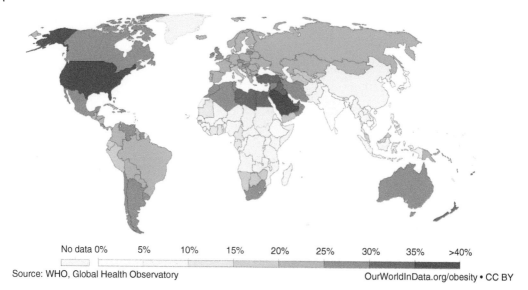

Source: WHO, Global Health Observatory   OurWorldInData.org/obesity • CC BY

Figure 12.1   Share of adults that are obese. Obesity is defined as a body-mass index (BMI) equal to or greater than 30. BMI is a person's weight in kilograms divided by his or her height in metres squared. *Source:* Data taken from https://ourworldindata.org/grapher/share-of-adults-defined-as-obese / Our World in Data / Public domain.

United States was nearly $173 billion in 2019. Obesity is a major risk factor for type 2 diabetes (T2D; Chapter 13), cardiovascular disease and cancer.

This chapter considers some of the more common deficiency diseases. It also describes obesity and considers genetic and environmental factors that contribute to the development of obesity. 'Eating disorders', such as anorexia nervosa are also discussed.

Protein-Energy Malnutrition

Malnutrition is now relatively uncommon in resource-rich nations, but remains a major problem in some resource-poor countries, especially during food shortages brought about by natural disasters, such as drought, and exacerbated by military conflicts. The prevalence of underweight people globally is shown in Fig. 12.2, underlining its inverse geographical relationship with obesity.

Malnutrition, arising from insufficient food consumption, is commonly referred to as *protein-energy malnutrition*. However, it is now understood that the cause is not primarily a lack of dietary protein, but rather a general insufficiency of food energy intake. While specific amino acids essential for nutrition – those the body cannot produce in adequate amounts – are required, undernutrition typically does not lead to deficiencies in these amino acids. When food intake is inadequate, protein is used for energy, rather than for growth or tissue repair. In individuals with low-fat reserves, insufficient energy intake leads to weight loss, with breakdown of the body's tissue proteins. The body mass index (BMI) is a useful measure of the degree of protein-energy malnutrition in adults (Table 12.1), as well as obesity (see later). BMI is calculated using the formula:

$$\text{BMI (kg/m}^2) = \text{weight}\left(\text{kg}\right) \div \text{height}^2\left(\text{m}^2\right)$$

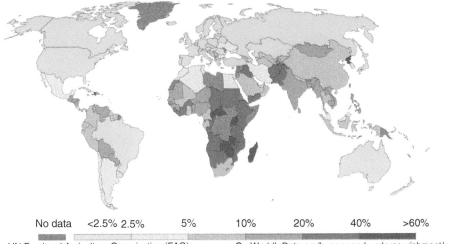

Figure 12.2   Share of the population that is undernourished. This is the main Food and Agriculture Organization (FAO) of the United Nations hunger indicator. It measures the share of the population that has a caloric intake that is insufficient to meet the minimum energy requirements necessary for a given individual. *Source:* Data taken from https://commons.wikimedia.org/wiki/File:Share_of_the_population_that_is_undernourished,_OWID.svg / Our World in Data / Public domain.

Table 12.1   Definitions of 'underweight' and 'malnutrition' using BMI.

Category	BMI
'Desirable'	20–25
Underweight	18.5–19.9
Moderate protein-energy malnutrition	17–18.4
Moderately severe protein-energy malnutrition	16–17
Severe protein-energy malnutrition	<17

Marasmus and Kwashiorkor

These are extreme forms of protein-energy malnutrition:

Marasmus is an inevitable consequence of a long-term shortage of food, caused by, or exacerbated by, famine. Sufferers exhibit extreme emaciation with no adipose reserves, muscle wastage and eventually loss of protein from critical organs such as the liver, heart and kidneys (although these are preserved as long as possible). The catabolism of protein is associated with reduced synthesis of protein in general and hence immune responses are impaired with a consequential increase in the risk of infections. Intestinal mucosal cell regeneration is drastically reduced, resulting in impaired absorption (the intestinal mucosa becomes secretory rather than absorptive under these conditions). Diarrhoea is a major problem in marasmus sufferers and worsens malnutrition. It can also result in difficulties when re-feeding is attempted.

Kwashiorkor is only observed in children. Children suffering from severe protein-energy malnutrition exhibit the same characteristic features as in marasmus, but in addition, they suffer from ascites, which produce misleading abdominal enlargement and limb puffiness which, from a distance, give the illusion of plumpness. The lack of sufficient protein intake leads to a decrease in oncotic pressure in the blood vessels. This reduction causes fluid to leak from the blood vessels into the interstitial spaces, including the abdominal cavity, resulting in ascites. Additionally, the liver may become fatty and enlarged. The colour and texture of the hair and skin are also affected. At one time, it was believed that the additional severity of kwashiorkor resulted from a lack of protein in the diet, even if total energy intake was adequate. This is now known not to be the case, but it is still not clear what additional factors result in kwashiorkor, rather than simply marasmus, although it is likely that general food deficiency coupled to reduced intake of antioxidants (such as vitamins C and E) are involved.

Micronutrient Deficiencies

Whilst protein-energy malnutrition is a major issue in resource-poor countries, deficiency of individual nutrients, especially of micronutrient *vitamins* and *minerals*, even if overall energy intake is adequate, is also a problem. The most common micronutrient deficiencies are:

Vitamin A Deficiency

Vitamin A deficiency is the leading cause of blindness in children worldwide, affecting approximately 14 million children under school age. This deficiency is particularly detrimental to vision because vitamin A, also known as retinoic acid or retinol, is a crucial component of rhodopsin, the visual pigment. Severe deficiency of vitamin A is referred to as *xerophthalmia*, characterised by a spectrum of ocular disorders including, initially night blindness, followed by abnormal dryness of the conjunctiva and cornea which can ultimately become ulcerated and result in blindness.

At a cellular level, retinoic acid, by interacting with its nuclear receptors, plays a pivotal role in controlling cell differentiation and turnover. The primary source of vitamin A is carotenoids – pigments found in many coloured vegetables. Beta-carotene, one of these carotenoids, is provitamin A, converting into vitamin A when broken down in the intestinal mucosa. Unlike vitamin A, which can be toxic at high levels, beta-carotene is much less toxic. Beyond its role as a precursor to vitamin A, beta-carotene, along with other carotenoids, also serves as a potent antioxidant. This property suggests its potential in the prevention of several disorders, including cancer, cardiovascular disease and diabetes.

Vitamins B1 and B2 Deficiencies

Deficiency of these vitamins is a problem in Africa and much of Asia. Vitamin B1, or *thiamin*, is required as a co-enzyme in carbohydrate metabolism. Many tissues are affected by thiamin deficiency, the most common result (coupled with relatively high carbohydrate intake) being *beriberi*, which is characterised by peripheral nerve damage and muscle weakness. Thiamin deficiency can also result in CNS damage (Wernicke-Korsakoff syndrome), especially in alcoholics.

Vitamin B2, or *riboflavin*, is involved in redox reactions in fuel metabolism. Deficiency results in skin disorders, *hyperaemia* (excess blood in the vascular beds of organs), mouth and throat oedema, and *cheilosis* (swollen, cracked lips).

Vitamin D Deficiency

Vitamin D is primarily produced when sunlight interacts with 7-dehydrocholesterol in the skin, producing cholecalciferol (vitamin D3). Cholecalciferol serves as a precursor for a hormone called calcitriol (1,25-dihydroxyvitamin D), which is synthesised by the kidney's 1α-hydroxylase after vitamin D undergoes 25-hydroxylation in the liver. In some cases, a synthetic form of vitamin D2, derived from ultraviolet exposure of the plant sterol, ergosterol, is used as a dietary supplement and can also contribute to the synthesis of calcitriol in a similar manner.

Calcitriol increases levels of calcium transport proteins, known as calbindin-D proteins, in the gastrointestinal tract, resulting in the increased uptake of calcium. This is the only mechanism by which the body can increase its calcium stores. Deficiency of vitamin D, due mainly to inadequate exposure to sunshine (and when dietary supplementation is absent), results in the under-mineralisation of bones, causing *rickets* in children and *osteomalacia* in adults. It is estimated that approximately 1 billion people worldwide have vitamin D deficiency. There is also increasing evidence for a protective role for vitamin D in infectious diseases, especially viral respiratory diseases, including COVID-19.

Other Vitamin Deficiencies

Other deficiencies include those of vitamin B6, B12 (resulting in pernicious anaemia, see Case Study CS23) and folic acid (see also Chapter 7). Folate supplementation during pregnancy decreases the likelihood of neural tube defects, such as spina bifida. *Scurvy* arises from a deficiency in vitamin C (ascorbic acid), while higher dosages of this vitamin may have beneficial antioxidant effects.

Iodine Deficiency

Iodine is essential for synthesis of the thyroid hormones, T3 and T4. In regions distant from the sea, where soil leaching in upland limestone areas prevents iodine absorption by food plants, enlargement of the thyroid gland, known as *goitre*, is common. The thyroid gland enlarges in an effort to increase iodine intake, but this may not be sufficient to produce enough thyroid hormone, leading to hypothyroidism.

Iron Deficiency

Iron deficiency results in anaemia because of the absolute need for iron in the *haem* moiety of haemoglobin. Anaemia is more common in women because of the loss of iron in menstrual blood. In some developing countries, intestinal parasites cause loss of blood in the faeces, resulting in anaemia. Iron deficiency anaemia is discussed in detail in Chapter 7.

Appetite Disorders

Disorders associated with the regulation of appetite and energy expenditure are increasing worldwide. In resource-rich populations where food is plentiful and relatively inexpensive, the incidence of obesity has been steadily rising, making it the most common chronic disease. However, there has also been a marked increase in other diseases associated with defective control of appetite, such as *anorexia nervosa* (an intense fear of eating despite marked weight loss) and *bulimia nervosa* (a related disorder involving normal or excessive 'binge' eating followed by self-induced vomiting before nutrition can be fully absorbed). Perhaps the cultural preoccupation with body image and weight gain is exacerbated by the obesity problem so that any fat is conceived to be 'bad'. Treatment for disorders associated with appetite is often difficult and relapse is frequent. However, recent advances in the understanding of the genetics of obesity and in appetite regulation at the molecular level, discussed below, may hold the key to the development of future therapies.

Obesity

Obesity can be crudely defined as an excess of adipose tissue. However, this begs the question, 'what exactly is an excess?' The presence of some adipose tissue is not only the norm but is necessary for normal function. 'Normal' percentages of body fat vary with age and biological sex (Fig. 12.3). Women have a higher normal adiposity than men. Levels of body fat substantially below these values can have significant physiological consequences – women who are anorexic, or elite athlete women with very low

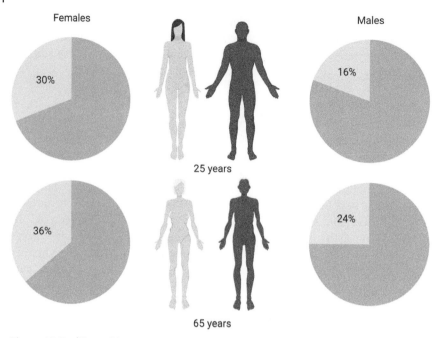

Females   Males

30%   16%

25 years

36%   24%

65 years

Figure 12.3 'Normal' body fat in males and females as a percentage of bodyweight. *Source:* Created with BioRender.com.

adiposity, are often infertile (amenorrhoeic) because some body fat is necessary for normal reproductive endocrine function. As we shall see later, adipose tissue secretes hormone-like *adipokines*, such as *leptin*, which can influence the hypothalamic-pituitary-gonadal endocrine axis, among their other widespread effects.

Because of its detrimental health effects, including the higher risk of various co-morbidities, such as T2D, cancer and cardio-vascular disease, obesity can be broadly defined as an excess of adiposity that increases the risk of disease and death. Extensive epidemiological data has established a connection between raised BMI and increased disease risk. For instance, the likelihood of a person developing T2D surges significantly as their BMI surpasses $30\,kg/m^2$, escalating to around 10 times greater than when the BMI is below $25\,kg/m^2$. Notably, during the COVID-19 pandemic, a high BMI was associated with a higher risk of severe respiratory disease and death. Using this type of evidence, BMI values associated with increased morbidity and

Table 12.2 Definition of overweight and obesity based on BMI.

Category	BMI (kg/m^2)
Underweight	<20
Healthy weight	20–24.9
Overweight	25–29.9
Obese	30–40
Severely (morbidly) obese	>40

mortality have been used to define overweight and obesity (Table 12.2).

Although BMI is widely used to define obesity, excess adiposity alone is not always linked to a higher risk of disease – the distribution of body fat is also a factor. *Central* (so-called visceral or abdominal) fat is more closely associated with co-morbidities than *peripheral* (subcutaneous) fat. Also, a high muscle mass and low amounts of fat, for example, in athletes and bodybuilders, can give a high BMI. Therefore, BMI should only be used as a crude surrogate marker of adiposity. Other measures, such as waist-hip ratio (WHR; the circumference of the waist divided by that of the hips), may be more accurate.

Apples and Pears: Differences in Body Fat Distribution and Disease Risk

The two major fat depots, visceral and subcutaneous, give rise to two distinct body shapes, often referred to as 'apple' and 'pear' (Fig. 12.4). 'Apple' obesity is male-pattern (*android*) central obesity involving visceral adipose tissue. This type of obesity also occurs in women after the menopause. 'Pear' obesity is female-pattern (*gynoid*) peripheral obesity, in which fat is deposited under the skin, especially in the hip and thigh area.

Compared to subcutaneous adipocytes, visceral adipocytes are more responsive to anabolic hormones, such as insulin (although this declines with insulin resistance) and catabolic hormones, such as adrenaline. This underlines the normal role of the visceral adipose depot as a short-term energy store, in contrast to subcutaneous adipose tissue, which is a longer-term store (e.g. to support the increased energy requirements of pregnancy). Pathological accumulation of long-term abdominal visceral fat is strongly associated with T2D and cardiovascular disease. In contrast, there is little correlation between the risk of these diseases and the size of subcutaneous fat depots. This is

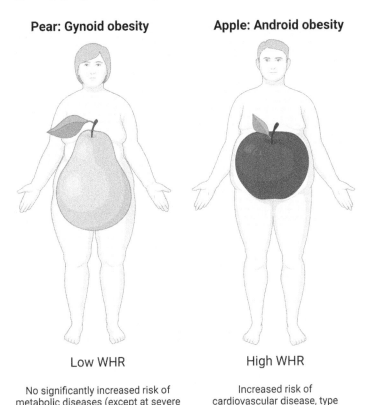

Pear: Gynoid obesity **Apple: Android obesity**

Low WHR High WHR

No significantly increased risk of metabolic diseases (except at severe end of obesity spectrum)

Increased risk of cardiovascular disease, type 2 diabetes

Figure 12.4 Body fat distribution – apples versus pears. Male pattern 'android' apple-shaped distribution of body fat reflects the accumulation of central (visceral) adipose tissue, which is associated with an increased risk of cardiometabolic diseases (e.g. type 2 diabetes and cardiovascular disease). In contrast, the female pattern 'gynoid' pear-shaped (subcutaneous) distribution is not associated with an increased risk of these disorders, except at markedly high BMI (e.g. severe obesity). *Source:* Created with BioRender.com.

particularly the case for fat in the hip and thigh region, although recent evidence suggests that more abdominal subcutaneous fat is linked to an increased risk of cardiometabolic disorders. Thus, BMI alone is not sufficient to categorise risk – it is important to also measure WHR, which is substantially increased in visceral obesity, but not in subcutaneous obesity.

Causes of Obesity

The cause of obesity appears, at first sight, to be rather simple and is an example of the first law of thermodynamics – the law of the conservation of energy – as such it can be considered as an 'inverse Micawber' effect:

As Charles Dickens had his character Micawber say in the novel, 'David Copperfield': *'Annual income twenty pounds, annual expenditure nineteen nineteen and six, result happiness. Annual income twenty pounds, annual expenditure twenty pounds ought and six, result misery'.*

Obesity is thus the inverse of Micawber's fiduciary dilemma, since excess energy 'income' over 'expenditure' results in excessive adiposity.

Energy is normally measured in joules (J) or kilojoules (kJ). In the case of food, kilocalories (kcal – often referred to as *Calories*) are still used. If energy intake, in the form of the chemical energy contained in food, exceeds energy expenditure (required for heat-generation-*thermogenesis*, muscular activity, metabolic and other physiological processes), then the excess energy is stored. Energy is stored either as glycogen (polymeric glucose), or, mainly, as triacylglycerol (triglyceride, TAG). TAG is favoured because it is a denser energy store (energy content 37.7 kJ/g or 9.0 kcal/g) compared to glycogen (16.7 kJ/g or 4.0 kcal/g; although the water of crystallization of glycogen, which is at a ratio of 2 g of water for every 1 g of glycogen, makes this even less dense). However, food intake is regulated by a combination of behavioural and neuroendocrine

factors. The body can adjust its energy expenditure to match intake, allowing for a decrease in expenditure during scarcity and an increase during abundance. Although changes in metabolic rate do not entirely explain shifts in energy expenditure, certain processes, like the modulation of non-shivering thermogenesis (carried out by *beige* adipocytes in adult humans), can impact energy expenditure. The concept of a *set-point* for body weight is evident, as weight loss leads to a reduction in metabolic rate, which in turn encourages a return to the original weight.

Genetic changes in humans due to evolution typically occur over extended periods, making it improbable that modern obesity trends are driven by recent genetic alterations. Instead, the surge in obesity, especially in the 20th and 21st centuries, is primarily attributed to environmental shifts, notably the increased availability of cheap, energy-dense, processed and highly palatable food in resource-rich regions, coupled with a sedentary lifestyle. Nonetheless, robust evidence underscores a significant genetic component of obesity, complemented by epigenetic effects, the latter of which can swiftly and heritably alter gene activity, potentially explaining how factors such as decreased nutrition availability can influence gene expression linked to body fat storage and eating behaviours. Several hypotheses have been advanced to explain the modern epidemic of obesity. They include:

- The *thrifty gene hypothesis* is based on the high incidence of obesity and diabetes in hunter-gatherer populations. In terms of human evolution, these populations recently moved from a traditional lifestyle, which supposedly incorporated periods of scarcity ('famine') interspersed with periods of abundance ('feast'), to a more 'Western' lifestyle of food abundance. This hypothesis suggests that selective pressures of the traditional lifestyle selected for genes which promote fat storage for use during food shortages. However, this hypothesis does not entirely account for variations in adiposity. Moreover, GWAS studies (Chapter 10) have not been able to identify the genes responsible.

- The *thrifty phenotype hypothesis* suggests that the increased prevalence of metabolic diseases observed in post–World War II generations could be attributed to the epigenetic re-programming of children whose mothers underwent significant food shortages during and immediately after the wartime period. It suggests that inadequate calorie intake during pregnancy led to adaptations in the baby, such as a predisposition to muscle insulin resistance and hyperinsulinaemia. These adaptations were meant to anticipate a nutrient-deficient environment after birth, but when that scarcity failed to manifest in the wealthier world, it resulted in obesity and diabetes. However, this explanation fails to account for the recent continuing rise in obesity and related conditions.

Several models have since been constructed to explain both the variation in adiposity between individuals, which appears to have a strong genetic component, and the recent rise in obesity, which would seem to be associated with the environment, in particular modern food availability and sedentary lifestyles. The two most prominent models are:

- The *energy-balance model*, which may initially seem grounded in the first law of thermodynamics, but which actually centres on the brain's pivotal role in regulating body weight. The brain orchestrates food intake by integrating external cues, such as available nutrition, with internal feedback from organs. Through a complex interplay of neural, hormonal and metabolite inputs, structures like the hypothalamus, brainstem and basal ganglia unconsciously control hunger and food preferences, guided by changing energy needs, which are in turn influenced by environmental factors such as food availability and types. Hormones like glucagon-like peptide-1 (GLP-1) and ghrelin, alongside neural signals via the vagal system, govern short-term eating patterns. Meanwhile, long-term signals, exemplified by the effects of leptin (see below), contribute to varying energy intake. As a result, short-term energy intake versus expenditure balance can fluctuate greatly over days, but the brain's influence on energy balance extends across broader timeframes. The energy balance model posits that the global rise in obesity primarily stems from the proliferation of inexpensive, energy-dense (rich in fats and simple carbohydrates), heavily processed foods, often aggressively marketed and deficient in vital nutrients, such as protein and fibre.

- The *carbohydrate-insulin model* proposes that excessive carbohydrate consumption, especially simple carbohydrates, leads to heightened insulin secretion and subsequent hyperinsulinaemia. This, in turn, supports fat storage as adipose tissue at the expense of non-adipose tissues, such as skeletal muscle. However, this model is probably an oversimplification since it assumes dietary carbohydrate is the sole metabolic source, and that insulin is the exclusive driver of fat accumulation, even though other factors contribute to fat storage. While the carbohydrate-insulin model forms the basis for numerous 'low carbohydrate' diets targeting weight loss, it is likely an incomplete framework and such diets are unlikely to be inherently superior to others.

The Genetics of Obesity

The earliest evidence of a genetic component to obesity was based on studies of *monozygotic* (identical) twins. Such twins, even when separated at birth and raised in very different family environments, develop similar bodyweights and susceptibility to obesity. Studies estimate that 75% of BMI is attributable to genetics. The Genetics of Obesity Study (GOOS), pinpointed around 15 genes associated with obesity. The Study Into Lean and Thin Subjects (STILTS), focused on naturally thin individuals with a BMI of $\leq 18\,kg/m^2$, identified variations in many of the same genes. Collectively, these studies demonstrate that genetic factors influencing adiposity contribute to some of the variation observed within the general population.

Monogenic Obesities

While several rare syndromic monogenic conditions (e.g. Prader-Willi syndrome, Chapter 10) are characterised by obesity, it is the non-syndromic monogenic obesities, especially those linked to the leptin-melanocortin system, that have provided the most significant insights into the genetics of obesity.

Leptin and Leptin Receptor Mutations

Some of the earliest attempts to identify obesity-related genes involved studies of monogenic obesity in animal models – the most notable being the genetically obese *ob/ob* mouse (Fig. 12.5). Researchers employed the labour-intensive and time-consuming approach of positional cloning to isolate and identify the *ob* gene. This breakthrough was met with considerable enthusiasm and media attention. It was demonstrated that the normal

Figure 12.5   An obese mouse and a healthy mouse; the mouse on the left has more fat stores than the mouse on the right. Human Genome wall for SC99 on ornl.gov. The original uploader was Bigplankton at English Wikipedia https://commons.wikimedia.org/wiki/File:Fatmouse.jpg.

version of the *Ob* gene in mice encodes a protein they dubbed *leptin* (derived from the Modern Greek 'leptos', meaning thin). The gene is now commonly referred to as the *Lep* gene. In the *ob/ob* mouse, a mutation in this gene either halts transcription or yields a 'nonsense' transcript that cannot generate a functional protein. As a result, these mice have an absolute deficiency of leptin. Mice heterozygous for the abnormality, carry just one copy of the mutated gene, and consequently produce enough leptin to prevent the development of obesity.

The *Lep* gene is primarily expressed in adipose tissue. As adipose mass enlarges, leptin secretion increases. Leptin functions as a signalling molecule, conveying information about the size of the adipose depot to specific areas of the brain that regulate appetite and energy expenditure, particularly within the arcuate nucleus of the hypothalamus (Fig. 12.6). Leptin was one of the first of a group of molecules, now known as *adipokines* (adipose-derived hormones and cytokines), to be identified. The significance of leptin in appetite control is evident in the *ob/ob* mouse, which exhibits excessive eating (hyperphagia) and is hyperinsulinaemic, insulin-resistant, diabetic and has a thermogenic defect, leading to an inability to maintain body temperature in colder environments. Treating *ob/ob* mice with recombinant leptin normalises their eating behaviour, body weight and metabolism.

Another genetically obese mouse, the 'diabetic' (*db/db*) mouse, carries a mutation in the leptin receptor gene (*Lepr*). The *db/db* mouse cannot express the long form of the leptin receptor, which acts as a type of cytokine receptor to activate JAK/STAT signalling. The *db/db* mouse exhibits the same phenotype as the *ob/ob* mouse, mirroring the traits of another rodent with a *Lepr* mutation, known as the 'Fatty' or Zucker (*fa/fa*) rat. Unlike the *ob/ob* mouse, mice with *Lepr* mutations have extremely elevated levels of circulating leptin and are unresponsive to both endogenous and injected leptin.

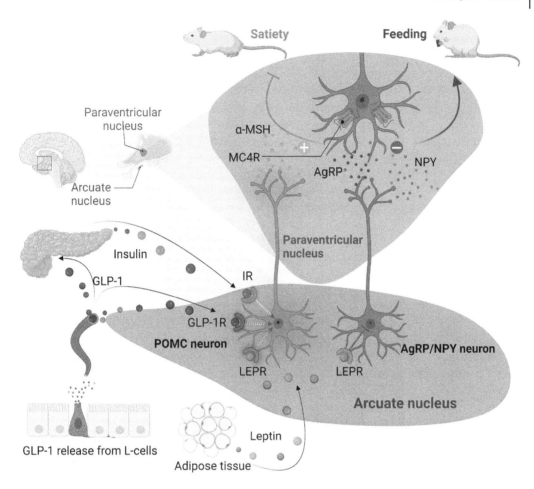

Figure 12.6 Leptin, insulin and GLP-1 play significant roles in appetite regulation. The arcuate nucleus in the hypothalamus contains two critical neuron populations: POMC (pro-opiomelanocortin) neurons and AgRP/NPY (agouti-related peptide/neuropeptide Y) neurons. Produced by adipose tissue, leptin activates POMC neurons, leading to the release of α-MSH (a POMC cleavage product). On the other hand, leptin inhibits AgRP/NPY neurons. The inhibitory action of leptin on AgRP/NPY neurons leads to a reduction in the synthesis and release of AgRP and NPY. Insulin is produced by the pancreas in response to elevated blood glucose. Insulin activates POMC neurons, further promoting the appetite-suppressing effects of α-MSH. GLP-1 (glucagon-like peptide-1) is produced by L-cells in the intestines in response to food intake. GLP-1 has both peripheral effects (e.g. it enhances insulin secretion) and central (brain) effects. GLP-1 receptors are expressed on POMC neurons and their activation promotes POMC activity and further contributes to appetite suppression. Neurons in the paraventricular nucleus express the melanocortin-4 receptor (MC4R). Activation of MC4R by α-MSH (from POMC neurons) leads to decreased food intake and increased energy expenditure. On the other hand, AgRP acts as an antagonist (or inverse agonist) to the MC4R, promoting appetite and NPY is a potent appetite stimulant. Thus, by inhibiting the activity and release of these neuropeptides, leptin plays a role in reducing food intake. *Source:* Created with BioRender.com.

Leptin influences neuronal circuits within the brain that regulate feeding and metabolism primarily through two types of neurons; those expressing POMC (proopiomelanocortin) or those expressing AgRP (agouti-related protein) (Fig. 12.6). In POMC-expressing neurons, leptin increases the release of αMSH (alpha-melanocyte-stimulating hormone), an anorexigenic peptide that inhibits feeding, and reduces the release of neuropeptide Y (NPY), which promotes food intake. In AgRP-expressing neurons, leptin downregulates the release of AgRP. This effect is crucial because AgRP typically blocks the receptor for αMSH, thereby counteracting the anorexigenic signal.

Leptin also exerts an indirect influence on insulin secretion by modulating neural pathways and directly engaging with insulin-producing cells in the pancreas. Conversely, insulin promotes the generation of leptin within adipocytes. This intricate interplay creates a feedback loop between the two hormones, contributing to the orchestration of energy homeostasis. Additionally, following food consumption, GLP-1 is generated by L-cells of the intestines and serves as an *incretin hormone* (a hormone released from the gastrointestinal tract in response to the intake of food), promoting insulin secretion and engaging with receptors in the hypothalamus to induce feelings of satiety. This concept forms the foundation for the development of GLP-1 receptor agonists as appetite suppressants in the therapeutic management of obesity. Leptin also regulates reproductive function in both animals and humans. For instance, female *ob/ob* mice and women with leptin deficiencies are often infertile (providing an explanation for amenorrhea in very thin females). In both species, affected males also exhibit delayed puberty.

Early investigations into both lean and obese individuals revealed a consistent correlation between body weight and blood levels of leptin but did not provide evidence of leptin deficiency or abnormalities in leptin receptors. This led to the concept of *leptin resistance* to explain why obese individuals consume food despite elevated leptin levels, although the specific mechanisms responsible are not known. However, mutations in the leptin gene do occur in humans, although they are rare (accounting for about 1% of cases of severe childhood obesity). Extremely obese children with a mutation in the leptin gene exhibit an intense drive to eat; treatment of these children with recombinant leptin reduces their appetite and weight.

Leptin also has a role in anorexia nervosa. Anorexic individuals appear to lose their normal feeding instincts. This would predictably happen if leptin levels were inappropriately high. However, this is not the case, as levels of leptin are very low in anorexic individuals, consistent with reduced adipose tissue mass. Low levels of circulating leptin in anorexia, paradoxically, lead to decreased appetite and reduced hunger cues, potentially contributing to food restriction. Additionally, the low leptin levels disrupt metabolism, slowing energy expenditure and aggravating weight loss challenges.

MC4R and Other Monogenic Forms of Obesity

A type of monogenic obesity seen in a strain of mice called the agouti (*Ay/Ay*) mouse, is caused by mutations that affect the function of αMSH, and its receptor, known as melanocortin receptor 4 (MC4R; Fig. 12.6). In these mice, the mutated agouti protein, instead of regulating pigment colour through the MC1R receptor, mimics the effects of AgRP, acting as an antagonist of the MC4R receptor. As a result, the mice exhibit a yellow colouration due to altered pigmentation and obesity due to disrupted feeding patterns.

In humans, mutations in MC4R, POMC and PCSK1 (prohormone convertase 1/3, which cleaves POMC to generate α-MSH) cause monogenic obesity. These mutations disrupt the normal regulation of appetite, energy expenditure and metabolism, leading to a genetic predisposition to obesity. Mutations in the MC4R gene are relatively common and account for about 5% of cases of early-onset (childhood) obesity and approximately 1% of all cases of obesity. These mutations may also be associated with changes in pigmentation, such as red hair. However, unlike the agouti mouse, obesity and red hair are not necessarily linked in humans. Notably, gain-of-function mutations in MC4R have been identified in about 6% of the human population and are associated with a reduced risk of obesity.

Polygenic Obesity

GWAS has identified around 1000 gene loci that are associated with obesity, although together they only account for around 5–6% of the total genetic influence on obesity – a considerable 'heritability gap' of unidentified genes still exists. One of the best-established genes is FTO (fat mass and obesity-associated protein), identified first in a GWAS of T2D. Individuals homozygous for specific variant single nucleotide polymorphisms (SNPs) are on average 3 kg heavier than those people homozygous for the protective variant. The FTO protein has enzymatic properties related to the demethylation of RNA; FTO is highly expressed in the hypothalamus and affects expression of genes encoding several feeding-regulating peptides, including downregulation of the orexigenic galanin-like peptide. FTO also plays a role in regulating adipogenesis. A large majority of genes identified in polygenic obesity are expressed in the brain and may have a role in the control of food intake.

Epigenetics and Obesity

Research, exemplified by the *Dutch Famine Cohort Study*, examining generations affected by wartime famine exposure, has highlighted the role of epigenetics in obesity and its related conditions, and has led to the so-called *epigenomic hypothesis* of obesity. In support of this, expression of many of the genes implicated in obesity are regulated by epigenetic processes. For example, changes in DNA methylation of the promoter region of the leptin gene are associated with alterations in adiposity and BMI. Epigenetic modifications in the adiponectin gene influence obesity-related traits. DNA methylation of the INS gene (which encodes insulin) and the IRS1 gene (which encodes insulin receptor substrate 1) are associated with insulin resistance and metabolic dysfunction. Changes in histone modifications, including methylation and acetylation, have been observed in response to dietary factors and can influence the expression of genes involved in adipogenesis and appetite control, such as POMC and NPY. Non-coding RNAs, including miRNAs and long non-coding RNAs, can also impact the expression of genes relevant to obesity. For example, IRS1 is downregulated by miR221, which in turn contributes to insulin resistance.

Crucially, DNA methylation and histone modifications can be inherited through both spermatozoa and oocytes, suggesting that the traits associated with obesity can be passed from either parent to their offspring. This has significant implications for our understanding of the heritability of obesity and related conditions.

Treatment of Obesity

At first glance, treating obesity might seem straightforward – controlling diet and increasing physical activity to reduce energy intake and raise expenditure. However, reality paints a more complex picture. The human body possesses a remarkable ability to uphold a stable 'set point' for body weight over extended periods. While extreme weight-reduction measures, like very low-calorie diets, can succeed in the short run (weeks/months), their long-term failure rate approaches 100%, as the body's physiology adjusts to restore its established equilibrium. Therefore, while genetics and epigenetics contribute to adiposity to a certain extent, the body is adept at adapting metabolically to maintain this 'set point'. Virtually any diet can lead to short-term weight loss, yet the most effective and enduring results stem from scientifically grounded diets that induce gradual and consistent reductions in adiposity. This is due to two primary factors:

- First, dramatic weight losses often stem from water loss. A brief examination of weight loss thermodynamics supports this. Thus, a claimed loss of roughly 7 kg of fat in a week implies a negative energy balance of $7000 \, g \times 9 \, kcal/g$ (fat) $= 63\,000 \, kcal/week$ or $9000 \, kcal/day$. This seems highly improbable given that the body typically expends about 2000–2500 kcal/day unless engaged in extreme physical activity.
- Second, a gradual weight loss of 0.5–1.0 kg per week, achieved through a well-balanced diet (ideally combined with additional exercise), decreases the tendency of the body to conserve energy stores around a 'set point'. This approach is more sustainable over the long term and may adjust the body's natural equilibrium.

Due to the inherent and unyielding nature of the body's ability to maintain a set point for body weight, which tends to increase with age, traditional dietary approaches to weight loss often prove ineffective. In such cases, alternative treatments may be required. Surgical interventions, such as gastric bypass or gastric banding, can offer success in severe cases of obesity, but the ultimate goal of obesity treatment, akin to a 'philosopher's stone', would be a pharmaceutical solution – a pill designed to make you lean.

Pharmaceutical endeavours to inhibit food absorption from the gut, exemplified by drugs, such as orlistat, that inhibit fat digestion and absorption, showed limited effectiveness when combined with diet and exercise. Moreover, they often also have undesirable side effects, such as oily stools (*steatorrhoea*). Sibutramine, another drug designed to suppress appetite by inhibiting the re-uptake of the neurotransmitter serotonin, caused severe side effects that prompted its withdrawal in 2010. Rimonabant, introduced in 2006, generated excitement due to its apparent effectiveness as an appetite suppressant. Acting as an antagonist to the CB1 cannabinoid receptor, it blocked the action of endocannabinoids responsible for stimulating appetite. Rimonabant was linked to serious psychiatric side effects and was withdrawn in 2008.

Following the discovery of leptin, Amgen paid $20 million to Rockefeller University to explore its potential as a drug. However, most obese individuals do not suffer from leptin deficiency but rather exhibit resistance to leptin. Consequently, for most obese individuals, high doses of leptin have little or no impact on weight loss. In contrast, leptin treatment has transformed the management of the rare leptin deficiency disorders.

Ongoing research focuses on identifying new drugs that can influence appetite, particularly by modulating the leptin-MSH-MC4R pathway. For instance, setmelanotide, an MC4R agonist, has demonstrated effectiveness in treating monogenic obesities caused by defects in the LEPR, POMC, or PCSK1 genes.

Most recently, a series of novel drugs initially developed for treating patients with T2D were shown to be effective in managing obesity and have received regulatory approval for such use. They include GLP-1 agonists, such as exenatide, liraglutide and semaglutide (see also Chapter 13). Semaglutide, in particular, demonstrates a remarkable average weight loss of up to 12% after 6 months of once-weekly injections. These injectable peptide medications work by reducing appetite through the activation of GLP-1 receptors (GLP-1R) in the hypothalamus.

Summary

Nutritional disorders range from deficiencies, including undernourishment – a major problem in resource-poor countries, to so-called 'diseases of affluence' – which is a growing problem among resource-rich nations. This chapter has described the most common of these disorders, paying particular attention to obesity. It has also discussed recent advances in our understanding of the genetics of obesity and considered new therapeutic approaches being developed to treat obesity.

Key Points
1) Obesity is the commonest, and most rapidly increasing, nutritional disorder in resource-rich countries. 2) Protein-energy malnutrition and micronutrient deficiency are still common in resource-poor countries. 3) Obesity has genetic and environmental components and is linked, especially in the case of 'abdominal' obesity, to an increased risk of type 2 diabetes, cancer and cardiovascular disease. 4) The study of monogenic obesity disorders has elucidated the role of the adipokine, leptin and the hypothalamic melanocortin system, in the regulation of feeding. 5) GWAS of polygenic obesity has identified over 1000 genes, each making a very small contribution to the overall risk. The majority of these genes influence feeding control mechanisms in the brain.

Further Reading

Bender D.A. & Cunningham S.M.C. (2021) *Introduction to Nutrition and Metabolism* (6th edn). CRC Press, New York.

Hall K.D., Farooqi I.S., Friedman J.M. *et al.* (2022) The energy balance model of obesity: beyond calories in, calories out. *The American Journal of Clinical Nutrition* 115(5): 1243–1254. https://doi.org/10.1093/ajcn/nqac031.

Trang K. & Grant S.F.A. (2023) Genetics and epigenetics in the obesity phenotyping scenario. *Reviews in Endocrine & Metabolic Disorders* 1–19. https://doi.org/10.1007/s11154-023-09804-6.

Farooqi I.S. (2021) Monogenic human obesity syndromes. *Handbook of Clinical Neurology* 181: 301–310. https://doi.org/10.1016/B978-0-12-820683-6.00022-1.

Friedman J.M. & Halaas J.L. (1998) Leptin and the regulation of body weight in mammals. *Nature* 395: 763–770.

13

Diabetes

Introduction: The Pandemic of Diabetes

The prevalence of type 2 diabetes (T2D) has reached pandemic proportions globally, with 537 million affected adults (20–79 years), representing 10% of the world's adult population. Current projections indicate that this figure will rise to 783 million by the year 2045 and 1.3 billion by 2050, affecting almost 10% of the world's adult population. Globally, 966 billion US dollars is currently being spent on healthcare for patients with diabetes, with countries like the United Kingdom allocating 10% of their healthcare expenditure towards the treatment of diabetes and its associated conditions. Given the projected increase in diabetes prevalence, the economic burden will continue to grow. This increasing prevalence is especially noticeable in resource-rich countries but is becoming an increasing concern in resource-poor countries. Epidemiological data over the past few decades indicate that the number of individuals affected by diabetes, primarily T2D, roughly doubles

The Biology of Disease, Third Edition. Edited by Paul G. Murray, Simon J. Dunmore, and Shantha Perera.
© 2024 John Wiley & Sons Ltd. Published 2024 by John Wiley & Sons Ltd.
Companion website: www.wiley.com/go/murray/biologyofdisease3e

every 10 years (Fig. 13.1). This increase in T2D aligns with the surge in obesity rates, a condition strongly associated with it (Fig. 13.2); projections indicate there will be ~1.12 billion obese individuals (and 2.16 billion overweight individuals) by 2045. The surge in diabetes is not limited to T2D, as the incidence of Type 1 diabetes (T1D) is also increasing.

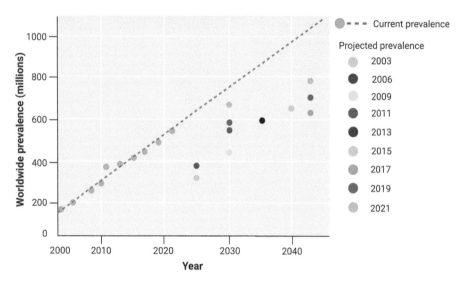

Figure 13.1 Projected worldwide prevalence of diabetes. Data from IDF Atlases of Diabetes 2006–2021 indicating that previous projections underestimated the likely future prevalence of diabetes. Current prevalence in green circles. Other coloured circles are projections that were made in the year indicated. *Source:* Created with BioRender.com.

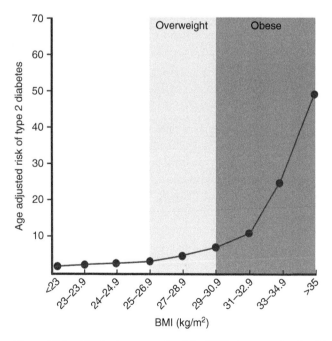

Figure 13.2 Obesity increases the risk of T2D. An example of epidemiological evidence showing that obesity is associated with an increased risk of T2D. Graph based on data from Chan J.M., Rimm E.B., Colditz G.A. *et al.* (1994) Diabetes Care **17**: 961–969. *Source:* Created with BioRender.com.

The History of Diabetes Mellitus

Diabetes was named '*diabetes mellitus*' by medieval physicians to distinguish it from the unrelated disease '*diabetes insipidus*', an endocrine disorder caused by a deficiency of posterior pituitary-derived antidiuretic hormone (ADH). The word '*diabetes*' is Greek meaning '*a siphon*' and refers to the copious urination observed by ancient Greek physicians (Aretaeus of Cappadocia in the 2nd century AD remarked on its rarity, which contrasts with its current prevalence). Medieval physicians tasted the patient's urine and classified the disease as 'mellitus' (honeyed) if it tasted sweet, or 'insipidus' (tasteless) if it did not; the presence of sugars in the urine providing a clue to the underlying pathogenesis of diabetes mellitus. In this chapter, we will employ the commonly used term 'diabetes' to mean diabetes mellitus.

Diabetes was first described by the ancient Egyptians in the Ebers papyrus c.1550 BC when it was a rare disease. The current high incidence could be linked to the advent of farming around 12 000 BC, which transitioned societies from a 'hunter-gatherer' lifestyle to settled living. The elevated prevalence of metabolic syndrome and diabetes in modern populations has been attributed to this shift away from a hunter-gatherer lifestyle and is referred to as the '*thrifty gene hypothesis*' (see also Chapter 12). An alternative, so-called, '*thrifty phenotype hypothesis*' (Chapter 12) is based on the increased prevalence of obesity and diabetes among the children of parents who endured periods of enforced caloric restriction, such as occurred in the United Kingdom during, and for some years after, World War II. This hypothesis suggests that the babies of this generation, who tended to be underweight because of maternal undernutrition, developed metabolic adaptations that allowed adequate post-natal growth. This in turn, led to an increased propensity to develop obesity and T2D. However, this hypothesis cannot explain the continuing rise in the prevalence of T2D in the face of more recent parental over-nutrition. Consequently, the focus has shifted to a potential role for epigenetics in explaining the enduring adaptations to undernutrition. While a deeper exploration of the epigenetic effects on obesity is provided in Chapter 12, the core idea is that periods of starvation/severe undernutrition induce epigenetic changes in gene expression that may be inherited by the children of those exposed. These alterations may include those that contribute to a predisposition to T2D.

The *Dutch Famine of 1944–1945*, also known as the '*Hunger Winter*' provides a unique historical circumstance that has contributed substantially to our understanding of epigenetic changes associated with malnutrition and its impact on metabolic diseases, such as obesity and T2D. During this famine, parts of the Netherlands were cut off from food supplies, leading to widespread malnutrition. Individuals who were *in utero* during the famine, known as the *Dutch Famine Birth Cohort*, have been extensively studied to understand the long-term health effects of prenatal exposure to malnutrition. Children born to mothers who were malnourished during pregnancy were found to be more susceptible to metabolic disorders like obesity and T2D and had different methylation levels of key genes, such as IGF2 and LEP, in late adulthood compared to their sex-matched siblings without famine exposure. Evidence suggests that these epigenetic changes can be passed down to subsequent generations. Research from the same cohort also indicated that the timing of the malnutrition during pregnancy could lead to different outcomes; malnutrition during early pregnancy having different effects compared to malnutrition later on, affecting different organ systems and metabolic pathways.

Ancient Indian physicians are credited with the identification of the two distinct forms of diabetes. They found that one form affected young individuals who were thin and this type progressed rapidly, often leading to death. The other form affected older people who were more obese and its onset was more gradual. We now refer to these two variants as type 1 diabetes (T1D) and type 2 diabetes (T2D). There are other recognised types of diabetes, including *gestational diabetes* and *monogenic diabetes*, which are discussed below and listed in Table 13.1.

Although diabetes is characterised by elevated blood glucose levels, the precise definition of hyperglycaemia has changed over the years. The presence of glucose in the urine is an early sign of diabetes and only occurs when the blood plasma level exceeds the kidney's reabsorption threshold of ~12 mmol/L, a figure that varies between individuals. Diabetes is diagnosed when the *fasting plasma glucose* concentration is ≥7.0 mmol/L, a figure set by the World Health Organisation in 2000. The hyperglycaemia is caused by failure of the insulin-secreting β-cells in the pancreatic islets to produce and secrete adequate amounts of insulin to maintain circulating glucose homeostasis, caused by autoimmune destruction of β-cells in T1D, or β-cell dysfunction with loss through apoptosis combined with insulin resistance, in T2D. Insulin resistance alone is not sufficient to cause diabetes, a fact underscored by the presence of insulin resistance (sometimes severe) in obese subjects who have not yet developed (and in many cases will never develop) T2D.

Insulin and the Control of Metabolism

Insulin is of central importance in maintaining glucose homeostasis and functions in post-absorptive metabolism (i.e. after a meal) as the key controller of blood glucose levels. Insulin reduces blood glucose by promoting uptake into key insulin-sensitive tissues, primarily the liver, skeletal muscle and adipose tissues. It stimulates the storage of glucose as glycogen in skeletal muscles and liver and increases its conversion to triacylglycerol in the liver and adipose tissues, as well as its storage in adipose tissues. Insulin secretion

must be intimately coupled to the prevailing circulatory glucose concentration. The failure of appropriate insulin release (together with increased levels of the counter-regulatory hormone, glucagon) and/or the loss of sensitivity to insulin (insulin resistance) leads to hyperglycaemia.

Insulin Synthesis

Insulin is initially synthesised in β-cells as a precursor protein, known as *pre-proinsulin*. The nascent polypeptide chain attaches to the endoplasmic reticulum via the hydrophobic signal 'pre' sequence, which is cleaved off. This is *proinsulin* which is cleaved by endopeptidases (PC1/3 and PC2) in the Golgi and the membrane-bound insulin secretory granules to release the final insulin molecule and the connecting C-peptide. When insulin is secreted, C-peptide is co-released in equimolar quantities. C-peptide has a longer circulating half-life than insulin (30 minutes for C-peptide versus 6 minutes for insulin) and is used as a laboratory marker of insulin secretion (Fig. 13.3).

A β-cell is classified by its function – the synthesis and secretion of insulin. Ultrastructurally, β-cells can be identified by the presence of electron dense-core insulin granules surrounded by a clear peripheral mantle; the granules are approximately 300 nm in size, with approximately 10 000 per β-cell. Mature β-cells express certain distinctive genes and transcription factors; for example, PAX4, PDX1, MAFA, MAFB, DLK1, SIX2/3, ID1, IAPP (which encodes amylin), UCN3, OLIG1, which are highly (or exclusively) expressed in human β-cells.

Insulin Secretion

The Islets of Langerhans (often simply referred to as 'islets') represent ~1–2% of the cells of the pancreas. Islets are composed of endocrine cells (Fig. 13.4A, B) that release their hormones into the portal venous

Table 13.1 Characteristics of the main types of diabetes.

Characteristic	Type 1 diabetes	Type 2 diabetes	Latent autoimmune diabetes of adulthood (LADA)	Gestational diabetes	Monogenic diabetes
Onset	Rapid; early (child/young adult) rare late onset 'LADA'	Insidious; most frequently older (>40 years) – but increasingly in young adults & children	Adults >25 years	During pregnancy	Early (birth onwards)
Proportion of diabetes patients	~5%	~95%	10%*	~5% of pregnancies	~1%
Insulin sensitivity	Sensitive	Resistant	Sensitive	Resistant	Sensitive
Genetic tendency	Slight (mainly HLA haplotype-monozygotic twins c.30% concordant)	Strong (monozygotic twins c95% concordant)	Moderate (33% of LADA have relatives with T2D)	Strong	Wholly (autosomal dominant: MODY1-14: Genes include GK, HNF1α, HNF 4α; HNF1β; subunits of SUR/KATP)
Obese?	No (may be once treated)	Usually (or history of)	No	Frequently	No
Islet cell antibodies (autoimmune)	95% positive at onset	No	Yes	No	No
Insulin secretion/ beta cell function	Very low/absent	Initially high, progressively failing and insufficient	Yes, for at least 6 months after diagnosis	High but Insufficient to low	Normal but inappropriate to insufficient
Treatment	Insulin (injection; inhaled(?); islet transplant, etc.)	Diet/exercise; metformin, SGLT2i, GLP-1/ DPP-4 inhibitors; insulin/ GLP-1 analogue, sulphonylureas, TZD	Insulin once LADA diagnosis is made	Diet/exercise; metformin; insulin	Depends on severity/gene: e.g. GK = none/diet; HNF1α sulphonylureas/insulin

* LADA may be misdiagnosed as T1D or T2D; therefore, this percentage is included also in the proportion of people with either T1D or T2D.

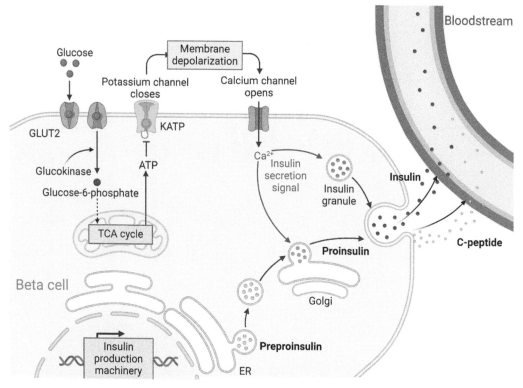

Figure 13.3 Glucose uptake and insulin secretion in β-cells. Glucose is transported into β-cells via glucose transporters (GLUT2). Glucokinase converts glucose to glucose-6-phosphate, which then triggers ATP (adenosine triphosphate) production (through multiple biochemical pathways not shown here for simplicity), which subsequently leads to the closure of ATP-sensitive potassium channels (KATP), depolarizing the cell membrane. This depolarization opens voltage-gated calcium channels, causing an influx of calcium ions (Ca^{2+}). The increased calcium concentration stimulates the exocytosis of insulin-containing vesicles. Meanwhile, pre-proinsulin, the precursor molecule, is translocated into the endoplasmic reticulum (ER), where the signal peptide is cleaved off, forming proinsulin. Proinsulin undergoes folding and disulphide bond formation in the ER and is transported to the Golgi apparatus. There, proinsulin is packaged into secretory vesicles. Enzymes within these vesicles cleave off the connecting peptide (C-peptide), leaving behind mature insulin and C-peptide. Upon stimulation by elevated calcium levels, these vesicles fuse with the cell membrane ("exocytosis"), releasing insulin and C-peptide into the bloodstream in equimolar amounts. C-peptide serves as a marker of insulin secretion. TCA cycle, tricarboxylic acid cycle. *Source:* Created with BioRender.com.

system, which then travel directly to the liver. The liver is a major target tissue for insulin and glucagon and, due to this anatomical arrangement, is exposed to the highest concentrations of these hormones. β-cells secrete insulin in response to elevated blood glucose levels (as well as other metabolic and neuroendocrine signals), α-cells secrete glucagon (which is counter-regulatory to insulin and secreted in response to falling glucose levels), δ-cells secrete somatostatin, ε-cells produce ghrelin and pancreatic polypeptide (PP)-cells

(sometimes known as γ cells) secrete pancreatic polypeptide. β-cells constitute ~60% of islet cells in humans and ~90% in mice. Each islet is a highly vascularised and highly innervated unit with a well-defined structure. The remaining 98% of pancreatic cells are acinar exocrine cells releasing digestive lipases and proteases.

Glucose-induced insulin secretion is controlled by glucokinase, an enzyme that is only expressed by β-cells and hepatocytes (Fig. 13.3). Unlike the hexokinase enzyme expressed by

(A)   (B)

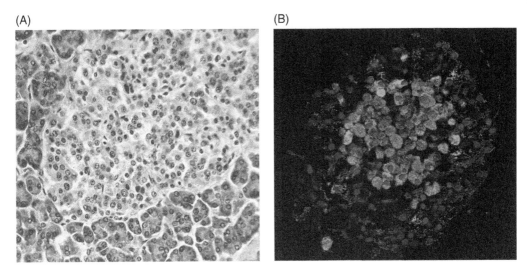

Figure 13.4   Pancreatic Islets. (A) H&E stain showing a high-power magnification of an Islet of Langerhans. The Islets are the endocrine cells of the pancreas and only compose about 2% of the cellular mass of the pancreas, 98% being acinar cells secreting digestive enzymes (proteases, lipases). Islet cells secrete peptide hormones via a capillary network into the portal vein. *Source:* Polarlys / Wikimedia Commons / CC BY 2.5. (B) Islet of Langerhans isolated from rat pancreas. Laser scanning confocal microscope image. 63×, oil immersion. objective. Nuclei stained blue with DAPI, Insulin (β-cells) stained green with anti-insulin dye-conjugated antibodies, Glucagon (α-cells) stained red with anti-glucagon dye-conjugated antibodies. *Source:* Masur / Wikimedia Commons / CC BY 2.5.

most cells, glucokinase has a much higher affinity for glucose, which is in the physiological blood glucose range. This means that a rise in blood glucose above the normal level (which is then reflected by a rise in intracellular glucose because of transport by the constitutively expressed glucose transporter of the β-cell, GLUT2) results in increased catabolism of glucose in the β-cell and a rise in the ATP/ADP ratio which depolarises the cell membrane by closing an ATP-sensitive potassium channel (KATP). This allows Ca^{2+} influx through a voltage-dependent calcium channel, which then stimulates the exocytosis of insulin granules.

Insulin Action

Insulin binds to cell-surface receptors on target cells in insulin-sensitive tissues (primarily liver, adipose tissue, skeletal and cardiac muscle). Insulin released following food intake, stimulates the uptake of glucose into these target tissues and increases glycogen synthesis in the liver and skeletal muscle. Insulin also stimulates lipogenesis and inhibits lipolysis in the

liver and adipose tissue. These effects occur through a complex signalling pathway initiated by insulin binding to the insulin receptor (IR) (Fig. 13.5). The IR is a heterotetrametric structure composed of two alpha and two beta subunits. Upon insulin binding to the extracellular alpha subunits, a conformational change activates the tyrosine kinase activity of the intracellular beta subunits. Each beta subunit autophosphorylates the adjacent one, amplifying the receptor's activity. This initiates a downstream signalling cascade that phosphorylates insulin receptor substrates, such as insulin receptor substrate 1 (IRS1) and IRS2. These substrates serve as docking platforms for various signalling molecules and adaptors, initiating a range of cellular events, including the recruitment of glucose transporter type 4 (GLUT4) glucose transporters to the cell membrane, which facilitates glucose uptake. Insulin also activates the mammalian target of rapamycin (mTOR) pathway, a central hub for cell growth and protein synthesis. Activation of mTOR stimulates anabolism, promoting protein synthesis and inhibiting protein degradation.

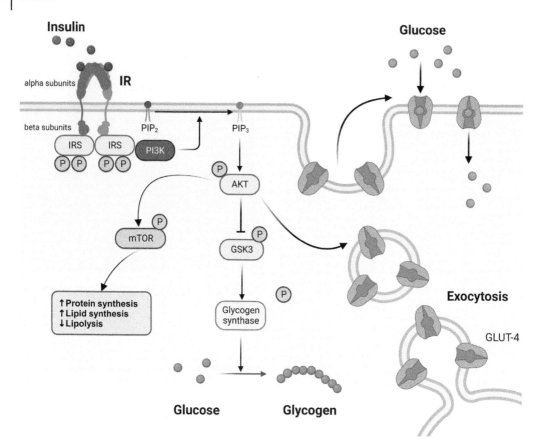

Figure 13.5 A simplified view of insulin signalling. Insulin signalling is a complex process that involves a series of molecular events triggered by the binding of insulin to its receptor on the surface of target cells. This signalling pathway plays a critical role in regulating glucose metabolism, cell growth and various physiological processes. One of the key downstream effects of insulin signalling is the activation of the AKT pathway, which in turn influences other important molecules such as glycogen synthase kinase 3 (GSK3), mammalian target of rapamycin (mTOR) and glucose transporter 4 (GLUT4). Insulin binds to the insulin receptor (IR) on the cell surface, leading to a conformational change in the receptor, which activates its intrinsic tyrosine kinase activity. The activated insulin receptor autophosphorylates tyrosine residues. Insulin receptor substrate (IRS) proteins are recruited to the phosphorylated tyrosine residues on the receptor. IRS proteins provide docking sites for phosphatidylinositol 3-kinase (PI3K). PI3K is activated and converts phosphatidylinositol 4,5-bisphosphate (PIP2) into phosphatidylinositol 3,4,5-trisphosphate (PIP3). PIP3 recruits and activates AKT, which phosphorylates and inhibits glycogen synthase kinase 3 (GSK3). Inhibition of GSK3 promotes glycogen synthesis by preventing GSK3 from phosphorylating glycogen synthase, allowing it to convert glucose into glycogen for storage. Insulin signalling also activates the mechanistic target of rapamycin (mTOR) pathway. mTOR plays a central role in regulating cell growth, protein synthesis and other cellular processes in response to nutrient availability and growth factors. AKT activation also leads to the translocation of glucose transporter 4 (GLUT4) to the cell surface. GLUT4 allows glucose to enter the cell, reducing blood glucose levels. Many of insulin's non-metabolic effects, including changes in gene expression and cell growth, are mediated by a separate pathway (not shown for simplicity) involving a Src homology 2 (SH2) domain of the receptor which, via interaction with growth factor receptor-bound protein 2 (Grb2) and Son of Sevenless (SOS), activates the mitogen-activated protein kinase (MAPK) pathway. *Source:* Created with BioRender.com.

Classification of Diabetes

T1D is almost invariably an autoimmune disease (also known in this case as type 1A diabetes). Rare variants of T1D exist that are of unknown aetiology and do not apparently have an autoimmune basis. These forms are referred to as *idiopathic* or type 1B diabetes.

T2D is strongly associated with obesity and insulin resistance, and usually arises in older adults, though it is increasingly seen in younger adults and even in children (which is probably linked to higher levels of obesity in these age-groups). *Gestational diabetes* is also associated with obesity and is essentially a form of T2D occurring in pregnancy, which may or may not persist after delivery. *Monogenic diabetes*, as the name implies, is a single-gene disorder. Monogenic diabetes usually follows an autosomal dominant pattern of inheritance, its severity depending upon which one of the 14 or so genes associated with it, are involved. There are other rarer forms of diabetes, including *mitochondrial diabetes* which is maternally inherited and often associated with deafness (see later).

Initial Presentation and Diagnosis of T1D and T2D

The signs and symptoms of diabetes are listed in Table 13.2. T1D usually presents more acutely with a short period of days or weeks of increasing severity of clinical symptoms, and at presentation, it is unusual for diabetes-related long-term complications to be present (see below); however, T1D patients may present with diabetic ketoacidosis (DKA; see below). Rarely, DKA can present in younger-onset T2D. Patients presenting with T2D may have had the disease for several years prior to its diagnosis and may be found on screening, coincidently, or because symptoms were precipitated by an event which increases insulin resistance, for example, a urinary tract infection. Consequently, newly presenting T2D patients may have both microvascular and macrovascular complications, discussed below.

Diagnostic criteria for all types of diabetes have been agreed upon by the WHO and the American Diabetes Association (ADA) (Table 13.3). *Fasting plasma glucose* (FPG; i.e. the circulatory blood glucose concentration) has been the traditional laboratory method for diagnosis, with a diagnostic threshold of $\geq 7.0\,\text{mmol/L}$ ($\geq 126\,\text{mg/dL}$).

Table 13.2 Signs and symptoms of diabetes.

	T1D	T2D
Symptoms	Polyuria – especially nocturia	Polyuria – especially nocturia
	Polydipsia – excessive thirst	Polydipsia – excessive thirst
	Lack of energy/fatigue	Lack of energy/fatigue
	Blurred vision	Blurred vision
	Polyphagia – constant hunger	Recurrent fungal infections (e.g. candidiasis)
	Unexpected weight loss	Tingling/numbness in extremities
		Slow-healing wounds
Clinical characteristics	Dehydration	
	Glycosuria	Glycosuria
	Muscle wasting	Obesity
	Ketonuria (frequent)	

Table 13.3 Diagnostic criteria for diabetes.

- Fasting plasma glucose (FPG) ≥7.0 mmol/L (126 mg/dL)

  or

- 2 hour plasma glucose ≥11.1 mmol/L (200 mg/dL) during an oral glucose tolerance test (OGTT) using a glucose load of 75 g

  or

- HbA1c ≥6.5% (48 mmol/mol)

  or

- Random plasma glucose of ≥11.1 mmol/L (200 mg/dL) in a patient with classic symptoms of hyperglycaemia or hyperglycaemic crisis

  or

- Where a state of hyperglycaemia is uncertain, diagnosis of diabetes requires two abnormal tests

Diagnosis requires a second confirmatory FPG on a separate occasion or an *oral glucose tolerance test* (OGTT). The non-diabetic, normal, reference range for FPG is 3.0–6.0 mmol/L. A value between 6.1 and 6.9 mmol/L is classed as *impaired fasting glycaemia* (IFG), which is considered to be *pre-diabetes*. Measures of random plasma glucose (taken most often when an urgent clinical situation requires a rapid assessment) can be indicative of diabetes, when greater than 11.0 mmol/L, but are highly influenced by recent food intake and are not diagnostic without a subsequent FPG measurement or OGTT (or unless the patient exhibits classic symptoms of hyperglycaemia or is hyperglycaemic crisis).

The OGTT involves the administration of a standard dose of glucose by mouth (75 g for adults) after an overnight (>8 hours) fast and measurement of plasma glucose before and 120 minutes afterwards, with intervening measurements occasionally taken. Fig. 13.6

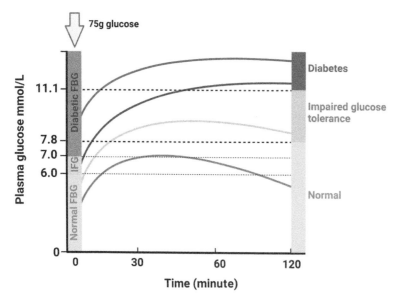

Figure 13.6 The oral glucose tolerance test (OGTT). A standard dose of 75 g glucose is given to overnight fasted adults (with the dose being adjusted for children and pregnant mothers). Blood glucose is sampled at time zero (fasting) and at 120 minutes. Intermediate time samples (e.g. 15, 30, 60 and 90 minutes) may also be taken if necessary. Time zero readings measure fasting plasma glucose (FBG); a value >7.0 mmol/L indicates diabetes, a value between 6.0 and 6.9 mmol/L is considered as impaired fasting glycaemia (IFG). The 120-minute post-glucose level will confirm diabetes (≥11.1 mmol/L), pre-diabetes (impaired glucose tolerance; IGT, 7.8–11.0 mmol/L) or non-diabetic status (<7.8 mmol/L). The upper red line shows an example of a patient with clear diabetes on FPG and confirmed by the 120 minutes level. The dark red line indicates a patient who has IFG based on FBG but who is confirmed to have diabetes by 120 minutes OGTT reading. The yellow line shows a case of an apparently normal FPG but with a 120 minutes glucose indicating pre-diabetes (IGT). The green line shows a normal subject who is non-diabetic on both FPG and 120 minutes post-OGTT criteria. *Source:* Created with BioRender.com.

shows the diagnostic threshold for diabetes, which is a 120-minute post-OGTT level of 11.1 mmol/L. A post-OGTT level between 7.8 and 11.0 mmol/L indicates *impaired glucose tolerance* (IGT), another indicator of pre-diabetes. It should be noted that although pre-diabetes confers a very significant risk of developing diabetes, it does not inevitably lead to the disease.

HbA1c, a measure of glycated haemoglobin, has been adopted as a diagnostic tool, in addition to its use in monitoring treatment of the disease. The basis of this test is the spontaneous reaction of glucose with amino-groups in proteins (an example of the so-called Maillard reaction) including intrachain lysine side-groups and amino-terminal amino acids (in HbA1c this is primarily the amino-terminal valine on the β-globin subunit). The following diagnostic criteria (established by the IDF and the WHO) apply:

- In individuals without diabetes, HbA1c accounts for around 5% (in old 'DCCT' measurement) of total haemoglobin (Hb). This gives a normal non-diabetic range for HbA1c of 31–41 mmol of HbA1c per mole of total Hb (5–5.9%). (HbA1c is currently measured in mmol/mol, according to the International Federation of Clinical Chemistry and Laboratory Medicine [IFCC] standard).
- Pre-diabetes is defined by an HbA1c level ranging from 42 to 47 mmol/mol (6–6.4%).
- Diabetes is diagnosed when the HbA1c level is 48 mmol/mol or higher (≥6.5%).

The Glycation Gap

When HbA1c levels are near the diagnostic threshold, inaccurate results can arise due to the presence of haemoglobin variants (such as HbS) or anaemia. However, these can usually be adjusted for and alternative diagnostic methods like FPG or OGTT can be used. The glycation gap (GGap; Fig. 13.7) is a discrepancy between measured HbA1c levels (which reflect intracellular glucose concentrations in the erythrocyte) and estimated HbA1c (based on glycaemia measurements using markers of extracellular glucose concentrations). In the

United States, a similar discrepancy indicator is termed the 'Haemoglobin Glycation Index' (HGI). GGap is calculated (in %) by subtracting the fructosamine-predicted HbA1c from actual measured HbA1c. In individuals with diabetes, a discrepancy of >1% occurs in at least 40% of patients and is consistent in magnitude and direction in the same individual over years. The GGap may impact on clinical treatment decisions as well as the risk of complications – diabetes patients with positive, compared with negative, GGap are at higher risk of microvascular and macrovascular complications and there is a U-shaped relationship between GGap (from negative to positive) and mortality. One potential explanation for the GGap is the relative activity of an intracellular deglycating enzyme, fructosamine-3-kinase, which is highly expressed in erythrocytes but only present at very low levels in the plasma and is three times more active in the erythrocytes of patients with a positive GGap compared to those with a negative GGap.

Pathophysiology of Type 1 Diabetes

T1D is a comparatively uncommon form of the disease that was conventionally thought to affect children and younger adults, though it is now known that onset can occur at any age (but with peaks in onset in young people). T1D is characterised by very low circulating insulin levels due to the targeted autoimmune destruction of β-cells. The rapid onset of disease becomes apparent once the residual insulin secretory capacity falls below ~30% and normoglycaemia can no longer be maintained. Evidence for an autoimmune basis for T1D includes:

- The occurrence of autoantibodies (AAbs) to islets and to specific β-cell antigens at or around the time of diagnosis (Table 13.4). These AAbs decline as the disease progresses, mainly due to the decline in β-cell numbers. The major autoantigens identified to date include anti-glutamic acid decarboxylase

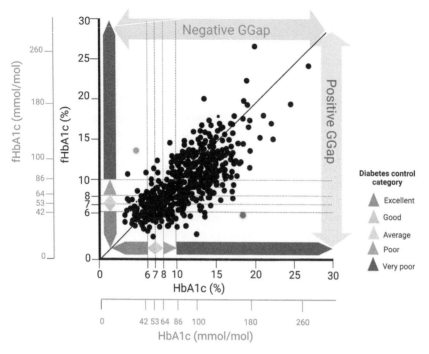

Figure 13.7 The Glycation Gap: The scatterplot visualises the relationship between estimated HbA1c (fHbA1c derived from fructosamine assays) and actual HbA1c levels in diabetes patients. By assessing HbA1c levels, diabetes control can be separated into distinct 'decision blocks': HbA1c >10% indicates very poor control; 8–10% denotes poor control; 7–8% represents average control; 6–7% signifies good control; and <6% is viewed as excellent control. Some patients exhibit a 'glycation gap' (GGap), which can shift their HbA1c readings, potentially impacting treatment decisions. For instance, a patient's treatment might change from a less aggressive approach like metformin or lifestyle changes to a more intensive one like insulin, or vice versa. Two examples of patients with negative or positive GGaps are shown as green and red dots respectively. *Source:* Created with BioRender.com.

Table 13.4 Diabetes-specific autoantibody tests.

Autoantibody	Abbreviation	Test	Specificity
Anti-glutamic acid decarboxylase autoantibodies	Anti-GAD65	Measures antibodies against glutamic acid decarboxylase in β cells	Present in 84% of patients with T1D
Insulin autoantibodies	IAA	Measures antibodies against the insulin molecule	Presence of IAAs is dependent on age and sex. IAAs are present in 81% of children <10 years with T1D, versus 61% in older patients. In patients <15 years, the presence of IAAs is similar in both sexes; in patients >15 years, the male:female ratio is 2:1
Insulinoma-associated-2 autoantibodies	IA-2	Detects antibodies specific for the enzyme, IA-2 phosphatase in β cells	Present in 58% of patients with T1D
Islet cell cytoplasmic autoantibodies	ICA	This test, measures the reaction between human islet cell antibodies and islet cell proteins from animal pancreas	Present in 70–80% of new-onset patients with T1D
Zinc transporter 8 autoantibodies	ZnT8Ab	This is a newer test that detects antibodies targeting zinc transporter 8 (ZnT8). Currently, this test is not readily available	Present in 80% of patients with T1D, with 99% specificity. Provides an independent measure of autoreactivity, as 25–30% of T1D patients negative for IAA, GAD65 and IA2 are ZnT8Ab positive

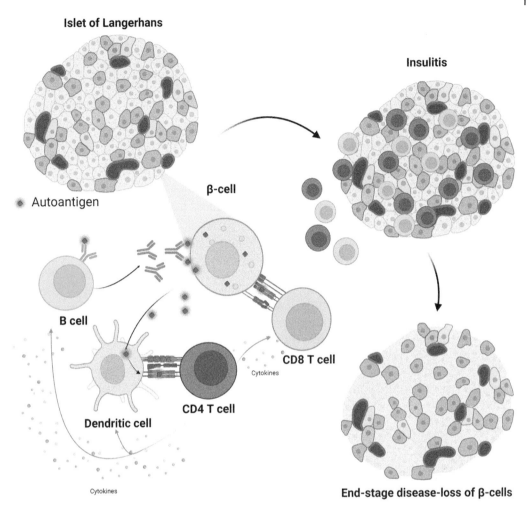

Figure 13.8 Simplified summary of the immunological basis of T1D. Antigen-presenting cells, including dendritic cells, present β-cell autoantigens such as GAD65 and insulin and promote the activation of β-cell-specific CD4+ and CD8+ T cells. The B cells react to being exposed to β-cell autoantigens by producing autoantibodies, which are islet specific and are present even before symptoms develop. T cell and inflammatory responses to the β-cells lead to insulitis (Islet inflammation) and cause damage and destruction of these insulin-secreting cells. Cytokines play a pivotal role in this process. Pro-inflammatory cytokines such as IL-1, TNF-α and IFN-γ are secreted by activated T cells and other immune cells. These cytokines amplify the immune response and contribute to the local inflammation within the islets. They can also directly induce β-cell apoptosis and disrupt the normal function of surviving β-cells, further exacerbating the loss of insulin secretion. The net result is a decline in function. *Source:* Created with BioRender.com.

autoantibodies (anti-GAD65), insulin auto-antibodies (IAA), insulinoma-associated-2 autoantibodies (anti-IA-2), islet cell cyto-plasmic autoantibodies (ICA) and zinc transporter 8 autoantibodies (anti-ZnT8A).

- In experimental systems, including animal models, such as the non-obese diabetic (NOD) mouse, β-cell-specific CD8 T-cells have been shown to destroy insulin-producing β-cells. CD4+ T-cells, also known

as helper T-cells (Chapter 6), amplify the immune response by activating other immune cells, including B-cells that produce the autoantibodies (Fig. 13.8). Lymphocytic infiltrates (*insulitis*) are rarely observed and only in occasional Islets of Langerhans in post-mortem examinations of individuals with newly diagnosed T1D (in contrast to the NOD mouse model where ~100% of islets have insulitis).

- The association of T1D with specific HLA haplotypes (see Chapter 6 and below). Certain haplotypes are associated with the earliest appearance of some autoantibodies, particularly anti-GAD65 and anti-insulin antibodies (see below).

Genetics and Environment in T1D

The concordance rate of T1D in monozygotic (identical) twins is relatively low (~30%, though recent longer-term studies have shown higher rates), which is much lower than in T2D. Genetic susceptibility to T1D is associated with specific HLA haplotypes, especially DR3-DQ2 and DR4-DQ8, which not only confer an increased risk, but also influence the timing of the appearance of autoantibodies; HLA-DR4-DQ8 and HLA-DR3-DQ2 are associated with the earlier detection of insulin autoantibodies and anti-GAD65 antibodies, respectively.

Environmental triggers are also apparently essential for the development of T1D. Human enteroviruses, especially coxsackie viruses, are significant triggers. This is because of the molecular similarity between Coxsackie B4 viral coat protein and a GAD65 autoantibody binding site. Other viruses implicated in the development of T1D include Rotavirus, Cytomegalovirus, Epstein-Barr virus, Parechovirus, Influenza virus, Parvovirus, Mumps virus, Rubella virus, and Human endogenous retroviruses. More recently, SARS-CoV-2 has been suggested to also trigger T1D.

Pathophysiology of Type 2 Diabetes

T2D is by far the most common type, accounting for ~95% of all diabetes. It is caused by the combination of insulin resistance and β-cell dysfunction/loss.

Insulin Resistance

Insulin resistance is a feature of obesity that is compensated for by hyperinsulinemia in non-diabetic obese individuals. T2D results when insufficient insulin can be produced to overcome the resistance. Insulin resistance starts primarily in skeletal muscle, developing later in adipose tissue and liver. There are multifaceted causes of insulin resistance, one factor being a decrease in cell-surface insulin receptors (insulin receptors may be internalised following insulin binding so that high concentrations of circulating insulin result in decreased numbers of receptors in the plasma membrane). Other factors include increased circulating FFA, ectopic triacylglycerol accumulation in muscle and liver, inflammatory processes and release of ROS, increased circulation of adipose tissue-derived detrimental adipokines, such as TNFα, and reduction in beneficial adipokines, such as adiponectin (discussed later, see also Chapter 12). A combination of all these factors causes a shift in the insulin-response curve to the right (Fig. 13.9). Insulin resistance is only rarely a consequence of genetic mutations affecting the insulin receptor or of its signalling pathway. The correlation between insulin sensitivity and insulin secretion follows a hyperbolic curve (Fig. 13.10). This relationship indicates that either heightened insulin secretion or improved insulin sensitivity, or both, can sustain normal glucose levels. However, the breakdown of these mechanisms eventually leads to IGT, with the potential to progress to T2D.

Failure of β-cells

β-cell failure, an essential prerequisite for the development of T2D, results from the loss of β-cell functionality, including secretion of partially processed pro-insulin, failure to respond to metabolic stimuli and loss of β-cells through apoptosis. Numerous obesity-associated factors contribute towards β-cell failure:

- *Inflammation:* Obesity often leads to a chronic, low-grade inflammatory state, which impairs β-cell function. For example, TNF-α and IL-6 can directly induce β-cell apoptosis.

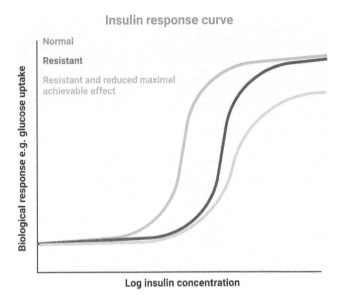

Figure 13.9 Insulin resistance. Insulin resistance may result from a decrease in receptor numbers or a post-receptor defect, which shifts the dose-response curve to the right, meaning that more insulin is required to achieve the same effect (red curve). A decrease in receptor numbers and/or a post-receptor defect in the insulin signalling pathway, can also lead to a decreased maximal response (orange curve). The latter is mostly observed in normal human insulin resistance. *Source:* Created with BioRender.com.

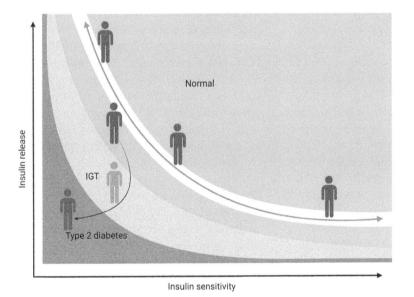

Figure 13.10 Hyperbolic relationship between insulin secretion and insulin sensitivity: The correlation between insulin sensitivity and the insulin response by β-cells is not linear, but forms a hyperbolic curve, making an evaluation of β-cell function contingent upon an understanding of both insulin sensitivity and the insulin response. Hypothetical zones outlining normal glucose tolerance (green), impaired glucose tolerance (IGT; yellow) and T2D mellitus (red) are depicted. With changes in insulin sensitivity, insulin secretion reciprocally increases or decreases to uphold normal glucose tolerance, leading to shifts 'up' or 'down' along the curve. Individuals at high risk of developing T2D see a transition from normal glucose tolerance to T2D, often passing through impaired glucose tolerance and effectively 'falling off the curve'. In many cases, these individuals veer from the curve even during normal glucose tolerance, indicating that β-cell function may be compromised before the onset of hyperglycaemia. Adapted from Kahn S., Hull R., Utzschneider K. *et al.* (2006) Nature **444**: 840–846. *Source:* Created with BioRender.com.

- *Raised FFA:* Elevated FFAs, particularly saturated fatty acids, can induce a state of *lipotoxicity*, in which the accumulation of toxic lipid metabolites in the β-cells interferes with their normal function, including insulin secretion.
- *Trans fatty acids:* Although still under study, there is growing evidence to suggest that dietary trans fatty acids may contribute to β-cell dysfunction, potentially exacerbating the effects of saturated fatty acids.
- *Metabolic stress:* Excessive nutrient supply can overwhelm the metabolic machinery of β-cells, causing metabolic stress that compromises their ability to secrete insulin.
- *Oxidative stress:* ROS can impair β-cell function.
- *Endoplasmic reticulum (ER) stress:* β-cells are metabolically highly active. Excessive metabolic demands can trigger the unfolded protein response (UPR) in an attempt to prevent the accumulation of unfolded/misfolded proteins in the ER; if the UPR is overwhelmed, ER stress results and can lead to apoptosis of β-cells.
- *Adipokines:* Detrimental adipokines can decrease β-cell function and/or number, whereas beneficial adipokines can have the opposite effect.
- *Amylin*, also known as islet amyloid polypeptide (IAPP), is a peptide hormone co-secreted with insulin by β-cells. While it plays a role in regulating gastric emptying and food intake, its accumulation and misfolding can lead to significant implications for β-cell health. In conditions such as T2D, prolonged hypersecretion of insulin and amylin can contribute to the formation of amyloid and its deposition within pancreatic islets. These amyloid deposits are composed of β-pleated sheet structures that aggregate into cytotoxic fibrils. The accumulation of these fibrils within β-cells can disrupt their normal function and lead to further cellular damage, apoptosis and reduced numbers of functional β-cells.

Genetics of T2D

T2D has a significant genetic component, demonstrated by almost complete concordance among identical (monozygotic) twins and a 10-fold-elevated risk for diabetes in the children of individuals with the condition. Nevertheless, it is important to acknowledge that the genetic makeup of contemporary humans is unlikely to have changed significantly from that of our ancestors spanning hundreds of thousands of years. Despite this, the prevalence of T2D has surged, implying a crucial role for environmental factors (and possibly of epigenetics, as discussed in the context of obesity in Chapter 12).

Genome-wide association studies (GWAS; Chapter 10) have identified multiple genetic susceptibility loci, underscoring the complex multifactorial basis of T2D. Most of the identified loci are associated with an insulin secretory deficit and only a minority are linked to insulin action/resistance, highlighting β-cell failure as a key factor which, when coupled with insulin resistance, precipitates diabetes.

Non-diabetic obesity is also characterised by insulin resistance, but the development of overt diabetes requires the presence of genetic risk factors, leading to failure of insulin secretion. Major genes so far identified include *TCF7L2*, which encodes a transcription factor involved in Wnt signalling. Single nucleotide polymorphisms (SNPs) in this gene are associated with a substantially increased risk of T2D (and gestational diabetes, see below), though the mechanism has yet to be elucidated. While multiple genes synergise to increase the risk of T2D, how they interact with environmental and dietary/lifestyle factors is complex and remains poorly understood. While obesity is often labelled a *modifiable risk factor*, and weight loss can indeed mitigate the risk and even induce diabetes 'remission', these effects might prove transient. Ultimately, the genetic component wields a potentially more powerful influence, steering individuals toward over-nutrition. Notably, obesity itself has a substantial genetic component (Chapter 12).

Role of the Gut Microbiome

A growing body of evidence supports the notion that abnormal changes in the gut microbiota, known as *dysbiosis*, contribute to the emergence of metabolic syndrome (discussed later and in Chapter 9) and T2D. Dysbiosis can be influenced by dietary factors (such as the low intake of dietary fibre or high-fat diets) as well as antibiotic usage. Dysbiosis can lead to fluctuations in metabolite levels, including short-chain fatty acids, which in turn, promote the development of insulin resistance and/or β-cell dysfunction.

Gestational Diabetes

Gestational diabetes mellitus (GDM) is a form of insulin-resistant diabetes (and so a form of T2D) that occurs in pregnancy. It is characterised by elevated blood glucose levels that develop for the first time during pregnancy in women who did not previously have diabetes. GDM affects around 14% of pregnant women globally, but prevalence varies, ranging from 7.1% in North Africa and the Caribbean, 7.8% in Europe, 10.4% in South America and Central America; 14.2% in Africa; 20.8% in South-East Asia, to 27.6% in the Middle East. It typically arises around the second or third trimester and is triggered by hormonal changes that affect how the body processes insulin. During pregnancy, the body becomes more resistant to insulin. In women with GDM, the body is unable to produce sufficient insulin to overcome this increased resistance, leading to elevated blood glucose levels. This can potentially pose risks to both the mother and the developing baby, which include:

- *Preeclampsia:* There is a higher risk of developing preeclampsia, a condition characterised by high blood pressure, and potential organ damage, during pregnancy.
- *Polyhydramnios:* in other words, the excessive accumulation of amniotic fluid.
- *Pre-term birth:* Gestational diabetes can increase the risk of premature birth by four- to fivefold.
- *Large birthweight:* Babies born to mothers with gestational diabetes are at risk of growing larger than average (*macrosomia*), which can complicate delivery and increase the risk of birth injuries.

- *Respiratory distress syndrome:* Babies born to mothers with gestational diabetes have a higher risk of respiratory distress syndrome.
- *Hypoglycaemia in the baby:* After birth, the baby may develop hypoglycaemia due to the excessive insulin production that occurs in response to high maternal blood glucose levels.
- *Risk of T2D:* A proportion of women affected by GDM may continue to have T2D following delivery (up to 20% in the USA) or else develop it later in life (up to 50% within 5 years).

GDM can also affect organogenesis in the foetus and rates of neonatal mortality are higher. Children of GDM mothers have an increased risk of obesity, insulin resistance and hypertension later in life. Managing GDM typically involves dietary modification, monitoring blood glucose levels, oral treatments such as metformin and if necessary, instituting insulin therapy. After delivery, blood glucose levels usually return to normal, but both the mother and the baby remain at increased risk of developing T2D later in life. Regular follow-up is important to monitor and manage risks.

The diagnosis of GDM uses the criteria of the International Association of Diabetes and Pregnancy Study Groups (IADPSG), which are considered to have a greater diagnostic accuracy (i.e. sensitivity and specificity) for GDM, than those defined by the WHO and American Diabetes Association. The IADPSG diagnosis is based on an FPG of ≥5.1 mmol/L and/or an OGTT 2-hour plasma glucose of ≥8.5 mmol/L. These criteria are not widely adopted; currently in the United Kingdom, an FPG of ≥5.6 mmol/L (100 mg/dL), or a 2-hour OGTT plasma glucose of ≥7.8 mmol/L (140 mg/dL), are used.

Less Common Types of Diabetes

Latent Autoimmune Diabetes of Adulthood

Latent autoimmune diabetes of adulthood (LADA) is a form of autoimmune diabetes that begins in adulthood (>25 years) and has an estimated 10% prevalence in people with diabetes. It has features of both T1D and T2D and is alternatively termed *Type 1.5 DM*. One important distinction from T1D is that LADA is slowly progressive and does not require insulin for glycaemic control for at least the first 6 months after diagnosis. In common with T1D, serum autoantibodies to β-cells are present. The majority of patients present with mild hyperglycaemia (a distinction from T1D) and are diagnosed and treated as T2D. Only later (perhaps years later) do these patients fail to control their hyperglycaemia on standard T2D oral drug regimens and require insulin.

A diagnosis of LADA should be considered if:

- The patient is >25 years
- Metabolic syndrome is absent; initial diagnosis being 'non-obese T2D'
- Hyperglycaemia is poorly controlled, despite using oral agents.
- There are co-existent autoimmune diseases (e.g. coeliac disease, autoimmune diseases of the adrenal or thyroid).
- The patient has elevated levels of pancreatic autoantibodies.
- C-peptide levels are low.
- There is only a moderate family history of T2D (33% of LADA patients have relatives with T2D).

The islets of patients with LADA are mainly infiltrated by macrophages, in contrast to the islets of patients with T1D, in which the infiltrate is predominantly composed of CD8+ T lymphocytes.

MODY and Other Forms of Monogenic Diabetes

Monogenic diabetes is the cause of 1–2% of cases and encompasses a diverse range of diabetes types caused by single-gene mutations that impact various aspects of glucose homeostasis, insulin secretion and β-cell function. The spectrum of monogenic diabetes is continually expanding as new genetic mutations are discovered.

MODY is an acronym for *maturity-onset diabetes of the young*, a form of monogenic diabetes occurring in children and young adults (Table 13.1). MODY bears little resemblance to T2D as there is no insulin resistance and, while some insulin secretory capacity is retained, insulin may not be secreted appropriately postprandially (after a meal). MODY is also distinct from T1D as there is no underlying autoimmune basis and it is not characterised by β-cell loss. MODY is typically inherited in an autosomal dominant pattern and is only rarely inherited recessively. Specific genes implicated in causing MODY include:

- GCK (MODY subtype 2), which encodes glucokinase.
- HNF1A (MODY subtype 3), which encodes hepatocyte nuclear factor-1 alpha. HNF1A is a transcription factor involved in β-cell development.
- HNF4A (MODY subtype 1), which encodes another member of the hepatocyte nuclear factor family, also critical for β-cell development.
- HNF1B (MODY subtype 5), a third member of the same family.
- INS (MODY subtype 10), which encodes insulin.

The severity of MODY depends upon the specific mutation. For example, people with MODY caused by GCK mutations may require no treatment or only dietary management, whereas those with either HNF1α or HNF4α mutations may need oral sulphonylureas or insulin. MODY should always be considered when there is:

- *A strong family history:* One of the hallmark features of MODY is its autosomal dominant inheritance pattern. This means that an affected individual has a 50% chance of passing the condition on to their offspring.
- *Low insulin requirement:* Many individuals with MODY have a significant amount of endogenous insulin production, leading to minimal insulin requirements, especially when compared to typical T1D.

- *Early age of onset:* While MODY can present at any age, it often manifests before the age of 25. An early age of onset combined with a strong family history and absence of typical T1D autoantibodies can be indicative of MODY.

In addition to MODY, other forms of monogenic diabetes exist, including:

- *Neonatal diabetes:* Mutations in the KCNJ11 gene (which encodes the Kir6.2 protein, a subunit of the KATP channel) and the ABCC8 gene (which encodes SUR1, a sulphonylurea receptor (SUR) subunit of the KATP channel) are associated with neonatal diabetes and are also found in some types of MODY. These mutations disrupt KATP channel function and insulin secretion.
- *Maternally-inherited diabetes and deafness (MIDD):* This rare form of diabetes is caused by mutations in mitochondrial DNA, which is inherited exclusively from the mother. These mutations impair cellular energy production, which in turn affects insulin secretion.

Exocrine Pancreas Diseases

Any disease process that damages the pancreas (pancreatitis, cystic fibrosis, haemochromatosis and partial/total surgical removal) can cause diabetes. These are forms of *secondary diabetes* (also known as Type 3c diabetes).

Drug-induced Diabetes

Secondary diabetes can also be caused by drugs. Drug-induced diabetes may be reversible if the drug is discontinued, though this is not always the case. Drugs that are well known to induce diabetes include corticosteroids, thiazide diuretics, beta-blockers, antipsychotics and statins.

Metabolic Memory (Legacy Effect)

The *metabolic memory* or *legacy effect* refers to the observation that achieving tight diabetes control early in the disease course can have lasting benefits in the later stages, even if diabetes

management at these later stages becomes less than ideal. This effect was noted in the Epidemiology of Diabetes Interventions and Complications (EDIC) study, which followed the Diabetes Control and Complications Trial (DCCT) of T1D patients. Despite the passage of years since the DCCT concluded, patients in the intensive treatment group continued to show reduced rates of both small blood vessel (microvascular) and large blood vessel (macrovascular) complications. A similar trend was observed in the United Kingdom Prospective Diabetes Study (UKPDS) of T2D patients; two decades after the study's end, participants from the intensive treatment group still exhibited fewer major cardiovascular incidents.

Aetiology of Chronic Diabetes-related Complications

The characteristic hyperglycaemia of diabetes arises either from the lack of insulin (in T1D), or from the relative deficiency (in the face of insulin resistance) of insulin (in T2D). Reduced insulin release, combined with normal or excessive secretion of glucagon from α-cells, causes many of the signs and symptoms of diabetes (Table 13.2).

Microvascular complications (neuropathy, nephropathy and retinopathy, see Case Study CS70) and macrovascular complications (e.g. cardiovascular disease and thrombotic stroke; see Table 13.5) arise in part from pro-inflammatory processes and the generation of ROS resulting from excessive uptake of glucose and fatty acids, and in part from the

Table 13.5   Complications of diabetes.

Microvascular	Macrovascular
Retinopathy	Cardiovascular disease
Nephropathy	Cerebrovascular disease
Neuropathy	Peripheral vascular disease
Erectile dysfunction	

non-enzymatic glycation of proteins. Glycation not only impairs the function of key proteins, but also generates advanced glycation end-products (AGEs), which in turn promote inflammation. In T2D (as in obesity) both insulin resistance and hyperinsulinaemia exacerbate vascular complications by promoting dyslipidaemia.

Microvascular complications are common in patients with T1D and T2D, and whilst macrovascular complications are more common in T2D patients, T1D patients do develop them as well.

While maintaining optimal diabetes control is crucial for preventing or decelerating the onset of diabetes-related complications, comprehensive patient monitoring is also important. For instance, regulating blood pressure minimises the risk of diabetic retinopathy and nephropathy. The dyslipidaemia often seen in diabetes patients presents as increased total cholesterol, raised levels of atherosclerotic low-density lipoprotein (LDL) cholesterol and triglycerides, alongside decreased levels of protective high-density lipoprotein (HDL) cholesterol. Statin therapy can mitigate these lipid complications. Thus, periodic medical checkups, including urine testing for albumin (an early indicator of kidney disease), and eye examination allow assessment of disease progression and informed treatment adjustments.

Metabolic Syndrome and the Link Between Obesity and Diabetes

Metabolic syndrome (which is also sometimes described as insulin resistance syndrome) may be defined as the co-occurrence of cardiovascular risk factors, including insulin resistance, obesity, atherogenic dyslipidaemia and hypertension. In clinical settings, specific criteria are used to diagnose metabolic syndrome (see Chapter 9). Metabolic syndrome is a complex and interconnected medical condition with an increased risk of developing cardiovascular diseases, T2D, and other health complications. It is not a disease itself, but rather a combination of several underlying health issues that together contribute to a higher risk of serious health problems. In the United States, it is estimated that approximately 45% of individuals over 50 years old have metabolic syndrome. Metabolic syndrome is considered a precursor to T2D.

Adipokines – A Potential Link Between Obesity, Diabetes and Vascular Disease

Adipokines are hormone-like molecules, sometimes referred to as 'adipocytokines', released by adipose tissue. The term adipokine is preferred over 'adipocytokine' because not all of these molecules exhibit cytokine activity. One of the earliest identified adipokines was TNFα, which was shown to contribute to insulin resistance in obesity. *Leptin*, discussed in detail in Chapter 12, is another well-known adipokine, with effects on insulin secretion and atherosclerosis. *Adiponectin* is a beneficial adipokine owing to its anti-inflammatory and anti-atherogenic properties, and its ability to enhance insulin secretion while reducing insulin resistance. Adiponectin is highly abundant in blood, constituting about 0.01% of plasma proteins and reaching levels as high as 30 µg/mL. Surprisingly, adiponectin secretion is higher in lean individuals and lower in obesity. This is because small adipocytes with less fat release more adiponectin, while bigger adipocytes in obese individuals release less. In contrast to adiponectin, most other adipokines are elevated in obesity and have harmful effects, for example by inducing insulin resistance and β-cell failure. Notable examples include resistin, apelin, chemerin, visfatin, omentin and adipsin. A comprehensive discussion of these and other recently described adipokines is outside the scope of this chapter.

Role of Non-coding microRNA in Diabetes

MicroRNAs (miRNAs) are a novel class of small non-coding single-stranded RNA molecules 18–24 nucleotides long that regulate gene expression at the posttranscriptional level. Evolutionarily conserved, miRNAs bind to the 3′ un-translated regions of messenger RNAs (mRNAs) and induce degradation of the mRNA or inhibit protein translation; their inhibitory actions lead to reduced target gene expression. miRNAs possess many critical regulatory functions in a wide range of biological processes, such as cell proliferation, differentiation, survival, apoptosis and the stress response. A single miRNA can modulate the expression and functions of hundreds of downstream target genes. In addition, the existence of feedback regulatory mechanisms between miRNA and their targets allows for amplification or inhibition of a specific signal. Hence, changes to only a handful of miRNAs can dramatically alter cell function. Emerging evidence suggests an increasing role for miRNA in both T1D and T2D with potential for their use as novel disease biomarkers. Altered glucose levels can influence the expression of miRNAs, which may, in turn, modulate pathways related to oxidative stress, inflammation and other cellular processes vital for maintaining glucose homeostasis. miRNAs also have key roles in modulating β-cell differentiation, function and survival.

Therapy for T1D and Its Complications

Insulin is the only option in T1D as there is an absolute insulin deficiency. The aim is to reproduce the normal physiological response to the ingestion of a meal (Fig. 13.11), which is composed of a first-phase response with rapid insulin release, followed by a more sustained second-phase response that maintains basal insulin levels between meals. This may be achieved by:

A *basal-bolus* regimen, also known as a *multiple daily injections (MDI) regimen*: This involves regular subcutaneous injections prior to meals and in the evening before bed. Typically, quick-acting insulin (acting for 1–2 hours, depending on the preparation) is used to replace the first phase response and may be injected before each of three meals (e.g. 8 a.m., 12 p.m. and 6 p.m.) with a longer-acting insulin (acting for 8–24 hours, depending on the preparation) injected at 9 p.m.

A *twice-daily mixed insulin regimen* (also known as a *biphasic insulin regimen*). This combines a quick-acting and a long-acting insulin injected twice daily (e.g. at 8 a.m. and 6 p.m.).

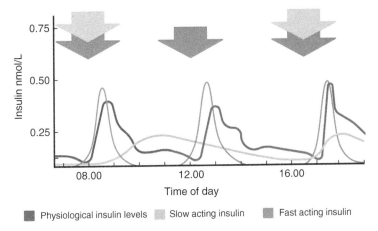

Figure 13.11 Physiological daily insulin levels and insulin injection targets. Normal daily physiological levels are shown in red and a target ideal 'mimicking' profile of insulin injection, which might be achieved using an appropriate regimen of slow-acting insulin (e.g. 'Lente') twice daily (blue dashed lines) and peri-prandial rapid-acting insulin (e.g. 'Lispro' – green dashed line). Note that there is a sustained physiologically low level of insulin secretion throughout the day and peaks at mealtimes, which show a typically biphasic profile with a rapid first phase and a slower second phase-declining as glucose level concentrations fall. Slow and fast-acting insulins may be administered separately or as a mixture to reduce the number of injections required. *Source:* Created with BioRender.com.

The morning dose provides coverage for breakfast with its quick-acting component and for lunch with its long-acting component. The evening dose provides coverage for the evening meal with its quick-acting component and covers basal insulin needs overnight with its longer-acting component. This regimen simplifies the injection routine, as it involves only two injections per day.

Subcutaneous insulin infusion (SCII) pumps, in which quick-acting insulin is given as a bolus before each meal, with a basal infusion between bolus doses (Fig. 13.12).

A hybrid closed loop system, which combines a continuous glucose monitoring (CGM) sensor with an insulin pump to automate the delivery of insulin. Using real-time glucose readings from the CGM, paired with a predictive algorithm, the system can adjust insulin infusion rates to maintain blood glucose levels within a target range. However, it is termed 'hybrid' because manual inputs are still required, especially during mealtimes, or when there is an anticipated sudden rise in blood glucose. CGM on its own is becoming an increasingly important method of monitoring diabetes in a non-invasive way with the added advantage that it can be monitored by patients using a smartphone or PC and the physician can easily calculate average blood glucose and glycaemic variability.

Inhaled insulin was developed but the first delivery system that was marketed ('Exubera') was withdrawn because of a number of issues, including poor uptake due to the bulky inhaler and concerns about respiratory system side-effects including a slightly increased risk of lung cancer. A newer inhaled insulin product ('Afrezza') was approved by the FDA in the USA in 2014 and is still marketed there for T1D and T2D but has a relatively limited uptake. Oral insulin is still in development.

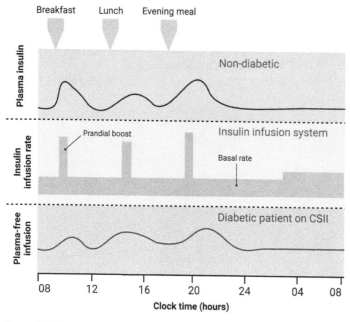

Figure 13.12 Insulin delivery by subcutaneous insulin infusion pumps. Insulin can also be delivered using subcutaneous insulin infusion (SCII) pumps in which quick-acting insulin is given as a bolus before each meal, with a basal infusion between bolus doses. Upper panel shows normal insulin levels in a non-diabetic individual. Middle panel shows the insulin infusion algorithm. Lower panel shows the insulin levels resulting from the SCII infusion and bolus doses. Insulin pump therapy is the most flexible insulin replacement therapy available, providing a constant rate of basal insulin, but the patient must activate boluses of insulin at mealtimes and at other times if blood glucose levels rise. *Source:* Created with BioRender.com

Cadaveric pancreatic islet cell transplant may be performed, but only in specialised centres, and often only for T1D patients with multiple diabetes-related complications. It is also restricted by the availability of donors. Patients receiving such a transplant must remain on lifelong immunosuppressive therapy to prevent rejection of the transplanted islet cells. This immunosuppression carries an increased risk of infection and malignancy. The use of transplants of stem-cell derived β-cells is still under development.

Monitoring insulin therapy to titrate insulin doses is undertaken by the patient using self-monitoring blood glucose measurements on plasma from a pinprick before and after meals. CGM devices are used that measure subcutaneous glucose. The HbA1c level is another crucial metric. It provides an indicator of average glucose control over a span of 12 weeks. Higher HbA1c levels point towards suboptimal diabetes management and a heightened risk for microvascular complications, such as retinopathy and nephropathy (Fig. 13.13).

Insulin deficiency in uncontrolled T1D not only has major metabolic consequences such as hyperglycaemia, increased protein catabolism and increased lipolysis (Fig. 13.14), but can progress to diabetic ketoacidosis (DKA, see below). Achieving good glucose control not only involves maintaining optimal blood glucose levels around mealtimes, but also simultaneously avoiding the risks of intensive management that could cause hypoglycaemia. However, treatment objectives vary depending on age and the existence of co-morbidities. For instance, elderly patients might prioritise avoiding hypoglycaemia, which could lead to dangerous falls, while someone recently diagnosed with T1D might aim for more aggressive glucose management to reduce the risk of future complications. In the DCCT study, compared to patients receiving the standard therapy, those randomised to receive intensive blood glucose control had reduced levels of retinopathy and nephropathy, but they also experienced more hypoglycaemic events (Fig. 13.15).

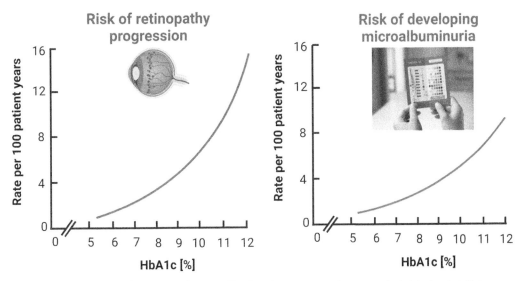

Figure 13.13 The risk of microvascular complications increases as glucose control deteriorates. The Diabetes Control and Complications Trial (DCCT) was a large trial of T1D patients randomised to either conventional treatment or intensive insulin treatment for six and half years. The risk of microvascular complications increased as glucose control deteriorated, measured by increasing levels of glycated haemoglobin (HbA1c). *Source:* Hxa098020 / Wikimedia Commons / CC BY-SA 4.0. Created with BioRender.com.

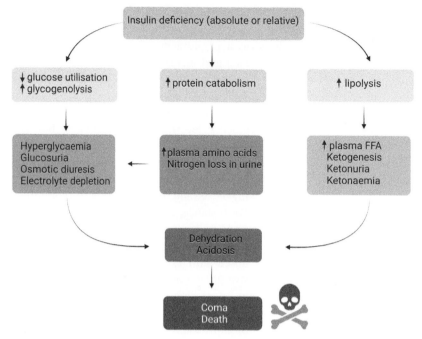

Figure 13.14 Effects of uncontrolled T1D. Uncontrolled T1D can have severe consequences. When there is an absolute lack of insulin, as in the case of T1D, or a relative insulin deficiency, which can arise from additional health conditions like infections or heart attacks that increase insulin resistance, a cascade of metabolic events ensues. This can ultimately lead to diabetic ketoacidosis, a dangerous condition that, if left untreated, can be fatal. *Source:* Created with BioRender.com.

Dose Adjustment For Normal Eating (DAFNE) is a structured education program designed for individuals with T1D. The primary aim of DAFNE is to teach individuals how to adjust their insulin doses to match their carbohydrate intake and lifestyle, allowing them more flexibility and better blood glucose control. In the DAFNE course, participants receive education about carbohydrate counting, insulin dose adjustment, as well as hypo- and hyperglycaemia management.

Treatment of T2D and Its Complications

The management of T2D is multifaceted and involves a combination of pharmacological interventions and lifestyle modifications; diet and exercise form the bedrock of management. Structured education programs, like Diabetes

Education and Self-Management for Ongoing and Newly Diagnosed (DESMOND), offer patients guidance and support in managing their condition. For those aiming to reduce obesity, extreme dietary interventions such as clinically supported very-low-calorie diets (approximately 800 kcal/day for several months) have demonstrated potential in inducing T2D remission. However, such stringent diets are challenging for patients to sustain over the long term. Pharmacologically, both oral and injectable medications help regulate blood glucose levels, typically reducing HbA1c by at least 0.5%. Each drug class operates differently, allowing for combination therapies to optimise individual patient outcomes:

Metformin is a well-established and frequently prescribed medication for the treatment of T2D. Metformin works by suppressing gluconeogenesis and glycogenolysis in the liver, which in turn reduces the release of

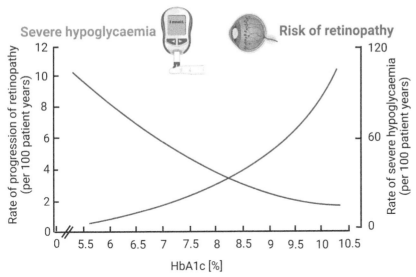

Figure 13.15 Intensive insulin therapy increases the risk of hypoglycaemia. In the DCCT study, increased glucose control (measured by HbA1c) reduced the risk of retinopathy (red line), but increased the risk of hypoglycaemia (blue line). *Source:* Created with BioRender.com.

glucose into the bloodstream. It is often the first line of treatment for T2D due to its efficacy, safety profile and low risk of causing hypoglycaemia compared to some other diabetes medications. In addition, it may increase glucose usage in the muscles. Metformin also results in lower serum lipid concentrations. Gastrointestinal side effects may limit its use or maximal dosage and it is contraindicated in patients with decreased renal function. In the UKPDS trial, metformin treatment reduced all-cause mortality from diabetes and its benefits in terms of reduced macrovascular complications were maintained on follow-up.

Sodium-glucose co-transporter-2 (SGLT2) inhibitors are oral agents and include dapagliflozin and empagliflozin. They inhibit SGLT2, which is responsible for 90% of glucose reabsorption in the renal proximal tubules. SGLT2 inhibitors induce glycosuria with associated loss of energy from the body and weight loss. Side effects include genitourinary infections and an increased risk of DKA, including a form known as *euglycaemic DKA*, in which ketone levels are high, but blood glucose levels are normal. SGLT2 inhibitors are less effective

and contraindicated in patients with renal impairment. Empagliflozin and other SGLT2 inhibitors are now frequently recommended for T2D patients who also have cardiovascular disease or heart failure, as they not only help in managing blood glucose levels but also offer cardio-renal protection.

Incretin therapies are based on the action of the incretin hormones, GLP-1 (glucagon-like peptide-1) and GIP (glucose-dependent insulinotropic peptide) released from the L and K cells of the small intestine, respectively, in response to food intake. They enhance insulin secretion, lower glucagon levels, slow gastric emptying, and promote glucose control and feelings of satiety. They include: *DPP-4 inhibitors* (gliptins; e.g. sitagliptin, vildagliptin and saxagliptin), which act on the enzyme, dipeptidyl peptidase 4 (DPP-4), which is responsible for degrading GLP-1 and GIP. The main action of GLP-1 is to stimulate glucose-dependent insulin release from the pancreas. DPP-4 inhibitors have few side effects and do not cause hypoglycaemia, but they should not be used in patients with pancreatitis. Linagliptin can be used in renal failure. *GLP-1 analogues*, given as subcutaneous daily or weekly injections, include exenatide, liraglutide and semaglutide. They have modified structures that make them resistant to DPP-4 degradation. Like GLP-1, they stimulate glucose-dependent insulin release from the pancreas. In addition, they slow gastric emptying, decrease post-prandial glucagon release and reduce food intake. Liraglutide and semaglutide are also licensed for weight loss in obese people without diabetes. They do not cause hypoglycaemia, but have gastrointestinal side effects, and should also not be used in patients with pancreatitis. Liraglutide, semaglutide and dulaglutide are also recommended for patients with established atherosclerotic cardiovascular disease as they have been shown to reduce adverse cardiovascular outcomes, including mortality. The newly approved dual GLP1/GIP analogues, known as *twincretins*, are administered as weekly subcutaneous injections. Currently, tirzepatide is the sole representative of this class, though more are being developed. They can lead to better blood glucose control and significant weight loss, and do not cause hypoglycaemia. However, they have gastrointestinal side effects and are not recommended for patients with pancreatitis.

Second-line therapies *Sulphonylureas* are oral agents that include glibenclamide (long-acting) and glipizide (short-acting). They stimulate insulin release by binding to a subunit of the KATP channel in β-cells. Common side effects include weight gain and hypoglycaemia. Caution is needed when using them in patients with renal disease. *Meglitinides*, including nateglinide, work in a similar way to sulphonylureas, but bind to a different site on the KATP channel.

Thiazolidinediones (TZDs), also known as 'glitazones', are oral medications. Pioglitazone is the only member of this drug class available for use in Europe. TZDs primarily work by activating the PPARγ nuclear receptor, particularly in adipose tissue and muscle, which improves insulin sensitivity. Pioglitazone also reduces liver inflammation and fibrosis in patients with T2D who also suffer from non-alcoholic liver disease. However, these drugs can cause fluid retention, which poses a significant risk for patients with heart failure.

In patients with T2D, insulin therapy might become necessary as the disease advances due to the progressive decline in β-cell function. Additionally, during certain illnesses or times of stress, the body may experience heightened insulin resistance, thereby temporarily increasing the demand for insulin. In such situations, daily or even twice-daily, injections of longer-acting insulin might be prescribed. This insulin regimen is typically combined with oral glucose-lowering medications to achieve optimal glycaemic control.

Large-scale clinical trials (such as the ADVANCE, VADT and ACCORD trials) have underscored the importance of maintaining optimal glucose control in T2D patients. These trials revealed a marked reduction in microvascular complications, including retinopathy and nephropathy, due to effective glucose control. However, in the ACCORD study, which pursued an aggressive diabetes control strategy targeting an HbA1c level of 6.5% or below (42 mmol/mol), there was an increase in both hypoglycaemic episodes and mortality. A possible explanation for this is that certain biochemical and haematological parameters might not revert to baseline even 24 hours post a hypoglycaemic incident, potentially elevating the risk of cardiovascular events. The long-term data from the UKPDS trial supports the concept of metabolic memory, described earlier. It suggests that establishing good diabetes control early on, can reduce the risk of macrovascular complications, particularly cardiovascular events, in subsequent years (Fig. 13.16).

Blood glucose monitoring is typically not required in T2D patients, unless the patient is on insulin, though there is increasing use of CGM devices. Medical practitioners often rely on HbA1c measurements done every 3–6 months in clinic to guide therapy.

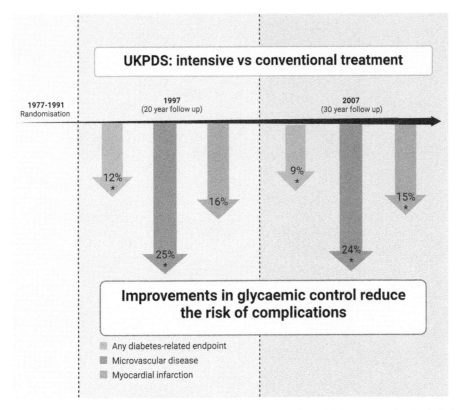

Figure 13.16 Long-term benefits of early intervention in T2D. UKPDS was a randomised clinical trial that investigated the effects of intensive diabetes management on T2D. The study showed that intensive glucose control led to statistically significant reductions in overall diabetes-related outcomes and a marked decrease in microvascular complications, as well as myocardial infarctions. A 30-year follow-up revealed that despite the decline in glucose control within the intensively managed group, the benefits, including reduced incidences of heart attacks, persisted, illustrating the phenomenon of metabolic memory. *Source:* Created with BioRender.com.

Hypoglycaemia

Hypoglycaemia is an acute event and the most common complication of treated diabetes, being more frequent in T1D than in T2D, and associated with significant morbidity, mortality and poor quality of life. Hypoglycaemia tends to occur in T2D patients being treated with insulin, sulphonylureas or meglitinides. When blood glucose drops below 3.9 mmol/L (70 mg/dL), individuals may experience symptoms such as sweating, shakiness, palpitations, hunger and dizziness. Further decrease can lead to confusion, seizures and unconsciousness (hypoglycaemic coma). Severe hypoglycaemia

occurs when the individual requires external assistance to administer carbohydrates, glucagon, or other necessary interventions. As blood glucose goes below 3.9 mmol/L, the body's protective responses kick in, which can vary from person to person. These responses, which aim to reverse the hypoglycaemia, cause a surge in catecholamines, cortisol and glucagon. This can lead to low potassium levels and cardiac arrhythmias. Simultaneously, these reactions might intensify oxidative stress, hinder cardiovascular reflexes (lasting up to 16 hours), raise inflammatory signs, stimulate platelets and add to blood vessel damage (Fig. 13.17). Moreover,

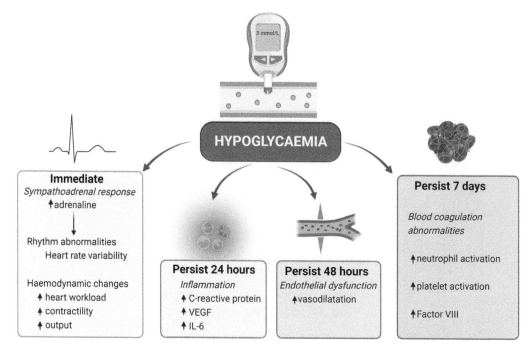

Figure 13.17 Pathophysiological consequences of hypoglycaemia. Hypoglycaemia triggers a cascade of physiological responses. The immediate reaction involves the activation of the sympathoadrenal system, which can be reversed once the hypoglycaemia is treated. However, hypoglycaemia also has more prolonged effects. For example, it can provoke inflammatory responses and activate platelets, and the complement and clotting systems. Unlike the rapid counter-regulatory response of the sympathoadrenal system, these alterations might take days to revert to their baseline levels. This lingering disruption raises concerns, as it may play a role in precipitating cardiovascular events.
VEGF, vascular endothelial growth factor; IL-6, interleukin 6. *Source:* Created with BioRender.com.

those with heart diseases or on insulin therapy are at risk of heart problems linked to hypoglycaemia. Treating hypoglycaemia typically involves consuming glucose (like tablets or juice), glucagon injections to promote glucose release from the liver or in severe cases receiving glucose intravenously in hospital. Hypoglycaemia can also occur more rarely in individuals without diabetes from a range of causes such as insulin-secreting tumours (insulinoma), PHHI (persistent hypoglycaemic hyperinsulinaemia of infancy, a genetic condition), liver or kidney disease and accidental or deliberate administration of insulin or sulphonylureas. Hypoglycaemia caused by overproduction of insulin can in some cases be successfully managed using somatostatin analogues (such as octreotide).

Diabetic Ketoacidosis

Diabetic ketoacidosis (DKA) is a medical emergency caused by a deficiency in insulin, either absolute or relative. While often the first symptom of T1D, DKA can also manifest in both T1D and T2D patients due to stressors such as significant illnesses (like a heart attack), infections, or certain medications. DKA usually presents with symptoms that include excessive thirst, frequent urination and *ketotic* (acetone-smelling) breath. If left untreated, this can escalate to rapid and deep (Kussmaul) breathing, nausea, lethargy, coma and even death (Fig. 13.14). Insulin deficiency results in increased gluconeogenesis, glycogenolysis and decreased glucose utilisation contributed to by the counter-regulatory hormones (glucagon, cortisol, growth hormone and catecholamines) leading to hyperglycaemia. The high glucose levels cause an osmotic diuresis leading to dehydration that may result in decreased kidney function and electrolyte loss. Simultaneously, lipid breakdown is accelerated, resulting in the generation of free fatty acids, which are converted in the liver to ketone bodies, namely β-hydroxybutyrate, acetoacetate and acetone. This results in metabolic acidosis. During acidosis, potassium is lost from muscle cells and is excreted in the urine, causing potassium deficiency. Diagnostic signs of DKA include elevated blood glucose, a blood pH under 7.3, low serum bicarbonate and high total plasma ketones. These imbalances are mirrored by the detection of glucose and ketones in urine tests. Effective DKA management involves rehydrating the patient, replenishing potassium levels and administering insulin.

Summary

This chapter has summarised the major features of diabetes, with a special focus on T1D and T2D. T2D is a rapidly growing pandemic, which is having a profound impact on global morbidity and mortality. The importance of insulin and its deficiency are discussed in the context of the pathways that regulate insulin secretion and activity. The chapter has also described the autoimmune basis of T1D and the major mechanisms underpinning the development of T2D, which involve the complex interplay between insulin resistance and impaired insulin secretion from pancreatic β-cells. It has also considered how diabetes is diagnosed and summarised treatment approaches, including the rapidly growing armoury of drugs available to manage the condition.

Key Points

1) Diabetes is defined as a group of chronic metabolic disorders characterised by elevated levels of blood glucose (hyperglycaemia), either due to insufficient insulin production, reduced sensitivity to insulin or a combination of both.

2) The main types of diabetes are:
   - T1D, which is primarily an autoimmune disease.
   - T2D, which is by far the most common, and which has a complex pathogenesis involving the interplay between multiple genes (polygenic) and environmental factors, resulting in the combination of insulin resistance and relative insulin deficiency (β-cell failure). T2D is associated with obesity and is a major global health problem of pandemic proportions.
   - Gestational diabetes, which is closely related to T2D and affects a significant number of pregnant women with potential complications for both mother and baby.
   - Latent autoimmune diabetes of adulthood (LADA), thought to be a variation of T1D.
   - Monogenic forms of diabetes (e.g. MODY and neonatal diabetes), which in most cases are autosomal dominant disorders involving mutations in genes involved in insulin secretion.

3) The accurate diagnosis of diabetes is dependent upon several tests including fasting plasma glucose, glycated haemoglobin and oral glucose tolerance tests. Glycated haemoglobin is also used for long-term diabetes monitoring.

4) The signs and symptoms of diabetes, and its complications (both microvascular and macrovascular), arise primarily from the high blood glucose levels leading to the glycation of proteins, inflammation and oxidative stress.

5) T1D is treated by subcutaneous injectable insulin or by insulin pumps.
6) The treatment of T2D, gestational diabetes and the monogenic forms of diabetes includes diet and exercise, oral therapies (including established drugs, such as metformin, sulphonylureas and pioglitazone, and newer ones, such as SGLT2 inhibitors and DPP-4 inhibitors). New injectable treatments include GLP-1 receptor agonists; insulin may also be required in some cases.
7) Hypoglycaemia and diabetic ketoacidosis are acute complications that require urgent treatment.

Further Reading

Holt R.I.G. & Flyvbjerg A. (eds) (2024) *Textbook of Diabetes* (6th Edn). Wiley-Blackwell, Chichester.

Katsarou A., Gudbjörnsdottir S., Rawshani A. *et al.* (2017) Type 1 Diabetes mellitus. *Nature Reviews Disease Primers* **3**: 17016. https://doi.org/10.1038/nrdp.2017.16.

Noble J.A. & Erlich H.A. (2012) Genetics of Type 1 Diabetes. *Cold Spring Harbor Perspectives in Medicine* **2**: a007732. https://doi.org/10.1101/cshperspect.a007732.

Galicia-Garcia U., Benito-Vicente A., Jebari S. *et al.* (2020) Pathophysiology of Type 2 Diabetes Mellitus. *International Journal of Molecular Sciences* **21**: 6275. https://doi.org/10.3390/ijms21176275.

Naylor R., Knight Johnson A. & del Gaudio D. (2018) Maturity-onset diabetes of the young overview. In Adam M.P., Ardinger H.H., Pagon R.A. *et al.* (eds) *GeneReviews®*. University of Washington, Seattle.

Bonnefond A., Unnikrishnan R., Doria A. *et al.* (2023) Monogenic diabetes. *Nature Reviews. Disease Primers* **9**(1): 12. https://doi.org/10.1038/s41572-023-00421-w.

Nayak A.U., Singh B.M. & Dunmore S.J. (2019) Potential clinical error arising from use of HbA$_{1c}$ in diabetes: effects of the glycation Gap. *Endocrine Reviews* **40**(4): 988–999. https://doi.org/10.1210/er.2018-00284.

14

Toxicology and Environmental Diseases

Introduction

Toxicology is routinely described as the science of poisons, which invites the question-what is a poison? A familiar answer lies in the famous quotation '*Alle ding sind gifft ... allein die dosis macht das ein ding kein gifft ist*' ('*Everything is a poison ... it is only the dose that makes it not a poison*'; Paracelsus, 1493–1541). More specifically, a poison is any chemical with the potential to cause harm, even death, to an organism. Such a compound, by interacting with the organism, may alter homeostasis to the extent that normal function is compromised. Poisons may be drugs, food additives, environmental pollutants, industrial chemicals, natural toxins, or household products. The definition of toxicology may be extended to also include the effects of physical agents, such as radiation or heat. Toxicology embraces several overlapping disciplines:

- *Environmental toxicology* examines the effects of toxins on ecosystems and the environment. It involves the study of how pollutants and contaminants impact organisms, populations, communities and ecosystems. Environmental toxicologists assess the potential risks to environmental health and work towards mitigating the adverse effects of pollutants on ecosystems.
- *Medical toxicology* focuses on the diagnosis, management and treatment of poisoning and adverse effects caused by exposure to toxins. Medical toxicologists play a critical role in emergency medicine, providing expertise in evaluating and managing toxic exposures and overdoses.

The Biology of Disease, Third Edition. Edited by Paul G. Murray, Simon J. Dunmore, and Shantha Perera.
© 2024 John Wiley & Sons Ltd. Published 2024 by John Wiley & Sons Ltd.
Companion website: www.wiley.com/go/murray/biologyofdisease3e

- *Analytical toxicology* involves the development and application of analytical methods and techniques to detect and quantify toxins and their metabolites in biological and environmental samples.
- *Forensic toxicology* investigates the presence of toxins in legal cases. Forensic toxicologists analyse biological samples from post-mortem examinations, as well as living individuals, to determine the role of toxic substances in accidents and crimes.
- *Regulatory toxicology* assesses the safety and potential risks of chemicals and substances. Regulatory toxicologists establish guidelines and regulations for the use and disposal of chemicals. Their work involves risk assessment, hazard identification and the development of safety standards.
- *Genetic toxicology* investigates the ability of toxins to cause damage to DNA and induce mutations. It involves studying the genotoxicity of substances to assess their potential to cause long-term effects, such as cancer. Genetic toxicologists use various assays to evaluate the genotoxic potential of chemicals.

Factors Affecting the Outcome of Exposure to Noxious Stimuli

Before any noxious stimulus can inflict its toxic effect, it must gain access to the body. For example, exposure to *xenobiotics* (foreign substances) can occur by intentional ingestion, occupational or environmental exposure and through accidental or intentional poisoning. With some exceptions, the most likely routes of entry are through the skin, lungs, or gastrointestinal tract. For a given stimulus, exposure can be defined by the means of entry, how many exposures have taken place, the concentration of the toxin and the timescale of repeated exposures. The ease of entry, and subsequent distribution, depends on the physicochemical properties of the toxic substance, in particular the degree of water and lipid solubility; in general, lipid-soluble substances move through cell membranes with greater ease

than more ionic substances. Distribution is also influenced by the extent of binding to endogenous macromolecules. *Acute exposure* to a substance usually refers to a single application of that substance to an organism, resulting in a toxic effect which occurs within minutes or hours of exposure. *Chronic exposure* is repeated delivery over time, for example, leading to accumulation of the xenobiotic.

The concentration of a xenobiotic is influenced by its metabolism, which mostly takes place in the smooth endoplasmic reticulum of the liver. Initial chemical reactions are hydrolysis, reduction or oxidation and are referred to as *phase I reactions*. They are usually catalysed by mixed function oxygenase enzymes, including cytochrome P-450. Phase I reactions are often, though not always, followed by *phase II reactions* which result in the formation of conjugates with glucuronyl, sulphate or methyl groups. Consequently, the xenobiotic is rendered less toxic, and is often more water soluble and thereby readily excreted in the urine. Metabolism, and subsequent excretion, influence how long the xenobiotic remains in the body. The general rule is that if the rate of excretion does not exceed the rate of entry, then accumulation will occur, which increases the toxic potential of a given xenobiotic. Some xenobiotics may be harmless unless metabolised to toxic forms in the body.

Some xenobiotics cause mutations, in which case they are described as *mutagens*. In some cases, exposure can result in an increased risk of cancer, in which case the substance is referred to as a *carcinogen*. If the xenobiotic is harmful to the embryo or developing foetus, then it is described as a *teratogen*.

Table 14.1 gives examples of mechanisms through which selected noxious stimuli exert their pathological effects.

Types of Environmental Exposures

The environment we inhabit has a profound impact on our health, influencing our physical, mental and social well-being. There are

Table 14.1 Examples of mechanisms of toxicity.

Cellular target	Example
Excitable membranes	Tetrodotoxin, a potent neurotoxin carried by pufferfish, blocks sodium channels. Both sensory and motor nerves are affected. Death may occur due to skeletal paralysis.
ATP production	Fluoroacetate occurs naturally in at least 40 plants in Australia, Brazil and Africa. It inhibits the Krebs cycle and deprives the cells of ATP.
Biomolecules	Carbon monoxide (CO) is a highly toxic gas. When inhaled, it has a high affinity for iron in the haem molecule of haemoglobin, forming carboxyhaemoglobin (COHb). This affinity is stronger than that of oxygen, leading to the preferential binding of CO to haemoglobin.
Sulphydryl groups	Lead and other heavy metals exert their toxic effects partly by binding to sulphydryl groups in proteins.
Calcium homoeostasis	Cadmium interferes with calcium signalling. One of the mechanisms involves the inhibition of calcium ATPases, which disrupts the normal extrusion of calcium ions, leading to an accumulation of calcium within the cell.
Nucleic acid synthesis	Cytarabine is a drug commonly used in cancer chemotherapy. It is structurally similar to cytidine. When cytosine arabinoside is taken up by cells, it is converted to cytosine arabinoside triphosphate, which competes with dCTP for incorporation into DNA during replication.
Lipids	Carbon tetrachloride causes severe liver damage, including necrosis. The mechanism of liver toxicity involves the formation of reactive intermediates, particularly the trichloromethyl peroxy radical, which leads to covalent binding to microsomal lipids and direct interaction with membrane phospholipids.

numerous environmental agents that can cause disease. They are discussed below under several broad headings and include air pollutants, industrial and occupational agents, household substances and drugs.

Air Pollution

Air pollution, resulting from industrial emissions, motor vehicle exhaust or the burning of fossil fuels, for example, poses significant health risks. Air pollution causes or exacerbates respiratory diseases such as asthma, chronic obstructive pulmonary disease (COPD) and lung cancer. The WHO estimates that outdoor air pollution, contributes to approximately 4.2 million premature deaths worldwide each year. Important air pollutants include:

Fine Particulate Matter (PM2.5)

PM2.5 refers to airborne particles with a diameter of $2.5\,\mu m$ or smaller. These particles can be generated by combustion processes, such as the burning of fossil fuels, vehicle emissions,

power generators and residential heating. Natural sources, such as wildfires, also contribute. The composition of PM2.5 varies widely and can include organic compounds, metals, sulphates and nitrates. Owing to their small size, PM2.5 particles can remain suspended in the air for long periods. PM2.5 are of particular concern because they can bypass physical defences in the upper respiratory tract to penetrate deep into the lungs when inhaled. The WHO guidelines recommend that the annual average concentration of PM2.5 should not exceed 10 micrograms per cubic metre ($\mu g/m^3$). Constituents of PM2.5 activate immune cells, causing the release of inflammatory mediators, the ensuing inflammation contributing to tissue damage, impaired lung function and chronic respiratory conditions.

Nitrogen Oxides

Nitrogen oxides (NOx), including nitrogen dioxide (NO_2) and nitrogen monoxide (NO), are respiratory irritants. NOx can form nitric acid (HNO_3) contributing (together with sulphur dioxide which makes sulphuric acid) to the formation of so-called 'acid rain'. Acid rain can damage forests by leaching essential nutrients from the soil and releasing toxic metals, making trees more susceptible to disease and trauma. Trees act as carbon 'sinks', absorbing more carbon dioxide than they emit. Deforestation releases this stored carbon back into the atmosphere, contributing to global warming. Short-term exposure to NOx causes coughing, wheezing and chest tightness. Prolonged exposure causes reduced lung function, increases the risk of respiratory infections and worsens symptoms of pre-existing respiratory conditions, particularly asthma. NOx are primarily generated by combustion processes, for example combustion of natural gas or nitrogen-rich fuels. Lightning strikes are a natural source of NOx because they convert nitrogen and oxygen into NOx. It is estimated that lightning produces 8.6 million tonnes of NOx per year, in contrast to NOx emissions from fossil fuel combustion, estimated at 28.5 million tonnes per year.

Volatile Organic Compounds

Volatile organic compounds (VOCs) encompass a wide range of chemicals emitted from various sources, including vehicle exhaust, industrial emissions and solvents. Short-term exposure to VOC can cause: eye, nose and throat irritation; headaches; dizziness and nausea. Long-term exposure to certain VOC, such as benzene, and other solvents, increases the risk of cancer and long-term respiratory and neurological disorders.

Ozone

Ozone (O_3) occurs naturally in the Earth's upper atmosphere, known as the ozone layer, where it plays a crucial role in absorbing the sun's harmful ultraviolet radiation. However, ozone can also be formed closer to ground level through chemical reactions involving NOx and VOC. Ground-level ozone, often referred to as

tropospheric ozone, is a major component of air pollution and is a highly reactive gas and potent respiratory irritant. It forms in urban and industrial areas where there are high levels of NOx and VOC emissions, such as from vehicle exhaust, industrial emissions and chemical solvents. Sunlight breaks down nitrogen dioxide (NO_2) into nitrogen monoxide (NO) and an oxygen atom. The oxygen atom then reacts with another oxygen molecule (O_2) to form ozone (O_3). Because this process requires sunlight, it occurs primarily under sunny conditions.

Carbon Monoxide

Carbon monoxide (CO) is a colourless, odourless and tasteless gas produced when fuels are burned incompletely due to insufficient oxygen supply. Common sources of CO include the combustion of gasoline, diesel, natural gas, coal and wood, in various applications such as vehicles, heating systems, stoves, fireplaces and generators. When CO is inhaled, it enters the bloodstream and binds to haemoglobin with a greater affinity than that of oxygen, resulting in the formation of carboxyhaemoglobin (COHb). COHb reduces the oxygen-carrying capacity of the blood. The health effects of CO depend on its concentration and the duration of exposure. Even at low levels, CO exposure can cause headaches, dizziness, nausea, fatigue, confusion and shortness of breath. Prolonged exposure to higher concentrations of CO can lead to more severe symptoms, including impaired vision and coordination, mental confusion, loss of consciousness and even death. CO is particularly dangerous in enclosed or poorly ventilated spaces where it can accumulate to high concentrations without detection (e.g. in the absence of CO monitors).

Lead

Lead is a toxic heavy metal that can be emitted into the air as a pollutant, primarily from industrial processes and the combustion of

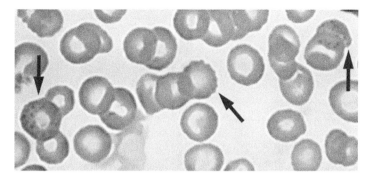

Figure 14.1   Lead poisoning: Peripheral blood film from an anaemic 53-year-old geophysicist who complained of fatigue and constipation. The blood film of his asymptomatic wife showed similar changes, but that of his 12-year-old son was normal. Basophilic stippling of this patient's red cells (arrows) suggested lead poisoning. Despite repeated inquiry, however, no source for such poisoning emerged. The plot thickened when blood lead concentrations were substantially elevated in both the man and his wife but normal in their son. Continued sleuthing ultimately unveiled the culprit-cocktail glasses decorated with lead-based paint. The husband and wife drank from these glasses daily, but their son never drank from them. Washing the glasses by machine presumably caused leaching of lead salts. Chelation therapy for the two adults returned their hematologic findings to normal and the patient became asymptomatic. Moral: Basophilic stippling of red cells can be the first, best, or only clue to lead poisoning. *Source:* Herbert L / Wikimedia Commons / CC BY 2.0.

lead-containing fuels. While air emissions have been significantly reduced in recent decades due to the phasing out of leaded fuels (although lead is still used in some aviation fuels, specifically in avgas (aviation gasoline), which is used in piston-engine aircraft), lead is still widely used in the manufacture of lead-acid batteries for motor vehicles, paints, pigments, solder, glassware, ammunition, cosmetics and traditional medicines. According to the WHO's 2021 update on the public health impact of chemicals, lead exposure remains a significant contributor to the global burden of disease. In 2019, nearly half of the 2 million lives lost to known chemical exposures were attributed to lead. Globally, lead exposure is estimated to account for 21.7 million disability-adjusted life years (DALYs). It is responsible for approximately 30% of the global burden of idiopathic intellectual disability, 4.6% of the global burden of cardiovascular disease and 3% of the global burden of chronic kidney disease.

The primary mechanism behind the toxic effects of lead is its ability to bind sulphydryl groups and disrupt the function of enzymes by displacing essential metals, such as calcium, iron or zinc. One important enzyme target is delta-aminolevulinic acid dehydratase (δ-ALAD), which is important for the biosynthesis of haem, leading to abnormalities in red blood cell formation and function (Fig. 14.1).

Lead exposure can cause severe and permanent effects on brain and nervous system development in young children and can lead to hypertension and kidney damage in adults. Pregnant women exposed to high levels of lead are at risk of premature birth, miscarriage and stillbirth.

Industrial and Occupational Exposures

According to the International Labour Organization (ILO), every year, over 2.3 million women and men die from occupational injuries or diseases. Out of the total deaths, over 350 000 are due to accidents and almost 2 million are due to work-related diseases caused by occupational hazards and exposures. The ILO estimates that annually there are around 160 million cases of non-fatal work-related diseases. In the United States, the cost of occupational diseases and injuries represents around 4% of global GDP, equivalent to approximately 3.94 trillion USD annually. Occupational cancers constitute a significant component of this global burden; the WHO estimates that occupational carcinogen exposure accounts for approximately 8% of all cancers. Respiratory diseases related to occupational exposures are also a significant concern. The ILO estimates that more than 334 million workers suffer from asthma, COPD and other respiratory conditions caused by, or exacerbated by, workplace factors. We briefly discuss some examples of occupational exposures below.

Asbestos

Asbestos is a naturally occurring mineral that has been used widely due to its desirable properties, including heat resistance and insulating capabilities. However, the inhalation of asbestos fibres can lead to severe lung conditions, including asbestosis and different forms of lung cancer including adenocarcinoma and mesothelioma. Archaeological evidence shows that asbestos was used as early as the Stone Age. Large-scale mining and usage of asbestos began in the late 19th century when it became popular as a building and fireproofing material (Fig. 14.2). However, by the 1970s, the adverse health effects of asbestos became more widely recognised, leading to its prohibition and regulation in many countries. Despite these regulations, many older buildings constructed before the 1980s still contain asbestos. Moreover, many countries continue to support its use as a building material. As a result, asbestos mining continues, with Russia currently being the top producer. The slow-developing nature of asbestos-related diseases means that the consequences of exposure can take several decades to manifest. Globally, the number of deaths attributed to asbestos-related diseases is estimated at 100 000 per year.

Figure 14.2 Example of asbestos cement siding and lining on a post-war temporary house in Yardley, Birmingham, UK. Nearly 40 000 of these structures were built between 1946 and 1949 to house families after World War II. *Source:* John M / Wikimedia Commons / CC BY-SA 2.0.

There are several forms of asbestos, including chrysolite (or white asbestos) and crocidolite (or blue asbestos). Crocidolite asbestos, composed of extremely fine sharp fibres, is considered the most hazardous type. It is the physical shape and dimensions of the asbestos fibre that accounts for its toxicity. When inhaled, the fibres resist digestion by macrophages, leading to the death of some of the macrophages and the release of pro-inflammatory molecules. This attracts more lung macrophages and fibroblasts, which cause interstitial fibrosis leading to asbestosis (Fig. 14.3). *Pneumoconiosis* is the general term for this class of interstitial lung disease caused by the inhalation of dust and other fine particulates (see also Case Study CS69).

Coal and Silica Dust

'Black lung disease' is a type of pneumoconiosis caused by the long-term inhalation of coal dust. Also known as coal workers' pneumoconiosis (Case Study CS69), it primarily affects coal miners and others working in coal-related industries. It is characterised by the progressive accumulation of coal dust, leading to inflammation, fibrosis and, in severe cases, necrosis. The development of coal workers' pneumoconiosis begins with a milder form called *anthracosis*, which is relatively common and usually asymptomatic. Anthracosis is found to some extent in urban dwellers due to general air pollution. However, prolonged exposure leads to more serious forms of the disease. *Simple coal workers' pneumoconiosis* and *complicated coal workers' pneumoconiosis*, also known as progressive massive fibrosis (PMF), are the more advanced stages. PMF is characterised by extensive fibrosis. Coal miners exposed to coal dust may also develop industrial bronchitis, defined as a persistent productive cough for at least 3 months per year, for a minimum of 2 years.

Inhalation of silica dust, particularly crystalline silica in occupations such as mining, construction, foundries and sandblasting can expose workers to high levels of silica, leading to *silicosis*, a type of pneumoconiosis, Silicosis is similar in its pathology to coal miner's pneumoconiosis with early-stage disease (Fig. 14.4) eventually progressing to extensive lung fibrosis.

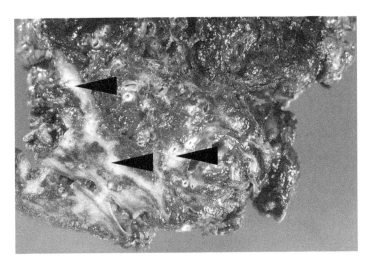

Figure 14.3 Asbestosis. Extensive fibrosis of pleura and lung parenchyma (white areas; shown by black arrowheads). *Source:* Yale Rosen / Wikimedia Commons / CC BY-SA 2.0.

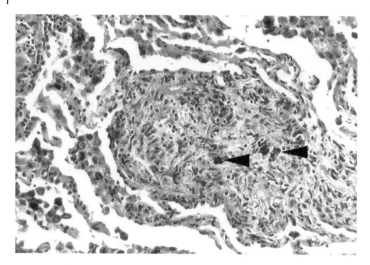

Figure 14.4 Silicosis. The earliest lesion of silicosis consisting of a small nodule containing predominantly macrophages with some anthracotic pigment present (black arrowheads). *Source:* Yale Rosen / Wikimedia Commons / CC BY-SA 2.0.

Benzene

The International Agency for Research on Cancer (IARC) classifies Benzene as a Group 1 carcinogen ('known to be carcinogenic to humans'). Exposure increases the risk of several haematological malignancies, including acute myeloid leukaemia, myelodysplastic syndrome and acute lymphoblastic leukaemia. Benzene causes DNA strand breaks and chromosomal damage. Exposure to benzene can occur through tobacco smoke, exhaust from motor vehicles and industrial emissions. Benzene can also be ingested or absorbed through skin contact.

Aromatic Amines

Workers exposed to occupational aromatic amines have an increased risk of bladder cancer. Historically, aromatic amines, such as benzidine and beta-naphthylamine, were extensively used in the dye industry, leading to a significant increase in bladder cancer incidence among workers. Painters also have an elevated risk of bladder cancer due to their potential exposure to these compounds found in certain paints and solvents. Similarly, workers in rubber tyre factories were exposed to a variety of chemicals, including aromatic amines, during the tyre production process. The carcinogenic potential of aromatic amines arises from their metabolic activation in the body, which forms compounds that can bind to DNA. While many industries have discontinued or heavily regulated the use of specific aromatic amines, the legacy of past exposures and continued risks underscore the importance of protective measures, continuous monitoring and medical surveillance for workers in these and other at-risk professions.

Pesticides

Pesticides can cause both acute and long-term health effects. Acute effects include pesticide poisoning, which can be a medical emergency. Long-term exposure is linked to birth defects, foetal death, neurodevelopmental disorders, cancer and neurological illnesses such as Parkinson disease. According to the Stockholm Convention on Persistent Organic Pollutants (2001), 9 of the 12 most dangerous and persistent chemicals were pesticides. Exposure occurs through various routes, including in the workplace or at home, through air and water, soil contamination, or as residues in food. Almost all humans have had some level of exposure to pesticides. *Pesticide drift* (the unintentional movement of sprayed pesticides away from the intended target area during application) can be a significant source of exposure to the general public. Moreover, pesticide residues can persist in the environment for extended periods of time, for example, dichlorodiphenyltrichloroethane (DDT), which has been banned in the United States since 1972.

Different classes of pesticides have different mechanisms of action. Organochlorines (OCP) or persistent organic pollutants (POP), for example, accumulate in fatty tissues and disrupt hormonal homeostasis, leading to adverse effects on reproductive tissues and can increase the risk of cancer. Pesticides can also affect cell receptors, ion channels, signalling pathways, DNA methylation and histone modifications, resulting in changes in gene expression and cellular function.

Foodstuffs

There are numerous examples of the adverse effects of unwanted substances in foodstuffs. Some are discussed below:

Methanol

Methanol is highly toxic. Ingesting as little as 10 mL can cause permanent blindness by destroying the optic nerve. The median lethal dose is 100 mL. Methanol is toxic through two mechanisms. First, it acts as a central nervous system depressant similar to ethanol poisoning, leading to effects on the CNS that can be fatal. Second, methanol is metabolised in the liver to form formaldehyde, which is further metabolised to formic acid, Formic acid inhibits mitochondrial cytochrome c oxidase, which is involved in the electron transport chain of cellular respiration, disrupting the process by which cells generate energy from oxygen,

causing cellular hypoxia and metabolic acidosis. Methanol poisoning usually occurs due to the contamination of drinking alcohol. Methanol is commonly used as a denaturant additive for ethanol manufactured for industrial purposes, making the resulting 'methylated spirit' unsuitable for consumption, thereby exempting it from excise taxation. Globally, there have been numerous instances of methanol poisoning. Many of these have occurred in India, which has a thriving 'moonshine' industry; methanol-tainted batches have killed over 2000 people in the last 3 decades.

Methylmercury

Methylmercury is produced by microbes in aquatic systems such as lakes, rivers and the open ocean. Anaerobic bacteria in the sediment, particularly in reservoirs and wetlands, are the primary contributors to methylmercury production. Natural sources of inorganic mercury, including volcanoes, forest fires, ocean volatilization (the process by which substances, such as gases or volatile compounds, are released from the surface of the ocean into the atmosphere) and weathering of mercury-bearing rocks, indirectly contribute to the production of methylmercury. Anthropogenic (arising as a result of human activity) sources, such as burning waste containing inorganic mercury and fossil fuels such as coal, also release mercury into the environment. Methylmercury bioaccumulates in aquatic food chains, resulting in higher concentrations at each step. Bacteria convert inorganic mercury to methylmercury, which is then transferred from bacteria to plankton, macroinvertebrates, herbivorous fish and finally to piscivorous (fish-eating) fish. Humans are exposed to methylmercury mainly through the consumption of fish and other aquatic species.

When ingested, methylmercury is readily absorbed by the gastrointestinal tract and transported throughout the body. It binds to proteins and peptides containing cysteine, mimicking methionine. Methylmercury is not easily eliminated and has a half-life of about

50 days in the blood. *In utero* exposure can lead to subtle developmental deficits in children, including loss of IQ points, language and memory impairments, and attention deficits. In adults, methylmercury exposure is linked to an increased risk of cardiovascular disease and autoimmune disorders.

Severe cases of methylmercury poisoning have been documented. For example, industrial waste release in Minamata, Japan and the Basra poison grain disaster in Iraq, resulted in mass poisonings. However, concerns focus mainly on the more subtle effects caused by the moderate levels of methylmercury exposure from dietary fish consumption. Methylmercury also causes diminished reproductive success in fish, fish-eating birds and mammals in contaminated aquatic ecosystems.

Nitrosamines

Nitrosamines are a class of chemical compounds that are carcinogenic only after their metabolic conversion into alkylating agents, which can modify DNA and induce mutations. Approximately 90% of the 300 nitrosamines tested have been shown to be carcinogenic in different experimental systems. A common source of nitrosamines is tobacco products, but they can also be formed in certain foods through the reaction of nitrites (used as food preservatives) and secondary amines. Secondary amines are produced during the degradation of proteins in food. The intake of nitrites and nitrosamines through diet has been associated with an increased risk of gastric cancer, oesophageal cancer and nasopharyngeal carcinoma.

Mycotoxins

Mycotoxins are toxic secondary metabolites produced by fungi. Typically, they are associated with fungal contamination of crops. Different fungal species can produce multiple mycotoxins and the same mycotoxin can be produced by several species. Aflatoxins are produced by *Aspergillus* spp. and are found on commodities grown in tropical and subtropical regions, including peanuts, maize and spices. Aflatoxin exposure increases the risk of hepatocellular carcinoma. Ochratoxin, produced by *Penicillium* and *Aspergillus* spp. is found in contaminated food and beverages. Citrinin, produced by various *Penicillium* and *Aspergillus* spp., causes 'yellowed rice disease'. Citrinin has various toxic effects, including nephrotoxicity and hepatotoxicity.

Household Exposures

Bleach, dishwasher powder, kettle descalers and drain cleaners are examples of household products that have toxic properties, usually due to their corrosive nature. Accidental poisoning can occur, particularly in young children. However, more severe effects are seen in cases of intentional poisoning. Drinking household bleach can lead to severe burns in the mouth and oesophagus, which can cause oedema of the pharynx and larynx. When bleach enters the stomach, it reacts with hydrochloric acid and releases hypochlorous acid and chlorine gas. Chlorine gas, if inhaled, is toxic to the lungs. Kettle descalers, which often contain formic acid, are also highly corrosive.

Drugs and Drug Abuse

Drug use and abuse, with substances such as marijuana, cocaine and methamphetamine, as well as the misuse of prescription drugs and illicit opioids, is rising exponentially. Drug use can result in dependency and addiction, as well as an increased risk of injuries and accidents. Here we discuss a few examples.

Paracetamol

Paracetamol, commonly available over the counter as an analgesic and antipyretic, is generally safe when used at therapeutic doses of 325–1000 mg. However, in cases of overdose, particularly in suicide attempts, serious intoxication can occur. Severe poisoning can lead to hepatic coma and, ultimately, death. The initial symptoms of paracetamol poisoning include nausea, vomiting, loss of appetite and abdominal pain. Within 2–4 days, there is an increase in liver enzymes and bilirubin levels, indicating liver damage. The liver damage associated with paracetamol overdose is characterised by centrilobular necrosis, which is the region of the liver most susceptible to toxic compounds.

The metabolism of paracetamol in the liver involves its conjugation to form paracetamol glucuronide or paracetamol sulphate. A smaller proportion (about 5%) is metabolised to mercapturic acid via glutathione. None of these metabolites are toxic themselves, but the conversion of paracetamol to mercapturic acid involves the formation of a reactive intermediate called *N*-acetyl-*p*-benzoquinone imine (NAPQI). In cases of overdose, the usual detoxification of NAPQI by glutathione becomes overwhelmed, resulting in the accumulation of NAPQI. NAPQI binds covalently to cysteine residues and can also conjugate with glutathione, deplete glutathione levels and be reduced back to paracetamol. Glutathione depletion and the subsequent oxidation of protein sulphydryl groups impairs cellular transport, particularly calcium-transporting proteins, leading to alterations in calcium homeostasis and Na^+-K^+-dependent adenosine triphosphatase (ATPase). Similar reactions may occur in the kidneys, leading to renal damage. Due to glutathione depletion and liver cell necrosis, the ability to metabolise paracetamol becomes increasingly compromised, resulting in a prolonged half-life of the drug. This exacerbates the toxicity unless intervention is initiated. In cases of severe intoxication, intensive supportive therapy along with the administration of sulphydryl-containing agents, particularly *N*-acetyl-cysteine, is essential.

Thalidomide

Thalidomide is a drug that was marketed in the 1950s and early 1960s as a sedative and treatment for morning sickness in pregnant women. However, it was later discovered to cause severe birth defects and other health problems in babies born to mothers who had taken the drug during pregnancy. The tragic consequences of thalidomide led to one of the biggest medical disasters in history. The drug was initially developed by Chemie Grünenthal in Germany and was later licensed to various pharmaceutical companies around the world, including The Distillers Company in the United Kingdom and Richardson-Merrell in the United States. Thousands of pregnant women took the drug. However, reports of abnormalities in newborn babies began to emerge, including limb deformities, organ malformations and other severe birth defects. In 1961, thalidomide was taken off the market due to widespread public pressure. It is estimated that thalidomide caused the death of approximately 2000 children and serious birth defects in more than 10 000 children worldwide. The thalidomide tragedy highlighted the need for stricter regulations and testing procedures for new drugs, particularly those intended for use during pregnancy. It also led to significant changes in drug approval processes and increased awareness of the potential risks of medications on foetal development. Thalidomide is still used to treat certain medical conditions (e.g. myeloma, graft-versus-host disease and some skin disorders, e.g. complications of leprosy) but under strict regulations and with stringent precautions to prevent its use during pregnancy.

Fentanyl

Fentanyl is a powerful synthetic opioid analgesic primarily used for pain relief with a potency 50 to 100 times greater than morphine, making it especially valuable for managing pain in cancer patients and post-surgical cases. Fentanyl acts quickly, and even a modest amount can trigger an overdose. The side effects of pharmaceutical-grade fentanyl are akin to those found in other opioids, including potential addiction, confusion, respiratory depression (that can escalate to fatality if unchecked), drowsiness and hallucinations.

Combining fentanyl with alcohol or other drugs, such as cocaine and heroin, can intensify these adverse reactions. Naloxone (Narcan) can reverse the effects of fentanyl. Despite its medical importance, which led to its inclusion in the WHO's List of Essential Medicines, its misuse has surged alarmingly. The United States has seen a dramatic increase in synthetic opioid-related fatalities, predominantly attributed to fentanyl. Since 2018, fentanyl surpassed heroin as the primary cause of drug overdose deaths in the United States. The drug's ease of production and powerful effect mean it is now commonly found mixed with other narcotics or disguised as different medications, compounding its widespread illicit use.

Smoking

The harmful effects of smoking tobacco, in all its forms, have been recognised for centuries. In 1604, King James I wrote 'A Counterblaste to Tobacco', wherein he described smoking as a *'custome lothesome to the eye, hateful to the nose, harmful to the brain, dangerous to the lungs, and in the black and stinking fume thereof, nearest resembling the horrible stygian smoke of the pit that is bottomless'*. Observational studies in the late 1800s and early 1900s began to establish an association between tobacco use and various cancers, including lung cancer. From the 1950s onwards, large case-control studies and prospective cohort studies began to accumulate evidence of the causal relationship between smoking and lung cancer, cardiovascular diseases and other disorders. Richard Doll and Austin Bradford Hill, with Richard Peto, conducted the British Doctors Study, which provided strong evidence of the harmful effects of smoking. This work found that the risk of developing lung cancer was directly related to the number of cigarettes smoked per day. Heavy smokers were approximately 25 times more likely to develop lung cancer compared to non-smokers. The study also showed an increased risk of other types of cancer among smokers, including cancers of the mouth, throat, bladder, pancreas and kidney. It also found that smoking was a major risk factor for cardiovascular disease.

Cigarette smoke contains a complex mixture of thousands of compounds, many of which are responsible for its harmful effects. Among these, more than 60 are established carcinogens, including polycyclic aromatic hydrocarbons (PAHs), *N*-nitrosamines, aromatic amines, aldehydes, volatile organic hydrocarbons and metals. Efforts to reduce smoking rates and promote tobacco control measures continue worldwide. The recognition of smoking as a major public health concern has led to comprehensive tobacco control policies, including smoking bans, public health campaigns, and regulations on advertising and packaging (Fig. 14.5). However, smoking remains a leading cause of preventable deaths globally, with estimates suggesting that it is responsible for about 5 million deaths per year.

Alcohol

Alcohol consumption has both short-term and long-term effects on health. In the short term, alcohol can cause intoxication, leading to symptoms such as slurred speech, clumsiness and delayed reflexes. In the long term, heavy and prolonged alcohol use can have serious detrimental effects, increasing the risk of developing *alcohol use disorder* – the inability to control or stop drinking despite the negative consequences. Alcohol misuse can lead to malnutrition, chronic pancreatitis, alcoholic liver disease (including cirrhosis) and several types of cancer, including liver, oesophageal, oral and breast cancer. Alcohol also affects the central and peripheral nervous systems. The developing adolescent brain is particularly vulnerable to the toxic effects of alcohol, which can have long-lasting impacts on cognitive and emotional functioning. The Global Burden of Disease Study and other research have shown that long-term alcohol consumption, even at moderate levels, is associated with an increased

Figure 14.5 Health warnings on cigarette packaging: In its background, this image depicts an opened pack of cigarettes with its side-panel health warning to would-be smokers stating some of the ill effects attributed to smoking, and in the foreground, a ruby-coloured glass ashtray containing the butts of two cigarettes. Smoking harms nearly every organ of the body, causing many diseases and affecting the health of smokers in general, as well as those inhaling 'second-hand' smoke. *Source:* CDC/Debora Cartagena, / Wikimedia Commons / Public Domain.

risk of death. While there may be some apparent benefits for older women in reducing the risk of certain diseases, such as ischaemic heart disease and diabetes, these potential benefits are offset by an increased risk of death from breast cancer and other causes.

The Impact of Climate Change on Health

Climate change presents a complex set of challenges that directly and indirectly impact human health on a global scale. Extreme weather phenomena, including hurricanes, floods and prolonged droughts, not only cause immediate physical harm, but also lead to large-scale displacement of communities, causing significant psychological and mental health problems. For many regions, these natural disasters exacerbate pre-existing social and economic problems.

Changes in the climate have also altered the spread of vector-borne diseases. Previously temperate regions are now witnessing a surge in diseases such as malaria, dengue fever and Lyme disease due to warmer conditions becoming more hospitable for vectors such as mosquitoes and ticks. Altered rainfall patterns and the rapid melting of glaciers, which act as freshwater reserves for billions globally, feeding rivers and streams, cause changes in sea levels and compromise water quality. In turn, these factors not only increase the risk of water-borne illnesses, but impact on food security.

Climate change-induced environmental degradation also has profound effects on global health. As humans encroach further into wildlife habitats, owing to deforestation and urban expansion, delicate ecosystems can be disrupted. The resulting loss of biodiversity and the increased likelihood of contact between humans and wildlife can increase the emergence and spread of zoonotic diseases.

Summary

At the beginning of the chapter, toxicology was described as the study of poisons. The extent to which a given chemical can exert a deleterious effect occurs in a dose-dependent manner which is governed by the exposure and the distribution characteristics of the toxin. It also depends on the physicochemical properties of the agent and the susceptibility of the receiving organism. Toxic effects are described as acute, after a single exposure to a toxin followed by a rapid onset of action. Chronic effects are the result of repeated exposures over a period of time. Toxic effects develop from perturbation at the molecular level, progressing to altered cellular function, eventually giving rise to a disturbance of tissue/organ function. The resulting influence on homeostasis may range from mild discomfort to death. Different types of environmental exposures have significant impacts on human health, which are too numerous to discuss in full here. Climate change is already affecting human health, for example by changing the incidence and geographical spread of vector- and water-borne diseases and increasing the risk of the emergence of new diseases.

Key Points

1) The outcome of exposure to noxious stimuli is influenced by various factors such as the route of entry, concentration, frequency and duration of exposure, and the physicochemical properties of the toxic substance.
2) Metabolism and excretion play a crucial role in determining the effect of xenobiotics, with accumulation increasing their toxic potential. Some xenobiotics become harmful only after undergoing metabolic transformation.
3) Noxious stimuli can have different mechanisms of action, which include their ability to induce mutations, cause cancer and induce abnormalities in the developing foetus.
4) Climate change can cause extreme weather events, leading to physical injury, displacement and mental health challenges among affected populations. Changes in disease patterns, including the expansion of vector-borne diseases and the increased risk of water-borne diseases, are also consequences of climate change.

Further Reading

Detels R., Abdool Karim Q., Baum F., Li L. & Leyland A.L. (eds) (2021) *Oxford Textbook of Global Public Health* (7th edn). Oxford University Press, Oxford.

Klaassen C.D., Watkins J.B., Sams R.L. (eds) (2021) *Casarett & Doull's Toxicology: The Basic Science of Poisons* (9th edn). McGraw-Hill Education, New York.

Case Studies

Case Study 1   Five-Day History of Yellowing of the Skin, Fatigue and Flu²-like Symptoms

Clinical Presentation

A 35-year-old man presented to the Emergency Department with a 5-day history of yellowing of the skin, fatigue and flu²-like symptoms. He had also experienced tan-coloured bowel movements, dark urine and nausea. He reported no recent travel, insect or chemical exposures, or any known sick contacts. He denied consuming anything out of the ordinary during the prior week. On examination, the patient was alert and oriented and not distressed. Physical examination revealed yellow discoloration of the skin and eyes (scleral icterus), and a mildly distended abdomen with mild tenderness in the right upper quadrant.

Investigations

A CT scan revealed that the appearance of the gall bladder, liver, common bile duct and pancreas were all within normal limits. Laboratory investigations revealed elevated liver enzymes and bilirubin levels. Serum albumin levels were decreased and clotting time increased. The clinical history and laboratory results suggested acute hepatitis (Table CS1.1).

Diagnosis

Liver function tests are important in the evaluation of acute hepatitis as they can help determine the extent of liver injury and the type of hepatitis present. Elevated levels of liver enzymes, such as ALT and AST, indicate liver cell damage and inflammation, while increased bilirubin levels suggest impaired bile flow. The pattern and degree of elevations in liver function tests can also help distinguish between different types of hepatitis, such as viral hepatitis or drug-induced hepatitis. In addition, monitoring changes in liver function test results over time can be used to track the progression of the disease and evaluate the effectiveness of treatment. In a patient with acute hepatitis, the decrease in albumin levels is likely due to liver damage, causing decreased production of this protein by the liver. The liver plays a crucial role in synthesising many coagulation factors; the decreased synthesis of clotting factors explains the increased coagulation time (prothrombin time; PT) observed in this patient.

Given the apparent liver dysfunction, a hepatitis panel was ordered. The results of

The Biology of Disease, Third Edition. Edited by Paul G. Murray, Simon J. Dunmore, and Shantha Perera.
© 2024 John Wiley & Sons Ltd. Published 2024 by John Wiley & Sons Ltd.
Companion website: www.wiley.com/go/murray/biologyofdisease3e

Table CS1.1 Results of selected laboratory investigations.

Test	Result	Reference range
Alanine transaminase (ALT)	**455 U/L**	5–40
Aspartate transaminase (AST)	**448 U/L**	10–40
Alkaline phosphatase (ALP)	**118 U/L**	25–115
Bilirubin	**2.5 mg/dL**	<1.2
Gamma-glutamyltransferase (GGT)	**65 U/L**	11–50
Albumin	**29 g/L**	34–48
Prothrombin time (PT)	**21 seconds**	11–16

Values outside the reference range are in bold.

this panel showed no evidence of infection with Hepatitis B virus or Hepatitis C virus, but the patient did have detectable levels of anti-Hepatitis A virus immunoglobulin (IgM), which in the absence of a history of recent Hepatitis A vaccination suggests recent (acute) Hepatitis A virus infection. Follow-up testing for Hepatitis A RNA (nucleic acid) on the serum sample was positive, confirming acute infection.

Diagnosis: Acute Hepatitis A.

Discussion

Hepatitis A virus (HAV) is a non-enveloped picornavirus that contains a single-stranded positive-sense RNA genome encased in a protein shell. The virus is mainly transmitted via the faecal-oral route. Once ingested, the virus travels to the liver, where it replicates in hepatocytes and Kupffer cells. HAV exits the host cell by lysis, allowing virions to be released into the bile and stool. HAV spreads easily in conditions of poor sanitation and overcrowding, and food-borne outbreaks are common, especially caused by the consumption of shellfish cultivated in polluted water. HAV is resistant to detergents, acids, solvents and drying, and can survive for months in fresh and salt water. HAV infections are common in children in resource-poor countries.

In resource-rich countries, infection is often acquired during travel to high-incidence areas or through contact with infectious individuals. Lifelong immunity typically results following infection.

The virus can be detected in the blood (known as *viraemia*) and in the stools during the first few weeks of infection, usually around 10–12 days before symptoms appear (Fig. CS1.1). HAV-specific IgM antibodies are only present during an acute infection (or vaccination); they can be detected 1–2 weeks after the initial infection and remain in the blood for up to around 14 weeks. The presence of IgG antibodies after infection indicates that the acute phase of the infection has passed and the individual may be immune to further infection. Thus, immunity to HAV can be determined by testing for IgG antibodies.

Treatment and Prognosis

There is currently no known cure or specific treatment for Hepatitis A and most people recover; the risk of death from HAV infection is less than 0.5%. The recovery period may last for a few weeks to several months. The focus of treatment is on ensuring the patient's comfort and maintaining proper nutritional balance, including replenishing fluids lost due to vomiting and diarrhoea. *Fulminant hepatitis,*

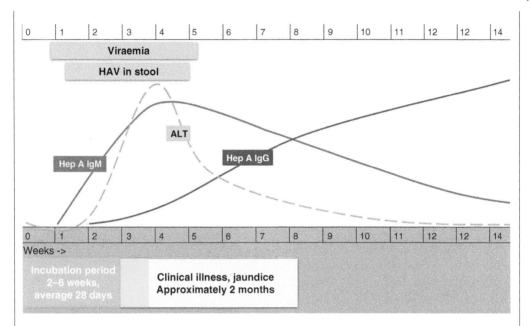

Figure CS1.1 Acute Hepatitis A infection. Timeline of the appearance of symptoms, virus detection and IgM and IgG antibodies in acute Hepatitis A infection. *Source:* https://microregistrar.com https://commons.wikimedia.org/wiki/File:Hepatitis_A_serology.jpg / Microregistrar / CC by 4.0.

also known as fulminant hepatic failure, is a rare, but severe and potentially life-threatening complication of acute hepatitis. It is defined as the development of severe acute liver injury with encephalopathy within 8 weeks of the onset of symptoms in a patient with no prior history of liver disease.

Questions

1   What is a single-stranded positive-sense RNA virus?

2   What are Kupffer cells and what do they do?

Answers on p. 531.

Case Study 2   Coughing Up Blood in a Retired Coal Miner

Clinical Presentation

A 65-year-old retired coal miner with a smoking history of 20–30 unfiltered cigarettes per day was admitted to hospital after coughing up blood (*haemoptysis*) on several occasions over the past 3 days. He described the amounts as ranging from a teaspoonful to a cupful. He had worked as a miner for 20 years, but had been forced to retire early due to worsening chronic

bronchitis and emphysema. He regularly produced copious amounts of white or yellowish sputum and required antibiotics for chest infections several times a year. On examination, he was slightly breathless when speaking and had clubbing of the fingernails. Examination of his chest confirmed the presence of features consistent with chronic bronchitis and emphysema, but no obvious signs of other acute disease.

He was not anaemic and all other aspects of his examination were normal.

Investigations

The results of laboratory investigations are shown in Table CS2.1. The haemoglobin concentration was within normal limits, suggesting that blood loss had been minimal. The chest X-ray revealed a rounded shadow in the left upper lobe of the lung, with a crescent of air above. This appearance is typical of a fungal ball in the lung, colonising an *emphysematous bulla* (a large, air-filled space or cavity that forms as a result of lung tissue destruction in emphysematous patients). Sputum culture showed a heavy growth of *Moraxella catarrhalis*, a commensal organism of the upper respiratory tract, which is often associated with colonisation or infection of the bronchi in patients with chronic obstructive pulmonary disease (COPD). A moderate growth of *Aspergillus fumigatus* was also isolated, which may indicate colonisation of the sputum, but is not a diagnostic finding when taken alone.

COPD often leads to impaired lung function, particularly reduced airflow during exhalation. This can result in reduced oxygen levels as well as inadequate removal of carbon dioxide from the bloodstream. As a consequence, carbon dioxide accumulates in the blood, leading to a condition known as *respiratory acidosis*. The increased carbon dioxide combines with water in the blood to form carbonic acid (H_2CO_3), which then dissociates into bicarbonate ions (HCO_3^-) and hydrogen ions (H+). The increase in H+ ions leads to a decrease in blood pH, making it more acidic.

Diagnosis

Apart from his episodes of haemoptysis, the patient's history and examination are entirely consistent with his chronic lung disease.

Table CS2.1   Results of laboratory investigations.

Test	Result	Normal range
White blood cell count	7.6×10^9/L	4.0–11.0
Haemoglobin	16.5 g/dL	13.0–18.0
Red cell count	5.9×10^{12}/L	4.5–6.5
Sputum: culture	**Heavy growth of *Moraxella catarrhalis*, moderate growth of *Aspergillus fumigatus***	—
Sputum: microscopy	**Moderate numbers of pus cells seen**	—
Blood gases		
Arterial pH	**7.32**	7.35–7.45
Bicarbonate	**31 mmol/L**	22–30
Partial pressure of oxygen (PaO$_2$)	**55 mmHg**	75–100
Partial pressure of carbon dioxide (PaCO$_2$)	**50 mmHg**	36–46
Spirometry		
FEV1/FVC ratio	**0.53**	>0.70

Values outside the reference range are in bold.

However, the chest X-ray findings, coupled with the isolation of *Aspergillus fumigatus* from sputum suggest the presence of a fungal ball (*aspergilloma*) colonising an old emphysematous bulla. The aspergilloma has eroded a pulmonary blood vessel, leading to the blood loss, which was coughed up.

Diagnosis: Aspergilloma eroding a pulmonary blood vessel in a patient with pre-existing COPD.

Discussion

COPD is characterised by persistent respiratory symptoms and airflow obstruction and can be caused by emphysema and/or chronic bronchitis. Indeed, patients with COPD often present with features of both emphysema and chronic bronchitis.

Chronic bronchitis is defined as a productive cough for at least 3 months in 2 consecutive years and is characterised by increased numbers of goblet cells, and enlarged submucosal glands. The excess mucus can narrow the airways, thereby limiting airflow and contributing to a decline in lung function.

Emphysema is a different pathological process characterised by abnormal and permanent enlargement of lung air spaces with the destruction of alveolar walls without any fibrosis and destruction of lung parenchyma with loss of elasticity. In the United Kingdom, the prevalence of COPD is approximately 2%, and the mortality rate is around 20 000/year. Spirometry measures the proportion of a person's vital capacity that they are able to expire in the first second of forced expiration (FEV1) to the full, forced vital capacity (FVC). If the FEV1/FVC is less than 0.7 then this indicates airway obstruction, such as COPD, in the appropriate clinical context. In restrictive lung disease, the FEV1 and FVC are usually equally reduced due to fibrosis or other lung pathology, resulting in a normal ratio.

Bacteria and viruses are a major cause of COPD exacerbations, whereas infections with fungi are less common. *Aspergillus* is a ubiquitous environmental mold which is found predominantly in decaying vegetable matter. Its spores (conidia; Chapter 4) can often be detected in the air, and their small size (2.5 – 3.0 µm) allows them to be inhaled deep into the respiratory tract. It is likely that all humans have been exposed to *Aspergillus* at some time, and many have continuing, frequent exposure, though few will become ill as a result. It is not surprising, therefore, that *Aspergillus* spp. may often be isolated from sputum samples without being of clinical significance. *Aspergillus fumigatus* is the species most frequently associated with human disease, while *Aspergillus flavus* is associated with infection in the immunocompromised host. There are three principal respiratory syndromes associated with *Aspergillus* infections:

1) *Allergic bronchopulmonary aspergillosis*: some patients who already suffer from asthma may develop hypersensitivity to *Aspergillus* spores, leading to varying degrees of lung damage.
2) *Invasive pulmonary aspergillosis:* profoundly immunocompromised patients may develop a life-threatening and rapidly progressive necrotising pneumonia which may be difficult to diagnose *ante-mortem* as sputum may not be produced for culture. Such patients may also develop aspergillosis at other sites, for example bone and sinuses.
3) *Pulmonary aspergilloma:* as illustrated in this case. Spores may be inhaled and grow to form a ball of hyphae inside (Fig. CS2.1). While some lesions may resolve without treatment, others may become secondarily infected leading to abscess formation or may erode into adjacent structures such as the pleural space or blood vessels resulting in haemoptysis.

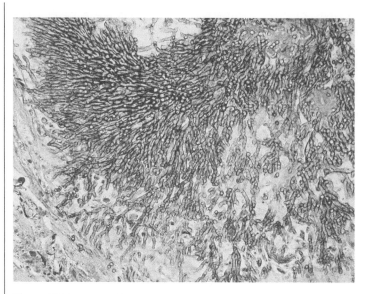

Figure CS2.1 Aspergilloma. Part of a colony forming an Aspergilloma. The fungal hyphae exhibit dichotomous 45 degree angle branching and septate typical of *Aspergillus* spp. *Source:* Yale Rosen / Wikimedia Commons / CC BY-SA 2.0.

In all of the pulmonary conditions described above, *Aspergillus* may be detectable in the sputum, but is not invariably present and will only aid diagnosis when considered as part of the overall clinical picture.

The innate immune response is important in controlling *Aspergillus* infections. Resident alveolar macrophages and epithelial cells recognise pathogen-associated molecular patterns (PAMPs; Chapter 6) on *Aspergillus* spores through pattern-recognition receptors (PRRs), including Toll-like receptors (TLRs), C-type lectin receptors and NOD-like receptors. This recognition leads to the production of proinflammatory cytokines and activation of inflammatory pathways. Alveolar macrophages directly phagocytose and kill *Aspergillus* conidia. Chemokines and cytokines attract other immune cells, including neutrophils and natural killer cells to the infected site. Neutrophils play a critical role in anti-fungal defence by phagocytosis, reactive oxygen species (ROS; see also Chapter 3) production and the formation of neutrophil extracellular traps (NETs; see also Chapter 5). The adaptive immune response against *Aspergillus* involves CD4+ T helper cells that differentiate into different subsets, including Th1, Th2, Th17, and Treg cells (see Chapter 6). Th1 cells enhance the anti-fungal activity of macrophages and neutrophils through the production of proinflammatory cytokines. Th2 cells are not typically protective during *Aspergillus* infection but are predominant in allergic bronchopulmonary aspergillosis.

Treatment and Prognosis

The prognosis of pulmonary aspergilloma is the same as the pre-existing lung disease, and conservative management is appropriate. Small amounts of haemoptysis will usually subside spontaneously. However, if complications arise, for example severe haemoptysis, lung abscess etc., surgical excision of the lobe of lung or drainage of the abscess may be indicated. Systemic anti-fungal therapy is often ineffective as the causative organism is protected within the cavity and inaccessible to systemically administered drugs. Percutaneous intra-cavitary anti-fungal therapy, which involves injecting anti-fungal drugs directly into the aspergilloma under radiologic guidance can be an option for patients who are not candidates for surgery. When anti-fungal drugs are indicated (especially in invasive disease occurring in the immunocompromised host), useful compounds include voriconazole, itraconazole or posaconazole.

A serological assay is approved by the US Food and Drug Administration (FDA) for the detection of galactomannan, a molecule found in the cell wall of *Aspergillus* spp. Serum galactomannan can often be detected a mean of 7–14 days before other diagnostic features become apparent.

Questions

1 Are *Aspergillus* virulence factors important?
2 What preventive measures might protect severely immunocompromised patients from aspergillosis?

Answers on p. 531.

Case Study 3 Fever, Cough, Myalgia and Shortness of Breath for the Past 5 days

Clinical Presentation

A 55-year-old man presented to the Emergency Department with fever, cough, myalgia and shortness of breath for the past 5 days. He had a history of hypertension and type 2 diabetes, which were well controlled with medication. He denied any recent travel history or exposure to sick contacts. On examination, his vital signs were as follows: temperature 38.6°C, blood pressure 141/86; heart rate 77 bpm, respiratory rate 25 bpm and oxygen saturation 92% on room air. Lung auscultation revealed bilateral crackles. He was admitted to the hospital for further management. The reverse-transcriptase-polymerase chain reaction (RT-PCR) test on a combined nasopharyngeal/throat swab was reported as 'SARS-CoV-2 RNA Detected'.

Investigations

A Chest radiograph (CXR) showed bilateral peripheral opacities in the lung fields. A CT scan of the chest showed ground glass opacities in both lung fields. Some consolidation was also seen. The results of laboratory and clinical tests are shown in Table CS3.1.

Diagnosis

The patient had elevated CRP and ESR, which indicate an inflammatory response. This is a common finding in patients with COVID-19 and can be used as a marker for disease severity and progression. *Ground glass opacities* (GGOs) refer to areas of the lung that appear hazy or cloudy on imaging. *Consolidation* refers to an area of the lung where the air spaces are filled with fluid or inflammatory material, making it difficult for air to flow through. This can lead to impaired oxygenation of blood, which can be life-threatening if not treated promptly. The combination of GGOs and consolidation in both lungs is a common finding in patients with severe COVID-19 pneumonia. D-dimer is a laboratory test that can be used to assess clotting function. Some patients with COVID-19 have a hypercoagulable state with microvascular thrombosis as a result of activation of the coagulation cascade by the infection.

Diagnosis: COVID-19 pneumonia.

Table CS3.1 Results of laboratory and clinical investigations.

Test	Result	Reference range
White blood cell count	**12.1 × 10⁹/L**	4.0–11.0
Differential white cell count		
Neutrophils	**80%**	40–75
Lymphocytes	**10%**	20–45
Monocytes	8%	2–10
Eosinophils	2%	1–6
Basophils	0%	1
Haemoglobin	13.5 g/dL	13.0–18.0
Platelets	178 × 10⁹/L	150–400
Sodium	140 mmol/L	135–146
Potassium	4.0 mmol/L	3.5–5.0
Creatinine	78 μmol/L	59–104
Total bilirubin	0.8 mg/dL	<1.2
ALT	30 U/L	5–40
AST	25 U/L	10–40
Prothrombin time	**18 seconds**	11–16
D-dimer	**750 ng/mL**	<500
Blood gases		
pH	7.36	7.35–7.45
PaCO₂	45 mmHg	36–46
PaO₂	**62 mmHg**	75–100
SaO₂	**92%**	>95
Inflammatory markers		
C-reactive protein	**25 mg/L**	<10
Erythrocyte sedimentation rate	**40 mm/hour**	<20

Values outside the reference range are in bold.

Discussion

Acute COVID-19 is biphasic. In most people, an early innate immune response transitions into a broadly effective adaptive immune response that controls the virus. However, a variable proportion of symptomatic patients require hospitalisation. Long-term inflammation in patients receiving supportive care in ICU can lead to pulmonary fibrosis, representing a third phase of the disease. A proportion of such patients recover but are at risk of long-term reduced lung function and early mortality. It has been suggested that severe lung injury in some patients is a consequence of hyper-activated innate immune responses, supported by elevated levels of circulating inflammatory cytokines, including IL-6, IL-8, IL-10, TNF-α, IFN-β, IFN-γ and IL-1α (Fig. CS3.1). Although subsets of inflammatory monocytes and macrophages are considered to be an important source of some of these cytokines, there is considerable

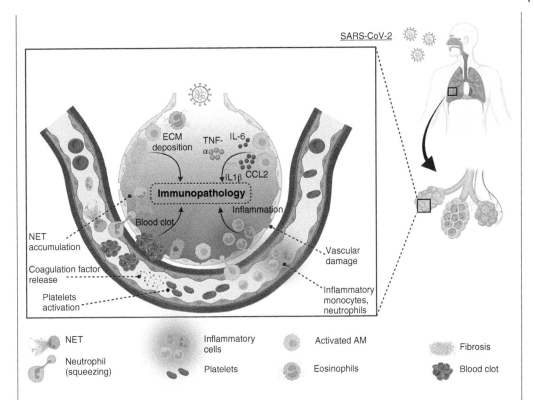

Figure CS3.1 Innate cell-mediated immunopathology in COVID-19: In COVID-19, the innate immune response is triggered by virus entry, resulting in inflammation, damage to blood vessels and clot formation. Alveolar macrophages (AM) and interstitial macrophages in lung tissues are among the first immune cells to respond to the virus. These cells release inflammatory molecules, which include TNFα, IL-6, IL-1β and CCL2. These molecules not only create a local inflammatory environment but also attract monocytes and neutrophils to the infection site. IL-1β induces the proliferation of fibroblasts that contribute to fibrosis. SARS-CoV-2 also promotes platelet and neutrophil activation and the secretion of pro-coagulants, leading to the formation of aggregates of white blood cells and platelets, and neutrophil extracellular traps (NETs, see also Chapter 5). This complex cascade of innate immune responses contributes to the pathophysiology of COVID-19. *Source:* https://onlinelibrary.wiley.com/doi/10.1002/jmv.28122 /John Wiley Sons.

uncertainty about the pathogenic contributions of other cell types, including the appearance in blood, of neutrophil precursors (as a consequence of emergency myelopoiesis) as well as dysfunctional neutrophils expressing PD-L1. Compared to other serious respiratory infections, such as influenza, the TNF-α/IL-1β inflammatory response dominates in COVID-19, and in severe cases, is associated with an accompanying robust, but delayed, type I interferon (IFN-I) response. In a mouse model of the SARS coronavirus,

an IFN-I response was shown to promote the accumulation of monocytes-macrophages and the production of pro-inflammatory cytokines, resulting in lethal pneumonia with vascular leakage and impaired virus-specific T cell responses.

In severe COVID-19, numerous studies have reported accompanying signatures of systemic dysfunction in adaptive immunity that include lymphopenia with markedly reduced numbers of peripheral T cells and natural killer cells. The diminished T cell pool in the blood of patients with active disease shows evidence of early CD4+ and CD8+ T cell activation with counter-regulation by the inhibitory checkpoint molecules, BTLA, Tim-3, PD-1, TIGIT and CD73, Th17-like and regulatory T cell populations, and non-conventional (but not classical anti-viral) Th1 cell polarisations. T cell receptor (TCR) sequencing has revealed that SARS-CoV-2-specific responses are partly driven by TCR clusters shared between patients and that TCR diversification is associated with immune control of SARS-CoV-2, as has been reported for other infections.

While respiratory pathology is the most frequent manifestation of SARS-CoV-2 infection, extra-pulmonary disease is also common, including thrombotic complications, myocardial dysfunction, acute renal injury, gastrointestinal symptoms, hepatocellular damage and neurological complications, prompting the definition of COVID-19 as a multi-system disorder. The mechanisms underpinning the development of extra-pulmonary disease include but are not limited to: (i) the presence of inflammatory cells in blood vessels suggesting microvascular dysfunction secondary to endothelial injury, (ii) systemic effects due to cytokines and (iii) direct infection of ACE-2-expressing cells in these organs (ACE-2 being the target for the spike protein to attach to in initial infection).

Most patients who have recovered from COVID-19 (or who have been vaccinated against SARS-CoV-2) have circulating SARS-CoV-2-specific CD4+ and CD8+ T cells. The key T cell subset involved is the CD4+ T helper type 1(Th 1) cell. Th1 cells reactive against Spike protein and several other key viral proteins have been found in recovered patients. SARS-CoV-2-specific CD8+ cells recognising Spike and other viral proteins have also been found in these patients. However, SARS-CoV-2-specific CD4+ Th1 cells demonstrated the strongest correlation with a good outcome and very low levels are associated with a strikingly high mortality. In contrast, CD8+ T-cells cells and antibodies are thought to be less relevant in preventing severe disease. However, the presence in an individual of high titres of circulating neutralising antibodies, from vaccination or past infection (or both), is correlated with protection against infection and severe disease.

Treatment and Prognosis

Oral dexamethasone (a corticosteroid) was commenced, as steroids have been shown in large randomised control trials to decrease mortality in patients with hypoxia due to COVID-19. Low molecular weight heparin was administered to reduce the risk of venous thrombo-embolism. Due to his worsening dyspnoea, the patient was treated with remdesivir. After 2 weeks, the patient's condition improved. After 3 weeks, all abnormal laboratory values returned to normal and he tested negative for SARS-CoV-2. GGO were markedly less on CT scanning. He was discharged and made an uneventful recovery.

Questions

1 Which drugs have been approved by the US Food and Drug Administration (FDA) for the treatment of COVID-19?
2 mRNA vaccines are a relatively new development and have been used extensively in the global fight against SARS-CoV-2, but what are they?

Answers on p. 532.

Case Study 4   Fatigue, Abdominal Pain and Yellow Discoloration of the Skin

Clinical Presentation

A 33-year-old man presented to his GP complaining of fatigue, abdominal pain and yellow discoloration of his eyes and skin for the past 2 weeks. He reported no significant past medical history and denied alcohol or drug use. On physical examination, vital signs were within normal limits. An abdominal examination revealed mild tenderness in the right upper quadrant and palpable enlargement of the liver. He was clearly jaundiced with a yellow colour to his skin and the sclera of his eyes.

Investigations

Laboratory findings included elevated levels of liver enzymes, increased bilirubin levels and a positive serum Hepatitis B surface antigen test (Table CS4.1).

Diagnosis

Elevated AST and ALT levels are indicative of liver inflammation, while elevated bilirubin levels suggest impaired bile flow and possible liver damage. In the context of the clinical findings and a positive Hepatitis B surface

Table CS4.1   Results of selected laboratory investigations.

Test	Result	Reference range
Aspartate transaminase (AST)	**228 U/L**	10–40
Alanine transaminase (ALT)	**252 U/L**	5–40
Bilirubin	**2.2 mg/dL**	<1.2
Hepatitis B surface antigen	**Positive**	—

Values outside the reference range are in bold.

antigen test, a diagnosis of acute Hepatitis B infection is confirmed.

Diagnosis: Acute Hepatitis B infection.

Discussion

The Hepatitis B virus (HBV) genome is comprised of partially double-stranded circular DNA. Important proteins of HBV include:

- *HBsAg (Hepatitis B surface antigen):* The first discovered HBV antigen, comprising small (S), medium (M) and large (L) proteins.
- *HBcAg (Hepatitis B core antigen)*, which is the main structural protein of the HBV nucleocapsid.
- *HBeAg (Hepatitis B envelope antigen).*

HBV is transmitted by infected bodily fluids, including blood and semen, for example, by unprotected sexual contact with an infected person, or sharing needles with people who inject drugs. Initial HBV infection may be asymptomatic or result in acute symptomatic hepatitis. Acute infection is characterised by elevation of serum liver enzymes, but most people clear the virus within 6 months. However, a small percentage of individuals with acute hepatitis may experience a worsening of their condition and develop acute liver failure, also known as *fulminant hepatitis*, which is a serious and potentially life-threatening condition. Acute liver failure is rare (1-6 cases per million in the United States and Europe); the rapid loss of function in acute liver failure is the result of massive hepatocyte necrosis. Managing acute liver failure requires hospitalisation in an intensive care unit, and without a liver transplant, mortality rates are around 50%. Hepatic encephalopathy is a complication of acute liver failure and can present with a wide range of symptoms, ranging from minimal changes in mental state, to coma. In normal functioning livers, ammonia, a by-product of protein metabolism, is converted into urea and then excreted by the kidneys. However, when liver function is compromised, ammonia levels rise. Ammonia is toxic to brain cells and can cause alterations in brain function and structure, as well as a disruption of the blood–brain barrier. The brain injury, which can be followed by oedema, can lead to raised intracranial pressure.

Extrahepatic manifestations are present in up to 10% of HBV-infected patients and include serum-sickness-like syndrome, acute necrotising vasculitis (polyarteritis nodosa), membranous glomerulonephritis and papular acrodermatitis of childhood (Gianotti-Crosti syndrome).

Although the Hepatitis B virus infects hepatocytes, it does not directly damage these cells. Instead, the tissue destruction is an indirect result of the immune response initiated by the host against the virus. In particular, cytotoxic T-cells (CD8+ T-cells) attack and kill the infected hepatocytes. The dead cells release pathogen-associated molecular patterns (PAMPs) and damage-associated molecular patterns (DAMPs) (Chapter 6). Both are recognised by the immune system as indicators of potential harm. PAMPs are found in pathogens. Examples include lipopolysaccharide (found in the cell wall of bacteria) or viral RNA. DAMPs are molecules released by damaged or dying cells. These molecules are not meant to be outside the cell and thus, when detected by the immune system, signal tissue damage or cellular stress. Examples include ATP, DNA, and heat shock proteins. Activation of the innate immune system by these molecules leads to the 'immunopathology'- effectively 'collateral damage', caused by the immune response to the virus. In chronic infection, this immune response fails to clear the virus and so repeated cycles of injury, healing and fibrosis ultimately lead to the development of cirrhosis.

The natural history of chronic Hepatitis B infection can be divided into five phases (Fig. CS4.1).

- *Phase 1 – Immune tolerant phase*: In the early phase, individuals typically have high levels of viral replication, but there is minimal immune response targeting the virus. As a result, there is little to no liver injury, and transaminase levels (ALT and AST) are often within the normal range. Blood tests are typically positive for HBsAg and HBeAg and for the IgM subclass of antibodies against Hepatitis B core (HBcAb). Hepatitis B virus DNA is also detectable in blood.
- *Phase 2 – Immune reactive phase*: This phase is characterised by active liver disease. There is ongoing high-level viral replication, but the immune response begins to target the infected liver cells. This leads to liver cell necrosis and inflammation, causing transaminase levels to rise. Blood tests remain positive for HBeAg.
- *Phase 3 – Immune control/inactive carrier phase:* In this stage, the host's immune response becomes more effective at controlling viral replication. Viral load decreases, and liver enzymes return to normal or are only mildly elevated. HBeAg levels become low or undetectable; as antibodies against Hepatitis B e (HBeAb) are now detectable. This phase is associated with a stable, non-progressing, liver disease state.
- *Phase 4 – Immune escape phase:* In some cases, the virus can undergo reactivation. Viral replication increases and transaminase levels rise again. Importantly, HBeAg remains undetectable or at very low levels. This phase is associated with renewed viral activity and liver inflammation.
- *Phase 5 – Virus clearance phase:* In a subset of individuals, the immune response becomes strong enough to clear the virus from the body. Viral antigens, including HBeAg and HBsAg, are absent. Antibodies

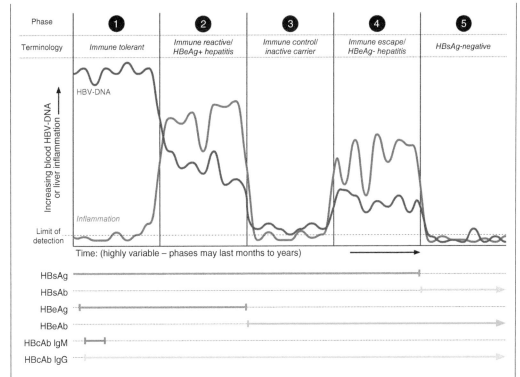

Figure CS4.1 Graph and line plot detailing the approximate trajectories of blood Hepatitis B virus DNA levels and of liver inflammation over time. Hepatitis B virus serology (HBsAg, Hepatitis B surface antigen; HBsAb, Hepatitis B surface antibody; HBeAg, Hepatitis B e antigen; HBeAb, Hepatitis B e antibody; HBcAb IgM, Hepatitis B virus core immunoglobulin M antibody; HBcAb IgG, Hepatitis B core antibody IgG) is also shown. The graph describes the five phases of chronic Hepatitis B infection as detailed in the 2017 European Association for the Study of the Liver guidelines. *Source:* https://commons.wikimedia. org/wiki/File:Hepatitis_B_virus_phases_of_chronic_infection.png/Gwilz/Public Domain.

to Hepatitis B surface antigen (HBsAb) become detectable, indicating immunity to the virus.

If Hepatitis B surface antibodies are present in an individual but there are negative Hepatitis B core antibody and negative Hepatitis B surface antigen tests, then this indicates previous vaccination for Hepatitis B, rather than prior infection.

It is important to note that progression through these phases can vary widely between individuals; not all individuals will progress through all phases, and some may remain in certain phases for extended periods. The phases are a general framework to

understand the natural history of chronic Hepatitis B infection, but individual cases can be more complex. Monitoring liver function, viral load and serological markers is essential to manage patients.

The progression of Hepatitis B infection depends on factors like age, immune response and viral load. Thus, the likelihood of developing acute hepatitis and progressing to chronic infection decreases as the age of the host increases. Infections acquired during infancy are typically asymptomatic and have a high likelihood of progressing to chronic infection, with over 90% of cases progressing in this way. Among children aged between 1 and 5 years, approximately 20% will develop chronic infection. However, for individuals older than 5 years, particularly adults, over 90% will typically clear the virus within 6 months.

Treatment and Prognosis

The main objective in treating patients with Hepatitis B is to prevent the progression to cirrhosis, liver failure or liver cancer. This normally involves a combined approach with viral load suppression and boosting the patient's immune response with immunotherapeutic interventions. Antiviral treatments, such as pegylated interferon (Peg-IFN) or nucleos(t)ide analogues, are often used.

Treatment for acute HBV is mainly supportive, and no specific therapy is available. For patients with chronic active Hepatitis B disease, treatment is recommended based on various algorithms. For example, the National Institutes of Health (NIH) recommends nucleos(t)ide therapy for patients with acute liver failure or cirrhotic patients with detectable HBV DNA and those with complications, such as cirrhosis or advanced fibrosis with positive serum HBV DNA, or reactivation of

chronic HBV during or after chemotherapy or immunosuppression. For HBeAg-positive patients with evidence of chronic HBV disease, treatment is advised when the HBV DNA level is at or above 20 000 IU/mL and when serum ALT is elevated for 3–6 months. For HBeAg-negative patients with chronic Hepatitis B disease, treatment is given when the HBV DNA load is >2000 IU/mL and the serum ALT is elevated for 3–6 months.

The patient was advised to avoid alcohol and to take some rest. He was scheduled for follow-up appointments to monitor his liver function.

Questions

1 What types of drugs are approved to treat Hepatitis B virus infections?
2 What is the underlying basis of Hepatitis B-associated serum-sickness-like syndrome?
3 Is there an increased risk of hepatocellular carcinoma (HCC) following Hepatitis B infection?

Answers on p. 532.

Case Study 5   Altered Mental Status, Right Facial Droop and Slurred Speech in a 31-year-old Woman

Clinical Presentation

A 31-year-old woman arrived at the Emergency Department with symptoms of altered mental status, right facial droop and slurred speech. She had a previous history of intravenous drug abuse until age 27 and now was seropositive for HIV. The patient denied headache, neck stiffness, vision changes, fever, abdominal pain, vomiting, or changes in urinary or bowel habits. She also denied recent alcohol or drug use. During the physical examination, the patient appeared confused and had difficulty following commands.

Investigations

A computed tomography (CT) scan revealed multiple areas of decreased density in both brain hemispheres. Further imaging by brain magnetic resonance imaging (MRI) showed numerous enhancing lesions scattered throughout the cerebral and cerebellar hemispheres. Her white cell count was elevated, and she had a raised erythrocyte sedimentation rate (ESR) and elevated C-reactive protein (CRP) level. In view of her HIV-positive status and the MRI findings, the differential diagnoses included cerebral toxoplasmosis, neurocysticercosis and CNS lymphoma. Initial analysis of cerebrospinal fluid did not show any significant abnormalities. However, subsequent testing for *Toxoplasma gondii* DNA revealed 2531 copies/mL. Serological tests for Epstein-Barr virus, Hepatitis B virus and Hepatitis C virus were negative. The patient's *Toxoplasma gondii* IgG levels were elevated (155 IU/mL; normal range: 0.0–7.1 IU/mL), while the IgM levels were within the normal range.

Diagnosis

Diagnosing toxoplasmosis involves serological testing, including detection of IgG and IgM antibodies. Serological tests such as the indirect fluorescent antibody (IFA) test or enzyme immunoassay (EIA) are commonly used. Neonates with suspected congenital toxoplasmosis may require additional testing, including PCR-based assays of amniotic fluid or other body fluids. Imaging studies such as CT or MRI of the brain are performed to evaluate CNS involvement.

Diagnosis: Toxoplasmosis of the CNS

Discussion

Toxoplasma gondii is a ubiquitous parasite that infects birds and mammals, including humans. It has a complex life cycle involving both sexual and asexual reproduction stages. Cats are the definitive hosts for *T. gondii* (Chapter 4). Sexual reproduction occurs in the cats' intestinal tracts, leading to the production of infectious oocysts in their faeces. The primary mode of transmission to humans is through the ingestion of oocysts, for example, through the consumption of undercooked meat of animals harbouring tissue cysts, food or water contaminated with cat faeces, handling cat litter or via blood transfusion or organ transplantation. Oocysts can also be inhaled when cat faeces dry and oocysts become aerosolised. Once ingested

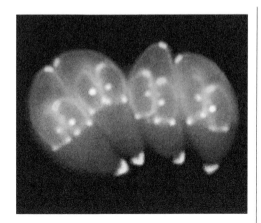

Figure CS5.1 *Toxoplasma gondii*, an obligate intracellular human parasite; tachyzoite stage. The parasite has a unique cytoskeletal apparatus that is used for invading host cells and for parasite replication. Shown here are images of *T. gondii* constructing daughter scaffolds within the mother cell. Green: YFP-α-Tubulin; bright yellow: mRFP-TgMORN1 (see Hu *et al.*, Figure 6A–C). *Source:* Ke Hu and John M. Murray. 2006 / from PUBLIC LIBRARY OF SCIENCE (PLOS).

(or inhaled), the oocysts develop into tachyzoites (Fig. CS5.1), which spread throughout the body and form tissue cysts in neural, eye and muscle tissues. These cysts can remain dormant for years, and contain bradyzoites, the dormant forms. The clinical manifestations of toxoplasmosis vary depending on the individual's immune status and the organ systems involved. Acute toxoplasmosis is usually asymptomatic or may present with flu'-like symptoms, lymphadenopathy and hepatosplenomegaly. CNS toxoplasmosis, commonly seen in immunocompromised individuals, particularly those with AIDS, presents as encephalitis with neurological symptoms, including altered mental status, seizures, focal deficits and ring-enhancing intracranial mass lesions (Figs. CS5.2 and CS5.3).

Treatment and Prognosis

Treatment of toxoplasmosis depends on the severity of the infection and the patient's

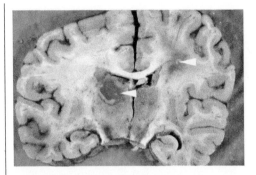

Figure CS5.2 Cerebral toxoplasmosis. This is a coronal slice of the brain of an HIV-positive adult with encephalitis caused by toxoplasmosis. This patient had received no therapy. The ventricles are compressed (by oedema). The foci of toxoplasmosis (arrowheads) are haemorrhagic and necrotic; the left basal ganglia are particularly affected. Generalised and focal symptoms such as seizures, hemiplegia and personality change occur in cerebral toxoplasmosis. *Source:* Wellcome Collection / https://wellcomecollection.org/works/e9xmj4hm/ items / last accessed November 20, 2023.

immune status. In immunocompetent patients with mild acute toxoplasmosis, treatment may not be necessary. However, for severe or disseminated infections, as well as in immunocompromised individuals, combination therapy with pyrimethamine and sulphadiazine, along with leucovorin to prevent bone marrow suppression, is the standard treatment. Alternative regimens may be used in patients who are allergic to sulphonamides or who are intolerant to certain medications.

Questions

1 Is there a potential evolutionary advantage to cats being the definitive hosts?
2 What is latent *Toxoplasma* infection, and does it have any clinical significance?

Answers on p. 533.

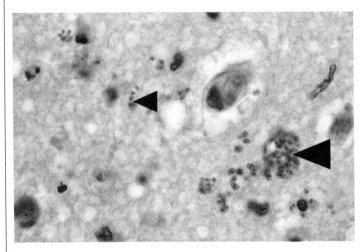

Figure CS5.3 Cerebral toxoplasmosis: microscopy. High-power view of *Toxoplasma* tachyzoites (small arrow head) in the brain shows numerous small blue parasites in the parenchyma. They may have emerged from the bradycyst seen towards the right-hand side (large arrow head). H&E stain. Creative Commons Zero (CC0) terms and conditions https://creativecommons.org/publicdomain/zero/1.0. *Source:* Brain: toxoplasmosis. SB Lucas / Wellcome Collection.

Case Study 6   Shortness of Breath and Vaginal Discharge

Clinical Presentation

A 28-year-old woman presented with a 3-day history of fever, shortness of breath on exertion and a dry cough. She denied chills, night sweats, nausea, vomiting, diarrhoea or abdominal pain. She had no headache, change in vision, muscle weakness or change in sensation. She had lost 10 pounds in the last 3 months despite eating what she described as a 'healthy' diet. She had noted a new vaginal discharge accompanied by itching, as well as some burning in her mouth when she drank coffee. She was not on any medication. She had used intravenous drugs from the age of 18–23 and, during this time, had unprotected sexual intercourse with multiple male sexual partners. In the last 2 years, she had one steady male partner with whom she continued to have a sexual relationship. There was no other significant past medical history, no allergies and all members of her family were in good health.

Investigations

On physical examination, the patient was thin, in moderate respiratory distress and febrile (temperature: 39.4°C). Blood pressure was 100/60. Pulse was regular at 100 b.p.m. Respiratory rate was 24/minute. Examination of the oral cavity revealed three discrete white patches on the hard palate but was otherwise normal. There was no apparent lymphadenopathy. Pulmonary examination revealed bilateral, diffuse end-inspiratory rales, but no rhonchi (wheezes) and no sign of consolidation or pleural effusion. Cardiovascular examination was normal. Abdominal examination showed no hepatomegaly or splenomegaly, and no masses or tenderness. Neurological examination revealed normal mental status, no muscle wasting and no focal weakness or sensory abnormalities. On pelvic examination, a thick white vaginal discharge was noted. There was no cervical tenderness and no cervical discharge. A cervical smear was performed.

Initial laboratory work-up included fourth-generation assays, which detected both HIV antibodies and HIV (p24) antigens in one test, which were positive. A CD4+ T-cell count revealed 150 cells/mm^3 (reference range, 800–1200 cells/mm^3). CD8+ T-cell count was 300 cells/mm^3 (reference range, 400–600 cells/mm^3). Microbiological cultures of blood were negative. The patient had a reduced haemoglobin level (9.8 g/dL; reference range 11.5–16.5) and a lower-than-normal white cell count (1.2 × 10^9/L; reference range 4.0–11.0). A chest radiograph showed bilateral interstitial infiltrates, no cavities, consolidation or pleural effusion. Microscopic examination of a sample obtained at bronchioalveolar lavage revealed the presence of *Pneumocystis jirovecii* organisms (Fig. CS6.1) and a PCR test on the fluid detected *Pneumocystis jirovecii DNA*. Microscopic examination of swabs taken from the oral cavity lesions and the vagina revealed the presence of fungal forms

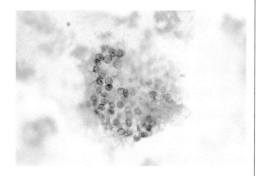

Figure CS6.1   High power view of *Pneumocystis jirovecii* obtained by bronchoalveolar lavage. This cluster of organisms with black-staining walls was found in the lavage of a patient with HIV/AIDS and lung disease. The individual pneumocystis organisms are clearly seen. Grocott silver stain. *Source:* Wellcome Collection / https://wellcomecollection. org/works/mm8y2psq / CC0 1.0 Universal.

consistent with *Candida albicans*. The cervical smear showed a grade 3 cervical intraepithelial neoplasia (CIN 3), reflecting infection with human papillomaviruses (HPV).

Diagnosis

The patient had *Pneumocystis jirovecii* pneumonia (Fig. CS6.2) and *Candida albicans* infection of the vaginal and oral cavities. An HIV antigen/antibody test was positive. Center for Disease Control (CDC) classification was used as a reference for clinical categorisation of HIV/AIDS stage (Table CS6.1). HIV infection causes acquired immune deficiency syndrome (AIDS). The risk factors associated with acquisition of the virus in this patient included multiple unprotected sexual partners and the use of intravenous drugs.

Diagnosis: AIDS with opportunistic *Pneumocystis jirovecii* and *Candida albicans* infections.

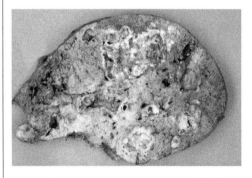

Figure CS6.2 *Pneumocystis* pneumonia (PCP). This is chronic pneumocystis pneumonia in a patient with AIDS. The lung shows consolidation and numerous areas of necrosis and cavitation. This is the effect of repeated attacks of PCP, with incomplete healing, fibrosis and cavitation. Note, the acronym for this clinical entity is PCP, not PJP – the organism responsible for infection was previously known as *Pneumocystis carinii* and the resulting clinical entity was first termed *Pneumocystis carinii* pneumonia. Nowadays, Pneumocystis pneumonia is still abbreviated 'PCP'. *Source:* Wellcome Collection / https://wellcomecollection.org/works/qngp3jkw / CC0 1.0 Universal.

Table CS6.1 Stage of HIV infection based on age-specific CD4+ T-cell count.

	CD4+ T-cell count/µL		
Stage	Age <1 year	Age 1–5 years	Age >5 years
1	≥1500	≥1000	≥500
2	750–1499	500–999	200–499
3	<750	<500	<200

Discussion

HIV is primarily divided into two major types: HIV-1 and HIV-2. HIV-1 is thought to have evolved from a Simian immunodeficiency virus (SIV) that initially infected chimpanzees before mutating and infecting humans. This jump from one species to another likely involved the virus evolving to escape the immune system's natural barriers. Particularly in Europe and the Americas, most HIV-1 infections can be traced back to a single strain from Africa that first spread to Haiti in the 1960s and later to other parts of the Western world.

HIV primarily infects CD4+ T-cells, which play a crucial role in coordinating the body's immune response (Chapter 6). HIV can also infect other types of immune cells, such as macrophages and microglial cells. HIV gains entry by binding its HIV envelope glycoprotein to the CD4 receptor on the surface of the target cell. This glycoprotein consists of two subunits: gp120 and gp41 (Fig. CS6.3). After gp120 binds to the CD4 receptor, a conformational change occurs that allows it to also bind to one of two chemokine co-receptors – CCR5 or CXCR4. This binding facilitates the fusion of the viral envelope with the cell membrane, allowing the viral genome to enter the host cell and begin the process of viral replication. Different strains of HIV preferentially use either CCR5 or CXCR4, and this has implications for disease progression and treatment. Once inside the

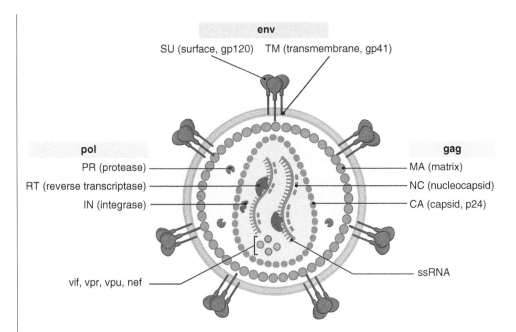

Figure CS6.3 Structure of the HIV virion. The viral particle, or virion, is composed of a viral genome made up of two identical strands of positive-sense, single-stranded RNA. This RNA genome is tightly packed within a conical capsid made of the viral protein, p24. The viral RNA encodes several genes that are crucial for the HIV life cycle. The three major genes are: *Env:* This gene encodes the envelope glycoproteins gp120 and gp41. Gp120 is responsible for binding to the CD4 receptor on the host cell, while gp41 facilitates the fusion of the viral envelope with the host cell membrane. *Pol:* This gene encodes the viral enzymes, reverse transcriptase (RT), integrase (IN) and protease (PR). Reverse transcriptase converts the viral RNA into DNA once the virus enters the host cell. Integrase helps incorporate this DNA into the host cell genome, and protease cleaves long protein chains into functional viral proteins during the maturation of new virions. *Gag:* This gene encodes the core structural proteins, such as the capsid (p24). There are also several other accessory and regulatory genes, such as viv, vpr, vpu and nef, that are involved in various stages of the viral life cycle, affecting processes that include viral replication, transcription and evasion of the host immune response. *Source:* Created with BioRender.com.

host cell, the HIV genomic RNA is transcribed into DNA, which integrates into the host genome. This integrated viral DNA (also known as a provirus; Chapter 4), is then transcribed back into viral RNA, allowing the virus to assemble new virions and spread to other cells. The process of converting RNA to DNA is prone to errors, which leads to a high mutation rate. This mutability provides the virus with opportunities to develop drug resistance and escape the host's immune responses.

HIV can exist in either the productive or the latent state in CD4+ T-cells. In a *productive infection*, new viral particles are produced,

while a *latent infection* is an infection without virion production. A subset of infected T-cells exist as memory T-cells that can reactivate the virus at later time points. Thus, HIV has evolved to survive in long-lived memory T-cells, which allow it to persist, potentially for the lifetime of the individual. A similar strategy is adopted by the Epstein-Barr virus, in which long-term persistence is mediated by infection of memory B-cells (Case Study 67).

Shortly after infection, there is a surge of viral replication that results in a rise in the amount of virus present in the blood (viraemia). Following this, the viral load decreases, coinciding with a rise in CD8+ T cells that can eliminate cells infected with HIV. This initial stage is often called the *primary infection* phase. After this, the disease enters a chronic stage, marked by a gradual reduction in the number of CD4+ T cells. This is accompanied by an increased risk of *opportunistic infections*. An opportunistic infection is infection by an organism that is ubiquitous in the environment, and which rarely establishes an infection in an immunocompetent host. There is also an accompanying hyperstimulation of some parts of the immune system characterised by elevated levels of circulating immunoglobulins and impaired ability to fight off intracellular pathogens, such as *Salmonella* spp. and *Mycobacterium tuberculosis*.

The earliest dependable indicator of HIV infection is the presence of HIV RNA, which can usually be detected 10–11 days post-exposure. This is typically followed by the detection of the p24 antigen, the most plentiful protein in HIV, around 4–10 days after HIV RNA first becomes detectable. Approximately 20 days post-exposure, levels of IgM antibodies start to increase, eventually transitioning to an IgG antibody response that can be identified between 2 and 6 weeks after the initial appearance of HIV RNA.

P. jirovecii infection of the lungs is one of the most common opportunistic infections seen in untreated HIV sufferers and may rapidly progress to pneumonia. The clinical features of *P. jirovecii* pneumonia include fever, a dry cough and progressive dyspnoea. On examination, the lungs often sound normal or end-inspiratory rales may be heard. Rales or 'crackles' are short explosive sounds that can be heard when collapsed small airways open during inspiration. Rales heard at the end of inspiration suggest fibrosis or the accumulation of fluid (oedema) within the lungs. Untreated patients with *P. jirovecii* pneumonia become incapacitated from dyspnoea and eventually die of hypoxaemia.

Vaginal and oral infections with the fungal organism *C. albicans* are also common in the immunocompromised host. The decline in the CD4+T-cell count is a useful indicator of the severity of immune dysfunction and is measured frequently in HIV-positive individuals. Clinically, the decline in immune function can be charted through several distinct stages (Table CS6.1).

The increased frequency of CIN lesions in HIV-positive females is well recognised. CIN refers to premalignant transformation of the squamous epithelium of the uterine cervix (Chapter 11). In CIN, the transformed cells are retained within the epithelium by the underlying basement membrane. CIN 3 is at the most severe end of the spectrum of these disorders and, if left untreated, may progress to invasive squamous carcinoma of the cervix. In invasive squamous carcinoma, the cells comprising the CIN lesion breach the basement membrane and spread into surrounding connective tissues causing local tissue destruction (Chapter 11). Extensive spread throughout the connective tissues and lymphatics has serious consequences for the patient and can lead to the blockage of vital pelvic organs. The severity of CIN in immunosuppressed females may be explained in part by the association of HPV with CIN. Certain subtypes of HPV (particularly subtypes 16 and 18) are causative agents of CIN. Immunosuppression allows outgrowth of HPV-infected tumour cells, the proliferation of which might otherwise be efficiently controlled by a competent immune system.

Treatment and Prognosis

Six classes of antiretroviral medications, each targeting different phases of the HIV life cycle, are currently approved for clinical use. The first-line treatment usually consists of a combination of drugs, including integrase strand transfer inhibitors (INSTIs), protease inhibitors (PIs), or non-nucleoside reverse transcriptase inhibitors (NNRTIs), along with a backbone of two nucleoside/nucleotide reverse transcriptase inhibitors (NRTIs). For patients with advanced HIV and low CD4+ T cell counts, additional prophylactic measures against opportunistic infections are crucial. Incomplete adherence to antiretroviral therapy (ART) can lead to drug resistance and virological failure. The lifelong administration of ART requires ongoing consideration of patient preferences and medical comorbidities to maintain safety and effectiveness. A small group of infected individuals, known as *HIV controllers*, can suppress viral replication without antiretroviral medications. Some of these controllers can keep the virus at almost undetectable levels, while others maintain low but detectable levels of the virus.

People infected with HIV are encouraged to lead a normal healthy life and to have all associated infections treated promptly. When treated early, *P. jirovecii* pneumonia responds well to therapy. The most common form of treatment is trimethoprim/sulfamethoxazole (TMP/SMX), which is also known as co-trimoxazole. These agents may also provide effective prophylaxis for HIV-positive individuals who have not yet developed *P. jirovecii* pneumonia. Many clinicians continue prophylaxis while the patient has a CD4+ T-cell count of 200/mm^3 or less. If the CD4 count increases above 200/mm^3, as a result of ART, some clinicians will discontinue prophylaxis. Drugs such as nystatin and fluconazole may be used to treat *C. albicans* infections. Treatment is also available for many other opportunistic infections such as cerebral abscesses due to *Toxoplasma gondii* (Case Study 5) and meningitis due to *Cryptococcus neoformans*. Patients who complete treatment for an opportunistic infection frequently remain on the same medication at a lower dose for the rest of their lives as prophylaxis. There is no specific treatment for HPV. Rather, the infected

cervical tissue must be removed through any number of approaches, including freezing the tissues, electrocautery or surgical removal. Vaccination against HPV is recommended post removal of the infected tissue.

It should be noted that significant progress has been made in reducing the transmission and impact of HIV through a variety of approaches, which include:

- *Condom use:* Male and female condoms are a cornerstone for prevention, significantly reducing the transmission rate when used consistently and correctly.
- *Voluntary medical male circumcision (VMMC):* This has been shown to reduce HIV infection rates in heterosexual men by approximately 59%.
- *Harm reduction:* This includes measures like needle exchange programs and opioid agonist therapy for people who inject drugs.
- *Pre-exposure prophylaxis (PrEP):* This involves taking antiretroviral drugs before potential exposure to HIV. It is highly effective, especially when adherence is high.
- *Post-exposure prophylaxis (PEP):* This is an emergency treatment started as soon as possible within 72 hours after a recent possible exposure to HIV.
- *HIV testing:* Modern HIV tests, including fourth- and fifth-generation tests, have increasingly shortened the 'window period' from infection to the time when infection is first detectable. This means earlier detection and treatment, which in turn reduces transmission.

The concept of *treatment as prevention* (TasP) is also important. Thus, initiating ART early leads to a 96% decrease in HIV transmissions between partners. This is because ART dramatically lowers HIV levels in blood and genital fluids. This is commonly summarised by the phrase *undetectable equals untransmittable*, or *U=U*, meaning that individuals with HIV levels suppressed below certain thresholds are extremely unlikely or unable to transmit the virus to others.

Questions

1 This patient presented with CIN 3, which is regarded as a precursor to invasive carcinoma of the cervix. Is the

incidence of other tumours increased in HIV-positive individuals?

2 AIDS-associated tumours are often treated less aggressively than the same tumours arising in immunocompetent individuals. Why do you think this is?

Answers on p. 533.

Case Study 7 Two-week History of Abdominal Pain, Fever and Vomiting

Clinical Presentation

A 45-year-old man presented to the Emergency Department with a 2-week history of abdominal pain, fever, loss of appetite and vomiting. He had also noticed a mass in the upper right quadrant of his abdomen. On physical examination, he appeared pale and distressed. His temperature was 38.1 °C, blood pressure was 130/80 mmHg and pulse was 100 beats per minute. On palpation, there was a firm, non-tender, mobile mass in the right upper quadrant. There were no signs of jaundice or hepatosplenomegaly.

Investigations

Blood tests (Table CS7.1) showed a raised white cell count (with increased neutrophils and reduced numbers of lymphocytes), elevated C-reactive protein and a mild elevation of liver enzymes.

Imaging studies, including ultrasound and CT scan of the abdomen, showed a large cystic lesion in the liver measuring 10 × 12 cm with daughter cysts within. Based on the clinical presentation and imaging studies, the patient was diagnosed with a hydatid cyst of the liver. A hydatid cyst is a parasitic infection caused by the tapeworm *Echinococcus granulosus*.

Diagnosis: Cystic echinococcosis.

Table CS7.1 Results of laboratory investigations.

Test	Result	Reference range
Haemoglobin	15.4 g/dL	13.0–18.0
White blood cell count	**13.2 × 10⁹/L**	4.0–11.0
Neutrophils	**79%**	40–75
Lymphocytes	**11%**	20–45
Monocytes	3%	2–10
Eosinophils	7%	1–6
Basophils	0%	1
C-reactive protein	**60 mg/L**	<10
Alanine transaminase	**46 U/L**	5–40
Aspartate transaminase	**47 U/L**	10–40
Alkaline phosphatase	**125 U/L**	25–115

Values outside the reference range are in bold.

Discussion

Echinococcosis is a human disease caused by different types of tapeworms from the *Echinococcus* genus. There are three types of importance in humans: cystic echinococcosis (caused by *E. granulosus*), alveolar echinococcosis (caused by *E. multilocularis*) and polycystic echinococcosis (caused by *E. vogeli* or *E. oligarthrus*). *E. granulosus* is widespread in several regions, including South America, Eastern Europe, Russia, the Middle East and China, where human incidence rates can be as high as 50 per 100 000. In some locations, such as slaughterhouses in South America, the prevalence of *E. granulosus* infection can be much higher. The parasite has two hosts in its life cycle (Fig. CS7.1): the primary (definitive) host and the intermediate host in which illness occurs. The adult form of the parasite lives in the intestines of primary hosts such as cats, dogs, wolves and foxes, causing only intestinal parasitosis but no organ disease. The adult parasite typically lives for around 5 months in the dog intestines and spreads millions of parasite eggs in the faeces. Humans become intermediate hosts, usually by consuming food contaminated

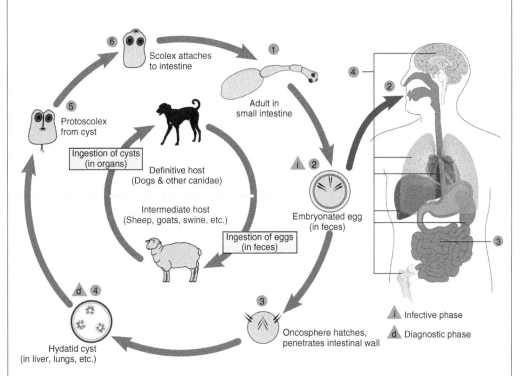

Figure CS7.1 Life cycle of the dog tapeworm, *Echinococcus granulosus*. Adult *E. granulosus* tapeworms inhabit the intestines of definitive (main) hosts, such as domestic dogs, wolves, foxes and other canids. The adult tapeworm (1) produces eggs that are released into the environment through the host's feces. (2) Intermediate hosts, including humans, become infected when they inadvertently ingest food, water, or vegetation contaminated with *E. granulosus* eggs. (3) Once ingested, the eggs release oncospheres (larval forms) within the small intestine. Oncospheres penetrate the intestinal wall, enter the bloodstream and are carried to various organs, most commonly the liver and lungs. (4) Oncospheres develop into hydatid cysts. These cysts grow gradually over time. Carnivores such as dogs become infected by consuming infected intermediate hosts or by ingesting eggs from contaminated environments. (5) In the dog's intestine, the cyst wall is digested, releasing protoscolices (larval stages) that (6) attach to the intestinal wall. The protoscolices develop into adult tapeworms in the definitive host's intestine. https://commons.wikimedia.org/wiki/File:CDC_Echinococcus_Life_Cycle.svg. Richfield, David (2014). 'Medical gallery of David Richfield'. WikiJournal of Medicine 1 (2). DOI:10.15347/wjm/2014.009. ISSN 2002-4436., CC BY-SA 4.0 <https://creativecommons.org/licenses/by-sa/4.0>, via Wikimedia Commons.

with the eggs. The liver is the most affected organ, accounting for 50–70% of cases, followed by the lungs (20–30%). After ingestion, oncospheres are released from the eggs and penetrate the intestinal wall, entering the portal venous system and spreading to various organs, including the liver and lungs. The oncospheres form cysts, which range in size from 1 to 15 cm in diameter and are typically unilocular. The immune system responds to the cyst by forming a calcified fibrous capsule around it, which is the structure that is most visible on imaging. As the cyst grows, it forms a combination of protoscolices (Fig. CS7.2) and daughter cysts, which appear grain-like on ultrasound imaging and are referred to as 'hydatid sand'. Cysts can remain asymptomatic for many years due to their slow growth. However, depending on the size and location of the cysts, they can eventually put pressure on nearby structures. Presenting symptoms include epigastric and/or right hypochondriac pain, nausea and vomiting. In 85–90% of cases, only one organ is affected, and more than 70% of patients have only a single cyst. Additional clinical manifestations may include biliary obstruction, cholangitis with biliary ruptures and portal hypertension.

Treatment and Prognosis

The patient was started on albendazole, a medication that kills the parasite. As the cyst was large and symptomatic, it was removed surgically. On examination, the cyst was found to contain multiple daughter cysts. The patient made an uneventful recovery and was discharged after a week of hospitalisation. He was advised to continue albendazole for 6 months to prevent recurrence.

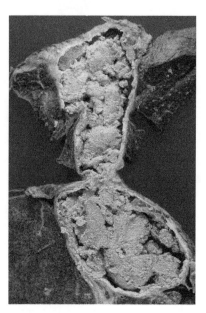

Figure CS7.3 Echinococcosis of the liver and lungs: Echinococcus cyst (so-called hydatid cyst) in the liver with the typical capsule (pericyst). A similar hydatid cyst is seen in the adjacent right lower lobe of the lung. The involvement of both organs is the result of trans-diaphragmatic extension of the liver cyst into the lower lobe of the right lung. *Source:* Yale Rosen / wikimedia Commons / CC BY-SA 2.0.

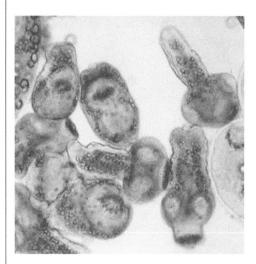

Figure CS7.2 Protoscolices liberated from a hydatid cyst. *Source:* CDC / Wikimedia Commons / Public Domain.

Complications of hydatid cyst disease include cyst rupture, either spontaneously or due to trauma. The rupture of a hydatid cyst can release antigens that trigger a severe allergic reaction, which may lead to anaphylaxis (Chapter 6). It can also increase the risk of secondary bacterial infections in the surrounding tissues or organs. Spread of cystic material within the peritoneal cavity can cause peritoneal seeding and peritonitis. Occasionally, ruptured hydatid cysts can cause pulmonary embolism. Additionally, if a liver cyst ruptures transdiaphragmatically, the contents can enter the pleural space. Another possible complication of a liver cyst rupture is the formation of a bronchial fistula. This occurs when the cyst communicates with the bronchial tree, leading to the release of cystic material into the lungs (Fig. CS7.3). Hydatid cysts can also exert pressure on surrounding tissues, including nearby blood vessels, leading to compromised blood flow. In the liver, this can be a cause of Budd-Chiari syndrome. Apart from mechanical compression, the inflammatory response to the parasite also activates the blood coagulation system, leading to thrombosis, which also contributes to the development of Budd-Chiari syndrome.

Questions

1   What is Budd-Chiari syndrome?
2   What type of drug is albendazole?

Answers on p. 533.

Case Study 8   Fever with Chills and Rigors

Clinical Presentation

A 28-year-old man presented to his local hospital with fever, headache, backache and nausea for 10 days duration. No regular pattern of fever was noted during the first 3–4 days, but recently the fever recurred every other day. A brief chill that progressed to rigors preceded each episode of fever that terminated after 3–6 hours with a bout of profuse sweating. The patient revealed that he had visited a malaria-endemic area the previous month; symptoms appeared about 2 weeks after returning home. No previous experience of similar symptoms was recorded. On examination, he looked ill and was pale, but was not jaundiced. Oral temperature was 100.4°F. He was conscious, rational and there was no neck stiffness. The spleen was just palpable. The liver was not enlarged.

Investigations

The raised erythrocyte sedimentation rate and C-reactive protein (ESR, CRP; Table CS8.1) indicated the presence of a disease process. Reduced haemoglobin, haematocrit and red cell count, but with other red cell indices within the normal range, indicated a normocytic-normochromic anaemia. The increased reticulocyte count supported a haemolytic condition. Malaria parasites in the blood film indicated a diagnosis of malaria due to *Plasmodium vivax* (Fig. CS8.2).

Diagnosis

The presence of malaria parasites in a blood smear confirms the diagnosis of malarial infection. A thickly spread blood film is useful for spotting the parasites, whereas thinly spread films enable accurate species identification and quantification of

Table CS8.1   Results of laboratory tests.

Test	Result	Reference range
Total white cell count	4.0×10^9/L	4.0–11
Red cell count	**3.47×10^{12}/L**	4.5–6.5
Erythrocyte sedimentation rate (ESR)	**48 mm/hour**	<15
Haemoglobin	**9.5 g/dL**	13–18
Haematocrit	**0.29 L/L**	0.40–0.54
C-reactive protein (CRP)	**35.2 mg/L**	<1
Mean cell volume	85.1 fL	82–92
Mean cell haemoglobin	27.3 pg	27–32
Platelet count	178×10^9/L	150–400
Reticulocyte count	**3.1%**	0.5–2.5
Blood film	**Malaria parasites seen** *Plasmodium vivax*-**parasitaemia = 0.24%**	
White blood cell differential count:		
Neutrophils	50%	40–75
Lymphocytes	**44%**	20–40
Monocytes	4%	2–10
Eosinophils	2%	1–6

Values outside the reference range are in bold.

the parasitaemia. It is recommended to make blood film examination on at least 3 consecutive days before malaria is excluded as the cause of fever in a probable case. Because of the acquisition of immunity to malaria in endemic areas, malaria parasites may be present in the blood without causing obvious symptoms.

In this patient, the clinical picture and the haemolytic, normochromic and normocytic anaemia are consistent with a diagnosis of malaria for which the identification of *Plasmodium vivax* parasites in a blood film is definitive evidence.

Diagnosis: Malaria due to *Plasmodium vivax* infection.

Discussion

Malaria is a mosquito-borne disease caused by a protozoan parasite of the genus *Plasmodium*. Several species of *Plasmodium* are mainly responsible for human disease: *P. vivax*, *P. falciparum*, *P. malariae*, *P. knowlesi*, and *P. ovale*. Malaria remains prevalent in many tropical and subtropical parts of the world. Globally, in 2020, there were a reported 241 million cases of malaria with 619 000 deaths: 77% of deaths occur in children <5 years old. Most infections are due to either *P. vivax* or *P. falciparum*.

The mosquito species primarily responsible for transmitting malaria belong to the genus *Anopheles*, which includes various species worldwide. Female *Anopheles* mosquitoes are the carriers of the malaria parasite and are known for their preference to bite humans during the night, particularly during dusk and dawn. They have a unique ability to detect human body heat, odour and carbon dioxide, which helps them locate potential hosts. *Anopheles* mosquitoes are found in various habitats, including areas with stagnant water, such as swamps, rice fields and standing pools. Efforts to control malaria involve targeting mosquito populations through measures such as insecticide-treated bed nets, indoor residual spraying (application of insecticide to the interior walls and surfaces of houses and other human dwellings), and larval control programs (e.g. altering or eliminating mosquito breeding sites, such as draining stagnant water, introducing natural predators of mosquito larvae, such as fish or copepods, or the application of chemical or biological larvicides).

Following the bite of an infected mosquito, the parasites first invade cells of the liver (Fig. CS8.1); they multiply and 6–12 days later, parasites are released and infect red blood cells (Fig. CS8.2), where they continue to multiply (Fig. CS8.3).

The clinical features of malaria are due to the blood stage parasites and the cyclical nature of their development, which results in the classical periodicity of symptoms. *P. vivax* infection is characterised by febrile paroxysms, that is, distinct episodes of fever associated with chills and rigors typically occurring every third day (tertian pattern). This periodicity is often not apparent in the initial stages of disease due to the asynchronous development of parasites. In addition, headache, myalgia, arthralgia, nausea and abdominal pain are frequently seen. Although distressing to the patient, clinical manifestations of *P. vivax* infections are usually medically uncomplicated and the disease is rarely fatal.

In *P. falciparum* malaria, fever periodicity is not as regular as with *P. vivax* and the paroxysms are less prominent. Any organ or system of the body can be affected, the brain and the circulatory system being two important sites, leading to severe, often fatal, complications which can manifest, for example, as cerebral malaria, severe anaemia, renal failure and severe hypoglycaemia. The anaemia that is common in malaria is due not only to the rupture of infected red cells, but probably also to destruction of un-infected red cells and bone marrow suppression. Mild enlargement of the spleen is common in acute malaria and is due to increased blood flow to the immunologically activated spleen causing congestion of the red pulp. Massive enlargement of the spleen can be seen in chronic infections in malaria-endemic areas. This once common, but now generally rare, condition is referred to as *tropical splenomegaly syndrome*.

Malaria should be suspected in all cases of fever in an endemic region and also in a patient with a travel history to a malaria-endemic area. Diagnosis of malaria in an area where there is no local transmission of the infection requires a high degree of clinical suspicion. Detailed history of clinical symptoms is important and travel history imperative, as would be a history of a recent blood transfusion. A patient's occupation may indicate the diagnosis, especially in areas of low malaria transmission, as outdoor workers are more likely to be exposed to mosquitoes. While a temperature chart may show the typical malaria fever pattern, fever with chills and rigors are seen in other conditions, for example viral infections, bacterial septicaemia and urinary tract infections. However, anaemia is infrequent in viral infections, elevated white cell counts with

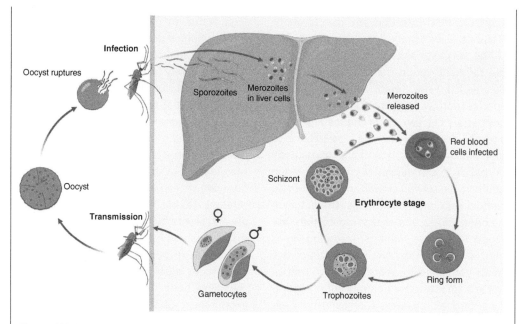

Figure CS8.1   Life cycle of parasites of the genus *Plasmodium* that are causal agents of malaria. *Mosquito (definitive host) stages:* Female *Anopheles* mosquitoes are infected with *Plasmodium* when they ingest the blood of an infected human host. The ingested blood contains gametocytes, the sexual stages of the parasite. Inside the mosquito's mid-gut, the male and female gametocytes fuse to form zygotes (not shown). Microgametes are the male gametes and macrogametes are the female gametes. The zygotes eventually transform into oocysts, which multiply through asexual division (sporogonic cycle) over a period of about 1–3 weeks. When mature, the oocysts rupture, releasing thousands of sporozoites. *Human (intermediate host) stages:* When the infected mosquito takes a blood meal from a human host, sporozoites enter the human bloodstream. *Liver stage (exoerythrocytic stage):* This is not shown here in detail, but sporozoites travel to the liver, where they invade hepatocytes. Inside hepatocytes, they undergo asexual multiplication, forming thousands of merozoites within each infected hepatocyte. Some merozoites differentiate into schizonts, which are structures that contain multiple merozoites formed through multiple rounds of asexual division. Schizonts eventually rupture, releasing merozoites. *Asexual blood stage (erythrocytic stage):* Within the red blood cells, merozoites transform into trophozoites, actively feeding on host haemoglobin and multiplying through multiple rounds of division. As the trophozoites mature, they develop into schizonts, which contain multiple merozoites. Eventually, the schizonts rupture, releasing a new wave of merozoites into the bloodstream, leading to the recurrent cycles of fever and clinical symptoms characteristic of malaria. This stage is responsible for the progression of the disease and the periodic episodes of fever and chills observed in infected individuals. When conditions become less favourable for asexual replication, such as changes in nutrient availability or other environmental cues, some trophozoites undergo a process called 'commitment' to gametocytogenesis. Committed trophozoites develop into gametocytes. When an infected mosquito takes a blood meal from an infected human host, it ingests both asexual blood stage forms and gametocytes present in the bloodstream, but only the gametocytes survive in the mosquito gut to continue its life cycle. *Source:* Created with BioRender.com.

polymorphonuclear leucocytosis character-ises bacterial septicaemia, and urinary tract infections will be marked by symptoms such as dysuria. Anaemia and splenic enlargement are common in haemolytic conditions but are also seen in leukaemia, though other haematological changes will be apparent in the latter condition. Reticulocytes appear in

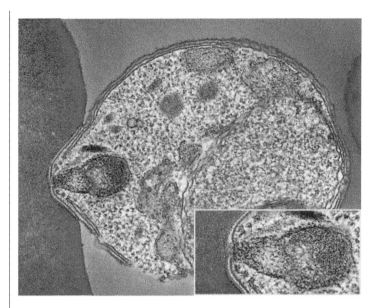

Figure CS8.2   Infection of red blood cell by malarial parasite. Colourised electron micrograph showing malaria parasite (right, blue) attaching to a human red blood cell. The inset shows a detail of the attachment point at higher magnification. *Source:* NIAID / Wikimedia Commons / CC BY 2.0.

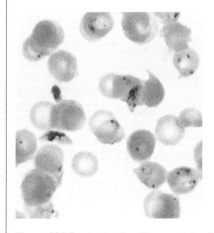

Figure CS8.3   A thin-film Giemsa-stained micrograph of ring-forms and gametocytes of *Plasmodium falciparum. Source:* TimVickers / wikimedia Commons / Public Domain.

larger than normal numbers because of the regeneration of red cells in the bone marrow, for example following haemolysis or haemorrhage.

Rapid diagnostic tests (RDTs) are frequently used in the diagnosis of malaria due to their ease of use as a bedside diagnostic. Unlike microscopy, these tests do not require expert technical skills and are relatively inexpensive. Molecular diagnosis, though high in sensitivity when compared with either RDTs or microscopy, is not routinely used in most endemic countries due to the high cost and the need for well-equipped laboratories.

Treatment and Prognosis

Prompt treatment of malaria with artemisinin-based combination therapy (ACT) is recommended in regions endemic to *P. falciparum*. Complete recovery occurs following effective treatment in uncomplicated malarial infections. Early diagnosis and effective treatment usually prevent severe and complicated disease in *P. falciparum* infections. Once complications develop, case fatality rates rise to around 10–20%. There is increasing concern about the emergence of drug-resistant malaria; artemisinin-resistant *P. falciparum* organisms are becoming established in Southeast Asia with reports of the emergence of resistant strains also in Africa. A malaria vaccine, known as RTS,S/AS01 (commercial name Mosquirix), targeting the circumsporozoite protein of *P. falciparum*, shows moderate protection (36% against the development of clinical malaria and 32% against severe malaria). New anti-malarial drugs and vaccines are in clinical development; the latter remains a priority if the WHO is to achieve its target of achieving a malaria vaccine with greater than 75% efficacy by 2030. Mosquito insecticide resistance is also a growing global issue that hampers attempts to reduce mosquito numbers.

Questions

1 Why do *P. falciparum* parasites cause a more severe form of malaria compared to other species?
2 What are the signs of treatment failure?

Answers on p. 534.

Further Reading

Jagannathan P. & Kakuru A. (2022) Malaria in 2022: increasing challenges, cautious optimism. *Nature Communications* **13**: 2678. 10.1038/s41467-022-30133-w.

Case Study 9 Fever, Vomiting and Neck Stiffness in a Child

Clinical Presentation

A 10-month-old boy presented with a short history of fever, persistent vomiting and marked drowsiness. On examination, he was febrile (temperature: 38.5°C), with photophobia and neck stiffness. Kernig sign (see below) was positive.

Investigations

A specimen of cerebrospinal fluid (CSF) was taken by lumbar puncture. The investigations shown in Table CS9.1 revealed an increase in the number of white cells (which were mainly neutrophils) and an increase in the amount of protein in the CSF (in viral meningitis, protein levels are usually <1.0 g/L. In contrast, acute bacterial meningitis is usually associated with much higher CSF protein concentrations). Gram stain showed the presence of Gram-positive diplococci in the CSF. The patient was started on ceftriaxone and vancomycin and was admitted to the paediatric intensive care unit for close monitoring

Table CS9.1 Results of CSF investigations.

Test	Result	Reference range
White cell count	**4600 × 10⁹/L (90% neutrophils)**	<5 × 10⁹/L (no neutrophils)
Glucose	**0.7 mmol/L** (blood glucose was 2.6 mmol/L)	CSF glucose levels are normally greater than two-thirds of blood glucose levels
CSF protein	**4.5 g/L**	<0.45

Values outside the reference range are in bold.

of cardio-pulmonary function and further management. By the second day of hospitalisation, culture of the CSF revealed the presence of *Streptococcus pneumoniae* sensitive to ceftriaxone and penicillin. Ceftriaxone is a third-generation cephalosporin that is highly penetrable into the meninges, eyes and inner ear.

Diagnosis

The early clinical signs of bacterial meningitis may include fever, depression and disturbance of the level of consciousness and vomiting. Other features which are more specific but appear later, include photophobia and neck stiffness. The triad of neck stiffness, photophobia and headache, is a sign of meningeal irritation. A specific clinical test for meningitis is Kernig sign. This is elicited by attempting to straighten the knee while the hip is held flexed. This manoeuvre pulls the spinal nerves where they penetrate the meninges, and therefore stretches the meninges themselves. Because the meninges are inflamed in meningitis, this is exquisitely painful and the muscles go into spasm to prevent it from happening.

Diagnosis: Bacterial meningitis.

Discussion

Meningitis is inflammation of the meninges. It can be caused by infections with bacteria, viruses, protozoa, or fungi. Viral meningitis typically has a benign clinical outcome unless the brain itself is also affected (viral encephalitis), when brain damage can ensue. Fungal and protozoal meningitis are most often seen in individuals who are immunosuppressed or immunodeficient, and in this context, are essentially opportunistic infections. Bacterial meningitis is a medical emergency and rapid diagnosis is critical.

Host factors that increase susceptibility to meningitis include the following:

- *Splenectomy/congenitally absent/functional asplenia:* Individuals who have had a splenectomy (removal of the spleen), are born without a spleen, or have a non-functioning spleen are at an increased risk of severe infections caused by encapsulated bacteria, including *Haemophilus influenzae* type b (Hib), *Streptococcus pneumoniae* and *Neisseria meningitidis*. The spleen plays a critical role in the immune response against these bacteria, so its absence or reduced function can lead to a decreased ability to clear these infections.
- *Diabetes and alcoholism:* Individuals with diabetes and those who abuse alcohol are at an increased risk of infections, particularly with *Streptococcus pneumoniae*.
- *Altered cell-mediated immunity:* Immunocompromised individuals, such as those receiving immunosuppressive therapy, or those living with HIV/AIDS, are at an increased risk of infections, with *Listeria monocytogenes* and *Cryptococcus neoformans*.
- *Fracture or bony defect of the skull:* These defects can serve as entry points for bacteria, particularly, *Streptococcus pneumoniae*.
- *Inherited defects in the late complement components:* Can increase the risk of recurrent invasive infections, particularly those caused by *Neisseria meningitidis*.
- *Pregnancy:* Pregnant women have an increased risk of certain infections, including *Listeria monocytogenes*, which can lead to meningitis. Listeriosis can be particularly concerning during pregnancy due to the potential for harm to the developing foetus.

70% of cases of bacterial meningitis occur in childhood (Table CS9.2). Three organisms, *Haemophilus influenzae*, *Neisseria meningitidis* and *Streptococcus pneumoniae*,

Table CS9.2 Typical bacterial pathogens associated with meningitis by age.

Age group	Organism
All ages	*Streptococcus pneumoniae*
	Neisseria meningitidis
	Haemophilus influenzae type b – in the 'pre-school' child – now rare, unless no Hib vaccine
	Mycobacterium tuberculosis (especially in high-prevalence countries)
Neonates	Group B *Streptococcus*
	Escherichia coli and other aerobic Gram-negative bacilli
	Listeria monocytogenes
Elderly	*Streptococcus pneumoniae*
	Listeria monocytogenes

are responsible for most cases. All these organisms have a carbohydrate (polysaccharide) capsule that is essential for virulence. The ability to mount an immune response to this capsule is an important protective mechanism. Antibodies to carbohydrate antigens, unlike antibodies directed to many other protein antigens, are thymus-independent (i.e. they do not require T-cell help). Moreover, the immune responses to carbohydrate antigens mature relatively late in development. Therefore, children under 2 years of age are particularly susceptible to infections with the organisms listed above.

In most developed countries, meningitis due to *H. influenzae* has now virtually disappeared due to the use of the so-called Hib vaccine (*H. influenzae B*); for example, in the United States, the introduction of the Hib conjugate vaccine has resulted in a decrease in the incidence of Hib disease of more than 98%.

In the United Kingdom, babies are also now routinely vaccinated with a type of pneumococcal vaccine known as the pneumococcal conjugate vaccine (PCV) as part of their childhood vaccination programme. This vaccine targets *Streptococcus pneumoniae* and is also given to adults over 65 years or to those with a pre-existing medical condition. However, after an initial promising decline in the incidence of pneumococcal meningitis following the introduction of vaccination, there has been a resurgence of cases, seemingly due to the proliferation of bacterial serotypes not targeted by the existing vaccine. These data indicate that meningitis caused by *S. pneumoniae* will continue to pose a major health challenge, at least until newer vaccines become available.

People all over the world are at risk of meningitis. However, the highest burden of disease is found in a region of sub-Saharan Africa, known as the 'meningitis belt' where there is a high risk of epidemics of both meningococcal and pneumococcal meningitis (Fig. CS9.1). Dry and dusty conditions during the dry season between December

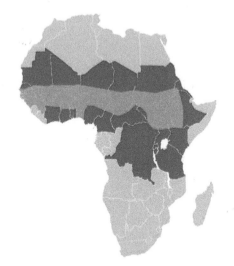

Figure CS9.1 African Meningitis Belt 2016. Orange – African meningitis belt. Red – countries within the meningitis belt and areas prone to outbreaks of meningitis epidemics. *Source:* CC BY-SA 4.0 <https://creativecommons.org/licenses/by-sa/4.0>, via Wikimedia Commons. https://commons.wikimedia.org/wiki/File:African_meningitis_belt_-_2016.JPG.

and June, as well as crowded living conditions, are thought to contribute to the increased incidence of bacterial meningitis in this region.

The decreased CSF glucose level observed in patients with bacterial meningitis is a characteristic feature. Initially attributed to bacterial glucose metabolism, a more significant contributor is the transition of neurons and glial cells from aerobic to anaerobic metabolism. Anaerobic respiration, which does not require oxygen but yields less energy, consumes more glucose. This shift intensifies glucose consumption. Infections trigger inflammation, causing fluid accumulation (oedema) and blood vessel compression, impairing blood supply. Microcirculation clotting exacerbates this issue. Under reduced oxygen conditions, cells switch to anaerobic metabolism, increasing glucose consumption and lactate production. Consequently, lower CSF glucose and elevated CSF lactate levels are associated with poor prognosis in bacterial meningitis.

Much of the tissue damage in bacterial meningitis is the result of the inflammatory response, rather than a direct effect of the organism. Lipopolysaccharide (LPS, endotoxin; Chapter 4) is a structural component of the cell wall of Gram-negative bacteria and is a potent stimulator of the inflammatory response. Gram-positive organisms have a similar component known as lipoteichoic acid. These cell wall molecules stimulate host cells (particularly macrophages) to secrete cytokines, including TNF-α and IL-1. These cytokines act on a wide variety of cells to initiate inflammation leading to increased capillary permeability (with associated local oedema), an ingress of neutrophils and stimulation of the complement and coagulation cascades (see also Chapters 5 and 8). The resulting cerebral oedema can, in turn, lead to an increase in intracranial pressure and impairment of cerebral blood flow.

Treatment and Prognosis

Prompt diagnosis is essential in bacterial meningitis. The Meningitis Research Foundation can help parents spot the warning signs of meningitis, which are divided into 'red symptoms' (high-risk, take immediate action), and 'amber symptoms' (medium-risk, seek professional medical advice) and are summarised at: https://www.meningitis.org/meningitis/check-symptoms.

Antibiotics are the cornerstone of the management of bacterial meningitis. Empiric therapy should be given immediately if meningitis is suspected and administered intravenously, at least during the acute stages of the illness, since vomiting, impaired consciousness and possible shock, make oral administration impossible. Many antibiotics do not cross the blood–brain barrier effectively, and thus have very poor penetration into the CSF. Dexamethasone (a corticosteroid) is an adjunct treatment. If given early, that is at the same time as the initial dose of antibiotic, dexamethasone can reduce the release of pro-inflammatory cytokines and limit the inappropriately severe host inflammatory response. Steroids should be administered prior to, or with, the first dose of antibiotics in suspected bacterial meningitis:

- In Hib meningitis, steroids reduce overall mortality and the incidence of deafness.
- In adults, with pneumococcal meningitis, steroids may reduce the risk of death.
- In tuberculous meningitis, steroids are used to reduce fibrosis and the risk of secondary hydrocephalus.

Despite modern antibiotic therapy, overall mortality in patients with bacterial meningitis remains around 10%. Up to 30% of survivors of bacterial meningitis are left with long-term sequelae, including around 10% who will be deaf due to direct cochlear damage; there is a small canal (the cochlear aqueduct) connecting the cochlea directly with CSF, and bacteria can pass directly to the inner ear and damage it, even in the early stages of the disease.

Questions

1 In some cases of bacterial meningitis, there is marked clinical deterioration within hours of starting antibiotic therapy. Can you suggest why this is?

2 What are the differences in the immune response generated against T-cell independent (TI) antigens, compared with T-cell dependent (TD) antigens?

Answers on p. 534.

Case Study 10   Sudden Onset of Severe Pain and Swelling in the Left Leg

Clinical Presentation

A 76-year-old man, who was a k/c/o ('known case of') diabetes, hypertension, ischaemic heart disease and end-stage renal disease (ESRD) on dialysis, presented to the Emergency Department (ED) with severe pain and swelling in his left leg. He reported that yesterday he noticed a small area of painful redness on his calf, which he thought was probably just an insect bite, but got worried when it quickly progressed to a larger, tender area with increasing pain and erythema. On presentation in the ED, he had a fever with chills and was feeling generally unwell.

On physical examination, the patient was unwell and in pain, the left leg had an area of induration measuring around 4 cm in diameter with surrounding warmth and erythema. Notable swelling was appreciated (extending beyond the area of visible redness) with no fluctuance. It was extremely tender to palpation. Crepitus (a crackling sensation) was appreciated on palpation. The skin was tense and shiny, and small bullae were also present. The patient's vital signs were unstable, with a heart rate of 120 beats per minute, blood pressure of 94/63 mmHg and a temperature of 102°F.

Investigations

The results of laboratory investigations are shown in Table CS10.1.

Table CS10.1   Results of laboratory investigations.

Test	Result	Reference range
White blood cell count	**22.5 × 10⁹/L**	4.0–11.0
Haemoglobin	**9.5 g/dL**	13.0–18.0
Platelet count	**450 × 10⁹/L**	150–400
Neutrophil count	**19.5 × 10⁹/L**	2.0–7.5
Lactate	**4.5 mmol/L**	0.5–2.0
C-reactive protein (CRP)	**280 mg/L**	<10
Blood culture	**Positive for Group A Streptococcus**	Not detected

Values outside the reference range are in bold.

Diagnosis

The elevated white blood cell count and neutrophil count indicated an active infection. Secondary thrombocytosis (increased platelet count) can also occur in response to infection. The patient also had a low haemoglobin level indicating anaemia, likely of chronic disease. CRP, a marker of inflammation or infection, was significantly elevated. Blood culture results came back the following day and were positive for Group A *Streptococcus*. Given the clinical presentation of the patient, the severity of the patient's symptoms and the rapid progression of the infection, a diagnosis of necrotising fasciitis was suspected. Pain out of proportion to examination findings is highly suggestive of necrotising fasciitis. Necrotising fasciitis is a clinical diagnosis; physical examination, laboratory tests and imaging studies helped to confirm this. Laboratory tests alone are not sufficient to diagnose the condition, as all markers are non-specific. Imaging may show extending oedema along fascial planes and/or soft tissue gas (which is highly suggestive of the diagnosis). CT/MRI has a higher sensitivity compared to X-ray but might take time to arrange and is not usually the first modality of choice in an emergency situation. X-ray findings are usually normal early on; hence, are not useful in ruling out the diagnosis. Ultrasound maybe done by the bedside in clinically unstable patients.

The patient was immediately started on broad-spectrum antibiotics and was taken to the operating room for urgent surgical debridement. During surgery, extensive necrosis of the fascia and muscle was noted, requiring aggressive surgical exploration and debridement. The patient was taken to the intensive care unit for further management.

Diagnosis: Necrotising fasciitis.

Discussion

Necrotising fasciitis is a rapidly progressing and potentially life-threatening infection of the deep fascia and surrounding soft tissues. It typically presents with the sudden onset of severe pain, redness, swelling and tenderness at the affected site. The skin over the affected area may appear shiny and tight, and there may also be blisters. Systemic symptoms such as fever, malaise and elevated white blood cell count, are also common presenting signs. As the infection progresses, the affected tissues become necrotic with gas formation, giving the characteristic 'flesh-eating' appearance. Early diagnosis is important as prompt surgical debridement and antibiotics are crucial for improving outcomes. Early diagnosis is aided by high clinical suspicion, as early necrotising fasciitis is difficult to differentiate from cellulitis. In an immunocompromised patient, it is especially important to look for pain out of proportion to physical findings and tenderness that extends beyond the erythematous area, as these are early signs that may be often missed. When clinical suspicion exists (severely unwell patients with cellulitis-like signs) but the signs are not severe, it may be sometimes advisable to perform an exploratory incision to confirm the diagnosis, as delaying surgery by over 24 hours significantly increases mortality.

Necrotising fasciitis can be caused by both aerobic and anaerobic bacteria. The most common causative organisms are: *Streptococcus pyogenes* (Group A *Streptococcus*), *Staphylococcus aureus*, *Klebsiella pneumoniae*, *Escherichia coli* and other Gram-negative bacilli. Necrotising fasciitis can also be *polymicrobial*, meaning it can be caused by mixture of different types of bacteria. Antibiotics are crucial in the management of patients; the primary goals being to eradicate the infecting bacteria, prevent the development of sepsis and reduce the risk of surgical complications. It is important to choose antibiotics that are effective against the specific bacterial species causing the infection, as determined by blood culture results. However, broad-spectrum antibiotics are used initially, followed by more targeted therapy once the results of culture and sensitivity testing are known. In general, the following broad-spectrum antibiotics are recommended for the treatment of necrotising fasciitis: Beta-lactams- this class of antibiotics includes the penicillins (e.g. nafcillin or oxacillin) and cephalosporins (e.g. ceftriaxone or cefotaxime). They are effective against a wide range of bacteria, including *Streptococcus pyogenes* and *Staphylococcus aureus*; Fluoroquinolones, including ciprofloxacin and levofloxacin, which are effective against many Gram-negative bacteria, including *Klebsiella pneumoniae* and *Escherichia coli*; Carbapenems, including imipenem and meropenem, effective against a wide range of Gram-negative and Gram-positive bacteria. Metronidazole can be used for the treatment of infections with anaerobic bacteria. In some cases of necrotising fasciitis caused by group A Streptococcus (GAS) infections, the use of IVIG (intravenous immunoglobulin) may be considered as an adjunctive treatment. IVIG is a preparation of antibodies collected from the blood of donors.

In addition to antibiotics, prompt surgical debridement of infected tissue is also a crucial component of the management of necrotising fasciitis. Surgical intervention is performed to remove all infected tissue, including any necrotic or gangrenous tissue, to prevent the spread of the infection and promote healing. Multiple debridements are often necessary. The management of necrotising fasciitis also involves supportive care, such as fluid and electrolyte management and pain control.

The pathogenesis of necrotising fasciitis involves the complex interplay between host factors and bacterial virulence factors. Certain host factors increase the risk of developing necrotising fasciitis. They include type 2 diabetes, a compromised immune system (e.g. in cancer and HIV infection), obesity, alcoholism, intravenous drug use, and open wounds or skin ulcers. Bacterial virulence factors that play a critical role in the pathogenesis of

necrotising fasciitis include: *Exotoxins* – certain bacterial strains produce exotoxins that damage host tissues and contribute to the development of tissue necrosis. *Enzymes*: bacterial enzymes, such as hyaluronidase and collagenase, facilitate tissue destruction. *Biofilm formation:* bacteria can form biofilms, which protect them from host defences and antibiotics, making them more resistant to treatment. Once the bacteria gain access to the subcutaneous tissues, they rapidly multiply and release toxins that cause extensive tissue destruction. The bacteria and host immune response further contribute to tissue destruction through the release of inflammatory cytokines and the formation of thrombi that impede blood flow to the affected tissues. As a result, tissue necrosis and liquefaction occur. Gas produced by the bacteria may accumulate in the tissues, causing the classic crepitus. Complications of necrotising fasciitis include sepsis and secondary infections.

Treatment and Prognosis

Over the next few days, the patient remained in a critical condition, requiring aggressive antibiotic therapy, fluid resuscitation, haemodynamic support and close monitoring. Despite aggressive debridement, the patient's condition continued to deteriorate and the patient was offered an above-knee amputation as a life-saving measure, which he refused despite being made aware of the risk of death. He subsequently developed sepsis and multi-organ failure. The family was consulted and the patient was made comfortable. He passed away 2 days later.

Questions

1   Why do some patients with necrotising fasciitis develop disseminated intravascular coagulation (DIC)?
2   What is responsible for the formation of the 'gas' in necrotising fasciitis?

Answers on p. 534.

Case Study 11   SARS-CoV-2 Infection in a Pregnant Woman

Clinical Presentation

A 26-year-old pregnant woman telephoned her GP at 28 weeks gestation with a cough that had been present for the past 5 days. She reported having tested positive for COVID-19 three days prior to this and had been self-quarantining at home. She was now worried about her baby because of recent articles in the press about the complications of pregnancy in mothers who test positive for SARS-CoV-2, and because she had noticed reduced foetal movements in the last few days. On examination in the surgery, the patient was noted to have a persistent cough but otherwise appeared healthy. However, foetal heart monitoring revealed a tachycardic foetal heart rate of 180 beats per minute. Blood tests and a nasal swab for PCR testing were taken.

Investigations

The patient was admitted to the local hospital and the neonatology team consulted for management of a potential high-risk pregnancy. Obstetric ultrasound revealed a small-for-gestational-age foetus with absent end-diastolic flow in the umbilical artery and abnormal placental thickness and vascularity. The results of laboratory investigations (Table CS11.1) revealed that the patient had raised levels of C-reactive protein and D-dimer. Nasal swab was positive for SARS-CoV-2 by PCR test.

Diagnosis

The patient received corticosteroids and an emergency caesarean section was performed due to concerns of foetal distress. A premature infant boy was delivered and admitted to the neonatal intensive care unit for management of respiratory distress syndrome. Macroscopically, the placenta showed yellow-tan patchy areas with apparent fibrin deposition and small infarcts (Fig. CS11.1). The pathologist examined the placenta, which showed chronic histiocytic intervillositis (CHI), massive peri-villous fibrin deposition (MPVFD)

Table CS11.1   Results of laboratory investigations.

Test	Result	Reference range
White blood cell count	9.0×10^9/L	4.0–11.0
Haemoglobin	14.5 g/dL	11.5–16.5
Platelet count	200×10^9/L	150–400
Blood gases		
pH	7.45	7.35–7.45
$PaCO_2$	40 mmHg	36–46
PaO_2	85 mmHg	75–100
SaO_2	98%	>95%
D-dimer	**800 ng/mL**	<500
C-reactive protein	**25 mg/L**	<10

Values outside the reference range are in bold.

and trophoblast necrosis (TN) (Fig. CS11.2). Immunohistochemistry for detection of SARS-CoV-2 encoded spike (Fig. CS11.3) and nucleocapsid proteins was positive, showing staining within the syncytiotrophoblast layer. The histological findings were consistent with COVID-19-associated placentitis, at the start of the third trimester of pregnancy.

Diagnosis: COVID-19-associated placentitis.

Discussion

When a pregnant woman contracts a viral infection, it can negatively affect both the outcome of the pregnancy and the long-term health of the developing foetus, even into adulthood. The harmful effects may not only result from direct infection of the placenta, but also from the systemic effects of the virus outside the placenta. The placenta serves as a critical barrier between the mother and foetus, protecting the foetus from infection while avoiding immune rejection of the semi-allogeneic foetus. Structurally, the placenta is made up mostly of foetal villi lined by a bilayer of cytotrophoblast and fused syncytiotrophoblast cells,

Figure CS11.1 Gross appearance of COVID-19 placentitis. The placenta has been sliced to show its abnormal appearance, with the pale areas (shown as white arrows on the right-hand side) resembling massive perivillous fibrinoid deposition with the involvement of a significant volume of placental parenchyma.

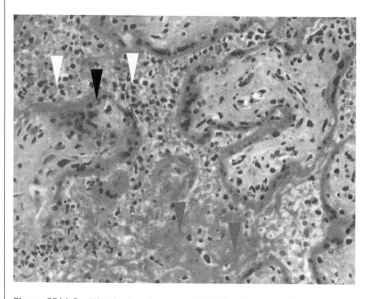

Figure CS11.2 Histological features of COVID-19 placentitis. Haematoxylin and eosin stained image of the affected placenta showing prominent intervillous histiocytic infiltrates (white arrowheads), fibrin deposition (red arrowheads) and trophoblast necrosis (black arrowhead). The features are typical of COVID-19-associated placentitis.

known collectively as the trophoblast layer. These villi, containing foetal blood, invade the mother's uterine decidua and are surrounded by the intervillous space, where maternal blood flows. Various mechanisms are at work at the maternal-foetal interface to prevent and limit viral infections, such as interferon responses, proinflammatory cytokine expression and the presence of maternal innate cells in the decidua, and foetal-derived macrophages, known as Hofbauer cells, in the villous stroma.

In April 2021, a report from County Cork, Ireland described a temporal cluster of six stillbirths and one miscarriage from pregnant women with COVID-19. Upon examination of the placentas from the stillborn foetuses, it was found that they were infected with

Figure CS11.3 Immunohistochemistry for SARS-CoV-2. Staining for spike protein was present in the syncytiotrophoblast layer (brown staining).

SARS-CoV-2 and had severe damage due to fibrin deposition, intervillositis and necrosis. In May 2021, a study in England analysing a national database of 342 080 pregnant women, 3527 of whom had COVID-19, found higher rates of foetal death among women infected with SARS-CoV-2 than among uninfected mothers. In November 2021, the US Centers for Disease Control and Prevention confirmed this association in a population-based study, demonstrating that pregnant women with COVID-19 have a greater risk of stillbirth than uninfected women. Although the increased risk of serious obstetric complications in pregnancy might be due to the systemic effects of severe respiratory disease, most cases of COVID-19 placentitis resulting in severe obstetric outcomes occurred in women with mild or asymptomatic COVID-19 respiratory infection. COVID-19 placentitis is characterised by three pathological features:

- *Chronic histiocytic intervillositis:* This is characterised by the presence of chronic inflammation in the spaces between the placental villi (intervillous spaces). The inflammation is typically marked by the accumulation of histiocytes (tissue macrophages).
- *Massive peri-villous fibrin deposition:* This describes an excessive accumulation of fibrin (the end-product of the coagulation cascade, see Chapter 8) around the placental villi (hence the term peri-villous).
- *Trophoblast necrosis:* In which trophoblasts undergo a form of cell death known as necrosis (Chapter 3).

These pathologies are usually widespread in the placenta, leading to destruction of the placenta and resulting in foetal demise in most cases. The role of SARS-CoV-2 in causing COVID-19 placentitis is not clear, as only about half of cases show evidence of placental infection. The extensive destruction of the placenta that follows COVID-19 placentitis hinders the delivery of sufficient oxygen and nutrients to the foetus and this, rather than direct virus infection of foetal tissues, is the cause of foetal death.

Treatment and Prognosis

The mother's recovery was uneventful. Her baby boy had to remain in the neonatal intensive care unit due to immaturity. Over the next few weeks, the baby received respiratory support from a ventilator, oxygen therapy and medications to support lung development. The baby gained weight and was gradually weaned off the ventilator. The baby was discharged from the hospital 4 weeks after delivery and continues to thrive. This was an unusual case history as, regrettably, COVID-19 placentitis almost always causes intrauterine death/stillbirth. It is noteworthy that most reported cases of SARS-CoV-2 placentitis occur in mothers who were not vaccinated against SARS-CoV-2.

Questions

1 What is the D-dimer and why is it raised in COVID-19?
2 What is C-reactive protein and what do raised levels mean?
3 Why was the patient given corticosteroids?

Answers on p. 535.

Case Study 12   Short History of Fever, Chills and Confusion

Clinical Presentation

A 72-year-old man with a history of hypertension and type 2 diabetes was brought to the Emergency Room accompanied by his wife who reported that he had developed fever and chills for the past 2 days on a background of a cough productive of yellowish-green sputum which had been present for at least a week. He initially took acetaminophen and rested, but his symptoms did not improve. His hypertension and type 2 diabetes were managed with lisinopril and metformin, respectively. The rest of his past medical history was not significant. In the last few hours, his wife noted that he was 'not himself' and appeared confused and drowsy. She called an ambulance and the patient was taken to hospital.

Investigations

Upon arrival, the patient was found to be febrile with a temperature of 38.4°C, tachycardic with a heart rate of 120 beats per minute and hypotensive with a blood pressure of 80/40 mmHg. He appeared confused and disoriented. Capillary refill was 6 seconds. His lungs had coarse crackles bilaterally, and his abdomen was soft and non-tender. There was no evidence of skin rash or petechiae. Laboratory tests revealed leucocytosis and an elevated blood lactate. Oxygen support was given. Blood samples were taken (including two sets of blood cultures), a urinary catheter was attached to monitor the hourly urine output, and after that, he was started on empiric broad-spectrum antibiotics, intravenous

Table CS12.1   Results of selected laboratory tests.

Test	Result	Reference range
White blood cell count	16.0×10^9/L	4.0–11.0
Lactate	4.2 mmol/L	0.50–2.0
Blood culture	Positive for *Klebsiella pneumoniae*	Negative
Creatinine	178 µmol/L	59–104
Urea	8.2 mmol/L	2.5–6.7
Sodium	136 mmol/L	135–146
Potassium	3.6 mmol/L	3.5–5.0
Bicarbonate	20 mmol/L	22–30
Random blood glucose	11.1 mmol/L	3.5–8.9

Values outside the reference range are in bold.

fluids and vasopressors. The results of laboratory tests are shown in Table CS12.1.

Diagnosis

The patient was admitted to the intensive care unit. His blood cultures flagged positive with Gram-negative bacilli on Gram stain the following day and subsequently grew *Klebsiella pneumoniae*. The patient had a raised white cell count, which was consistent with an infection. He also had raised lactate levels. Measuring serum lactate levels is routine in the management of sepsis. High serum lactate reflects tissue hypoperfusion and may revert back to normal quickly after adequate fluid resuscitation. In sepsis, the body's oxygen delivery to the tissues may become impaired due to a combination of factors, including decreased cardiac output and impaired blood flow to vital organs. This can lead to an accumulation of lactate in the blood, known as *lactic acidosis*. Lactate levels are a good indicator of outcome in sepsis; persistent elevation of lactate levels, despite resuscitation and treatment, suggests the need for further intervention. Lactic acidosis causes a decrease in serum bicarbonate concentration that is similar in magnitude to the increase in the lactate concentration. Urea and creatinine are markers of kidney function that are raised in this patient. Urine output is a good reflection of kidney function and/or intravascular volume depletion. Hence, hourly urine output is usually monitored in in-patient settings, which is why catheterisation is usually done upon presentation. In septic shock, the kidneys can be affected acutely due to decreased intravascular volume depletion. Some studies show that patients with acute kidney injury (AKI) evoked by sepsis have a worse prognosis than patients with non-septic AKI.

Diagnosis: Septic shock.

Discussion

Sepsis is a potentially life-threatening complication of an infection that can affect anyone of any age. Sepsis occurs when there is a dysregulated immune response to an infection, leading to systemic inflammation and organ damage. *Septic shock* is a subset of sepsis with circulatory and cellular/metabolic dysfunction associated with a higher risk of mortality. Septic shock occurs when a patient's blood pressure drops dramatically, despite receiving adequate fluid therapy, necessitating treatment with vasopressor drugs (Table CS12.2). Sepsis is more common in the very young, the elderly, or those with a weakened immune

Table CS12.2 Distinction between infection, sepsis and septic shock.

Infection	A clinical syndrome based on symptoms and signs caused by pathogenic organisms, which may or may not be identified.
Sepsis	Acute dysfunction of one or more organ consequent to infection.
Septic shock	A vasopressor or inotrope requirement to maintain mean arterial pressure (MAP) ≥65 mmHg despite adequate fluid resuscitation which has been triggered by infection.

Table CS12.3 Definitions of organ dysfunction.

System	Altered function
Central nervous system	Acutely altered mental status
Respiratory	New need for oxygen to achieve saturation >90% (note: this is a definition, not the target)
Cardiovascular	Systolic BP <90 mmHg or mean arterial pressure (MAP) <65 mmHg or systolic BP >40 mmHg below the patient's normal systolic BP. Lactate ≥4.2 mmol/L
Renal	Creatinine >170 μmol/L, or urine output <500 mL/24 hours, despite adequate fluid resuscitation
Haematological	Platelets $<100 \times 10^9$/L. Petechial or purpuric rash
Liver	Bilirubin >32 μmol/L

system. Internationally, approaches to sepsis management, based on early recognition of sepsis with resuscitation and timely referral to critical care, have reported reductions in mortality from septic shock in the order of 20–30%.

Organ dysfunction is identified if one or more of the abnormal findings listed in Table CS12.3 are diagnosed (either on the blood tests sent as part of Sepsis 6, or persisting clinically after the bundle has been administered).

Infection, sepsis and septic shock are clinical diagnoses. There is no one test that will confirm the presence of infection, sepsis or septic shock to the exclusion of other diagnoses. Rather, the suite of symptoms and signs along with the results of investigations need to be considered against the differential diagnoses and a clinical management decision made. It is important to note that while the identification of an organism is valuable in guiding targeted antimicrobial therapy, blood cultures are only positive in 40–55% of cases of sepsis. Hence, a negative culture does not preclude the diagnosis of sepsis.

Early identification and management of sepsis is critical to improve patient outcomes. This involves having systems in place to recognise sepsis so that patients are managed in a timely manner. Sepsis management includes prompt administration of antibiotics, control of the infection source (e.g. removal/debridement of infected tissues, abscess drainage, etc.) and supportive measures, such as fluid replacement and vasopressor therapy.

Sepsis is a global health concern affecting all age groups, with respiratory and gastrointestinal bacterial infections being the most common causes. In 2017, sepsis was responsible for an estimated 11 million deaths globally, accounting for almost 20% of all deaths with an age-standardised mortality rate of 148 per 100 000 population. Thirty-day survival for patients with sepsis admitted to critical care units is only two out of three. Sepsis survivors also have a higher risk of re-hospitalisation and death, with almost half requiring re-hospitalisation within a year and one out of six not surviving the first year.

Sepsis is characterised by an imbalanced immune response resulting in persistent immune cell activation and injury. The immune mechanisms initially activated to provide protection become harmful due to excessive inflammation associated with a concomitant immune suppression. Damage-associated molecular patterns (DAMPs; Chapter 6) contribute to immune stimulation in sepsis by activating pattern recognition receptors (PRRs). This creates a vicious cycle of sustained immune activation and dysfunction. Patients in intensive care can develop a chronic critical illness called *persistent inflammation, immunosuppression and catabolism syndrome* (PICS). PICS is characterised by prolonged hyperinflammation, immune suppression, dysregulated myelopoiesis, muscle wasting and cachexia.

The term 'cytokine storm' was first introduced in early preclinical studies, which demonstrated that experimental animals exposed to bacteria, or their products, experienced a strong systemic release of proinflammatory cytokines. Although it is now widely accepted that acute systemic challenge models may have little relevance for human sepsis, there is still a consensus that uncontrolled activity of proinflammatory cytokines contributes to injury in sepsis. Neutrophils contribute to hyperinflammation in sepsis by releasing proteases and reactive oxygen species (ROS; Chapter 3), as well as neutrophil extracellular traps (NETs), which are composed of a network of chromatin fibres containing antimicrobial peptides and proteases, such as myeloperoxidase, elastase and cathepsin G (Chapter 5). While NETs can trap and subsequently kill bacteria, excessive NETosis during sepsis can be harmful through various mechanisms, for example by inducing intravascular thrombosis and multiple organ failure.

The coagulation and complement systems are closely linked and have evolved from shared ancestral pathways. Complement activation can release C3a and C5a, which recruit and activate leucocytes, endothelial cells and platelets (Chapter 5). While complement activation is important for innate immunity, uncontrolled activation can cause tissue injury and organ failure. Activation of the coagulation system is also part of the innate immune response to invading pathogens. In sepsis, the coagulation system becomes unbalanced, resulting in a tendency toward thrombosis in the microvasculature. Disseminated intravascular coagulation (DIC) is a severe manifestation of sepsis-associated coagulopathy and can also cause bleeding due to the consumption of clotting factors, anticoagulant proteins and platelets. Tissue Factor, constitutively expressed by perivascular cells and induced on endothelial cells, monocytes and macrophages by microbial agents and inflammatory mediators, is the primary initiator of blood coagulation in sepsis. Moreover, neutrophil elastase and cathepsin G can stimulate coagulation and thrombus growth by enhancing Tissue Factor and FXII-dependent coagulation and locally breaking down the coagulation inhibitor, tissue factor pathway inhibitor (TFPI; Chapter 8). Platelets contribute to hyperinflammation and thrombosis in sepsis by attracting leucocytes to sites of infection, forming complexes with neutrophils and enhancing their ability to kill pathogens. NETs also promote platelet adhesion, activation and aggregation – crucial for the development of sepsis-induced intravascular coagulation in mouse models of sepsis. Complement factors can also activate coagulation proteases, and vice versa. For example, Factor VIIa, Factor Xa, thrombin and fibrin can convert C3 and C5 into C3a and C5a, respectively, while C5a and the membrane attack complex (MAC; Chapter 5) can stimulate the expression of Tissue Factor on endothelial cells. C5a can also disrupt the endothelial glycocalyx and further facilitate clotting.

Immune suppression contributes to sepsis risk. Moreover, immunosuppression is implicated in the increased susceptibility of patients with sepsis to secondary infections. In sepsis, there is also excessive apoptosis of CD4+ T-cells, CD8+ T-cells, B-cells, natural killer (NK) cells and follicular dendritic cells. The immunosuppressive milieu is further exacerbated by increased numbers of regulatory T (Treg) cells and myeloid-derived suppressor cells (MDSCs). Recent studies also implicate the activation of immune checkpoints in sepsis-associated immunosuppression. Accordingly, therapeutics targeting immune checkpoints hold promise to reverse sepsis-induced immunosuppression and preserve host immunity against primary and secondary infections.

Sepsis is a complex syndrome with significant heterogeneity between patients. This is one of the reasons why immune modulatory trials have mostly failed in sepsis patients, although attempts have been made to stratify patients more likely to benefit from specific therapies based on certain biological mechanisms, for example, by selecting patients with high baseline IL-6 levels, low expression of HLA-DR on circulating monocytes and/or low lymphocyte counts.

Treatment and Prognosis

The *Sepsis Six* is a bundle of medical approaches created to lower the mortality rate in patients with sepsis. It was developed by the UK Sepsis Trust in 2006, based on international guidelines resulting from the Surviving Sepsis Campaign, to provide healthcare professionals with a practical tool for delivering essential care quickly and consistently. Evidence published by the UK Sepsis Trust in 2011 demonstrated that using the Sepsis Six led to a 50% decrease in mortality, shorter hospital stays and fewer days spent in intensive care. The Sepsis Six consists of 'giving three' and 'taking three' – all to be delivered within one hour of the initial diagnosis of sepsis.

Give:

- Titrate *supplementary oxygen* to achieve oxygen saturation of 94–96% (88–92% in patients with chronic lung disease).
- *Broad-spectrum antibiotics:* Give antimicrobials as per local antimicrobial guideline based on the site of infection, whether community or healthcare acquired, and the patient's allergy status. Assess requirement for source control.

- *Intravenous fluid resuscitation:* Patients who present with hypotension should receive isotonic crystalloid within 1 hour of presentation with vasopressors started in patients who are fluid unresponsive. Patients with hypoperfusion should receive fluid to restore perfusion using a bolus and review technique. Give isotonic crystalloid, reassessing frequently. Boluses may be amended based on clinical context.

Take:

- *Blood cultures:* Take blood cultures using aseptic (non-touch) technique prior to giving antimicrobials unless this leads to a delay >45 minutes. If a central venous catheter is *in situ*, take blood cultures through that line. Take other specimens as indicated by history and examination, for example influenza swabs, wound swabs, sputum, urine, etc.
- *Blood tests:* Point of care lactate (venous or arterial), full blood count, renal profile, liver profile ± coagulation screen. Other tests and investigations as indicated.
- Assess *urinary output* as part of volume/perfusion status assessment. For patients with sepsis/septic shock, start hourly fluid balance charts.

Early effective fluid resuscitation is crucial for stabilisation of sepsis-induced tissue hypoperfusion or septic shock. Sepsis-induced hypoperfusion may manifest as acute organ dysfunction and/or

Table CS12.4 Definitions of empiric and targeted/definitive therapy.

Empiric therapy	Initial therapy started in the absence of definitive microbiologic pathogen identification.
Targeted/ definitive therapy	Therapy targeted to a specific pathogen (usually after microbiologic identification). Targeted/definitive therapy may be mono- or combination therapy, but is not intended to be broad-spectrum.

decreased blood pressure and increased serum lactate. Having been initially started on empiric therapy, this should subsequently be modified once microbiology culture and antimicrobial susceptibility results are available (Table CS12.4).

The choice of empiric antibiotic therapy is usually determined by the following factors:

- *Source of the infection:* Community or healthcare acquired.
- *Site of the infection:* Based on the history and examination of the patient, for example respiratory, abdominal, genito-urinary, device or catheter-related, cellulitis, central nervous system, bone or joint, or unknown.
- *Patient considerations:* Patient factors that need to be taken into consideration and may influence the choice of empiric antimicrobial therapy include:
  - Previous infection or colonisation with a multidrug-resistant organism, such as methicillin-resistant *Staphylococcus aureus* (MRSA), extended-spectrum β-lactamase (ESBL) producing/Gram-negative organism.
  - Recent antimicrobial therapy.
  - Current outbreaks.
  - Recent infections in close contacts.
  - Recent travel or hospitalisation and/or residence in another country.
  - Allergy status.

The initial empiric antibiotics prescribed for sepsis are dependent upon the above factors, but usually cover both Gram-positive and Gram-negative pathogens. For Gram-negative empiric cover, a broad-spectrum carbapenem (e.g. meropenem and imipenem) or extended-range penicillin/β-lactamase inhibitor combination (e.g. piperacillin/tazobactam) is usually used. However, several third- or higher-generation cephalosporins can also be used, especially as part of a multidrug regimen (e.g. alongside an aminoglycoside). The specific regimen can and should be modified by the anatomic site of infection, if it is apparent, and by knowledge of local antimicrobial resistance patterns.

However, increasing rates of carbapenem resistance in *K. pneumoniae* have been reported in recent years, leading to the use of alternative antibiotics, such as colistin, tigecycline and aminoglycosides. Given the current evidence, suspected or confirmed infection with carbapenem-resistant Enterobacterales (e.g. *K. pneumoniae)* should be treated with a combination of at least two antimicrobial agents likely to be effective.

The choice of antimicrobial agents should take account of the type of carbapenemase (e.g. KPC; *Klebsiella pneumoniae* carbapenemase, OXA; oxacillinase, another type of carbapenemase) and the most recent susceptibility test result of the CPE (carbapenemase-producing Enterobacterales) isolate from that person (if known). In people in septic shock, the addition of a third antimicrobial is often appropriate. For example, for treatment of KPC/OXA infection, a combination of ceftazidime-avibactam with tigecycline and colistin, or if an NDM/VIM/IMP carbapenemase, a combination of colistin, tigecycline and fosfomycin may be appropriate. (NDM/VIM/IMP refer to different types of carbapenemase enzymes; NDM is New Delhi metallo-beta-lactamase, VIM is Verona integron-encoded metallo-beta-lactamase and IMP is imipenemase).

Despite aggressive resuscitation, the patient's blood pressure remained low and he required increasing doses of vasopressors. He developed acute respiratory distress syndrome (ARDS) and was intubated and mechanically ventilated. The patient was moved to the ICU. He also developed acute kidney injury with no urinary output despite dialysis. Unfortunately, despite aggressive treatment, the patient's condition deteriorated. He developed multi-organ failure. His family made the difficult decision to withdraw life-sustaining therapies and he passed away peacefully in the ICU.

Questions

1 What is the role of TNF-α in sepsis?
2 What are pathogen-associated molecular patterns (PAMPs)?

Answers on p. 535.

Case Study 13   Fever, Confusion and Breathlessness

Clinical Presentation

A 70-year-old man with a history of diagnosed prostatic hypertrophy was admitted to hospital with fever, confusion, oliguria and tachypnoea, which had developed over the past 12 hours. On examination, he appeared slightly breathless, drowsy and was unable to give a coherent history of events. He was pale with cold extremities, with a cap refill of >5 seconds, and a core temperature of 38.5°C. His heart rate was regular but increased at 100 beats per minute and his blood pressure was lower than normal at 90/60 mmHg. Examination of his abdomen revealed an enlarged bladder and rectal examination confirmed the presence of prostatic hypertrophy.

Investigations

The results of initial laboratory investigations are shown in Table CS13.1. Although the total white cell count was low and the proportion of neutrophils was towards the lower end of normal; this may occur in some states of immunosuppression, but also in overwhelming infections. The urea was raised, consistent with mild renal failure or dehydration, but was most likely caused by the chronic obstruction of urine flow secondary to prostatic hypertrophy. Elevated creatinine also reflects kidney injury. Urine microscopy showed an increased white cell count and bacteria, indicating a potential urinary tract infection (UTI).

Table CS13.1 Results of laboratory investigations.

Test	Result	Reference range
Total white cell count	**3.6×10^9/L**	4.0–11.0*
Red cell count	5.0×10^{12}/L	4.5–6.5
Haemoglobin	14.2 g/dL	13.0–18.0
Platelets	210×10^9/L	150–400
Differential blood film		
Neutrophils	51	40–75
Lymphocytes	40	20–45
Monocytes	6	2–10
Eosinophils	3	1–6
Monocytes	0	1
Urea	**10.2 mmol/L**	2.5–6.7
Sodium	142 mmol/L	135–146
Potassium	4.1 mmol/L	3.5–5.0
Arterial blood gases		
pO_2	81 mmHg	75–100
pCO_2	**31 mmHg**	36–46
Bicarbonate	**34 mmol/L**	22–30
Urine microscopy		
White blood cell count	**>100/mL**	<5
Red blood cell count	**20/mL**	<5

* In sepsis, a WBC count of less than 4×10^9/L is considered leukopenia. Additionally, it is noted that even if the WBC count falls within the normal range, the presence of greater than 10% immature forms (often referred to as 'bands' or 'bandaemia') among the white blood cells can also be indicative of an abnormal response to infection, which may occur in sepsis. Values outside the reference range are in bold.

Diagnosis

The patient was pyrexial, confused, tachypnoeic and hypotensive, with a low neutrophil count. He had evidence of a UTI and a degree of urinary tract obstruction as indicated by the palpable bladder, which would predispose him to infection. The findings are typical of sepsis, and the most likely cause of a UTI is Gram-negative bacteria, though Gram-positive (enterococcal) infection is also possible. The following day, the urine cultured *Escherichia coli* and blood cultures were positive with Gram-negative bacilli on Gram stain. It took another day for blood cultures to grow the same *E. coli*.

Diagnosis: *E. coli* bloodstream infection secondary to UTI in a patient with partial urinary tract obstruction.

Discussion

Gram-negative bacteria form a major component of the normal human microbial flora. Apart from the anaerobic bacteria, *E. coli* predominates among the Gram-negative flora of the large intestine. By virtue of its fimbriae, *E. coli* frequently colonises mucosal surfaces, but rarely becomes invasive. Several strains possess specific P fimbriae, which bind to the P blood group antigen present in red blood cells and epithelial cells of the urinary tract. These *uropathogenic* strains possess a variety of other virulence factors and are the most common cause of UTI. *E. coli* bacteraemia is most commonly associated with UTI, particularly where there is an obstruction to the flow of urine.

Most of the clinical features associated with Gram-negative sepsis result from cytokine-mediated responses to the presence of bacterial endotoxin (LPS; as discussed in Case Study 9). Exposure to endotoxin releases many cytokines, the most important of which is TNFα. This cytokine activates the complement, coagulation, fibrinolytic and kinin pathways (Chapter 5) to the detriment of the host.

Initial signs of Gram-negative sepsis include fever, hyperventilation, relative neutropenia and changes in the mental state, but, if left unchecked, complications rapidly occur, including hypotension, disseminated intravascular coagulation (simultaneous features of bleeding and clot formation) and multi-organ failure. It is important that samples of blood, urine and material from other possible sites of infection are taken for culture as early as possible, ideally before the start of antimicrobial therapy.

Treatment and Prognosis

The treatment of Gram-negative sepsis has three principal components: (i) primary antimicrobial therapy, (ii) general supportive measures, and (iii) treatment of the factor(s) predisposing to infection. In most patients with sepsis, the initial choice of antimicrobial agent (empiric antimicrobial) is made 'blind' because the results of culture or antimicrobial susceptibility are usually not available at that time. A 'best guess', or an empiric treatment in medical terms, is therefore made (Case Study 12). Usually, a broad-spectrum antibiotic or combination of antibiotics, covering a wide variety of Gram-positive and Gram-negative organisms, will be chosen. This initial therapy may be modified once culture results are known, which is usually 24–48 hours later. It is important that the agent chosen achieves good levels at the site of infection. Intravenous administration of the antibiotic is therefore required. It is also important to consider local epidemiology of antimicrobial resistance in Gram-negative organisms when choosing empiric antibiotics. Antimicrobial resistance in Gram-negative pathogens has increased significantly over the last decade with increasing rates of extended-spectrum β-lactamase (ESBL), multi-drug-resistant organisms (MDRO) resistant to multiple antibiotic classes/families and carbapenemase-producing organisms detected.

General measures are essential in supporting the patient while the infection is brought under control. Their precise nature and duration will depend on the condition of the patient. The 'Sepsis Six' is the name given to the steps designed to reduce mortality in patients with sepsis. Since 2011, use of the Sepsis Six has led to a significant reduction in mortality, a decreased length of stay in hospital, and fewer days in intensive care. It consists of 'giving three' and 'taking three' – all to be delivered within one hour of the initial diagnosis of sepsis (for details, see Case Study CS12).

Treating the factors predisposing the patient to Gram-negative sepsis can usually wait until the acute infective episode is over and the patient is in better condition. Examples of such treatments include prostatectomy for obstructive prostatic hypertrophy and improved control of type 2 diabetes where appropriate.

The prognosis of a patient with Gram-negative sepsis depends largely on their age and general condition – the very young and very old, and those with significant underlying disease having a much higher mortality (50–60% in some studies) even with appropriate treatment. A worse outcome is also seen in those who have already developed complications or where appropriate treatment is delayed. When confused about whether a patient is septic or not, consider if he or she has any risk factors from the history/medical investigations, and approach the patient accordingly.

Questions

1   Although *E. coli* is a common cause of both hospital and community-acquired Gram-negative sepsis, *Pseudomonas aeruginosa* frequently causes life-threatening infections in immunocompromised patients. Why is this?

2   Why is *E. coli* the most common cause of UTI?

Answers on p. 536.

Case Study 14   Recurrent Epigastric Pain with Abnormal Liver Function Tests

Clinical Presentation

A 52-year-old male publican was admitted to hospital with severe epigastric pain radiating to the back. The pain had developed with increasing intensity over 6–8 hours and the patient had also experienced retching and vomiting. He had been seen in the Outpatient Department 3 months previously for investigation of recurrent epigastric pain and steatorrhoea (the passage of bulky greasy stools due to fat malabsorption). The patient reported experiencing similar episodes of dull epigastric pain (though less intense) within the past 6 months. Typically, the abdominal pain was worse 30 minutes or so after eating, lasted for 3–4 days, was accompanied by nausea, and could be eased slightly by sitting forward. He would try to manage it with painkillers at home but would occasionally present to his GP, who would also prescribe him painkillers without further workup. The patient also reported a weight loss of 5 kg over the last 6 months. Abdominal CT revealed pancreatic calcification, diffuse enlargement of the pancreas and ductal dilation. Faecal testing revealed high faecal fat content and low faecal elastase-1 (a pancreatic enzyme), suggesting pancreatic insufficiency. On admission to hospital, the patient's blood pressure was 175/95 mmHg. Laboratory results are shown in Table CS14.1.

Diagnosis

The most likely cause of the acute onset of epigastric pain in this case is acute exacerbation of chronic relapsing pancreatitis. However, it is important to exclude other unrelated conditions for which surgery might be indicated.

Damage to the exocrine pancreas results in the release of pancreatic enzymes into body fluids, so the very high serum amylase is indicative of acute pancreatitis. However, serum amylase levels also rise in perforated peptic ulcer, acute intestinal obstruction and acute biliary obstruction, but peak levels are usually less than five times the upper range of normal. In acute pancreatitis, serum amylase usually increases to at least five times the upper reference range within 24 hours. Other pancreatic enzymes, such as trypsin and lipase, may also be increased in patients with acute pancreatitis. In this case, the serum amylase was approximately eight times the upper reference range. High levels of bilirubin and liver enzymes together with red cell macrocytosis indicate liver damage, again possibly due to alcohol abuse.

Diagnosis: Chronic relapsing pancreatitis, with liver damage due to alcohol abuse.

Discussion

Chronic pancreatitis is characterised by ongoing inflammation of the pancreas, which can lead to structural changes, particularly fibrosis and narrowing of the ducts. This persistent damage results in a permanent decline in both the digestive (exocrine) and hormone-producing (endocrine) functions of the pancreas. In regions like the United States and Europe, the occurrence of chronic pancreatitis is estimated to be 0.2–0.6%, with around 7–10 new cases/100 000 people. A variety of factors contribute to the development of chronic pancreatitis; chronic alcohol use accounts for more than half the cases, followed by tobacco use. Furthermore, physical obstruction, whether from strictures in the pancreatic ducts, tumours or congenital conditions, such as *pancreatic divisum* (a congenital anomaly in which a single pancreatic duct is not formed, but remains as two distinct dorsal and ventral ducts), can also cause chronic pancreatitis. Common symptoms include upper abdominal pain, nausea and vomiting, steatorrhoea, weight loss, and in

Table CS14.1   Results of laboratory investigations on initial presentation.

Test	Result	Reference range
Total white cell count	**12.6 × 10⁹/L**	4.0–11.0
Haemoglobin	14.2 g/dL	13.0–18.0
Haematocrit	0.42	0.40–0.54
Mean cell volume (MCV)	**100 fL**	82–92
Platelet count	180 × 10⁹/L	150–400
Erythrocyte sedimentation rate	20 mm/hour	<20
Blood film		
WBC differential		
Neutrophils	**79%**	40–75
Lymphocytes	**16%**	20–45
Monocytes	4%	2–10
Eosinophils	1%	1–6
Basophils	0%	1
Biochemistry		
Random blood glucose	4.1 mmol/L	3.5–8.9
Total bilirubin	**30 μmol/L**	<21
Aspartate transaminase (AST)	**68 U/L**	10–40
Alanine transaminase (ALT)	**115 U/L**	5–40
Gamma glutamyltransferase (gamma-GT)	**190 U/L**	11–50
Serum amylase	**1600 U/L**	<220

Values outside the reference range are in bold.

some patients, the development of diabetes. Chronic pancreatitis can have an hereditary component in some cases, with a small group of patients showing a genetic predisposition (e.g. people harbouring mutations in PRSS1 and SPINK1 genes. Additionally, individuals with cystic fibrosis [CFTR mutations; Case Study 43]) commonly develop chronic pancreatitis at a younger age due to thick mucus which blocks the pancreatic ducts, impeding its exocrine function and causing pancreatic inflammation.

Acute pancreatitis, on the other hand, is a condition characterised by a sudden inflammation of the pancreas, typically resulting from direct injury, and often leading to temporary pancreatic inflammation. In some severe instances, a potent, and sometimes life-threatening, multi-system inflammatory reaction can develop in a significant fraction of those diagnosed. The two primary triggers for acute pancreatitis are gallstones and alcohol.

The most widely used criteria for diagnosing and classifying chronic pancreatitis are based on a combination of clinical, morphological, functional, and aetiological factors. Several systems exist, but the TIGAR-O and

Cambridge classifications are among the most recognised.

The **TIGAR-O classification**, which breaks down the aetiology and risk factors for chronic pancreatitis, is based on the following categories:

- **T**oxic/metabolic factors (like alcohol and tobacco use).
- **I**diopathic (cause unknown).
- **G**enetic factors.
- **A**utoimmune conditions.
- **R**ecurrent and severe acute pancreatitis.
- **O**ther causes (e.g. obstructions).

The **Cambridge classification** is a morphological classification based on computed tomography (CT) or endoscopic retrograde cholangiopancreatography (ERCP) findings. It classifies the severity of chronic pancreatitis based on the presence of ductal changes (e.g. dilation and presence of stones) and parenchymal changes (e.g. atrophy and calcification). In brief, the stages are:

- Normal (no changes).
- Mild (some ductal changes).
- Moderate (pronounced ductal changes).
- Severe (parenchymal changes).

Treatment and Prognosis

The patient was treated conservatively with bed rest, analgesia PRN (*pro re nata*, which means 'as needed') intravenous (IV) fluids, IV omeprazole (a proton pump inhibitor), and bowel rest (NPO; from the Latin *nil per os*, which means 'nothing by mouth'). He was eventually commenced on a low-fat diet. His acute condition resolved and he was discharged on a high-protein and high-calorie diet, analgesics, and pancreatic enzyme replacement therapy (PERT) containing a mixture of digestive enzymes, including proteases (to break down proteins), amylases (to break down carbohydrates), and lipases (to break down fats). On further direct questioning, the patient admitted to excess alcohol

intake, and he was counselled to abstain. He was discharged from hospital and eventually referred to a psychiatric clinic and to self-help groups when it became apparent he was continuing to drink heavily.

The patient continued to drink heavily, began to neglect himself and eventually lost his job. He attended his general practitioner's surgery every 3 months to collect a prescription for vitamins and on each occasion, he would claim to be in control of his drinking. Meanwhile, he was occasionally vomiting material that resembled 'coffee grounds' (typically indicating the presence of digested blood in the vomit; when blood mixes with gastric acid in the stomach, it takes on a dark, grainy appearance reminiscent of coffee grounds), which he did not report to his doctor. One day, the patient collapsed at home after vomiting fresh blood and an emergency admission to hospital was arranged. On admission, haematemesis due to upper gastrointestinal bleeding was confirmed, and gross abdominal ascites were noted. The patient was stabilised with IV fluids and blood to achieve haemodynamic resuscitation. Antibiotic prophylaxis and IV terlipressin were administered. Terlipressin is a synthetic analogue of vasopressin, a naturally occurring hormone that regulates water balance in the body by acting on vasopressin receptors in the kidneys. Terlipressin has a more prolonged duration of action than vasopressin and has important vasoconstrictive effects on the splanchnic circulation (the blood flow to the digestive organs) and is used to reduce portal hypertension. An endoscopic examination revealed oesophageal varices, which were managed with endoscopic ligation. The white cell count was raised (summary of laboratory results on this occasion is shown in Table CS14.2), and a differential count revealed neutrophilia (higher than normal neutrophil count). The haemoglobin concentration and red cell count were low, and coagulation tests were requested. The prothrombin

Table CS14.2   Laboratory results on the patient's second visit to hospital.

Test	Result	Reference range
Total white cell count	**11.5×10^9/L**	4.0–11.0
Red cell count	**3.9×10^{12}/L**	4.5–6.5
Haemoglobin	**9.6 g/dL**	13.0–18.0
Mean cell volume (MCV)	**108 fL**	82–92
Platelet count	**120×10^9/L**	150–400
WBC differential		
Neutrophils	**81%**	40–75
Lymphocytes	**14%**	20–45
Monocytes	3%	2–10
Eosinophils	1%	1–6
Basophils	1%	0–1
Biochemistry		
Random blood glucose	3.6 mmol/L	3.5–8.9
Total bilirubin	**33 μmol/L**	<21
Aspartate transaminase	**75 U/L**	10–40
Alanine transaminase	**132 U/L**	5–40
Gamma glutamyltransferase	**221 U/L**	11–50
Serum amylase	**631 U/L**	<220

Values outside the reference range are in bold.

time was prolonged (17 seconds, reference range 11–16). The activated partial thromboplastin time was also prolonged (60 seconds, reference range 26–35). Vitamin K and fresh frozen plasma were administered to correct the coagulation defect and the bleeding was eventually controlled.

The patient was discharged from hospital, but he died at home 6 weeks later. At post-mortem examination, the cause of death was confirmed as a massive bleed from oesophageal varices. The liver was found to be enlarged to a weight of 2165 g (normal weight 1200–1400 g). On slicing, the liver was rather yellow coloured and greasy in texture with pale areas of obvious fibrosis (Fig. CS14.1). Histological examination with

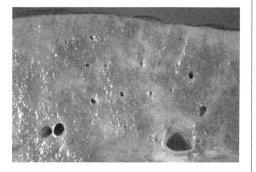

Figure CS14.1   Gross pathology of alcoholic liver cirrhosis. The cut surface shows areas with diffuse pallor due to a dense network of scar tissue. *Source:* Centers for Disease Control and Prevention/ Dr. Edwin P. Ewing / Wikimedia Commons / CC0 1.0.

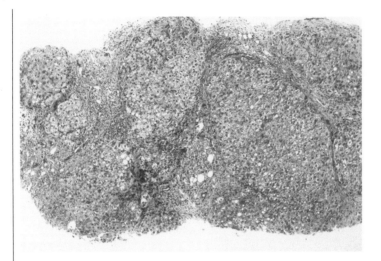

Figure CS14.2 Micrograph of a liver sample with trichrome stain showing cirrhosis as a nodular texture surrounded by fibrosis (collagen is stained blue). Courtesy of Ed Uthman, Houston, TX, USA. *Source:* Ed Uthman. / Wikimedia Commons / CC BY 2.0.

a trichrome stain showed bands of fibrosis surrounding nodules of hepatocytes, typical of cirrhosis (Fig. CS14.2). Thus, while chronic pancreatitis was the initial presenting sign, the underlying cirrhosis was clinically more significant in this case. At first presentation, the patient may have been in *compensated cirrhosis*, which refers to a stage of cirrhosis in which the liver is still able to perform its essential functions, despite the presence of significant scarring and damage. In other words, the liver can compensate for the loss of healthy tissue. Patients with compensated cirrhosis may not display noticeable symptoms. In contrast, *decompensated cirrhosis* may have developed in this patient over time. Decompensated cirrhosis is a more advanced stage in which the liver is no longer able to compensate for the damage. It is characterised by the development of complications which include ascites, portal shunts, hepatic encephalopathy, hepato-renal syndrome, variceal bleeding and jaundice.

Alcohol abuse accounts for substantial worldwide morbidity, causing a range of medical, psychological and sociological problems (see also Chapter 14).

Questions

1 What is the cause of red cell macrocytosis in liver disease?
2 Why was the prothrombin time prolonged in this patient?
3 What are the likely causes of gastrointestinal bleeding in patients with alcoholic liver disease?

Answers on p. 536.

Case Study 15   Memory Loss and Confusion in a 70-year-old Woman

Clinical Presentation

A 70-year-old woman presented at her GP with a 6-month history of increasing forgetfulness, difficulty with everyday tasks, and confusion. Her husband reported that she forgets appointments, misplaces items around the house and has difficulty following conversations. He also said she experiences difficulty with finding words and trouble with planning and organising tasks. She can perform basic self-care tasks independently, but requires assistance with more complex tasks such as managing medications and finances. She also appears to be socially withdrawn and less interested in her usual activities. Physical examination was unremarkable, with no focal neurological deficits. The Mini-Mental State Examination (MMSE) score was 20 out of 30, which suggests moderate cognitive impairment.

Investigations

Laboratory investigations were performed to exclude any underlying medical conditions that may be contributing to the patient's cognitive decline. A complete blood count, electrolyte panel, thyroid function tests, liver function tests, and vitamin B12, were within normal limits. An autoantibody screen to exclude vasculitis was negative. An MRI of the brain showed generalised cortical atrophy consistent with Alzheimer disease.

Diagnosis: Alzheimer disease.

Discussion

Alzheimer disease (AD) is a progressive neurodegenerative disease, characterised by the accumulation of both senile plaques and neurofibrillary tangles (Fig. CS15.1) within neurons of the cerebral cortex, and is associated with loss of the neurons. Neurofibrillary tangles are abnormal intracellular fibrils made up of tau protein, while senile plaques are hallmark lesions of AD brains that are composed of the extracellular accumulation of amyloid fibrils and activated glial cells, including astrocytes (Fig. CS15.2). Amyloid β (Aβ) deposition is highly specific to AD and represents one of the earliest pathological changes. Aβ exists as various species, including monomers and soluble aggregates of varying sizes (e.g. oligomers and protofibrils), as well as insoluble fibrils in plaques. Important pathogenic mechanisms in AD include glutamate-mediate neurotoxicity and acetylcholine deficiency due to neuronal injury, both of which cause cognitive decline.

AD is a multifactorial disorder, meaning its development involves interactions between genetic and environmental factors. While most AD cases are sporadic, there are some cases in which genetics play a significant role. There are three main genes directly related to the development of familial AD, a rare type of AD that is inherited in an autosomal dominant pattern. These genes are amyloid precursor protein (APP), presenilin 1 (PSEN1) and presenilin 2 (PSEN2). Mutations in these genes can cause the accumulation of Aβ protein in the brain, leading to the formation of senile plaques. Several other genes are linked to an increased risk of late-onset AD, the most common form of the disease. The most well-known of these is the apolipoprotein E (APOE) gene. There are three common variants of this gene – APOE ε2, APOE ε3, and APOE ε4. Carrying one or two copies of the ε4 variant increases the risk of developing AD, whereas the APOE ε2 variant is protective and the APOE ε3 variant is risk neutral.

Other risk factors play a role in the predisposition to AD. These include midlife hypertension, orthostatic hypotension (postural hypotension – a sudden drop in blood pressure when a person changes their position from lying down or sitting to standing up), diabetes, midlife obesity, head trauma, depression and stress. Moreover, studies suggest a link between AD and obstructive sleep apnoea

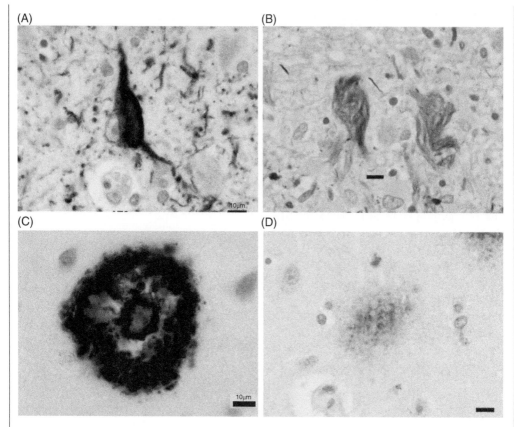

Figure CS15.1 Alzheimer disease histology. Microphotographs of Alzheimer disease brain sections showing intracellular (A) and ghost (B) neurofibrillary tangles, and neuritic (C) and diffuse (D) plaques. Immunohistochemical staining for phosphorylated tau with AT8 (A) and amyloid-β (C, D), and Gallyas-Braak silver staining (B). *Source:* Renpei Sengoku. 2019 / Reproduced from John Wiley & Sons.

(OSA). Treatment of OSA may improve AD biomarkers. For that reason, current available data recommend screening for sleep-related disorders in patients with cognitive impairment.

Treatment and Prognosis

AD typically progresses through several stages. While there is some variation between individuals, the main clinical stages are as follows:

- Preclinical stage, characterised by the presence of biomarkers associated with AD, such as abnormal levels of Aβ and tau proteins in the brain, but no outward symptoms or cognitive impairments.
- Mild cognitive impairment (MCI) stage, in which there are mild cognitive deficits noticeable to the individual and/or their

family and friends, but which do not significantly interfere with daily activities. MCI can be caused by a variety of factors and not all cases of MCI progress to AD.

- *Early-stage AD:* In this stage, cognitive deficits become more pronounced and interfere with daily activities. Individuals may have difficulty with memory, word-finding and other cognitive functions. They may also experience mood changes and behavioural symptoms.
- *Middle-stage AD:* Cognitive deficits become more severe and individuals may require assistance with activities of daily living, such as bathing and dressing. Behavioural symptoms may become more pronounced, and individuals may experience hallucinations, delusions and other psychiatric symptoms.

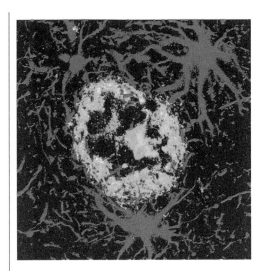

Figure CS15.2 Amyloid plaque (green) surrounded by reactive astrocytes (red) in the cerebral cortex of an AD patient. Confocal laser scanning microscopy. 3D reconstruction. *Source:* Rauhtopaz / Wikimedia Commons / CC BY 4.0.

- *Late-stage AD:* There is severe cognitive impairment and functional disability. Individuals may be unable to communicate, have difficulty swallowing, and may require round-the-clock care.

AD is an incurable disease. Treatments fall into three groups; 1) supportive care/lifestyle alterations, 2) treatments for conditions linked to dementia, and 3) anti-dementia drugs to maintain function and slow down disease progression. Anti-dementia drugs include: *Acetylcholinesterase inhibitors* (AChEIs; e.g. rivastigmine, galantamine, donepezil) which amplify the concentration of acetylcholine in the synaptic gap by reversibly inhibiting cholinesterase; *NMDA receptor antagonists* (e.g. memantine) that decrease glutamate-induced calcium mediated excitotoxicity, and; *Aducanumab*, an amyloid β monoclonal antibody that works by reducing Aβ plaque deposition. The patient was started on rivastigmine and referred to a memory clinic for further evaluation and management. Her husband was counselled on how to provide support and care for his wife. The patient attended regular follow-up appointments and her husband reported some improvement in her memory and cognitive function with treatment. However, as described above, AD is progressive, and her husband was advised to plan for long-term care as the disease advances.

Questions

1   What is the basis of the Mini-Mental State Examination (MMSE)?
2   Are there any new targeted therapies for Alzheimer disease?

Answers on p. 537.

Case Study 16   Fatigue and Severe Joint Pain in a 62-year-old Woman

Clinical Presentation

A 62-year-old woman was admitted to hospital with a 2-year history of fatigue and progressively more severe joint pain involving the elbows, wrists, knees and ankles, along with neuropathy characterised by sensory and motor deficits that included tingling sensations, numbness, and weakness in the extremities. Prior to the onset of her symptoms, she had been healthy and enjoyed walking regularly as a member of her local hiking club. She had been prescribed naproxen and tramadol hydrochloride for the pain. Rheumatoid arthritis was suspected, and the patient started on a trial of steroids. However, there was no improvement and she went back to her GP. On examination, the GP observed the patient to be pale, with pain that was particularly severe in the left knee, which was swollen and tender. There were palpable lymph nodes in both axillae.

Investigations

On admission, laboratory tests revealed she had a normocytic normochromic anaemia with a haemoglobin level of 9.8 g/dL and renal impairment as evidenced by a raised serum creatinine of 210 μmol/L. As part of the anaemia work-up, she underwent a bone marrow biopsy, which revealed 65% plasma cells including immature and multinucleate forms. A number of the plasma cells also showed intracytoplasmic inclusions (known as Russell bodies) and intranuclear inclusions (known as Dutcher bodies). The plasma cells expressed CD138 and by immunohistochemistry showed staining only for lambda light chain and not for kappa light chain. Serum protein electrophoresis revealed a large peak of monoclonal immunoglobulin and serum-free light chain (SFLC) analysis revealed an abnormal ratio with a large increase in lambda light chains compared to kappa light chains. These findings were consistent with multiple myeloma. However, skeletal imaging showed no distinctive myeloma-related bone abnormalities.

Diagnosis

The clinical picture and laboratory diagnosis were consistent with multiple myeloma. The joint pain could be a consequence of amyloidosis involving the joints. Bone marrow trephine staining was not conclusive for amyloid involvement. Therefore, a fat pad and lymph node biopsy were performed. Haematoxylin and eosin staining of sections revealed the presence of pink amorphous extracellular material consistent with amyloid deposits. These deposits stained with Congo red and exhibited characteristic apple-green birefringence under polarising microscopy (Fig. CS16.1A and B). The patient underwent a specialised serum amyloid protein (SAP) scintigraphy that confirmed organ involvement with amyloid. In this type of scan, SAP is labelled with a radioactive substance. When injected into the patient's bloodstream, the labelled SAP binds to amyloid deposits in organs and tissues, which can identify amyloid in various organs. The scan is semi-quantitative, meaning it not only detects the presence of amyloid but can also provide an estimate of the amount present. Alongside this, the SFLC assay is another diagnostic tool that measures the ratio of light chains in the blood, and is useful in the diagnosis and monitoring of patients with myeloma (see also Case Study 62).

(A)   (B)

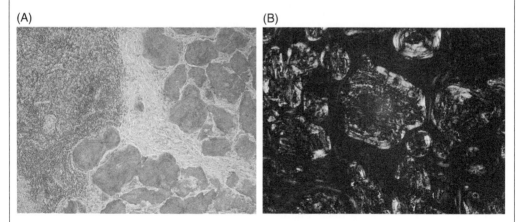

Figure CS16.1   (A) Amyloidosis in a lymph node stained with Congo red: The amyloid deposits are strongly congophilic when viewed before white light. *Source:* Ed Uthman. / wikimedia Commons / CC BY-SA 2.0. (B) Amyloidosis in a lymph node stained with Congo red, showing apple-green birefringence under polarizers. *Source:* Ed Uthman. /Wikimedia Commons / CC BY-SA 2.0.

Diagnosis: Amyloidosis in a patient with multiple myeloma-Light chain (AL) amyloidosis.

Discussion

Light chain (AL) amyloidosis can occur as a complication of multiple myeloma. Myeloma and related, so-called *plasma cell dyscrasias* (blood disorders affecting plasma cells), are discussed in Case Study 62. In the present case study, we focus our discussion on the nature and causes of amyloidosis.

Amyloidosis is the pathological deposition of an insoluble proteinaceous substance between cells in various tissues and organs of the body. Amyloid literally means 'starch-like' and was so-called because it was thought to be made up mainly of starchy carbohydrate. Amyloidosis results from the misfolding of a protein from its physiological tertiary structure into a more linear shape dominated by 'beta-pleated' sheets, which form

long fibres, known as fibrils. The fibrils are resistant to degradation. They replace normal tissues causing pressure atrophy and there is evidence that they, or their precursors, may also be directly toxic to cells. The net effect is compromised normal function of the tissues in which the deposits occur. One of over 35 different proteins is implicated in fibril formation. Amyloidosis can be either hereditary or not, localised or systemic, with different organ involvement and, accordingly, prognosis. Amyloid P component (AP) is a glycoprotein that is an additional component of amyloid that is always present and is bound to the amyloid fibril itself. Amyloid also contains carbohydrates called glycosaminoglycans, which affect the rate at which amyloid proteins accumulate, as well as their toxicity.

Amyloidosis is classified on the nature of the protein that forms the fibrils (Table CS16.1). The most common forms of

Table CS16.1 Examples of types of amyloidosis based on the nature of the protein involved.

Category	Associated disease	Major fibril protein	Precursor protein
Systemic amyloidosis			
Immunocyte dyscrasias with amyloidosis (primary form)	Multiple myeloma	AL	Immunoglobulin light chains
Reactive systemic amyloidosis (secondary form)	Chronic inflammatory conditions	AA	SAA
Haemodialysis-associated form	Chronic renal failure	$A\beta_2M$	$\beta2$-Microglobulin
Hereditary amyloidosis			
Familial Mediterranean fever		AA	SAA
Familial amyloidotic neuropathies		ATTR	Transthyretin
Localised amyloidosis			
Senile cerebral	Alzheimer disease	$A\beta$	APP
Endocrine	Medullary carcinoma of thyroid	A-Cal	Calcitonin
	Type II diabetes	AIAPP	Islet amyloid peptide
	Isolated atrial amyloidosis	AANF	Atrial natriuretic factor

amyloid are: the AL (amyloid light chain) type, which forms as a consequence of the excess immunoglobulin light chains derived from the malignant plasma cells in multiple myeloma; the AA (amyloid-associated) form, caused by the overproduction of acute phase proteins in chronic inflammatory conditions; the Aβ amyloid type found in the cerebral lesions of Alzheimer disease (see also, Case Study 15); and the ATTR form, caused by the misfolding and aggregation of transthyretin (TTR) protein into amyloid fibrils. Clinically, amyloidosis is also separated based on whether a disease is localised or systemic. The localised forms of amyloidosis include amyloidosis caused by the overproduction of calcitonin in medullary thyroid cancer.

Amyloid can be deposited in synovial tissues and the cartilage and tendons around the joints leading to joint pain. Joint involvement is more common in AL amyloidosis than in other forms of the disease. Amyloid deposition in joints can occasionally appear clinically as rheumatoid arthritis.

Treatment and Prognosis

As multi-organ involvement by amyloid can be present at diagnosis, up to 40% of patients will die from amyloidosis in the first year, but the overall outcomes are heterogeneous and require risk stratification. The updated Mayo Clinic 2012 scoring system uses measurement of SFLC ratios, in addition to serum brain natriuretic protein and troponin levels, to group patients into low- and high-risk categories.

Treatment of AL amyloidosis mirrors treatments for myeloma as the aim is to eradicate the malignant plasma cell clone. Combination triplet and quadruplet therapies, including monoclonal anti-CD38 antibodies, are highly effective and autologous peripheral blood stem cell transplant can improve outcomes in selected patients. New monoclonal antibodies against the amyloid protein have been developed and are currently undergoing evaluation in clinical trials (e.g. Birtamimab). Standard first-line triplet therapy involves the proteasome inhibitor bortezomib. As this patient was limited by a significant pre-existing neuropathy, a common side-effect of bortezomib, this drug was contraindicated and Daratumumab, in combination with lenalidomide and dexamethasone (DRd) was prescribed. There was a resolution of joint symptoms and normalisation of anaemia, renal impairment, serum immunoglobulins and SFLC ratios, after which time she was maintained on lenalidomide alone. This patient did not have cardiac involvement and was low risk (score=1) using the Mayo 2021 staging system and she remains in remission 3 years after her original diagnosis.

Questions

1 What other drugs might be used to treat amyloidosis?
2 What is the mode of inheritance of hereditary ATTR amyloidosis?

Answers on p. 537.

Case Study 17   Abdominal Pain, Nausea and Vomiting in a Young Woman

Clinical Presentation

A 22-year-old woman with no previous medical history of note was taken to the Emergency Department by her husband following a 2-day history of abdominal pain and nausea. The pain was initially peri-umbilical but, within a few hours, shifted to the right lower quadrant. She also complained of vomiting twice. Over the last 12 hours, the pain intensified. On physical examination the patient had a low-grade fever (37.9°C). There was evidence of '*guarding*' (voluntary tensing of the abdominal wall muscles on examination to protect inflamed organs within the abdomen), but not of '*rigidity*' (involuntary tightening of the abdominal muscles in response to touching). Examination for rebound tenderness was equivocal.

Investigations

A pregnancy (β-hCG) test was negative, as was urine microscopy. Erect chest and supine abdominal X-rays showed no abnormalities. An abdominal CT scan showed thickening of the appendix with 'hazy' outlines. Laboratory findings showed a neutrophilic leucocytosis and an elevated CRP and ESR (Table CS17.1). There were significant numbers of 'band' neutrophils – the nucleus of less mature neutrophils is not segmented but has a 'band' or rod-like shape. Taken together, these results are suggestive of an acute inflammatory process. The patient underwent laparoscopic surgery to remove the appendix. On receipt of the appendix in the histopathology laboratory, the pathologist described a swollen reddened appendix, the lumen of which contained yellow purulent fluid and a fecalith 1 cm in diameter. A macroscopic image of an inflamed appendix is shown in Fig. CS17.1A. Microscopic examination of the specimen by haematoxylin and

Table CS17.1   Results of laboratory investigations.

Test	Result	Reference range
Blood		
Total white cell count	$\mathbf{14.2 \times 10^9/L}$	4.0–11.0
Red cell count	$5.5 \times 10^{12}/L$	3.8–5.8
Haemoglobin	**11.2 g/dL**	11.5–16.5
Platelet count	$190 \times 10^9/L$	150–400
Erythrocyte sedimentation rate	**22 mm/hour**	<20
C-reactive protein	**15 mg/L**	<10
WBC differential		
Segmented neutrophils (%)	**73**	37–70
Band neutrophils (%)	**9**	3–5
Lymphocytes (%)	**15**	20–45
Monocytes (%)	2	2–10
Eosinophils (%)	2	1–6
Basophils (%)	0	1

Values outside the reference range are in bold.

eosin (H&E) stain showed abundant neutrophils in the muscular layer of the appendix with necrosis of the mucosal layer and haemorrhage extending into the serosal fat. A trichrome stain (Martius Scarlet Blue; MSB) revealed fibrin deposition on the serosal surface. The H&E also revealed cross sections of the parasite, *Enterobius vermicularis*. Fig. CS17.1B shows this parasite cut in cross section.

Diagnosis

The clinical presentation and histology were consistent with acute appendicitis. The classical physical finding is the McBurney sign, in other words, deep localised tenderness in the right lower quadrant located two-thirds

(A)

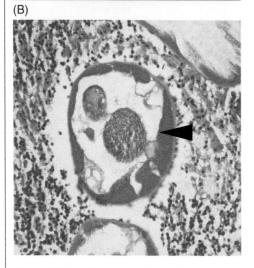

(B)

Figure CS17.1   (A) Acute inflammation: swollen appendix with exudates on the surface (more towards the tip). *Source:* Department of Pathology, Calicut Medical College / Wikimedia Commons / CC BY-SA 4.0. (B) Pinworms (enterobiasis; arrowhead) in the lumen of the vermiform appendix: surrounded by inflammatory cells: incidental finding in an acutely inflamed appendix. *Source:* Ed Uthman. / Wikimedia Commons / CC BY 2.0.

along a straight line from the umbilicus to the right anterior superior iliac spine. This sign is present in over 90% of patients and correlates with the most common anatomical location of the appendix. An atypical location may give rise to different physical examination findings, including a positive psoas (retrocaecal appendix) or obturator sign (appendix in the pelvis). Differential diagnosis includes mesenteric lymphadenitis (especially in children and young adults),

ectopic pregnancy, salpingitis, *mittelschmerz* (German for middle pain – owing to pelvic bleeding during ovulation), ovarian disease, Meckel diverticulum, terminal ileitis and urinary tract infection. Imaging by abdominal ultrasound and CT are important in establishing the diagnosis before surgery. For example, imaging can rule out mesenteric lymphadenitis, which is characterised by the presence of multiple enlarged mesenteric lymph nodes.

Diagnosis: Acute appendicitis.

Discussion

Acute appendicitis is often due to the presence of luminal obstruction caused by a fecalith. Obstruction leads to stasis and bacterial colonisation causing mucosal necrosis and acute inflammation, further exacerbating the increased intraluminal pressure, in turn compromising venous outflow. The reduced blood flow can lead to further necrosis with tissue breakdown and ultimately perforation and/or abscess formation. Neutrophil infiltration is a prominent sign typical of the acute inflammatory response. Early in the disease course, aerobic organisms predominate, whereas later, mixed infections are more common. *Enterobius vermicularis* (pinworm) is the most common parasite of the lower gastrointestinal tract. Although its role in the development of appendicitis is unclear and it may be an incidental finding in this case.

Classically, the pain of appendicitis is initially referred to the epigastric or umbilical region because the increased size of the appendix leads to the stimulation of afferent visceral nerve fibres at the T8–T10 spinal cord level. Because the pain is visceral, rather than somatic, in origin, it is referred pain and poorly localised. Vomiting, nausea and anorexia are common. Local tenderness over the appendix is often experienced later after the abdominal pain has subsided; this pain is the result of irritation of the parietal peritoneum, which stimulates somatic nerves.

Treatment and Prognosis

Serious complications of acute appendicitis include peritonitis and sepsis, so early surgical intervention is critical. The most common complication after laparoscopic appendectomy is wound infection, especially if the appendix is perforated or if pus is present. Patients may also present after surgery with a pelvic abscess. In this case, the patient made an uneventful recovery and returned to work 2 weeks after the operation.

Questions

1   What are the functions of neutrophils in the acute inflammatory response?
2   What is the definition of an abscess?
3   What is a 'positive psoas sign'?

Answers on p. 538.

Case Study 18   Recurrent Respiratory Infections, Chronic Cough and Worsening Breathlessness

Clinical Presentation

A 56-year-old female librarian reported to her GP with worsening symptoms of cough and dyspnoea for the past 1 week. For the last 5 years, she had been experiencing a chronic cough with mucopurulent sputum production, which was occasionally blood-streaked. The amount of sputum had increased significantly within the last week. She had been a heavy smoker for 25 years but stopped smoking 5 years ago. She had no history of asthma or allergies but had experienced recurrent chest infections for the past 10 years that required hospital admission on four occasions. When she was a child, she had two episodes of pneumonia requiring hospitalisation.

Investigations

On further examination, the sputum had a foul odour. Auscultation revealed wheezes (whistling sounds heard during expiration; caused by narrowed airways), coarse crackles ('popping' or 'bubbling' sounds that can be heard during inspiration, caused by fluid accumulation or mucus in the airways) in the right lower lung field, and rhonchi (low-pitched, rumbling sounds heard during both inspiration and expiration; caused by obstruction of the larger airways). The patient had a low-grade fever of 38.0°C. Laboratory investigations (Table CS18.1) revealed a raised white cell count with neutrophilia. The patient also had a normocytic normochromic anaemia as indicated by reduced haemoglobin levels and red cell count, but normal mean cell volume and mean cell haemoglobin (see also Chapter 7). The FEV1/FVC ratio was reduced. The initial clinical findings were suggestive of bronchiectasis. CT scan revealed dilated airways and the typical 'string of pearls' and 'signet ring' features of bronchiectasis (Fig. CS18.1).

Diagnosis

An FEV1/FVC ratio of 0.58 indicates an obstructive lung disorder. This means that the amount of air a person can exhale forcefully in 1 second (FEV1) is reduced compared to the total exhaled volume (FVC). In other words, the person is not able to empty their lungs as quickly as they should be able

Table CS18.1 Results of laboratory investigations.

Test	Result	Reference range
Haemoglobin	**8.9 g/dL**	11.5–16.5
White blood cell count	**14.5×10^9/L**	4.5–11.0
Red cell count	**3.5×10^{12}/L**	3.8–5.8
Mean cell volume	87.5 fL	82–92
Mean cell haemoglobin	28.2 pg	27–32
Differential white cell count		
Neutrophils	**77**	40–75
Lymphocytes	**19**	20–45
Monocytes	2	2–10
Eosinophils	2	1–6
Basophils	0	1
C-reactive protein (CRP)	**120 mg/L**	<10
Blood gas analysis		
pH	**7.30**	7.35–7.45
PaO_2	**67 mmHg**	75–100
$PaCO_2$	**52 mmHg**	36–46
Spirometry		
FEV1/FVC ratio	**0.58**	>0.70
Sputum culture	***Pseudomonas aeruginosa***	—

Values outside the reference range are in bold.

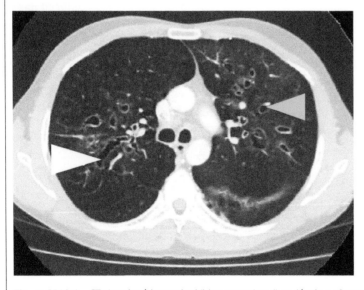

Figure CS18.1 CT showing 'signet ring' (blue arrowhead) and 'string of pearls' (white arrowhead) appearances, indicative of central bronchiectasis. Mucoid impaction and dilated bronchi are also seen. *Source:* Avani R. Patel / Wikimedia Commons / CC BY 3.0.

Figure CS18.2 Typical appearance of bronchiectasis at post-mortem (white arrowheads). *Source:* Yale Rosen / Wikimedia Commons / CC BY-SA 2.0.

to, which suggests obstruction of their airways. In general, an FEV1/FVC ratio of less than 0.70 is considered suggestive of obstructive lung disease. The imaging features in this case revealed the classic signs of bronchiectasis. A sputum culture showed growth of *Pseudomonas aeruginosa* sensitive to ciprofloxacin, indicating this is likely to be an infective exacerbation of bronchiectasis.

Diagnosis: Infective exacerbation of bronchiectasis.

Discussion

Bronchiectasis is a chronic lung condition characterised by irreversible dilation and destruction of usually medium-sized components of the bronchial tree, with loss of the muscular and elastic fibres that normally provide support to the airways (Fig. CS18.2). It is classified as an obstructive lung disease and is most commonly a focal process, involving a lobe, segment or sub-segment of the lung.

The pathogenesis of bronchiectasis is poorly understood, but is thought to involve a complex interplay of genetic, environmental, and immune factors. Bronchiectasis is usually caused by repeated episodes of respiratory infections. Immunodeficiency is a risk factor because it impairs the ability of the immune system to fight off respiratory infections. Bronchiectasis is also more common in cystic fibrosis sufferers. Cystic fibrosis causes a build-up of thick, sticky mucus in the airways that can lead to recurrent infections and inflammation. Some of the new drugs that have been developed to treat cystic fibrosis (Case Study 43) can partially reverse bronchiectasis in some cases (Fig. CS18.3). Other causes leading to chronic obstruction of the airways, such as from an inhaled foreign body or tumour, can also lead to bronchiectasis. Certain autoimmune disorders such as rheumatoid arthritis or Sjogren syndrome damage the airways and predispose to bronchiectasis. Bronchiectasis is also more common in patients with primary ciliary dyskinesia (PCD), a rare autosomal recessive disorder that affects the structure and function of cilia. In PCD, the cilia do not move properly, leading to poor mucus clearance.

Mnemonic: causes of bronchiectasis (CAPT Kangaroo has Mounier-Kuhn)

C: Cystic fibrosis or congenital cystic bronchiectasis (Williams-Campbell syndrome; cartilage in the bronchi is defective, leading to collapse of the airways).
A: Allergic bronchopulmonary aspergillosis (ABPA)
P: Post-infectious (most common)
T: Tuberculosis (granulomatous disease)
K: Kartagener syndrome (part of a larger family of PCD disorders)
M: Mounier-Kuhn syndrome

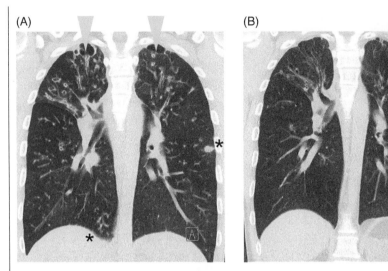

Figure CS18.3 Reversal of cystic bronchiectasis. This is a case of a 30-year-old male cystic fibrosis sufferer, who required non-invasive ventilation and supplemental oxygen whilst awaiting a lung transplant. He was commenced on the combination of elexacaftor, tezacaftor and ivacaftor (ETI), three different medications that target specific defects in the CFTR (cystic fibrosis transmembrane conductance regulator) protein. After 12 months of therapy, the patient's respiratory function had improved and non-invasive ventilation and supplemental oxygen were stopped. The patient was taken off the lung transplant waiting list. CT scans pre- and post-ETI are shown in A and B, respectively, with reductions in the peripheral cysts at both apices (blue arrowheads) and mucus plugging (asterisks) and reductions in bronchial wall dilatation and thickening.
Source: Courtesy of Middleton PG, Simmonds NJ.

The airway inflammation in bronchiectasis is primarily mediated by neutrophils. When the airways become inflamed or infected, neutrophils are recruited. However, in bronchiectasis, the neutrophil response can become excessive, leading to chronic inflammation and tissue damage. Neutrophils release a variety of pro-inflammatory cytokines, chemokines and proteases, such as elastase and cathepsin G, that cause damage to the bronchial epithelium and surrounding tissues. These factors can also stimulate the production of mucus and other secretions, which can further obstruct the airways. The overall effect is severely impaired clearance of secretions from the bronchial tree leading to pooling and stasis.

The patient had reduced oxygen levels in the blood (*hypoxaemia*) because the damaged airways impair the diffusion of oxygen from the air into the blood, leading to decreased arterial oxygen tension (PaO_2). *Hypercapnia* (increased arterial carbon dioxide tension; $PaCO_2$) is also present because the damaged airways impair the ability of the lungs to remove carbon dioxide from the blood. In bronchiectasis, air can be trapped in the lungs, which exacerbates impaired gas exchange. In this patient, the combination of hypoxaemia and hypercapnia could lead to respiratory failure and the need for supplemental oxygen therapy and mechanical ventilation. Regular monitoring of blood gas levels is important in the management of bronchiectasis to prevent the complications of respiratory failure.

Anaemia of chronic disease (ACD) is the most common form of anaemia seen in COPD patients, but other forms of anaemia can occur as well. The reason why anaemia develops in chronic diseases is multifactorial and involves several underlying mechanisms that can include:

- *Suppressed erythropoiesis:* Inflammation and chronic diseases can lead to the production of inflammatory cytokines, such as IL-6 and TNF-α. These cytokines interfere with the normal production of erythropoietin (EPO), a hormone that stimulates the bone marrow to produce red blood cells. As a result, the bone marrow's response to EPO is blunted, leading to decreased production of red blood cells.
- *Shortened red blood cell lifespan:* Chronic inflammation can lead to increased destruction of red blood cells (haemolysis) and decreased red blood cell survival. Inflammatory cytokines can affect red blood cell membranes and increase their susceptibility to premature destruction by the spleen.
- *Iron sequestration:* Inflammation can disrupt the normal regulation of iron metabolism. Iron is essential for haemoglobin synthesis. Hepcidin, which is released during inflammation, can inhibit iron absorption from the intestines and sequester iron within cells, reducing its availability for erythropoiesis.
- *Impaired iron utilisation:* Inflammation can also affect how iron is utilised by the developing red blood cells of the bone marrow. Even if iron stores are adequate, inflammatory signals can interfere with the incorporation of iron into haemoglobin molecules.

Chest X-rays, while often demonstrating abnormalities in bronchiectasis, are usually not sufficient for its precise diagnosis or grading. Features such as 'tram-track' opacities (parallel linear opacities indicative of thickened bronchial walls) suggest cylindrical bronchiectasis (uniform dilation of bronchi). In cystic bronchiectasis (saccular dilation of bronchi), air-fluid levels (lines revealing the presence of fluid in bronchi) can be seen. Increased bronchovascular markings and ring shadows (circular shadows from end-on views of dilated bronchi) are other radiographic findings. Computed tomography (CT) and high-resolution CT (HRCT) provide a better demonstration of the tissue architecture and can sometimes indicate underlying cause. Features of bronchiectasis best observed on CT or HRCT include:

- Bronchial wall thickening; bronchial walls are more distinct, and their thickness is better appreciated on CT.
- Lack of tapering of bronchi; in other words, the bronchi do not decrease in size as they progress peripherally.
- Bronchoarterial ratio alterations; the diameter of a bronchus should be approximately 0.65–1.0 times that of the adjacent pulmonary artery branch. A higher ratio (between 1 and 1.5) can be seen in some normal individuals. However, a ratio greater than 1.5 indicates bronchiectasis (though other conditions may show a similar increase).
- Mucoid impaction; visualised as mucus-filled dilated bronchi.
- Air-trapping and 'mosaic' perfusion; areas of trapped air which contrast with well-aerated lung.
- 'Tree-in-bud' pattern; representing small airway inflammation, seen as centrilobular nodules connected by linear branches.
- Specific signs (Fig. CS18.1), such as the 'tram-track' sign, 'signet ring' sign, 'string of pearls' sign and 'bunch of grapes' sign.

Recently, multidetector CT (MDCT) is favoured over HRCT due to its ability to produce thinner sections (⩽1 mm) with more precise details (HRCT typically achieves slices of only 1–1.5 mm). However, when MDCT is not accessible, HRCT remains the standard, yielding similar diagnostic signs.

Treatment and Prognosis

The patient was started on inhaled bronchodilators and nebulized hypertonic saline to help clear her airways. She was also prescribed a course of oral antibiotics (ciprofloxacin) for the *Pseudomonas* infection. Pulmonary rehabilitation was recommended to improve her exercise tolerance and respiratory muscle strength. She was advised to get yearly influenza and pneumococcal vaccinations. The patient's symptoms improved with treatment, and her cough and sputum production decreased. Her lung function and 6-minute walk distance also improved after completing pulmonary rehabilitation. However, she continued to have occasional exacerbations of her bronchiectasis, which were treated with antibiotics when needed. She is scheduled for regular follow-up appointments with her pulmonologist to monitor her lung function and manage her symptoms.

Careful control of bronchiectasis is important because there can be serious complications, including recurrent respiratory infections, which further damage the airways, leading to more severe bronchiectasis. In severe cases, bronchiectasis can lead to respiratory failure, a life-threatening condition in which the lungs cannot provide enough oxygen to the body. Chronic lung disease, including bronchiectasis, can cause the right side of the heart to work harder to pump blood to the lungs, leading to cor pulmonale associated with symptoms such as shortness of breath, fatigue and swelling of the legs.

Questions

1 What is Mounier-Kuhn syndrome?
2 Apart from poor mucus clearance in the lungs, can you predict what other abnormalities might exist in a patient with primary ciliary dyskinesia (PCD)?

Answers on p. 538.

Case Study 19   Six-month History of Increasing Abdominal Distention, Nausea and Fatigue

Clinical Presentation

A 65-year-old woman presented to her GP with a 6-month history of increasing abdominal distention, nausea and fatigue. She had a past medical history of congestive heart failure and hypertension, for which she was taking furosemide and lisinopril. She had no history of alcohol abuse or hepatitis. She reported no recent weight loss or jaundice. On physical examination, the patient had marked abdominal distention with diffuse tenderness and a positive *fluid wave* (gently tapping or pushing on one side of the abdomen causes a 'wave-like' movement of fluid shifting to the other side of the abdomen, indicative of the presence of fluid in the abdominal cavity and suggesting ascites). She also had peripheral oedema in her lower extremities.

Investigations

A transthoracic echocardiogram (TTE) revealed an ejection fraction (EF) of 30%, a dilated right atrium, a pulmonary artery pressure (PAP) of 40 mmHg and a dilated inferior vena cava. The EF is a measure of how effectively the heart is pumping blood. An EF of 30% suggests that the heart is not pumping as effectively as it should be (normal EF range, 50–70%). A PAP of 40 mmHg is elevated and suggests increased pressure in the pulmonary circulation. Dilation of the inferior vena cava suggests increased pressure in the right atrium of the heart, which is dilated.

Diagnosis

Based on the clinical findings and laboratory results (Table CS19.1), the patient was thought to have cardiac cirrhosis. This is a

Table CS19.1   Results of selected laboratory investigations.

Test	Result	Reference range
Total white cell count	10.6×10^9/L	4.0–11.0
Haemoglobin	12.2 g/dL	11.5–16.5
Haematocrit	0.48	0.40–0.54
Mean cell volume (MCV)	87 fL	82–92
Platelet count	180×10^9/L	150–400
Erythrocyte sedimentation rate	**35 mm/hour**	<30
Blood film		
WBC differential		
Neutrophils	74%	40–75
Lymphocytes	21%	20–45
Monocytes	4%	2–10
Eosinophils	1%	1–6
Basophils	0%	1
Biochemistry		
Random blood glucose	3.6 mmol/L	3.5–8.9
Albumin	**15.2 g/dL**	34–48
Total bilirubin	**31.5 μmol/L**	<21
Aspartate transaminase (AST)	**75 U/L**	10–40
Alanine transaminase (ALT)	**105 U/L**	5–40
Gamma-glutamyltransferase (gamma-GT)	**190 U/L**	7–32
Alkaline phosphatase	**162 U/L**	25–115
N-terminal pro-B-type natriuretic peptide (NT-proBNP)	**900 pg/mL**	<300

Values outside the reference range are in bold.

type of cirrhosis caused by long-standing congestive heart failure, which leads to liver injury and fibrosis. In this case, the elevated bilirubin, liver enzymes and low serum albumin levels are consistent with liver dysfunction, while the peripheral oedema and elevated BNP (brain natriuretic peptide) suggest congestive heart failure. Cardiac cirrhosis is associated with right-sided congestive heart failure and its aetiology reflects the multiple underlying causes of right-sided heart failure, including ischaemic heart disease, cardiomyopathy, valvular heart disease, primary lung disease and pericardial disease. While inferior vena cava thrombosis and Budd-Chiari syndrome share some pathophysiological features with cardiac cirrhosis, they are classified separately and not considered as causes of cardiac cirrhosis.

Diagnosis: Cirrhosis caused by long-standing congestive heart failure ('cardiac cirrhosis').

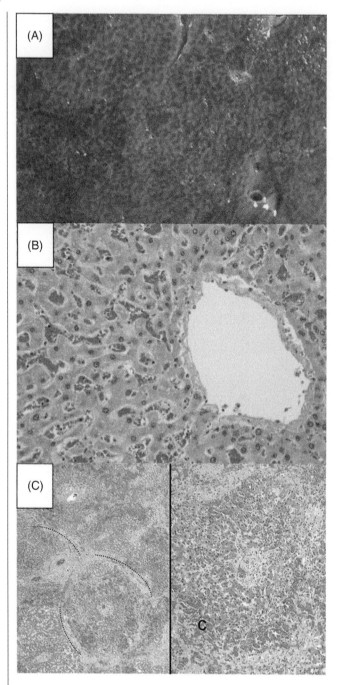

Figure CS19.1 Cardiac (congestive) hepatopathy. (A) Gross photograph of a liver with a classic 'nutmeg' appearance. Darker tissues are the congested hepatic venules and small hepatic veins. Paler areas are unaffected surrounding liver tissue. (B) Early passive congestion with sinusoidal dilatation (red blood cells are pink). (C) Microscopic images of congestive hepatopathy demonstrating marked pericentral congestion with bridging fibrosis (dotted lines). The higher magnification demonstrates a normal portal tract in the middle of the image with congestion (C) and fibrosis of the central-lobular region.
Source: Raouf E. Nakhleh M.D. 2017 / Reproduced from John Wiley & Sons.

Discussion

Cirrhosis is a progressive disease in which the liver is chronically injured and replaced by scar tissue, leading to impaired liver function. Cardiac cirrhosis is a type of cirrhosis that is caused by chronic right-sided heart failure. Although it is called cardiac cirrhosis, the condition does not always meet the precise pathological criteria for cirrhosis. Instead, the terms congestive hepatopathy and chronic passive liver congestion are often considered more appropriate. Nevertheless, the name cardiac cirrhosis has become a common convention.

In cardiac cirrhosis, chronic right-sided heart failure leads to increased pressure in the hepatic veins, which can cause the liver to enlarge and become congested. Over time, this congestion can cause liver cell damage and lead to the formation of scar tissue. At the cellular level, venous congestion hinders the efficient drainage of blood flow from sinusoids into terminal hepatic venules. Cardiac cirrhosis can present with the classical appearance of 'nutmeg liver' (diffuse mottling on imaging) due to ischaemia and fatty degeneration (Fig. CS19.1).

Treatment and Prognosis

It is important to treat and manage the underlying cause of cardiac cirrhosis, typically right-sided heart failure, to prevent progression of the disease and improve outcomes. Thus, management involves treatment of the underlying congestive heart failure with diuretics and angiotensin-converting enzyme (ACE) inhibitors or angiotensin receptor blockers (ARB). A low-sodium diet is also recommended to reduce fluid retention. In severe cases, liver transplantation may be considered.

The patient was started on a higher dose of furosemide and lisinopril and advised to follow a low-sodium diet. She was enrolled in a cardiac rehabilitation programme to provide structured exercise routines, dietary guidance and education to enhance heart function. She was also referred for further evaluation and management of her liver disease, including assessment of the extent of fibrosis using the Fibrosis-4 index (FIB-4). FIB-4 is a non-invasive tool used to assess the severity of liver fibrosis in patients with chronic liver diseases, such as cirrhosis. The FIB-4 index is based on four factors: age, platelet count, AST (aspartate transaminase) and ALT (alanine transaminase) levels. The FIB-4 index is calculated from a simple formula: (age × AST)/(platelet count × sqrt [ALT]). Higher FIB-4 scores indicate more advanced fibrosis and a worse prognosis for liver-related outcomes, such as liver failure, hepatocellular carcinoma and death.

Questions

1 Calculate the Fibrosis-4 index for this patient. You can use online tools. For example https://www.mdcalc.com/calc/2200/fibrosis-4-fib-4-index-liver-fibrosis
2 What is a FibroScan?

Answers on p. 538.

Case Study 20   Diarrhoea and Abdominal Distension

Clinical Presentation

A 40-year-old woman presented with fatigue and tiredness. On further history, she complained of palpitations and light-headedness. Her periods were regular and not heavy. She gave a 10-year history of intermittent diarrhoea that was diagnosed as irritable bowel syndrome associated with tiredness. The diarrhoea could reach up to five times daily, was described as creamy-white and was associated with flatulence and bloating. She was not on any medication. On examination, the patient was pale with marked ankle swelling and a distended abdomen. Her BMI was classified as underweight (Chapter 12). There was marked ankle swelling.

Investigations

The results of laboratory results are shown in Table CS20.1 and are consistent with a malabsorption syndrome with low serum albumin, anaemia with low folate and ferritin levels. Coeliac disease was suspected. Blood tests are used to identify autoantibodies

Table CS20.1   Results of laboratory investigations.

Test	Result	Reference range
Blood		
Haemoglobin	**10.1 g/dL**	11.5–16.5
Serum folate	**0.6 µg/L**	3.0–20.0
Red cell folate	**84 µg/L**	160–640
Vitamin B12	255 ng/L	160–760
Ferritin	**6 µg/L**	15–300
Albumin	**18 g/L**	34–48
Thyroid-stimulating hormone (TSH)	**40.8 mU/L**	0.4–4.0
Free thyroxine	9.8 pmol/L	9.0–24.0

Values outside the reference range are in bold.

associated with coeliac disease. The most common blood tests look for:

- *Anti-tissue transglutaminase antibodies (anti-tTG):* These antibodies are typically elevated in individuals with coeliac disease.
- *Anti-endomysial antibodies (anti-EMA):* EMA testing is highly specific for coeliac disease and is often used as a confirmatory test.

It is important to also measure total IgA levels at the same time. This is because the tTG test measures the levels of anti-tTG IgA antibodies. Some people may also have selective IgA deficiency, a genetic condition in which they produce lower than normal levels of IgA (see also, Chapter 6), meaning they can give a false negative result in the tTG test.

In this patient, both anti-tTG and anti-EMA antibody tests were positive. Upper gastrointestinal endoscopy was performed and biopsy samples taken from the jejunum.

Diagnosis

The jejunal biopsies showed subtotal villous atrophy, as well as a chronic inflammatory infiltrate within the epithelium and the lamina propria, consistent with coeliac disease.

Diagnosis: Coeliac disease.

Discussion

Adult coeliac disease is defined as a condition in which there is an abnormal jejunal mucosa, which improves when patients take a gluten-free diet, but relapses when gluten is reintroduced. The highest incidence of coeliac disease in the world is found in the west of Ireland, where 1 in 300 of the population is affected. In the United Kingdom, the incidence is around 1 in 2000. The clinical features of coeliac disease are common to all types of malabsorption and include weight loss, weakness and diarrhoea. The stools are pale, bulky and foul-smelling. In addition to deficiencies of iron and folate, there may also be hypocalcaemia due to vitamin D malabsorption; if long standing, this may result in *osteomalacia* (a decrease in calcified bone due to poor mineralisation).

Coeliac disease primarily affects the proximal small bowel. Loss of normal villi results in a mucosa, which appears flattened (Fig. CS20.1). There are increased numbers of intraepithelial lymphocytes and the lamina propria is infiltrated with T-lymphocytes, B-lymphocytes, plasma cells and antigen-presenting cells, like macrophages and dendritic cells.

In coeliac disease, the activation of a gluten-specific immune response is initiated by the interaction between gluten (an environmental trigger), and HLA-DQ2/8 (the main predisposing genetic factor). Transglutaminase 2 (TG2) serves as the specific autoantigen. Activated cytotoxic CD8+ intraepithelial T-cells display sustained activation of the Janus kinase-signal transducer and activator of transcription (JAK/STAT) pathway that transmits signals from various cytokines and growth factors. Sustained activation of intraepithelial T-cells contributes to the autoimmune-like attack on the intestinal epithelium, in turn leading to the release of inflammatory cytokines and the recruitment of additional immune cells, further exacerbating the tissue damage.

The patient had elevated levels of thyroid-stimulating hormone (TSH). Coeliac disease is associated with a higher risk of developing autoimmune thyroiditis. In response to damage to the thyroid gland, the pituitary gland releases more TSH to stimulate the thyroid gland to produce more thyroid hormones. The increased TSH levels are an attempt by the body to compensate for the reduced

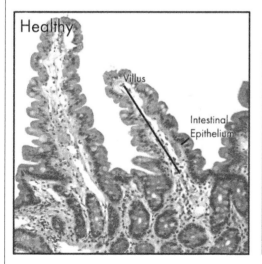
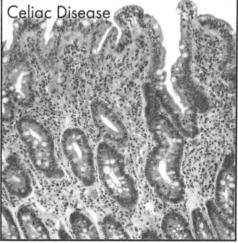

Figure CS20.1 Histopathology of villous atrophy in coeliac disease. H&E stain. Note, blunting of villi in coeliac disease. https://commons.wikimedia.org/wiki/File:Histopathology_of_villous_atrophy_in_celiac_disease.jpg GeneFood, CC BY 4.0 <https://creativecommons.org/licenses/by/4.0>, via Wikimedia Commons. *Source:* GeneFood / Wikimedia Commons / CC BY 4.0.

thyroid hormone production and maintain thyroid hormone levels within the normal range. In some cases, the thyroid hormone levels may still be within the normal reference range, but the TSH levels are increased. People with coeliac disease are also more likely to develop other autoimmune disorders, such as type 1 diabetes (Chapter 13). People with type I diabetes also have a higher risk of developing coeliac disease.

Treatment and Prognosis

A gluten-free diet combined with oral folic acid and iron supplementation, as well as nutritional supplements and an appropriate diet plan for weight restoration, produced a remarkable clinical response in this patient. Within 4 months, all the laboratory indices, including the serum albumin, had returned to normal. The thyroid deficiency was rapidly corrected by replacement therapy with thyroxine.

In recent years, there has been a significant focus on developing pharmacological treatments for coeliac disease to complement, or even eventually replace, existing dietary restrictions. This is because a strict gluten-free diet poses challenges in terms of both adherence and inadvertent gluten exposure. Moreover, the mucosal healing that normally occurs after dietary restriction does not happen in all patients with coeliac disease.

Questions

1 Can you explain the cause of anaemia in this patient?
2 Is there an increased risk of cancer in people with coeliac disease?

Answers on p. 538.

Case Study 21 Cough, Shortness of Breath and Fever in a 78-year-old Man

Clinical Presentation

A 78-year-old man with a history of hypertension and chronic obstructive pulmonary disease (COPD) presented to the Emergency Department with a cough, shortness of breath and fever for the past 3 days. He had smoked 30 cigarettes a day for the last 35 years. On examination, he was tachypnoeic with a respiratory rate of 26 breaths per minute (the normal respiratory rate for an adult at rest is 12–18 breaths per minute) and an oxygen saturation of 88% (normal SaO_2 is >95%). He had crackles in the right lower lung field, jugular venous distension (Fig. CS21.1) and pitting ankle oedema (fluid retention; applying pressure on the affected area causes a temporary indentation or 'pit' that slowly fills back up). There was also a parasternal heave (a palpable pulsation of the chest wall) and an enlarged and tender liver.

Investigations

The results of laboratory tests (Table CS21.1) showed a low pH (acidaemia), low PaO_2 (hypoxaemia) and high $PaCO_2$ (hypercapnia), leucocytosis and elevated levels of C-reactive protein, which may indicate an infection. The patient was also anaemic and there was evidence of renal impairment indicated by the raised blood creatinine and urea levels. A chest X-ray revealed consolidation in the right lower lobe with a mild pleural effusion. An ECG revealed right ventricular hypertrophy. Lung function tests revealed an FEV1/FVC ratio of 0.60. Sputum culture identified *Streptococcus pneumoniae*. Blood cultures were negative.

Diagnosis

This is bacterial pneumonia in a patient with pre-existing COPD. The right ventricular a common consequence of right-sided heart failure (see also Case Study 19).

The tachypnoea is likely due to the underlying COPD and pneumonia as well as right ventricular failure. Infection of the lungs causes inflammation, fluid accumulation and reduced oxygen exchange, leading to difficulty in breathing and an increased respiratory rate. The body's response to the reduced oxygen delivery is to increase the respiratory rate to improve oxygenation. Typically, tachypnoea results in a washout of CO_2, so an arterial blood gas analysis would be expected to show a low $PaCO_2$. However, our patient has elevated CO_2. This can be explained by one of the following: (a) COPD causes air trapping and CO_2 retention within the lungs and/or (b) the patient could be in hypercapnic respiratory failure (see below). The pitting ankle oedema is consistent with right ventricular failure; when the right side of the heart fails, blood can back up in the systemic veins, including those in the legs and feet, leading to swelling and fluid accumulation (Fig. CS21.2). The parasternal heave is caused by the right ventricle working harder to pump blood into the pulmonary circulation due to increased resistance or pressure in the pulmonary artery. The heave is typically felt in the region of the left sternal border.

Diagnosis: Community-acquired pneumonia and right ventricular failure.

Discussion

Pneumonia is classified by cause, the setting in which it was acquired, or its anatomical distribution:

Cause: Pneumonia may be; bacterial, most commonly *Streptococcus pneumoniae*, but also *Haemophilus influenzae*, *Legionella pneumophila* and *Staphylococcus aureus*; viral, for example, Influenza virus, Respiratory syncytial virus (RSV), SARS-CoV-2; fungal, including,

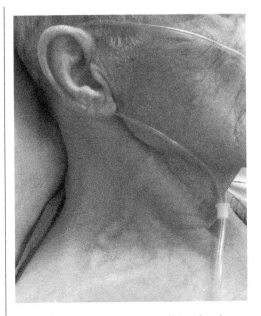

Figure CS21.1 Jugular venous distension. A person with congestive heart failure presented with an exceedingly elevated JVP. This is an edited version of the source image made for use in the 'Anatomist' iOS and Android app and shared here under the terms of the source image's Share Alike Creative Commons license. *Source:* James Heilman, MD / Wikimedia Commons / CC BY-SA 4.0.

hypertrophy and elevated JVP are likely due to right ventricular failure secondary to respiratory disease. In severe cases of COPD, low oxygen levels cause a rise in blood pressure in the arteries of the lungs, a condition known as pulmonary hypertension, placing additional workload on the right ventricle to pump blood through the lungs. As a result, the heart muscle weakens and right-sided heart failure can occur. When the right side of the heart fails to pump blood effectively, pressure in the right atrium and venous system can increase, leading to a backup of blood in the systemic veins, including the jugular veins in the neck. This can cause an elevation in the JVP. The tender and enlarged liver observed in the physical examination is consistent with hepatic congestion, which is

Table CS21.1   Results of selected laboratory investigations.

Test	Result	Reference range
White blood cell count	**$12.2 \times 10^9/\text{L}$**	4.0–11.0
Haemoglobin	**11.5 g/dL**	13.0–18.0
Platelet count	$200 \times 10^9/\text{L}$	150–400
C-reactive protein (CRP)	**25 mg/L**	<10
Sodium	140 mmol/L	135–146
Potassium	**3.2 mmol/L**	3.5–5.0
Bicarbonate	23 mmol/L	22–30
Creatinine	**189 μmol/L**	59–104
Urea	**7.8 mmol/L**	2.5–6.7
Bilirubin	0.8 mg/dL	<1.2
Alanine transaminase (ALT)	24 U/L	5–40
Aspartate transaminase (AST)	28 U/L	10–40
Alkaline phosphatase	70 U/L	25–115
Prothrombin time (PT)	12 seconds	11–16
Blood gases		
Arterial pH	**7.20**	7.35–7.45
PaO_2	**68 mmHg**	75–100
$PaCO_2$	**48 mmHg**	36–46

Values outside the reference range are in bold.

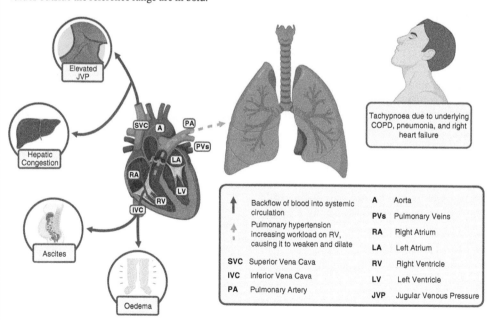

Figure CS21.2   The effect of backflow of blood into the systemic circulation secondary to right heart failure. Right heart failure leads to inefficient pumping of blood to the lungs. This causes an increase in pressure in the right atrium and venous system due to the blood pooling in the right side of the heart. The increased pressure in the systemic veins leads to an elevated JVP, hepatic congestion, ascites and oedema. Because of the impaired right heart function, less blood is pumped to the lungs, leading to reduced oxygenation of the blood. The body responds to the lack of oxygen by increasing the breathing rate (tachypnoea). This is the body's attempt to compensate for the reduced oxygen delivery to tissues. *Source:* Created with BioRender.com.

Histoplasma capsulatum, Cryptococcus neoformans and Aspergillus spp., or; parasitic, involving organisms such as *Toxoplasma gondii* or *Pneumocystis jirovecii*.

Setting: Pneumonia may be hospital-acquired pneumonia (HAP), ventilator-associated pneumonia (VAP) or community-acquired pneumonia (CAP).

Anatomical distribution: It can also be categorised as *lobar pneumonia*, which affects one or more lung lobes; *bronchopneumonia*, in which inflammation surrounds the bronchi or bronchioles—this form of pneumonia often appears more patchy and may affect multiple areas of both lungs; or, *interstitial pneumonia*, which affects the interstitial spaces between the alveoli.

Respiratory failure is a condition in which the respiratory system is unable to provide sufficient oxygen or remove enough carbon dioxide from the body. Respiratory failure can be acute or chronic and can occur due to various underlying causes. Acute respiratory failure is of sudden onset and is caused by pneumonia, pulmonary embolism or acute respiratory distress syndrome (ARDS). Chronic respiratory failure, on the other hand, is a long-term condition usually caused by underlying chronic lung disease, such as COPD or pulmonary fibrosis.

Hypoxaemic and hypercapnic respiratory failure are the two types of respiratory failure. Both can co-exist in some cases, particularly in patients with chronic lung diseases such as COPD.

Hypoxaemic respiratory failure occurs when there is a failure of oxygen uptake in the lungs, leading to a low level of oxygen in the blood (hypoxaemia). In this form of respiratory failure, the arterial oxygen tension (PaO_2) is less than 60 mmHg, while the arterial carbon dioxide tension ($PaCO_2$) is usually normal or low. Hypoxaemic respiratory failure is treated with supplemental oxygen therapy to improve oxygen levels in the blood.

Hypercapnic respiratory failure, on the other hand, occurs when there is an accumulation of carbon dioxide in the blood due to a failure of the respiratory system to remove enough carbon dioxide (hypercapnia). In hypercapnic respiratory failure, the $PaCO_2$ is greater than 45 mmHg, while the PaO_2 may be normal or low. In hypercapnic respiratory failure, mechanical ventilation is often necessary to support respiratory function and improve gas exchange. Non-invasive ventilation (NIV) is a viable and often preferred option for managing hypercapnic respiratory failure. NIV involves providing respiratory support without the need for an endotracheal tube or tracheostomy. Instead, a mask or other interface delivers positive pressure ventilation to the patient's airways. Two common types of NIV are Bi-level Positive Airway Pressure (BiPAP) and Continuous Positive Airway Pressure (CPAP). NIV is generally preferred over invasive mechanical ventilation, as it is associated with fewer complications and improved patient comfort.

The FEV1 (forced expiratory volume in 1 second) to FVC (forced vital capacity) ratio is used in pulmonary function testing. A normal FEV1/FVC ratio is greater than 0.70–0.75, which means that a person can forcefully exhale at least 70–75% of their total lung capacity within the first second of exhalation. In this case, the patient had an FEV1/FVC ratio of 0.60, which is lower than the normal range and typical for COPD.

Treatment and Prognosis

The patient was administered antibiotics for the pneumonia, diuretics for the right ventricular failure and non-invasive ventilation to correct his respiratory failure. Close monitoring of the patient's response to treatment, including repeat chest X-rays and measurement of oxygen saturation, is necessary to ensure adequate resolution of both conditions. The patient made a good recovery, but

8 months later, he was admitted to the hospital via the Accident and Emergency Department. He was cyanosed, breathless and febrile. A chest X-ray revealed right basal consolidation with the presence of a 2 cm mass in the hilar region. A biopsy of this mass revealed a well-differentiated squamous cell carcinoma (see Chapter 11 for more detail). CT scan showed multiple enlarged right supraclavicular, mediastinal and right hilar lymph nodes with compression of the trachea and oesophagus. Abdominal CT revealed a shadow in the right lower lobe of the liver with dilation of the intrahepatic bile duct and common bile duct consistent with liver metastases. The patient's performance score was 2 and he was started on chemotherapy using paclitaxel combined with cisplatin. The patient received four cycles of this therapy without any evidence of significant tumour reduction. 6 days after the last course of chemotherapy, the patient died suddenly at home. A post-mortem examination revealed the cause of death to be myocardial infarction.

Questions

1 Why do doctors have to be careful when giving oxygen to a patient with COPD?
2 Right heart failure occurs when the right side of the heart is unable to pump blood effectively. How does right heart failure lead to renal impairment:

Answers on p. 539.

Case Study 22   Abdominal Pain and Diarrhoea

Clinical Presentation

A 46-year-old woman complained of right-sided cramping abdominal pain and intermittent non-bloody loose stools for 8 weeks, averaging about 10 episodes per day. She had unintentionally lost 6 kg in weight over this period and had been feeling significantly more tired and fatigued. On examination, she was apyrexial (absence of fever), pale and had mouth ulceration. On digital rectal examination, perianal skin tags were noted. A tender mobile mass was felt in the right lower abdomen.

Investigations

The results of laboratory investigations are shown in Table CS22.1. The total leucocyte count was slightly higher than normal, with a mild neutrophilia. Erythrocyte morphology showed microcytic and hypochromic cells. The erythrocyte sedimentation rate (ESR) is very high, reflecting active inflammatory disease. Stool sample was negative for *Clostridium difficile*, a bacterium that can cause severe diarrhoea and other intestinal symptoms (often after antibiotic use). Ilecolonoscopy showed:

- Mucosal changes, including hyperaemia (increased blood flow) and friability (fragile mucosa that bleeds easily upon touch), indicative of inflammation.
- *Cobblestoning:* A term describing the appearance of the mucosa typically seen in Crohn's disease, in which there are areas of ulceration interspersed with normal or swollen mucosa, resembling cobblestones.
- *Aphthous ulcers:* Small, shallow, round ulcers found on the mucosal surface.
- *Skip lesions:* Intermittent areas of diseased and normal mucosa, which is characteristic of Crohn's disease.

Magnetic resonance enterography (MRE) and computed tomography enterography (CTE) are also useful tests to further evaluate the extent and nature of the disease.

Table CS22.1   Results of laboratory investigations.

Test	Result	Reference range
Total white cell count	**11.9 × 10⁹/L**	4.0–11.0
Haemoglobin	**9.3 g/dL**	11.5–16.5
Haematocrit	**0.30 L/L**	0.37–0.47
Mean cell volume	**73.0 fL**	82–92
Mean cell haemoglobin	**22.46 pg**	27.0–32.0
Platelet count	400 × 10⁹/L	150–400
Erythrocyte sedimentation rate	**103 mm/hour**	<20
White blood cell differential		
Neutrophils	**76%**	40–75
Lymphocytes	**17%**	20–45
Monocytes	5%	2–10
Eosinophils	2%	1–6
Basophils	0%	1

Values outside the reference range are in bold.

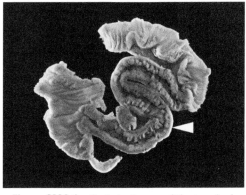

Figure CS22.1   Crohn's disease of the terminal ileum. Note the central region of thickening (white arrowhead). Licence: Attribution 4.0 International (CC BY 4.0) *Source:* Michael Frank / Wellcome Collection / CC BY 4.0.

Diagnosis: Crohn's disease.

Discussion

Each year, in most developed countries, Crohn's disease is diagnosed in approximately 5 people in every 100 000. The disease affects more women than men and often presents before the age of 30. It also tends to occur more frequently in certain families, suggesting genetic susceptibility in some cases. Typical symptoms of Crohn's disease and other inflammatory bowel disorders (notably ulcerative colitis), include colicky abdominal pain, diarrhoea, fever, weight loss and anorexia. Fistula formation is relatively common in Crohn's disease and may require surgical intervention. Changes associated with Crohn's disease may extend beyond the affected area indicated by radiography or laparotomy. It can be difficult to make a precise diagnosis of inflammatory bowel disease because the clinical features of Crohn's disease and ulcerative colitis can be similar (see below). The contribution of genetics and environmental triggers to the pathogenesis of Crohn's disease are discussed in Case Study 28.

Typical features of Crohn's disease include:

- *Bowel wall thickening:* A hallmark of Crohn's disease is thickening of the inflamed

Figure CS22.1 shows the characteristic thickening of the ileum wall in Crohn's disease.

Diagnosis

The mean cell volume (MCV) is below the reference range, suggesting that the red blood cells are smaller than normal, indicative of microcytic anaemia. The mean cell haemoglobin is also below the reference range, indicating that the amount of haemoglobin in the red blood cells is reduced, consistent with hypochromic anaemia. Iron-deficiency anaemia is typically microcytic, hypochromic. The combination of weight loss and diarrhoea suggests significant bowel pathology. Active inflammation or malignant disease are most likely to produce these results. Typical Crohn's disease was shown by ilecolonoscopy and imaging studies. The distal ileum is the most common site of presentation.

segments of the bowel, which can lead to strictures, causing obstructive symptoms.

- *Fistulae:* These are abnormal connections between the bowel and other bowel segments, or between the bowel and other organs (like the bladder or skin). Fistulae are more commonly associated with Crohn's disease than with other types of inflammatory bowel disease.
- *Abscesses:* These are essentially collections of pus, which can develop due to deep-seated infection, often associated with penetrating disease.
- *Skip lesions:* Crohn's disease is characterised by areas of diseased bowel interspersed with healthy segments, resulting in a 'skip' pattern.
- *Mesenteric fat proliferation:* This is increased fat deposition around the inflamed bowel segment, sometimes referred to as 'creeping fat'.
- *Lymphadenopathy:* Enlarged lymph nodes in the mesentery, reflecting the body's immune response to inflammation.
- *Bowel oedema:* Swelling or accumulation of fluid in the bowel wall can also be observed.
- *'Comb' sign:* This is due to engorged vasa recta (small blood vessels of the intestines) in the mesentery surrounding the inflamed bowel segments, giving an appearance reminiscent of a comb's teeth.
- *Presence of granulomas:* These are clusters of chronic inflammatory immune cells (Chapter 5) and are a typical histological feature of Crohn's disease.
- *Distension of bowel loops:* Sometimes, fluid-filled dilated loops can be seen, especially if there is a nearby stricture causing partial obstruction.

Treatment and Prognosis

The patient was admitted and IV fluids were started. Treatment goal in Crohn's disease patients is to establish remission (endoscopic, histological and clinical). The patient was commenced on high-dose corticosteroids. Initially, there was a good response to treatment. However, 3 months later, the patient returned with fever and right-sided abdominal pain. An abscess obstructing the right kidney was diagnosed and surgical drainage was performed. Crohn's disease can lead to various complications, including the formation of abscesses. The abscess obstructing the right kidney in this patient was most likely a complication of Crohn's disease; as described above, fistulae (abnormal connections between the intestine and other structures) can form, which can result in abscesses in adjacent organs or tissues. Moreover, the use of high-dose corticosteroids, while effective in inducing remission in Crohn's disease patients, can suppress the immune system, potentially making patients more susceptible to infections. This immunosuppression could have contributed to the development or worsening of the abscess.

Combination therapy using TNF-α inhibitors, such as infliximab or adalimumab, in conjunction with other immunomodulators, like azathioprine or methotrexate, is a favoured approach to treat moderate to severe Crohn's disease. However, not all patients can be given these combinations due to potential side effects, contraindications, or lack of response. In such cases, alternative treatments, including newer biologics, such as vedolizumab (a monoclonal antibody that binds to integrin α4β7, blocking GI tract-selective anti-inflammatory activity), or ustekinumab (a monoclonal antibody which blocks the pro-inflammatory cytokines, IL-12 and IL-23), can be considered. In practice, the choice of therapy is tailored to an individual's clinical presentation, disease severity and overall health profile.

Questions

1 Can you explain the haematological findings?
2 How does Crohn's disease differ from ulcerative colitis?

Answers on p. 539.

| Case Study 23 | Fatigue, Weakness and Inability to Concentrate |

Clinical Presentation

A 65-year-old man visited his GP complaining of fatigue and weakness for the past few months. He reported feeling tired even after getting enough sleep and had noticed a decline in his ability to concentrate. He denied any significant weight loss or fever. He also mentioned occasional numbness and tingling in his hands and feet. He had no history of recent illness or injuries. He reported a generally healthy lifestyle, including a good diet, with no smoking or excessive alcohol consumption. There was a history of hypertension, which was well-controlled with medication. Family history was significant for his sister and mother, both of whom had systemic lupus erythematosus. On examination, his blood pressure was 130/80 mmHg, heart rate 80 beats per minute, respiratory rate 16 breaths per minute, temperature 98.6°F (37°C). The patient appeared pale and fatigued. There was decreased sensation to light touch and pinprick in the distal extremities. Deep tendon reflexes were also diminished in the lower extremities.

Investigations

Blood tests revealed a reduced haemoglobin level (9.2 g/dL [reference range: 13.0–18.0 g/dL]). Mean corpuscular volume (MCV) was increased at 107 fL (reference range: 82–92 fL). White blood cell count and platelet count were within normal limits. A peripheral blood smear showed macro-ovalocytes and hyper-segmented neutrophils. Serum vitamin B12 level was low at 120 ng/L (reference range: 160–760 ng/L). Folate level was within normal limits.

Diagnosis

Vitamin B12, also known as cobalamin, is an essential water-soluble vitamin that plays a key role in DNA synthesis, red blood cell formation, nervous system function, and various metabolic processes in the body. Common causes of vitamin B12 deficiency include inadequate dietary amounts (e.g. in vegans), impaired absorption, age-related decreases in acid secretion, and pernicious anaemia, an autoimmune disorder characterised by vitamin B12 deficiency due to intrinsic factor deficiency. An intrinsic factor antibody test was positive.

Diagnosis: Pernicious anaemia.

Discussion

Pernicious anaemia is part of the clinical spectrum of autoimmune gastritis (AIG) and is considered a late-stage manifestation. It is characterised by megaloblastic anaemia, caused by vitamin B12 deficiency secondary to the autoimmune destruction of parietal cells in the stomach, leading to a deficiency of hydrochloric acid and intrinsic factor (IF), the latter of which is necessary for the absorption of vitamin B12. Macrocytosis is a common finding, with a MCV equal to or greater than 100 fL. However, normocytic anaemia can also be present if there is concurrent iron deficiency anaemia due to achlorhydria (the absence of stomach acid production, which can contribute to iron deficiency by impairing the absorption of dietary iron in the stomach). Iron deficiency may occur before or at the same time as a diagnosis of pernicious anaemia. A peripheral blood smear may contain macro-ovalocytes (enlarged red blood cells with an oval shape), hyper-segmented neutrophils (neutrophils with an abnormally high number of lobes in their nucleus) and *anisopoikilocytosis*. Hyper-segmented neutrophils, although often associated with megaloblastic anaemia, are not specific to pernicious anaemia. Folate levels should be measured to exclude macrocytic anaemia caused by folate deficiency.

Pernicious anaemia is a slowly progressing condition, and it may take 2–5 years before symptoms become apparent. Patients may not realise they have an underlying condition until the anaemia becomes severe. This is because compensatory mechanisms in the body, particularly in the cardiovascular and respiratory systems, help maintain adequate oxygen delivery, despite the deficiency. Frequent presenting symptoms include fatigue, lethargy, anorexia and weight loss. Neurological symptoms can manifest as confusion, difficulty in concentrating, memory loss and cognitive decline. Peripheral neuropathy is an early neurological manifestation of pernicious anaemia. It is caused by damage to the peripheral nerves, resulting in sensory abnormalities and motor dysfunction. The peripheral neuropathy associated with pernicious anaemia typically presents as a symmetric neuropathy, meaning that both sides of the body are affected similarly. The lower extremities are often more affected than the upper extremities. Common symptoms include a decrease in sensitivity to light touch, pinprick and vibration. Patients may also experience numbness, tingling, and abnormal sensations in their extremities. Patients with severe loss of position sense may demonstrate a positive Romberg test. Deep tendon ankle reflexes are commonly hypoactive or absent, indicating dysfunction of the reflex arc. However, more proximal reflexes may still be intact. In the later stages of disease, *subacute combined degeneration* (SCD) of the spinal cord may occur, leading to further neurological deficits, including limb weakness, *ataxia* (loss of coordination and balance), and visual disturbances. In some cases, the neurological findings can be present even in the absence of anaemia.

The detection of anti-IF antibodies has a sensitivity of 40–60% in identifying

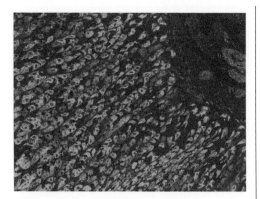

Figure CS23.1 Immunofluorescence pattern of gastric parietal cell antibodies. Antibodies in the serum of a patient with pernicious anaemia bind to a stomach section and are detected with a fluorescent (FITC) label. *Source:* Simon Caulton / Wikimedia Commons / CC BY-SA 3.0.

pernicious anaemia, with the rate of positivity increasing as the disease progresses. However, the specificity of anti-IF antibody testing is almost 100%. Anti-parietal cell antibodies are present in 90% of patients with pernicious anaemia (Fig. CS23.1), but are less specific than anti-IF antibodies. Combining both anti-IF and anti-parietal cell antibody testing significantly increases diagnostic accuracy. According to the British Committee for Standards in Haematology, patients with anaemia, neuropathy, or glossitis suspected of having pernicious anaemia should be tested for anti-IF antibodies, regardless of vitamin B12 levels. Anti-IF antibody testing is also recommended in patients with low serum vitamin B12 levels without anaemia, food malabsorption or other causes of deficiency to determine if they have an early or latent presentation of pernicious anaemia.

Treatment and Prognosis

The patient was given intramuscular injections of vitamin B12 (hydroxocobalamin) daily for 1–2 weeks. This was followed by weekly injections for 1–2 months

and then a monthly injection for maintenance. He was advised to have regular
follow-up visits to monitor his response to
treatment and ensure adequate vitamin
B12 levels.

Questions

1 What is the Romberg test?
2 What is anisopoikilocytosis?

Answers on p. 540.

Case Study 24   6-month History of Fatigue, Joint Pain and Abdominal Discomfort

Clinical Presentation

A 45-year-old man presented to his GP with
a 6-month history of fatigue, joint pain and
abdominal discomfort. He reported that he
had been feeling increasingly tired and weak,
with difficulty concentrating at work. He also
complained of pain in his fingers, knees and
ankles, which was worse in the morning but
improved throughout the day. He reported a
past medical history of hypertension and
was currently taking lisinopril. He had a family history of liver disease; his father having
died from cirrhosis at the age of 60. He
admitted to a history of heavy alcohol use
but had quit drinking 5 years ago. He denied
any recent blood transfusions or intravenous
drug use. On physical examination, the
patient was mildly jaundiced, with yellowing
of the sclerae and skin. His liver was palpable 3 cm below the right costal margin and
was firm to the touch, with an irregular edge.
He had mild tenderness on palpation of the
liver, but no palpable spleen or obvious
ascites. There were no stigmata of chronic
liver disease, such as spider angiomas or
gynaecomastia. His joints were tender to
palpation, with some swelling noted in his
fingers, knees and ankles. There was no
erythema or warmth over the joints.

Investigations

An abdominal ultrasound indicated that the
liver had a nodular contour consistent with
cirrhosis. Laboratory investigations, summarised in Table CS24.1, suggested iron overload. Serum iron is the amount of iron that is

Table CS24.1   Results of laboratory investigations.

Test	Result	Reference range
Haemoglobin	**12.2 g/dL**	13.0–18.0
Mean corpuscular volume (MCV)	**75 fL**	82–92
Iron	**56 μmol/L**	13–32
Ferritin	**340 nmol/L**	5.8–144
Transferrin saturation	**90%**	15–50%
Total iron-binding capacity (TIBC)	**25 μmol/L**	42–80
Aspartate transaminase (AST)	**102 U/L**	10–40
Alanine transaminase (ALT)	**100 U/L**	5–40
Alkaline phosphatase (ALP)	**150 U/L**	25–115
Bilirubin	**2.5 mg/dL**	<1.2

Values outside the reference range are in bold.

circulating in the blood bound to transferrin.
The patient's serum iron and ferritin levels
were significantly elevated compared to the
reference range, indicative of iron overload.
Total iron-binding capacity (TIBC) is a measure of the amount of iron that can be bound
to transferrin. The transferrin saturation is
calculated as the ratio of serum iron to
TIBC. The patient's transferrin saturation is
significantly elevated, indicating that a large
amount of transferrin is already bound to iron.

Diagnosis

Based on the clinical and laboratory findings, the patient was diagnosed with haemochromatosis, a disorder characterised by excessive accumulation of iron in various organs, including the liver, heart and pancreas.

Haemochromatosis can result from an inherited defect, in which case it is described as *hereditary (or primary) haemochromatosis*. When haemochromatosis occurs because of haematological disorders (e.g. β-thalassaemia, myelodysplastic syndrome), excessive iron supplementation, metabolic syndrome or chronic alcoholism, for example, it is referred to as *secondary haemochromatosis* (also called *haemosiderosis*).

Hereditary hemochromatosis is classified into four types based on age of onset, genetic cause and mode of inheritance. Type 1 and type 4, also known as ferroportin disease, typically present in adulthood with symptoms appearing between the ages of 40 and 60 in men and after the menopause in women. Type 2 haemochromatosis is a juvenile-onset disorder with symptoms often beginning in childhood. Iron accumulation causes decreased or absent secretion of sex hormones by age 20, leading to menstrual irregularities in affected females and delayed puberty or hormone-related symptoms in males. If left untreated, potentially fatal heart disease can develop by age 30. Type 3 haemochromatosis typically presents before age 30, with symptoms appearing at an intermediate point between types 1 and 2. Type 1 haemochromatosis is the result of mutations in the HFE gene on chromosome 6, and type 2 haemochromatosis is caused by mutations in either the HJV or HAMP gene. Mutations in the TFR2 gene cause type 3 haemochromatosis and mutations in the SLC40A1 gene cause type 4 haemochromatosis. The patient was referred for genetic testing and screening of his family members.

The patient tested positive for the C282Y mutation in HFE. This mutation leads to

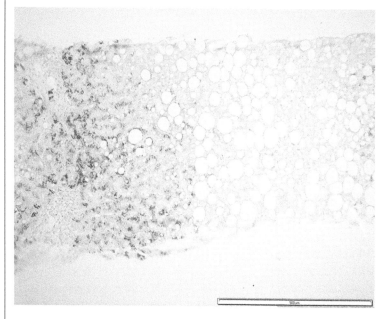

Figure CS24.1   Haemochromatosis: Hepatocyte iron accumulation (blue granules) with an acinar distribution pattern consistent with homozygous genetic haemochromatosis (stained with Perl Prussian Blue method). *Source:* Mathew, J et al. / Wikimedia Commons / CC BY 3.0.

the replacement of cysteine (C) with tyrosine (Y) at position 282. This is an autosomal recessive defect, so individuals must inherit two copies of this mutation (homozygous) to develop haemochromatosis. Liver biopsy is no longer necessary to diagnose and quantify iron overload and its distribution, as imaging and biological findings are sufficient. However, a liver biopsy can be useful in assessing complications such as fibrosis. Figure CS24.1 shows a Perl stain used to detect iron in a liver biopsy from a patient with haemochromatosis.

Diagnosis: Hereditary haemochromatosis.

Discussion

Iron homeostasis is maintained by regulating the concentration and biochemical forms of iron in the plasma. The main source of bioavailable iron for cells is plasma iron and its concentration is kept within normal limits to ensure adequate iron delivery to cells and to prevent iron-related disorders. Iron is bound to transferrin in plasma and stored bound to ferritin in tissues. Enterocytes and macrophages are the main sources of plasma iron, which is transported into the plasma through ferroportin, the only identified iron exporter in mammals (Fig. CS24.2). The activity of ferroportin is regulated by hepcidin, a peptide

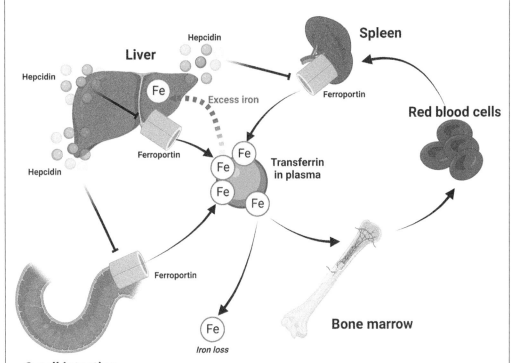

Figure CS24.2   Regulation of iron levels. Duodenal enterocytes absorb dietary iron. Iron is transported in blood by transferrin (% saturation indicates iron storage status). Hepatocytes store excess iron. Hepcidin is a peptide synthesised by hepatocytes. The sole known molecular target of hepcidin is the protein, ferroportin, which functions as a transmembrane conduit for the transfer of cellular iron to plasma. Hepcidin binds to ferroportin and promotes its degradation. The gene product of the HFE gene, known as human homeostatic iron regulator protein, or HFE protein (stands for High FE2+), regulates hepcidin synthesis. If the HFE gene is inactivated by mutation, then hepcidin levels are reduced. In the absence of hepcidin, ferroportin concentrations in enterocytes are increased, leading to enhanced transport of dietary iron into plasma. Excess plasma iron accumulates in organs in which iron uptake exceeds the rate of iron export. The liver is most commonly affected by iron overload due to the avid uptake of non-transferrin-bound iron by hepatocytes. *Source:* Created with BioRender.com.

secreted by hepatocytes. When hepcidin levels are low, ferroportin activity increases, resulting in an increase in plasma iron concentration and transferrin saturation. However, high transferrin saturation can lead to the appearance of non-transferrin-bound iron in plasma, which can be taken up by the liver and which is toxic to hepatocytes. Therefore, the control of hepcidin expression levels is crucial for the maintenance of systemic iron homeostasis. Excessive iron accumulation in cells leads to cellular toxicity, mainly due to the formation of reactive oxygen species (ROS; Chapter 3).

Treatment and Prognosis

The treatment of hereditary haemochromatosis involves regular phlebotomy, which removes excess iron from the body and reduces the risk of organ damage. The goal is to reduce the serum ferritin level while maintaining a healthy haemoglobin level. The frequency of phlebotomy depends on the severity of iron overload.

Questions

1   In some individuals, it may not be appropriate to treat haemochromatosis with regular phlebotomy. Are there any alternatives?
2   Why do women with hereditary haemochromatosis present later than men?

Answers on p. 540.

Case Study 25   Muscle Weakness and Numbness in the Legs for 3 months

Clinical Presentation

A 35-year-old female athlete presented to her GP with a 3-month history of muscle weakness and numbness in her legs. She reported that she was worried about her symptoms that started gradually at the beginning of her marathon training, but which had progressively worsened over the next 5 days. She also complained of an ongoing history of fatigue and difficulty with coordination and balance. One year prior to these new symptoms, the patient had developed insidious left eye pain and blurred vision, which lasted for 2 weeks but improved gradually. She had a past medical history of asthma and eczema but was otherwise healthy. She had no history of recent infections or other illnesses and denied any family history of neurological disorders. On examination, she was alert and oriented with no signs of distress. Her blood pressure was 120/80 mmHg, and her heart rate was 50 beats per minute.

Neurological examination revealed mild weakness in her left leg, with a positive Babinski reflex and decreased sensation to light touch in her right leg. A Babinski reflex is the upwards extension of the hallux (big toe) elicited through the stimulation of the lateral plantar aspect of the foot. In adults and children over 2 years of age, a positive Babinski reflex is generally a sign of damage to the corticospinal tract. The patient demonstrated difficulty when asked to perform the *tandem gait test* (heel-to-toe walking). The tandem gait test evaluates balance by asking the patient to walk in a straight line while the toe of one foot touches the heel of the other with each subsequent step. The patient also displayed *Lhermitte sign*, which is a brief, electric shock-like sensation that radiates down the spine and into the limbs when the neck is flexed forward. The sensation is often described as a jolt, buzz or vibration, and may last only a few seconds. Lhermitte sign is classically associated with multiple sclerosis (MS), though it may be found in other neurological conditions. The patient also showed decreased visual acuity when subjected to a letter chart examination.

Investigations

Magnetic resonance imaging (MRI) of the brain and spinal cord showed multiple contrast-enhancing T1 lesions and non-enhancing T2 hyperintensities in the white matter of the brain and spinal cord, consistent with MS (T1 with contrast shows active lesions while T2 hyperintensities signify areas of demyelination). A lumbar puncture was performed and analysis of cerebrospinal fluid (CSF) showed elevated levels of immunoglobulin G (IgG) with oligoclonal bands, indicating the presence of an inflammatory process in the CNS. *Neuromyelitis optica spectrum disorders* (NMOSD) and *myelin-oligodendrocyte glycoprotein antibody-associated disease* (MOGAD) were excluded from the diagnosis, as the patient did not have detectable antibodies to either aquaporin-4 (AQP4-IgG) or myelin-oligodendrocyte glycoprotein, respectively. Other blood tests investigating autoimmune disorders and infections were unremarkable. *A visual evoked potential (VEP)* test was indicated based on the clinical suspicion of MS. In this test, electrodes are placed on the scalp over the occipital cortex, the area of the brain responsible for processing visual information. The patient is then presented with visual stimuli, such as a flashing light or a checkerboard pattern, and the electrodes record the electrical responses generated in the brain. In this patient there was a delay in the time it takes for the visual evoked potentials to travel along the optic nerve and reach the visual cortex in the brain, suggesting demyelination of the optic nerve.

A diagnosis of MS can be difficult as each patient may present with a different set of signs and symptoms, depending on the location of neurological lesions (Table CS25.1). The McDonald criteria are a set of diagnostic guidelines that were first proposed in 2001, and updated most recently in 2017, to facilitate earlier and more accurate diagnosis of MS, particularly in cases in which clinical symptoms and diagnostic tests are inconclusive. The criteria require evidence of dissemination of lesions in both *time* and *space*. There are three main clinical forms of MS: relapsing-remitting, primary progressive and secondary progressive MS. Clinically isolated

Table CS25.1 Typical presenting signs and symptoms of multiple sclerosis.

Sign/symptom	Description
Muscle weakness	Difficulty moving limbs or grasping objects, or feeling weak or heavy
Numbness or tingling	Sensations of pins and needles, or loss of sensation in parts of the body
Fatigue	Feeling tired, even after rest or with minimal activity
Balance and gait	Difficulty maintaining balance or walking in a straight line, or falling easily
Muscle spasticity	Velocity-dependent increase in muscle tone in response to passive movement. This manifests as involuntary muscle contractions, spasms, or stiffness
Bowel and bladder	Difficulty controlling bowel or bladder function, or increased frequency or urgency to urinate
Cognitive impairment	Difficulty with memory, attention, or problem-solving, or changes in mood or personality
Pain	Pain or discomfort in different parts of the body, such as the face, limbs, or torso
Optic neuritis	Inflammation of the optic nerve, leading to pain, visual loss or colour vision changes
Trigeminal neuralgia	Pain in the face or jaw, often triggered by touch or movement
Diplopia	Double vision or blurry vision in one or both eyes
Vertigo	Dizziness, light-headedness or a sensation of spinning or motion

syndrome (CIS) is a term used to describe a first episode of neurological symptoms caused by inflammation or demyelination in the CNS. It is characterised by a single clinical episode lasting at least 24 hours, which is suggestive of MS but does not yet meet the criteria for a diagnosis of MS.

In *relapsing remitting MS* (RRMS), the most common form of MS, individuals experience periods of new or worsening symptoms, known as relapses, followed by periods of partial or complete recovery, known as remissions. For the purpose of diagnosing RRMS, dissemination in space is demonstrated clinically by two or more attacks with objective clinical evidence of two or more lesions. It may also be demonstrated if there were two or more attacks with objective clinical evidence of one lesion, plus historic evidence of a prior attack involving a lesion in a different location. Radiologically, dissemination in space is demonstrated by the presence of one or more T2 hyperintense lesions in at least two of four areas of the CNS (periventricular, juxtacortical or cortical, infratentorial and spinal cord). Dissemination in time, on the other hand, is established if there is a second clinical relapse occurring at least one month after the first attack. Radiologically, dissemination in time occurs when there are new T2 hyperintense or gadolinium-enhancing T1 lesions in comparison to a previous MRI, or when there are simultaneous gadolinium-enhancing and non-enhancing lesions in the same scan. Gadolinium is a contrast agent used in MRI to help distinguish active and inactive lesions. Gadolinium-enhancing lesions represent areas of active inflammation. Another way in which dissemination in time is established is through the presence of CSF-specific oligoclonal bands (abnormal bands of IgG seen on electrophoresis of CSF, but which are not observed in the blood). The presence of oligoclonal bands in the CSF indicates CNS inflammation and is an independent predictor of

conversion from CIS to clinically definite MS. Under the 2017 McDonald criteria, testing positive for oligoclonal bands can be sufficient to fulfil the criteria for dissemination in time, even if a patient only has evident damage at one time point.

Primary progressive MS (PPMS) is characterised by slowly worsening symptoms and disability from the start of the disease.

The diagnosis of PPMS differs from RRMS, requiring the presence of 1-year of disability progression, independent of clinical relapse, plus two out of the following criteria:

- One or more T2 hyperintense lesions characteristic of MS in one or more of the following brain areas (periventricular, cortical or juxtacortical, and infratentorial).
- Two or more T2 hyperintense lesions in the spinal cord.
- Presence of CSF-specific oligoclonal bands.

Secondary progressive MS (SPMS) is characterised by a gradual worsening of neurological symptoms and disability over time, independent of clinical relapses. Progression typically follows an initial period of RRMS. Unlike RRMS, which is based on the specific diagnostic criteria outlined above, the diagnosis of SPMS relies primarily on clinical judgement.

Diagnosis: Multiple sclerosis.

Discussion

MS is a chronic autoimmune demyelinating disease of the CNS with inflammatory and degenerative components. It is the most common demyelinating disease of the CNS and the leading cause of non-traumatic disability in young adults; being most common in those aged 20–35 years, with females having a three-times elevated risk compared to males. MS predominantly affects individuals of European ancestry and is less common among Native Americans, Asians, Blacks and Māori populations. Incidence rates rise with increasing distance from the equator, underscoring the potential role of vitamin D deficiency as a key risk factor. Other environmental risk factors include Epstein-Barr virus infection, smoking and obesity. Certain HLA alleles confer either an increased risk (e.g. HLA-DRB1*1501) or relative protection (e.g. HLA-A*02).

MS pathogenesis involves poorly understood interactions between these genetic and environmental factors, resulting in a dysregulated immune response against the myelin sheath, the protective covering of nerve fibres in the CNS (Fig. CS25.1). Peripheral autoreactive T cells, particularly CD4+ T-helper cells, become activated in response to antigen mimicry or other autoimmune triggering mechanisms (Chapter 6). The activated T-cells migrate to the CNS and recognise antigens presented locally by antigen-presenting cells. This recognition, combined with the effects of inflammatory mediators released by endothelial cells and other resident CNS cells, leads to a disruption of the blood-brain barrier (BBB). Once the BBB is compromised, there is an influx of other immune cells, including B-cells, macrophages and more T-cells, into the CNS. These immune cells release inflammatory mediators and cytotoxic substances, leading to demyelination and the formation of *focal plaques*, the hallmark lesions of MS, which are particularly evident around post-capillary venules. Plaques can be visualised on MRI or by staining of brain tissue postmortem (Fig. CS25.2). Over time, the repeated cycles of inflammation and demyelination lead to the accumulation of irreversible axonal damage and neuronal loss, resulting in progressive neurological disability. In addition to demyelination, MS is also characterised by other pathological features such as *astrogliosis*, which is the proliferation of astrocytes that form a glial scar (Fig. CS25.3), microglial activation, and the formation of lymphoid follicle-like structures in the cerebral meninges.

Although MS has conventionally been categorised into the distinct clinical types, described above, emerging evidence supports the idea that MS exists on a spectrum, with individuals displaying varied and overlapping pathological features at different times. The shift towards a progressive phase of the disease represents a transition from mainly acute, localised inflammation to more diffuse inflammation, neurodegeneration and a decline in repair mechanisms such as remyelination. As patients age, they become more vulnerable to CNS damage with reduced recovery capacity. Therefore, it might be more accurate to view the progression of MS as a continuum, shaped by the interplay between disease-causing processes and the body's reparative responses.

Treatment and Prognosis

The standard of care therapy for MS, currently an incurable disease, includes various disease-modifying therapies (DMTs) that aim to modulate immune reactivity and reduce disease activity. They include:

- *Interferon-beta*, a cytokine that can inhibit T-cell activity, helping to limit immune attack on the myelin.
- *Glatiramer acetate*, which is structurally similar to myelin, and acts as a 'decoy'.
- *Teriflunomide*, which inhibits the enzyme dihydroorotate dehydrogenase that is crucial for pyrimidine synthesis. By blocking DNA synthesis, teriflunomide reduces the proliferation of rapidly dividing cells, such as activated T-cells.
- *Dimethyl fumarate*, which has immune-modulatory properties based on its ability to divert cytokine production towards a Th2 profile.
- *Fingolimod, siponimod, ozanimod* and *ponesimod*, sphingosine 1-phosphate (S1P) signalling inhibitors, which cause lymphocyte retention in the lymph nodes, reducing the number of these cells available to cause damage in the CNS.

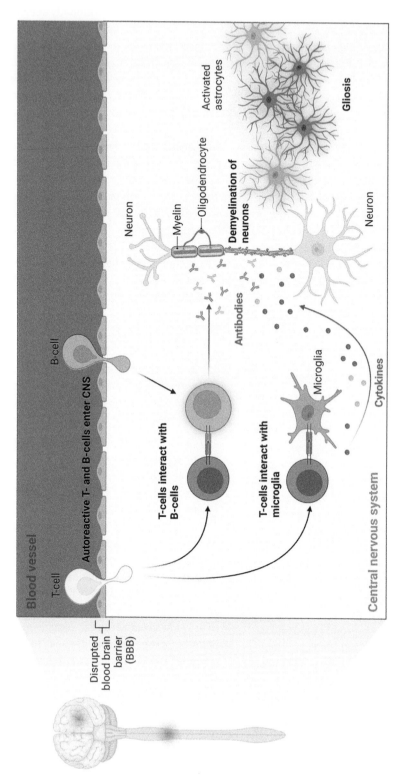

Figure CS25.1 Immune mechanisms in multiple sclerosis. Multiple sclerosis (MS) is an autoimmune disorder characterised by demyelination within the CNS. The precise aetiology remains unclear, but the immune pathogenesis involves a complex interplay between genetic predisposition and environmental triggers. In susceptible individuals, autoreactive T cells cross the blood–brain barrier (BBB) and interact with both B-cells and microglia causing persistent inflammation, destruction of the myelin sheath, axonal damage and formation of characteristic sclerotic plaques characterised by the presence of activated astrocytes, a process known as gliosis. Simultaneously, there is an impaired ability to repair or regenerate the damaged myelin. *Source:* Created with BioRender.com.

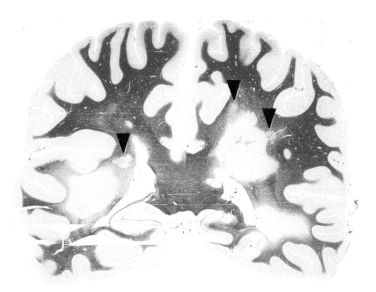

Figure CS25.2 Demyelination in multiple sclerosis. Arrow points indicate pale-staining pre-phagocytic areas which, on higher power magnification, have vacuolated myelin sheaths and apoptotic oligodendrocytes (not shown). Also prominent are old sharp-edged periventricular lesions and large subacute demyelinating lesions with demyelinated centres. Luxol fast blue-periodic acid Schiff staining; magnification, ×0.65. *Source:* Andrew P. D. Henderson MBBS et al. 2009 / Reproduced from John Wiley & Sons.

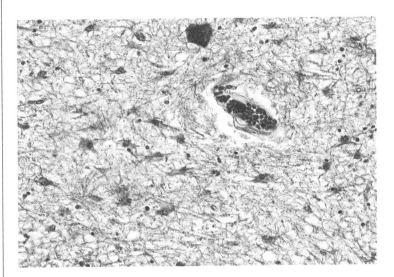

Figure CS25.3 Histopathology of gliosis. Gliosis refers to the proliferation of glial cells in the CNS, often as a reactive change associated with various types of neural injury. When stained with phosphotungstic acid-haematoxylin (PTAH), as shown here, the gliotic areas typically display intensified astrocytic cell bodies and processes, revealing the extent and pattern of astrocytic hypertrophy and hyperplasia in response to the underlying pathology.

- *Natalizumab*, a monoclonal antibody directed against α4 integrins expressed on the surface of leucocytes. By binding to these integrins, it prevents leucocytes from adhering to endothelial cells and migrating into the CNS.
- *Alemtuzumab*, a monoclonal antibody directed against CD52 that is expressed on T- and B-cells. Treatment leads to significant and prolonged depletion of T- and B-cells.
- *Ocrelizumab and ofatumumab*, which are CD20-targeting monoclonal antibodies, causing B-cell depletion.
- *Cladribine*, a purine analogue that is cytotoxic to T- and B-cells with a modest effect on neutrophils.

Symptomatic treatments and supportive care are also important components of MS management. These may include medications for *spasticity* (involuntary muscle contractions or spasms that cause stiffness and tightness in the muscles), pain, fatigue and bladder dysfunction, as well as physical and occupational therapy, rehabilitation programs, and psychological support.

The patient was started on glatiramer acetate injections. She was also referred to physical therapy for muscle strengthening and gait training. The patient was counselled on the importance of regular follow-up visits and the need for ongoing monitoring of her symptoms and disease progression. During the first 2 years following her diagnosis, she experienced three more relapses. Each episode of relapse resulted in progressive disability; at which time she was switched to natalizumab therapy. Because of the progressive nature of her illness, she took the difficult decision to retire from competitive sport and was referred to counselling therapy to help her cope with this.

Questions

1 What is *Uhthoff* phenomenon and what is its relevance to MS?
2 What is *rebound MS*?

Answers on p. 540.

Case Study 26 Three-week History of Fatigue, Itching and Jaundice

Clinical Presentation

A 42-year-old man presented to his GP complaining of fatigue, itching and jaundice of 3 weeks duration. He reported that the itching was worse at night and he had noticed some yellowing of his eyes and skin. He also complained of some abdominal discomfort and occasional fevers. The patient had a history of ulcerative colitis, for which he had been taking mesalamine. He had no significant family history of liver disease or any other medical conditions. On physical examination, the patient appeared pale and icteric (jaundiced; yellowing of the skin and the whites of the eyes due to elevated levels of bilirubin in the blood). His vital signs were within normal limits and he had no signs of dehydration. Abdominal examination revealed mild hepatomegaly with no palpable masses or tenderness. He had no stigmata of chronic liver disease such as spider angiomas, gynaecomastia or palmar erythema.

Investigations

Laboratory studies (Table CS26.1) showed elevated levels of the liver enzymes, alanine transaminase (ALT) and aspartate transaminase (AST), bilirubin and alkaline phosphatase. Viral hepatitis serology was negative.

Table CS26.1   Results of laboratory investigations.

Test	Result	Reference range
White blood cell count	6.5×10^9/L	4.0–11.0
Haemoglobin	**12.8 g/dL**	13.0–18.0
Platelet count	180×10^9/L	150–400
Alanine transaminase (ALT)	**200 U/L**	5–40
Aspartate transaminase (AST)	**250 U/L**	10–40
Bilirubin	**6 mg/dL**	<1.2
Alkaline phosphatase (ALP)	**350 U/L**	25–115
Serum albumin	**25 g/L**	34–48
Prothrombin time (PT)	**17 seconds**	11–16

Values outside the reference range are in bold.

Diagnosis

An abdominal ultrasound revealed a dilated common bile duct. Magnetic resonance cholangiopancreatography (MRCP) confirmed the presence of multifocal, short, annular strictures that alternate with normal or mildly dilated segments consistent with primary sclerosing cholangitis (PSC). Three criteria are used to diagnose PSC: (i) a chronically cholestatic serum biochemical profile; (ii) cholangiography that shows multiple areas of strictures and segmental dilations in the bile ducts inside and outside the liver (Fig. CS26.1) and (iii) compatible features, such as chronic cholangitis and periductal fibrosis, observed in a liver biopsy (Fig. CS26.2). However, if the patient presents with characteristic cholangiographic findings and fits the clinical context (such as a young male with inflammatory bowel disease and a cholestatic biochemical profile), liver biopsy may be unnecessary. It is important to remember that the initial presentation of PSC can be highly variable, depending on the disease stage and age of the patient. While some cases of PSC are incidentally diagnosed during blood work, others can present with advanced liver disease and/or concomitant cholangiocarcinoma. When investigating PSC, it is important to exclude secondary causes of sclerosing cholangitis, including autoimmune hepatitis, IgG4-related disease, infections (e.g. bacterial cholangitis) and certain medications or toxic exposures that can mimic primary sclerosing cholangitis. Children with PSC, in particular, may present with overlapping features of autoimmune hepatitis.

Diagnosis: Primary sclerosing cholangitis (PSC).

Discussion

PSC is a chronic liver and gallbladder disease that involves inflammation and scarring of the bile ducts. It can either cause no symptoms, or produce signs and symptoms of liver disease, such as yellowing of the skin and eyes, itching and abdominal pain. PSC is characterised by narrowing of the bile ducts due to scarring, which leads to a blockage of bile flow. Over time, this can result in cirrhosis of the liver and liver failure. PSC also increases the risk of various cancers, including hepatocellular carcinoma, gallbladder carcinoma, colorectal cancer and cholangiocarcinoma.

The underlying causes of PSC are still not fully understood, but it is widely accepted that the disorder is complex and multifaceted, with genetic, environmental and

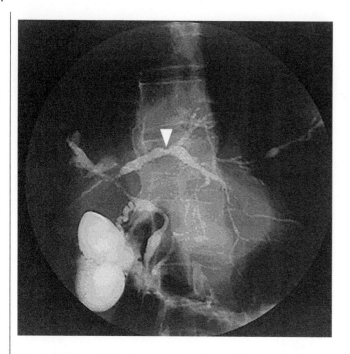

Figure CS26.1 Cholangiogram of a patient with primary sclerosing cholangitis. Typical cholangiographic findings show stricturing (example indicated by white arrowhead) of intrahepatic and extrahepatic bile ducts with dilatation of the areas either side. *Source:* Joy Worthington, Roger Chapman / Wikimedia Commons / CC BY 2.0.

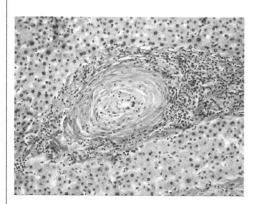

Figure CS26.2 Representative histology of small-duct PSC. Shown is the classic 'onion skin' fibrosis surrounding a small intrahepatic bile duct branch in PSC. *Source:* Jillian M. Cotter M.D & Cara L. Mack M.D. et al. 2017 / Reproduced from John Wiley & Sons.

immunological factors all potentially contributing. Cholangiocytes, the cells lining the bile ducts, are thought to be both the target and instigator of the disease. The environmental component of PSC is increasingly thought to involve the enteric microbiome and the enterohepatic circulation of microbial molecules, including lipopolysaccharide (LPS; Chapter 4). There is growing evidence linking the microbiome to PSC and animal models of microbial dysbiosis have demonstrated PSC-like hepatobiliary lesions. Genetic factors also appear to play a significant role in PSC; first-degree relatives of patients have a significantly elevated risk. Genome-wide association studies (GWAS; Chapter 10) have identified the human leucocyte antigen (HLA) gene family as the strongest risk locus for PSC.

The two most prominent diseases associated with PSC are inflammatory bowel disease (IBD), most commonly ulcerative colitis and colorectal cancer. IBD is often diagnosed before PSC, but the severity of bowel disease does not necessarily correlate with the severity of liver disease. The presence of IBD is associated with increased morbidity and mortality in PSC patients. PSC-IBD patients have a unique IBD phenotype characterised by pan-colitis with rectal sparing and back-wash ileitis, and a higher risk of post-colectomy pouchitis. PSC-IBD also confers a nearly fivefold increased risk of colorectal cancer compared to IBD alone, with colono-scopic surveillance recommended. PSC-AIH overlap syndrome, which meets the criteria for both PSC and autoimmune hepatitis (AIH), is also seen in some PSC patients and should be treated with immunosuppressive therapy.

Treatment and Prognosis

PSC tends to progress to end-stage liver disease, usually in a gradual manner. Without liver transplant, the median survival following a PSC diagnosis is approximately 15 years. However, patients with small-duct PSC generally have longer survival and a lower risk of developing cholangiocarci-noma than those with large-duct (i.e. classic) PSC. Prognostic models have been developed to predict survival in PSC; the revised natural history (Mayo) model being one of the most commonly used. However, these models have limitations in accurately predicting prognosis. Recently, serum ALP improvement or normalisation has emerged as a potential prognostic biomarker. Orthotopic liver transplant (OLT) is the only potentially curative treatment for end-stage PSC. OLT has high success rates, with 1-year and 5-year survival rates exceeding 90% and 80%, respectively.

The patient was referred to a hepatologist for further management of his liver disease and to discuss long-term management and liver transplant evaluation. The patient is currently undergoing regular monitoring with laboratory tests and imaging to monitor disease progression. He has also been advised to undergo regular colonoscopic surveillance.

Questions

1 What is backwash ileitis?
2 What is post-colectomy pouchitis?

Answers on p. 541.

Case Study 27   Shortness of Breath, Fatigue and Joint Pain

Clinical Presentation

A 45-year-old man presented to his GP with a 3-month history of shortness of breath, fatigue and joint pain. He reported feeling generally unwell and had lost weight without trying. He had a past medical history of hypertension and hyperlipidaemia, for which he was taking lisinopril and atorvastatin, respectively. He denied any recent travel or sick contacts. On physical examination, the patient appeared tired. Vital signs were stable. Lung auscultation revealed bilateral crackles in the upper lung fields. Cardiac examination was unremarkable. Joint examination revealed tenderness and swelling of the knees and ankles bilaterally. There was a rash on both legs.

Investigations

Laboratory results are provided in Table CS27.1. A chest X-ray revealed bilateral hilar lymphadenopathy and bilateral lung

Table CS27.1 Results of selected laboratory investigations.

Test	Result	Reference range
Angiotensin-converting enzyme (ACE)	**85 U/L**	10–70
Calcium	**2.81 mmol/L**	2.20–2.67
Erythrocyte sedimentation rate (ESR)	**35 mm/hour**	<15
C-reactive protein (CRP)	**24 mg/L**	<10
25-hydroxy Vitamin D	**20 nmol/L**	37–200

Values outside the reference range are in bold.

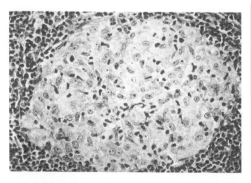

Figure CS27.1 Histopathology of sarcoidosis. The characteristic granuloma seen in sarcoidosis is non-caseating and consists of 'epithelioid' macrophages (pale cells in the centre) admixed with lymphocytes (darkly stained cells around the periphery). Multinucleate giant cell macrophages are not shown here, but may or may not be present. *Source:* Yale Rosen / Wikimedia Commons / CC BY-SA 2.0.

consolidations. A CT scan of the chest confirmed the presence of enlarged mediastinal and hilar lymph nodes, together with parenchymal nodular consolidations in the upper lobes bilaterally.

Diagnosis

The laboratory results showed that the patient had elevated ACE and CRP levels and a raised ESR. The patient was also hypercalcaemic with a low level of 25-hydroxy Vitamin D. Bronchoscopic transbronchial lung biopsy (TBLB) and endobronchial ultrasound (EBUS) guided transbronchial needle aspiration (TBNA) of a hilar lymph node were taken. Histopathological examination of the biopsy showed the presence of non-caseating granulomas consistent with sarcoidosis (Fig. CS27.1). Asteroid bodies (Fig. CS27.2), Schaumann bodies and birefringent crystalline particles were also present. A Ziehl-Neelson stain for acid-fast bacilli was negative.

Diagnosis: Sarcoidosis.

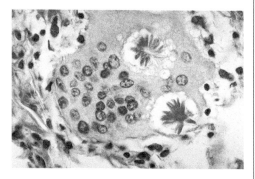

Figure CS27.2 Sarcoidosis – asteroid bodies. Asteroid bodies are stellate inclusions with numerous rays radiating from a central core that are present in the cytoplasm of giant cells. They may be seen in sarcoidosis and other granulomatous disorders. They are most frequently encountered in the giant cells of foreign body granulomas. Multiple asteroid bodies may be present in a single giant cell as seen in this image. *Source:* Yale Rosen / Wikimedia Commons / CC BY-SA 2.0.

Discussion

Sarcoidosis is a systemic granulomatous disease that can affect multiple organs. It commonly affects young and middle-aged adults, and frequently presents with bilateral hilar lymphadenopathy, pulmonary infiltration, and ocular and skin lesions. Any organ in the body can be involved. The exact cause of sarcoidosis is not known, but it is thought to result from an exaggerated immune response to one or more of a number of environmental or infectious agents in genetically susceptible individuals. Potential aetiological agents include mycobacteria and *Cutibacterium acnes* (previously known as *Propionibacterium acnes*). T-helper cells, particularly Th1 and Th17 cells, are thought to play a central role in the pathogenesis. These T-cells release cytokines that activate macrophages to form granulomas, the hallmark of the disease. Granulomas are clusters of immune cells, predominantly macrophages and lymphocytes, that form in response to chronic inflammation, causing tissue damage and fibrosis. The macrophages take on an epithelial morphology and are referred to as 'epithelioid' macrophages. Giant cell macrophages, which contain multiple nuclei, may also be present. The pathogenesis of sarcoidosis remains an active area of research, and further studies are needed to fully elucidate the underlying mechanisms of the disease.

ACE (angiotensin-converting enzyme) is a protein produced by cells in the lungs, blood vessels and other organs. ACE converts angiotensin I to angiotensin II, a potent vasoconstrictor that can increase blood pressure. ACE is also involved in the breakdown of bradykinin, a vasodilator that can lower blood pressure. In sarcoidosis, the granulomatous inflammation can stimulate the production of ACE by macrophages. Elevated ACE levels can be a marker of disease activity in sarcoidosis and may be used to monitor response to treatment. However, it is important to note that an elevated ACE level is not specific for sarcoidosis and can also be observed, for example, in tuberculosis. The patient's serum calcium level was also elevated, which in sarcoidosis can be caused by the granulomas, which produce vitamin D. This phenomenon is known as *granulomatous hypercalcaemia*. Granulomas produce excess vitamin D through the activity of 1-alpha-hydroxylase, which is present in macrophages and activated by inflammatory cytokines. 1-alpha-hydroxylase converts 25-hydroxy vitamin D, the inactive form of vitamin D, into its active form, 1,25-dihydroxy vitamin D. The excess 1,25-dihydroxy vitamin D stimulates the absorption of calcium from the gastrointestinal tract and the release of calcium from bones, leading to hypercalcaemia. The most common cause of granulomatous hypercalcaemia is sarcoidosis, but it is also found in other granulomatous diseases, such as tuberculosis, histoplasmosis or coccidioidomycosis.

Treatment and Prognosis

The patient was started on daily prednisone and referred to a pulmonologist for further management of his sarcoidosis. The treatment of sarcoidosis depends on the severity of the disease and the organs affected. In many cases, sarcoidosis resolves spontaneously and does not require treatment. However, in patients with persistent symptoms, treatment is aimed at suppressing the inflammation and preventing organ damage. Corticosteroids, such as prednisone, are often the first-line treatment and work by suppressing the immune system and reducing inflammation. Other immunosuppressive drugs (e.g. methotrexate, azathioprine and mycophenolate mofetil) and biologics (infliximab, adalimumab and rituximab) may also be used.

Questions

1 What is EBUS-TBNA?
2 What are Asteroid bodies and Schaumann bodies?

Answers on p. 541.

Case Study 28   Severe Abdominal Pain at the Site of a Recent Surgical Wound

Clinical Presentation

A 45-year-old man with a longstanding history of Crohn's disease affecting both the terminal ileum and colon underwent elective modified two-stage (Stage I: sub-total colectomy plus end ileostomy, Stage II: completion proctectomy plus J-Pouch Ileo-anal anastomosis without diversion loop ileostomy) proctocolectomy for medically refractory disease. The patient was being treated with steroids, infliximab (an antibody against TNFα) and azathioprine (an immunosuppressive drug that acts to inhibit purine synthesis and, in turn, the synthesis of white blood cells). Surgery was uneventful and initial postoperative recovery unremarkable. However, 5 days post-surgery the patient experienced severe abdominal pain at the site of the wound.

Investigations

On examination, the patient's abdominal contents were found to have partially extruded through the surgical wound. The wound was covered in saline-soaked gauze and the patient was given analgesics and started on broad-spectrum intravenous antibiotics. Prior to antibiotic therapy, swabs of the wound were sent for microbiology, but no evidence of infection was found. An urgent return to theatre was arranged for repair of the wound.

Diagnosis

Wound breakdown (*dehiscence*) typically occurs 5–8 days after surgery and is defined as a partial or total separation of the edges of a wound. It is a failure of proper wound healing. The incidence of wound dehiscence following abdominal surgery is between 0.4 and 3.5%. *Evisceration*, in which intra-abdominal organs herniate through the open wound, is a complication of complete wound dehiscence and requires immediate treatment.

Diagnosis: Wound dehiscence.

Discussion

Inflammatory bowel disease (IBD) is a group of chronic disorders of the gastrointestinal tract. They include Crohn's disease (see also Case Study 22), ulcerative colitis and IBD-unclassified (IBDU). Whereas the inflammation in ulcerative colitis is limited to the colonic/rectal mucosa, in Crohn's disease, transmural inflammation can affect both the small and large bowel. Non-caseating granulomas composed of 'epithelioid' macrophages are characteristic of Crohn's disease.

Crohn's disease has a familial basis; for example, siblings of affected individuals are 30 times more likely to develop the disease than the general population. Classical Crohn's disease is an example of a multifactorial disease in which multiple genetic (*polygenic*) and environmental exposures synergise to cause the disease. The contribution of individual genes in polygenic IBD is relatively small, in other words, each has low penetrance (Chapter 10). Over 30 genes have been shown to be associated with the development of Crohn's disease. They include LRRK2, XBP1 and the nucleotide-binding oligomerization domain-2 gene (NOD2) NOD2 encodes a so-called *pattern recognition receptor* (PRR; see also Chapter 6) that senses conserved fragments present in the cell wall of different types of bacteria and in response activates intracellular signalling pathways, for example NF-kB, to induce pro-inflammatory and anti-microbial processes. Mutations in NOD2 result in diminished phagocytosis of bacteria by macrophages.

Advances in genomics have identified rare monogenic disorders exhibiting IBD and IBD-like inflammation (so-called *monogenic IBD*) that are different from the more common polygenic form of IBD. Genes affected in monogenic IBD include IL10RA/B, XIAP, CYBB, LRBA and TTC7A. Most cases of monogenic IBD occur in children <6 years of age.

Monogenic IBD is classified into at least five groups on the basis of mechanism: (i) defects in the epithelial barrier, (ii) T-cell or B-cell defects, (iii) hyperinflammatory and autoinflammatory defects, (iv) phagocytic defects and (v) immunoregulatory (including IL-10) signalling defects.

Wound healing is a complex process involving a series of overlapping processes which begin with haemostasis, which is followed by the influx of inflammatory cells, the proliferation of cells and extracellular matrix to fill the wound and finally tissue remodelling. There are several underlying factors that can impair wound healing and therefore make wound dehiscence more likely. They include ischaemia, infection, diabetes, obesity or malnutrition, smoking and immunosuppression. It is likely that this patient's immunity was compromised by the immunosuppressive drugs (particularly the steroids) he was taking to treat his IBD.

Treatment and Prognosis

Prompt identification is important to prevent worsening dehiscence, limiting infection, and reducing the risk of other complications, such as sepsis (Case Study 12 and 13), or necrotising fasciitis (Case Study 10). Wound dehiscence after abdominal surgery is associated with mortality rates between 10% and 44%. In this case, the abdominal wound was successfully repaired, and the patient made a full recovery.

Questions

1   The presence of non-caseating epithelioid granulomas can be helpful in making a diagnosis of Crohn's disease. What are non-caseating epithelioid granulomas?
2   What is the difference between healing by first intention and second intention?

Answers on p. 542.

Case Study 29   Neutropenia and Recurrent Infection in a Patient with Rheumatoid Arthritis

Clinical Presentation

A 67-year-old woman was admitted to hospital with a severe chest infection having presented with breathlessness, a productive cough and mild confusion. She was diagnosed with rheumatoid arthritis (RA) 25 years ago and was under the care of a hospital rheumatology department. Despite her condition, she was leading an active life. She admitted to a chest infection 6 months earlier, though this episode did not require hospital admission. However, at that time, she was noted to be neutropenic.

Investigations

The results of laboratory investigations are shown in Table CS29.1. The patient was slightly anaemic and her mean corpuscular volume (MCV) was lower than normal. Although the total blood cell white count was

within the normal range, the differential count was abnormal; with a profound reduction in neutrophils and an increase in lymphocytes. On blood film examination, many cells had a large granular lymphocyte morphology.

Peripheral blood cells were also analysed by fluorescent-activated cell sorting (FACS). The results showed that around 85% of the cells were CD8+ T-lymphocytes and positive for CD57 (a marker of large granular lymphocytes). There was also an inverted CD4:CD8 ratio.

Diagnosis

The patient had a microcytic anaemia, that is the red blood cells are reduced in size (low MCV), which could indicate iron deficiency. Patients on non-steroidal anti-inflammatory drugs are at risk of losing iron through the gastrointestinal tract due to

Table CS29.1 Results of selected laboratory investigations.

Test	Result	Reference range
Blood		
Total white cell count	6.2×10^9/L	4.0–11.0
Red cell count	4.27×10^{12}/L	3.8–5.8
Haemoglobin	**10.6 g/dL**	11.5–16.5
Mean corpuscular volume	**74.4 fL**	82–92
Mean cell haemoglobin	**24.8 pg**	27–32
Platelets	343×10^9/L	150–400
Vitamin B12	260 ng/L	160–760
Serum folate	4.1 µg/L	3.0–20.0
Blood film		
Neutrophils	**9%**	40–75
Total lymphocytes	**85%***	20–45
Monocytes	3%	2–10
Eosinophils	2%	1–6
Basophils	1%	1

* Large granular lymphocytes comprised 70–80% of total lymphocytes (reference range 0–15%).
Values outside the reference range are in bold.

peptic ulceration. Alternatively, microcytosis is typical in anaemia associated with chronic inflammatory diseases such as RA, in which the anaemia does not respond to treatment with iron (Chapter 7). In the context of RA, a low neutrophil count in the absence of any other specific cause is likely to be Felty syndrome.

Diagnosis: Felty syndrome in a patient with a history of rheumatoid arthritis.

Discussion

Rheumatoid arthritis is a complex systemic autoimmune disease, of unknown aetiology, characterised by chronic inflammation of the joints, which in some cases, may lead to severe disability. The disease may start in the small joints, for example metacarpophalangeal (knuckle) joints of the hand and progress to the wrists and large joints, such as the knees and elbows. Typically, the inflammation of the joints manifests as erythema and swelling, which progressively gets worse; the patient suffering with pain, stiffness and restricted movement. Around 75% of RA patients are rheumatoid factor (RF) positive. The presence of RF is associated with a poorer prognosis and extra-articular manifestations, which may include subcutaneous nodules, cardiac involvement, vasculitis and Felty syndrome: in the latter case, patients tend to have high titres of RF. RFs are autoantibodies largely of the IgM isotype (but may also include IgG and IgA isotypes), which bind to the Fc region of IgG and cause immune complex formation.

The American College of Rheumatology (ACR)/European League Against Rheumatism (EULAR) classification criteria for RA aims to identify individuals with early RA. These criteria require the evaluation of four domains:

- *Joint involvement:* a score calculated based on the number of joints that are swollen (0–5 joints) and the number of joints that are tender (0–10 joints). Scoring is cumulative, with higher scores reflecting more joint involvement.
- *Serology:* scores based on the presence of RF and ACPA (anti-citrullinated protein antibodies)
- *Acute-phase reactants:* Considered elevated if the levels of C-reactive protein (CRP) or erythrocyte sedimentation rate (ESR) are above the upper limit of normal.
- *Duration of symptoms:* assessed and categorised as follows: (i) less than 6 weeks, (ii) 6 weeks to 3 months or (iii) more than 3 months.

A total score is calculated by summing the scores from each domain. Individuals are classified as RA if they score 6 or more.

Felty syndrome, often described as 'super rheumatoid' disease, is characterised by a triad of RA, splenomegaly and neutropenia, resulting in susceptibility to bacterial infections, though the presence of splenomegaly is not essential to make the diagnosis. In some cases, Felty syndrome may develop during a period when the symptoms of RA have subsided or are not present. In such cases, the presence of Felty syndrome may go unnoticed. On the other hand, in more rare instances, Felty syndrome may precede the development of RA. The exact relationship between Felty syndrome and RA is not fully understood, but it is believed that both conditions share underlying autoimmune mechanisms. However, the specific factors that lead to the development of Felty syndrome in some individuals, while others only experience RA, remain unclear.

Treatment and Prognosis

If the neutrophil count is not dramatically reduced and the patient is not suffering from infections, then there is no absolute requirement for treatment. However, if there is a profoundly low neutrophil count and/or recurring infections, then it is sensible to treat to prevent a life-threatening infection.

Disease-modifying anti-rheumatic drugs (DMARDs), such as methotrexate, sulfasalazine and leflunomide, are commonly used to treat the underlying RA in Felty syndrome. These medications reduce joint inflammation and slow the progression of joint damage. In cases of severe or refractory Felty syndrome, biological agents, like TNF-α inhibitors (e.g. etanercept and adalimumab) may be used in combination with DMARDs. Biologics target specific immune molecules involved in the inflammatory process, providing additional control over the disease. Short-term use of corticosteroids, such as prednisone, may be prescribed during flare-ups to reduce inflammation and alleviate symptoms. If the enlarged spleen causes significant complications, such as severe neutropenia or anaemia, surgical removal of the spleen (splenectomy) may be considered.

Questions

1 What is the cause of the neutropenia in patients with Felty syndrome?
2 What links Felty syndrome, large granular lymphocyte syndrome and neutropenia?

Answers on p. 542.

Case Study 30   Breathlessness in a Young Man

Clinical Presentation

A 23-year-old schoolteacher presented to his GP complaining of a 3-week history of shortness of breath, which was particularly troublesome after exercise. He was an occasional smoker and had a history of mild eczema and hay fever since childhood but had never suffered any serious illness. On examination, his doctor noticed mild conjunctivitis and small areas of chronic eczema on the back of his hands. Examination of his chest was unremarkable, except for a few scattered inspiratory rhonchi (wheezes) heard throughout his lung fields. Given these findings, his GP diagnosed asthma and prescribed a salbutamol inhaler with the advice that he should take two puffs whenever he felt short of breath. Two weeks later, after playing in a school cricket match, the teacher became breathless, despite taking puffs of the inhaler. He became increasingly wheezy and his chest 'felt tight'. A concerned colleague drove him to hospital. On admission to the Accident and Emergency Department he was barely able to speak because of his breathlessness. His pulse rate was 120 b.p.m. and on examination of his chest, the doctor commented that it sounded almost 'silent'. He was given nebulized salbutamol and ipratropium bromide and an intravenous dose of corticosteroids. He was admitted for observation and prescribed a 5-day course of oral steroids.

Investigations

A chest radiograph revealed hyperinflated lung fields but no areas of consolidation. Blood tests were within the normal reference range.

Diagnosis

The patient presented with acute severe asthma, which is a life-threatening condition. He required emergency treatment and admission to the hospital where he remained for several days. The important questions are: why did this happen to a previously healthy individual and could it have been prevented?

Diagnosis: Acute severe asthma.

Discussion

Although this patient suggested to his doctor that his breathlessness was a new symptom, it should have been tied in with the background history and clinical features, which clearly define an atopic tendency. The association with exercise was noted, but a more detailed history would have elicited that exercise in winter months was not associated with symptoms and therefore this was more likely to be associated with the patient's

recognised pollen allergy (hay fever). Had the seasonal nature of the symptoms been recognised (the cricket match took place in the English summer), a more appropriate therapeutic strategy might have been instituted, which would have prevented the need for hospital admission.

Adequate history taking is the key to identifying likely allergens in the atopic individual. In addition, the appropriate use of further investigations can be helpful. In this case, the patient was followed up at an immunology clinic and had further tests. Total serum immunoglobulin E (IgE) was raised at 544 U/L (normal reference range: 1.5–120 U/L) and pulmonary function

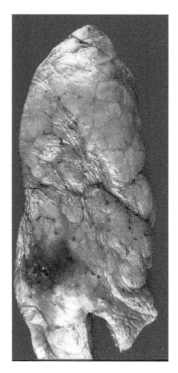

Figure CS30.1 Hyperinflated lung: Image shows a lung from a patient who died in status asthmaticus. The lung is hyperinflated.
Source: Yale Rosen / Wikimedia Commons / CC BY-SA 2.0.

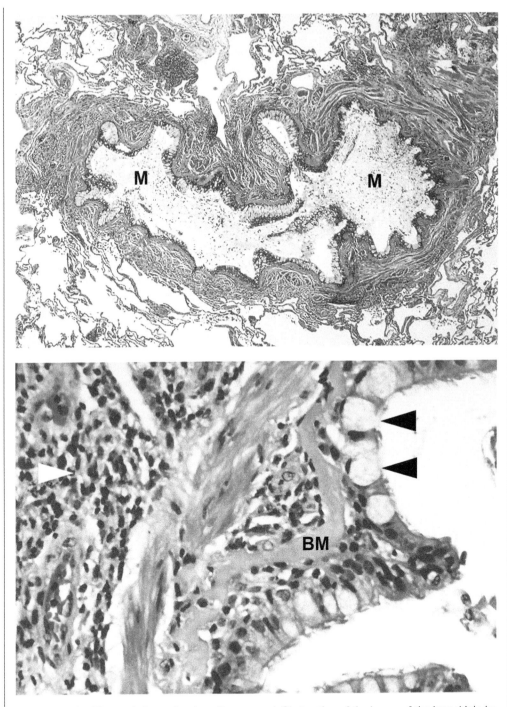

Figure CS30.2   Histopathology of asthma. Upper panel: Obstruction of the lumen of the bronchiole by mucoid exudate (M). Lower panel: Goblet cell metaplasia/hyperplasia (black arrows), epithelial basement membrane thickening (BM) and severe inflammation of a bronchiole (indicated by the presence of an infiltrate of inflammatory cells, white arrow). *Source:* Yale Rosen / Wikimedia Commons / CC BY-SA 2.0.

testing showed a moderate obstructive pattern, which reversed with β_2-agonists. The consultant who saw him performed skin prick testing, which demonstrated strong reactions to grass and tree pollens. Testing for allergen-specific IgE was not performed as it was not felt that this would add any useful information. Blood tests for allergen-specific IgE are only indicated when skin testing is not possible due to the presence of severe eczema, the need for antihistamine therapy, which cannot be withdrawn, or occasionally in young children.

Asthma is a type I hypersensitivity reaction. The mechanistic basis of type I hypersensitivity reactions is outlined in Chapter 6. When exposed to specific stimuli, in response, the bronchi undergo bronchospasm causing the airways to narrow, making it difficult to breathe. Bronchospasm is caused by contraction of the smooth muscles in the bronchial walls. Inflammation leads to further narrowing of the airways.

A hyperinflated lung (Fig. CS30.1) occurs when there is an increased volume of air in the lung. This can occur when the lungs are not able to fully exhale and can lead to changes in lung function and chest shape. During an asthma attack, the muscles around the airways contract, leading to their narrowing (bronchoconstriction). This narrowing makes it difficult for air to flow freely out of the lungs, leading to air being trapped within the alveoli. In response to the inflammation, the airways produce excess mucus. This thick mucus can further obstruct airflow. The inflamed airways also become swollen and sensitive. This inflammation can also lead to a loss of elasticity in the lung tissue.

The histopathology of asthma (Fig. CS30.2) is characterised by chronic inflammation of the airway walls, with infiltration of immune cells, such as eosinophils, lymphocytes and mast cells. There is also remodelling of the airway structures, including increased smooth muscle mass, thickened basement membrane, and hyperplasia and/or metaplasia of mucus-producing (goblet) cells; the inflammation stimulates epidermal growth factor receptor activation and IL-13 to induce both Clara cells (also known as 'club' cells, which are non-ciliated, secretory cells) and ciliated cells to transition into goblet cells. The excess mucus produced by the goblet cells contributes to airway narrowing and obstruction.

Treatment and Prognosis

Preventative therapy was instituted in this case. Inhaled disodium cromoglycate was prescribed as a mast cell stabiliser. This drug must be taken on a daily basis and acts by preventing the immediate release of mast cell mediators in response to allergen. In addition, the patient continued to carry his salbutamol inhaler and to take it prior to participating in any outdoor activities. This simple approach was adequate in controlling the teacher's symptoms, but in more refractory cases, regular long-acting β_2-agonists would be required with the possible addition of inhaled corticosteroids, usually as a combined inhaler.

Questions

1 What is status asthmaticus?
2 What is the reason for using corticosteroids and how do they work?

Answers on p. 542.

Case Study 31 Protein in the Urine and High Blood Pressure

Clinical Presentation

A 45-year-old man was found to have protein in his urine on dipstick testing at an employment medical. He was also hypertensive (blood pressure: 180/100). On questioning, he admitted to having had some swelling of his ankles at the end of the day for the last 6 months, but otherwise, there were no symptoms.

Investigations

The results of laboratory investigations are shown in Table CS31.1. Anti-nuclear antibodies were not detectable. The kappa/lambda ratio of serum-free light chains was normal and there was no paraprotein band seen on serum electrophoresis. Hepatitis B, Hepatitis C and HIV serology screens were all

Table CS31.1 Results of laboratory investigations.

Test	Result	Reference range
Blood		
Albumin	**27 g/L**	34–48
Sodium	138 mmol/L	135–146
Potassium	3.8 mmol/L	3.5–5.0
Urea	**11.2 mmol/L**	2.5–6.7
Creatinine	**189 μmol/L**	59–104
Estimated glomerular filtration rate (eGFR)	**50 mL/ minute**	>90
Fasting glucose	4.6 mmol/L	4.5–5.6
Cholesterol	**9.2 mmol/L**	3.5–6.5
Urine		
24-h urinary protein	**6.3 g/24 hour**	<0.15

Values outside the reference range are in bold.

negative. Antibodies to phospholipase A2 receptor (PLA2R) were detected in the blood. Renal ultrasound scan showed normal size and appearance of both kidneys. A renal biopsy was performed. Histological analysis showed diffuse thickening of the glomerular basement membrane (periodic acid Schiff stain, Fig. CS31.1 left panel). Immunofluorescence analysis showed granular deposition of IgG along the glomerular capillary loops (Fig. CS31.1, middle panel). On electron microscopy, there were sub-epithelial electron-dense deposits within glomerular basement membranes (Fig. CS31.1, right panel). Silver stain revealed 'spikes', separated by 'holes', in the glomerular basement membrane (Fig. CS31.2). The 'spikes' refer to the irregular thickening of the glomerular basement membrane. The 'holes' or 'sub-epithelial deposits' are the spaces between these spikes where immune complexes are deposited. These immune complexes consist of antibodies and antigens and contribute to the inflammation and damage seen in membranous glomerulonephritis. The biopsy findings were consistent with membranous nephropathy.

Diagnosis

This patient has nephrotic syndrome as indicated by high levels of protein in the urine (nephrotic range proteinuria; 300mg/dL of urinary protein, which equates to >3g/24 hour), low levels of protein in the blood (hypoalbuminaemia) and swelling (peripheral oedema). Nephrotic syndrome can occur as a result of immune-mediated damage to the glomeruli, which can be caused by systemic diseases, or diseases affecting only the kidney (*glomerulonephritis*). Other forms of renal disease such as chronic pyelonephritis and chronic interstitial nephritis may lead to proteinuria. However, in

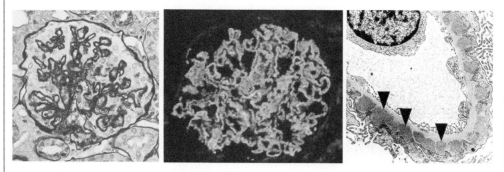

Figure CS31.1 Membranous glomerulonephritis. Left panel: Periodic acid-Schiff (PAS) staining highlighting the thickened glomerular capillary loops. Middle panel: Immunofluorescence showing granular deposition of IgG within the glomerular basement membranes. Right panel: Electron microscopy depicting sub-epithelial deposits (black arrowheads) within the glomerular capillary loop. *Source:* Courtesy of Prof. J. Charles Jennette.

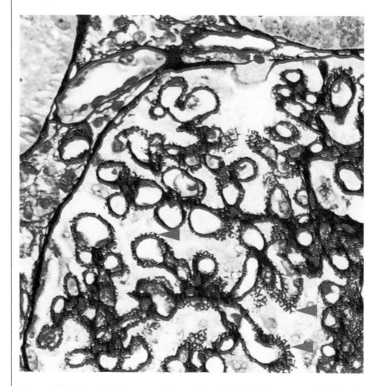

Figure CS31.2 Membranous glomerulonephritis. Higher magnification image of Jones Silver special stain highlighting the thickened glomerular capillary loops with 'spikes' and 'holes' (red arrowheads) along the glomerular capillary loops. *Source:* Courtesy of Prof. J. Charles Jennette.

tubulointerstitial disease, the degree of proteinuria will be much lower than what would be expected in a condition that directly targets glomeruli leading to glomerulonephritis. The absence of antinuclear antibodies in this case tends to rule out systemic lupus erythematosus as a cause of the renal disease. Diabetic nephropathy is unlikely due to a normal fasting glucose, and amyloidosis, a rare condition caused by excessive amyloid protein deposition in organs, is less likely as there was no detectable paraprotein in the blood. Positive PLA2R antibodies, along with the renal biopsy findings, are strongly supportive of a diagnosis of primary PLA2R-positive membranous nephropathy. Other types of glomerulonephritides causing proteinuria, including minimal change disease and focal segmental glomerulosclerosis, have no evidence of detectable antibodies in the blood and are diagnosed based on characteristic findings on renal biopsy.

Diagnosis: Membranous nephropathy.

Discussion

Membranous nephropathy is a common cause of nephrotic syndrome among adults. It is due to the deposition of antigen-antibody complexes (immune complexes) within the glomerular basement membranes. These immune complexes can be detected by immunofluorescence analysis or as electron-dense deposits on electron microscopy. There are two main types of membranous nephropathy. Primary membranous nephropathy, which is more common, and secondary membranous nephropathy, which is caused by another disease state or medication. About 70–80% of patients with primary membranous nephropathy have anti-phospholipase A2 receptor (PLA2R) antibodies which correlate with disease activity.

The deposition of immune complexes within glomeruli leads to glomerular inflammation, glomerular damage, including the deposition of collagen within glomeruli (glomerulosclerosis), and subsequent impairment of renal function. In membranous glomerulonephritis, the changes in the basement membrane manifest as the leakage of protein through the damaged basement membrane and into the urine. This patient has nephrotic range proteinuria sufficient to reduce the blood albumin levels below normal. The hypoalbuminaemia has in turn caused a reduction in the oncotic pressure of the blood, with leakage of fluid into the tissues resulting in the accumulation of fluid in the legs. The triad of proteinuria, hypoalbuminaemia and oedema is termed *nephrotic syndrome*. The patient showed evidence of impaired renal function with elevation of blood urea and creatinine and reduction in eGFR. The history of oedema for 6 months suggests the condition is not of recent origin. Hypertension has likely developed secondary to renal disease.

Hypercholesterolaemia can occur because of nephrotic syndrome. The liver is a major organ involved in the synthesis and regulation of cholesterol and lipoproteins. In nephrotic syndrome, the loss of proteins in the urine leads to a decrease in the oncotic pressure. As a compensatory mechanism, the liver responds by producing more proteins, including lipoproteins (such as very-low-density lipoprotein, VLDL) to maintain the balance of fluids between the blood and tissues. The balance between lipid metabolism and clearance is also disrupted in nephrotic syndrome. The loss of proteins in the urine affects the metabolism of lipoproteins, leading to an increase in the levels of low-density lipoprotein (LDL) cholesterol, often referred to as 'bad cholesterol' (Chapter 9).

Treatment and Prognosis

Supportive treatment is important in membranous nephropathy, including the use anti-hypertensive drugs when necessary (hypertension accelerates the decline of renal function). Cholesterol-lowering medication should be considered along with dietary salt and fluid restrictions and angiotensin blockade to reduce proteinuria. Diuretics can be administered either orally, or through intravenous injection, to promote the expulsion of water and decrease swelling. Anticoagulant treatment is indicated for patients with severe hypoalbuminaemia, due to the associated hypercoagulability and increased risk of venous thrombosis. This is, in part, a consequence of urinary losses of anti-coagulants such as antithrombin III, protein C and protein S.

Prognosis in membranous nephropathy is variable. Up to one-third of patients will enter spontaneous remission. This will often occur within 12 months of presentation and therefore, in the absence of severe manifestations, a 'watch and wait' approach is often employed. Patients who do not enter spontaneous remission have a variable risk of progressing to end-stage renal disease. The risk of progression depends on markers of severity, including renal function at presentation and severity of proteinuria and nephrotic syndrome. The decision to proceed to immunosuppressive treatment for membranous nephropathy is based on whether patients enter spontaneous remission, as well as risk factors, such as renal function, severity of proteinuria, severity of hypoalbuminaemia and PLA2R antibody titres. Patients at low risk may be managed with conservative measures without immunosuppressive treatment in the first instance. Patients at higher risk will usually receive immunosuppression with rituximab (a B-cell depleting agent that targets CD20), cyclophosphamide (a T-cell depleting agent) in combination with glucocorticoids, or a calcineurin inhibitor (T-cell targeting agent).

Questions

1 Hypercholesterolaemia is frequently encountered in patients with nephrotic syndrome. Why is it important to administer lipid-lowering drugs to this group of patients?
2 Why is serum creatinine a useful marker of renal function? Are there more accurate methods for the measurement of renal function?

Answers on p. 543.

Case Study 32 Persistent Coughing Up of Blood and Ankle Swelling

Clinical Presentation

A 24-year-old man with a 3-week history of persistent haemoptysis (the coughing up of blood) was admitted via casualty after coughing up obvious blood. He had passed only a small volume of blood-stained urine over the last 48 hours. On examination, the patient was hypertensive (blood pressure: 160/90 mmHg). There was some swelling of the ankles and lower legs, which he had only noted over the previous 2 days.

Investigations

The clinical features and laboratory results (Table CS32.1) established two immediate clinical problems:

- Haemoptysis.
- Significant impairment of renal function, indicated by the reduced urine output, raised blood urea and creatinine, and reduced estimated glomerular filtration rate.

Table CS32.1 Results of laboratory investigations.

Test	Result	Reference range
Blood		
Urea	**19.2 mmol/L**	2.5–6.7
Creatinine	**550 µmol/L**	59–104
Estimated glomerular filtration Rate (eGFR)	**12 mL/minute**	>90
Sodium	**148 mmol/L**	135–146
Potassium	**6.5 mmol/L**	3.5–5.0
Haemoglobin	**10.2 g/dL**	13.0–18.0
Albumin	**30 g/L**	34–48
Urine		
Protein (dipstick)	**2+***	0–trace
Blood (dipstick)	**2+**	0

* Indicates significant protein loss through the urine. Values outside the reference range are in bold.

A renal biopsy was performed. Histological analysis revealed a crescentic glomerulonephritis consistent with the clinical information, suggestive of a rapidly progressive glomerulonephritis. Crescentic glomerulonephritis is characterised by breakdown of the glomerular basement membrane and infiltration of neutrophils and macrophages in the extraglomerular space in Bowman capsule, which forms 'crescent' shapes, leading to the observed histological pattern (Figs. CS32.1 and CS32.2). Immunofluorescence analysis of the tissue showed the presence of IgG distributed in a uniform linear pattern along the glomerular basement membrane (Fig. CS32.3). A chest radiograph revealed diffuse interstitial and alveolar infiltrates.

Diagnosis

The presence of haemoptysis and glomerulonephritis is strongly suggestive of pulmonary-renal syndrome. This presents with a combination of pulmonary haemorrhage and glomerulonephritis as a

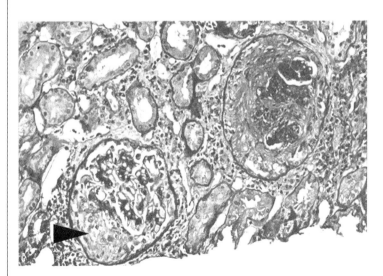

Figure CS32.1 Crescentic glomerulonephritis. Photomicrograph of a renal biopsy stained with periodic acid Schiff (PAS) showing crescentic glomerulonephritis (black arrowhead). *Source:* Nephron / Wikimedia Commons / CC BY-SA 3.0.

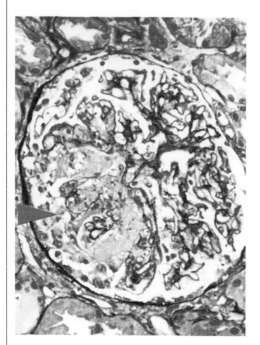

Figure CS32.2 Crescent formation. High magnification highlighting breakdown of the glomerular basement membrane and proliferation of inflammatory cells forming a crescent in the Bowman capsule (red arrowhead). *Source:* Courtesy of Prof. J. Charles Jennette.

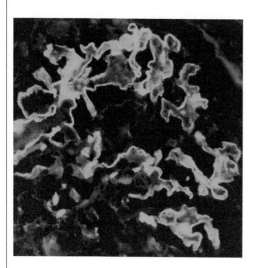

Figure CS32.3 Typical immunofluorescence picture of anti-GBM disease. Image shows bright linear IgG staining. *Source:* Courtesy of Prof. J. Charles Jennette.

manifestation of an underlying systemic autoimmune process. Pulmonary syndrome is usually due either to ANCA-associated vasculitis or anti-glomerular basement membrane (anti-GBM disease; also known as Goodpasture syndrome). Anti-neutrophil cytoplasmic antibodies (ANCA) were undetectable in this case, making ANCA associated vasculitis less likely. The detection of uniform and linear deposits of IgG within the glomerular basement membrane is a feature of anti-GBM disease. Anti-GBM disease is also characterised by the presence of anti-GBM auto-antibodies in the blood, which target antigens expressed in the basement membranes of capillary blood vessels of the kidney and lung.

Diagnosis: Anti-GBM disease.

Discussion

Anti-GBM disease is an autoimmune disease in which autoantibodies are directed towards the α chain of type IV collagen present within basement membranes of the lung and kidney. The autoantibodies, which are predominantly IgG, may be visualised within these basement membranes by immunofluorescence on tissue sections and are also detected in the serum of affected individuals.

The autoantibodies evoke a type II hypersensitivity reaction (Chapter 6) that is responsible for the characteristic inflammation and tissue damage. The pathogenic B- and T-cell epitopes are typically located within the structure of type IV collagen and are usually not easily accessible to the immune system. The precise stimulus for the formation of these antibodies is unclear, but it is hypothesised that damage to the basement membrane (e.g. by toxins or mechanical damage) exposes these epitopes and leads to loss of tolerance in susceptible individuals.

The clinical course of anti-GBM disease involves renal and lung involvement. Approximately 60–80% of patients develop rapidly progressive renal dysfunction leading to acute kidney injury (creatinine levels can double within a few days) together with haemoproteinuria (evidence of blood and protein on the urine dipstick, indicative of a glomerulonephritis). 50% of patients also present with haemoptysis, a symptom due to underlying pulmonary haemorrhage (Fig. CS32.4) secondary to alveolar damage; this can be life-threatening. Some patients may only present with renal involvement. Haematuria may be seen in many other urinary tract diseases and is therefore not specific. Hyperkalaemia, due to inadequate renal excretion of potassium associated with acute kidney injury, may produce life-threatening cardiac dysrhythmias. Oedema, often seen in acute kidney injury, is the accumulation of excess fluid in the interstitial space, presenting in this case as swollen ankles. In acute kidney injury, oedema is usually the result of inappropriate sodium and water retention, which increases extra-cellular fluid volume. Less commonly, if urinary protein losses are great, there may be a significant decrease in plasma protein concentration and consequent reduction in oncotic pressure. This may contribute to the oedema via the resultant passage of water out of capillaries and into the interstitial space.

Treatment and Prognosis

Treatment of anti-GBM disease aims to remove the pathogenic autoantibodies from the blood, reduce inflammation and stop further production of these autoantibodies. Treatment is usually a combination of daily *plasmapheresis* (plasma exchange) until the level of circulating autoantibodies is within the normal range (i.e. completely removed), alongside immunosuppressive therapy with high-dose glucocorticoids and intravenous cyclophosphamide. Dialysis support may be required for patients with severe acute kidney injury. The prognosis for patients with anti-GBM disease is favourable if an early diagnosis is made with studies showing 1 year survival of 80–90%. Renal outcome correlates with presenting creatinine, the need for dialysis and the extent of glomeruli involved with crescentic glomerulonephritis on renal biopsy at presentation. Patients requiring dialysis at presentation have a very low chance of recovering renal function.

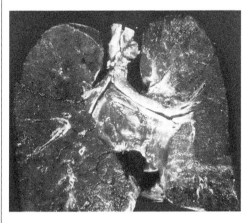

Figure CS32.4 Anti-GBM disease: Shows diffuse bilateral haemorrhage in both lungs. *Source:* Yale Rosen / Wikimedia Commons / CC BY-SA 2.0.

Questions

1 What are the likely causes of anaemia in this patient
2 The patient was hypertensive. How do you explain this?

Answers on p. 543.

Case Study 33   Diarrhoea, Jaundice and Rash

Clinical Presentation

A 27-year-old woman presented with diarrhoea and a rash 3 weeks after receiving an allogeneic hematopoietic stem cell transplant (HSCT). She was afebrile with a temperature of 37.2°C. Her heart rate was normal at 89 beats per minute and her blood pressure was normal at 115/85. On examination, she had a pink maculopapular rash covering approximately 40% of her body surface area and she was mildly jaundiced.

Investigations

The results of laboratory investigations are shown in Table CS33.1. The haemoglobin level was low, consistent with mild anaemia, but the total and differential white cell counts, and the platelet count, were increasing demonstrating good post-transplant haematological recovery. The bilirubin and alkaline phosphatase were moderately elevated. Her stool volume was measured at 0.5–1.0 litres per day.

Diagnosis

The clinical picture was consistent with a diagnosis of acute graft versus host disease (GvHD).

Diagnosis: Acute graft versus host disease.

Table CS33.1   Results of laboratory investigations.

Test	Result	Reference range
White cell count	4×10^9/L	4.0–11.0
Red cell count	**3.1×10^{12}/L (but increasing)**	3.8–5.8
Haemoglobin	**10 g/dL (but increasing)**	11.5–16.5
Haematocrit	0.37 L/L	0.37–0.47
Mean corpuscular volume	85 fL	82–92
Mean cell haemoglobin	28 pg	27–32
Platelet count	150×10^9/L	150–400
Urea	5.0 mmol/L	2.5–6.7
Urate	0.3 mmol/L	0.18–0.42
Bilirubin	**2.8 mg/dL**	<1.2
Alanine transaminase (ALT)	**52 U/L**	5–40
Aspartate transaminase (AST)	**43 U/L**	10–40
Alkaline phosphatase (ALP)	**118 U/L**	25–115
Blood film		
White cell differential		
Neutrophils	65%	20–75
Lymphocytes	24%	20–45
Monocytes	5%	2–10
Eosinophils	5%	1–6
Basophils	1%	1

Values outside the reference range are in bold.

Discussion

GvHD (see Chapter 6) is a relatively common complication of allogeneic HSCT. It occurs when donor-derived immune responses target the recipient's organs. Acute GvHD manifests in the first few weeks post-transplant and involves the skin, gut or liver. Chronic GvHD develops later after transplant and involves additional anatomical sites. The severity of acute GvHD is assessed by clinical stage and grade (Table CS33.2). In this case, the extent of the maculopapular skin rash, the degree of jaundice and the volume of liquid stools per day indicated Grade II acute GvHD.

Risk factors for acute GvHD include HLA antigen-mismatching between patients and donors, older patient age, use of female donors for male recipients, use of donors who have been allo-immunised due to previous blood transfusions or pregnancy and the type of GvHD prophylaxis used during the transplant process.

The underlying pathogenesis of GvHD is a type IV hypersensitivity reaction (Chapter 6), in which immunologically competent T-lymphocytes derived from the donor react against antigens expressed in recipient tissues (Fig. CS33.1). The pathophysiology of acute GvHD is conventionally described in three phases:

- *Initiation phase:* Prior to transplantation, recipients receive conditioning regimens involving chemotherapy and/or total body irradiation. This causes tissue damage, especially in the gastrointestinal tract, skin and liver. Damage to cells releases damage-associated molecular patterns (DAMPs). Bacteria, for example from the gut, can enter tissues because of a breakdown in the intestinal barrier, releasing pathogen-associated molecular patterns (PAMPs). These molecules are recognised by receptors, for example, Toll-like receptors (TLRs) and NOD-like receptors (Chapter 6), which then activate various

Table CS33.2 Clinical stage and grade in acute graft versus host disease.

Organ	Stage 0	Stage 1	Stage 2	Stage 3	Stage 4
Skin	No active rash	Maculopapular rash <25% BSA (Body Surface Area)	Maculopapular rash, 25–50% BSA	Maculopapular rash >50% BSA	Generalised erythroderma >50%, plus skin bullae and desquamation
Liver (bilirubin)	<2 mg/dL or normal liver function tests	2–3 mg/dL	3.1–6 mg/dL	6.1–15 mg/dL	>15
Upper GI tract	No or only intermittent nausea, vomiting or anorexia	Persistent nausea, vomiting or anorexia	—	—	—
Lower GI tract (adult)	No diarrhoea or <500 mL/day or <3 episodes/day	500–1000 mL/day, or 3–4 episodes	1000–1500 mL/day, or 5–7 episodes	>1500 mL/day, or >7 episodes	Severe abdominal pain ± ileus, bloody stools

Grade 0: No stage 1–4; Grade I: Stage 1–2 skin, without liver, upper or lower GI tract; Grade II: Stage 3 rash and/or stage 1 liver, and/or stage 1 upper GI tract and/or stage 1 lower GI; Grade III: Stage 2–3 liver and/or stage 2–3 lower GI, with stage 0–3 skin and/or stage 0–1 upper GI; Grade IV: Stage 4 skin, liver or lower GI involvement with stage 0–I upper GI.

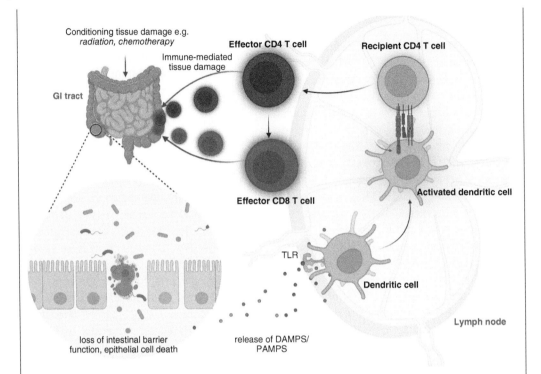

Figure CS33.1   The immune pathogenesis of acute GvHD. Prior to transplantation, patients undergo a conditioning regimen involving total-body irradiation and/or chemotherapy to eradicate the recipient's malignant cells and to create space in the bone marrow for the incoming donor cells. This regimen is toxic and can induce tissue damage, particularly in the intestines, leading to a compromised protective barrier function of the intestinal epithelium. Once this barrier is disrupted, bacteria that typically reside harmlessly in the intestine can move across the compromised barrier into the bloodstream and surrounding tissues. The ensuing tissue damage leads to the release of PAMPS and DAMPs, which are recognised by antigen-presenting cells (APC) such as dendritic cells. Donor-derived T cells recognise antigens presented by the APC leading to robust T cell activation. Once activated, these T cells proliferate and differentiate into effector CD4 T cells and activate donor CD8 T cells. As a result of this immune activation and attack on host tissues, patients can exhibit a range of symptoms depending on the organs affected. Common manifestations include skin rashes, diarrhoea and liver dysfunction. *Source:* Created with BioRender.com.

antigen-presenting cells (APCs), such as dendritic cells.

- *T-cell activation phase:* The activated APCs present host antigens to the incoming donor T-cells. When these T cells recognise disparities between their own and the host's MHC molecules, they become activated.
- *Effector phase:* Once activated, the donor T-cells proliferate and differentiate. Cytokines, such as TNF-α, IL-1 and IL-6, are released, which exacerbate the inflammation, resulting in tissue damage in target organs, particularly the skin, liver and gastrointestinal tract.

Treatment and Prognosis

Corticosteroids are used for first-line management of GvHD. For limited-stage cutaneous GvHD, corticosteroids are used topically. For more severe cases, including extensive cutaneous and/or gut and/or liver involvement, corticosteroids are given systemically. Immunosuppressive agents may be used, including calcineurin inhibitors (e.g. ciclosporin and tacrolimus), or mycophenolate mofetil (which inhibits de novo synthesis of purine nucleotides in immune cells). A proportion of cases do not respond to corticosteroids necessitating the use of other treatments, such as the oral JAK inhibitor, ruxolitinib, the anti-CD52 monoclonal antibody, alemtuzumab, anti-thymocyte globulin (ATG), the anti-TNF-α monoclonal antibody infliximab, or extracorporeal photopheresis (removing the patient's blood, exposing it to ultraviolet light and then returning it to the patient's body). Patients with severe GvHD that is refractory to corticosteroids have a high risk of mortality, often succumbing to infections that occur as a result of the profound secondary immunodeficiency resulting from immunosuppressive treatment.

Patients with severe acute GvHD are also more likely to develop chronic GvHD. In this case, the patient responded to systemic corticosteroids and ciclosporin, which were subsequently reduced and eventually stopped.

Questions

1 What is the effect of acute GvHD on immune reconstitution?
2 What are the possible late complications of allogeneic HSCT?

Answers on p. 543.

Case Study 34   Unexpected Death in a Young Man

Clinical Presentation

A 32-year-old man attended his local hospital following a referral from his general physician. He had complained of a retrosternal pain that appeared suddenly at 2 a.m. while he was attending a party. He had never been seriously ill and was active in sports. He worked as a representative of an industrial company and had no other significant medical history.

Investigations

Physical examination, including examination of the chest, was normal. Blood pressure was 130/80 mmHg, pulse rate 70/minute and regular, respiratory rate 12/minute. The electrocardiogram (ECG) was normal on admission. Laboratory investigations were normal. The patient continued to complain of chest discomfort and was kept on observation for suspicion of myocardial ischaemia or pericarditis, in spite of normal laboratory values. Later in the evening on the day of admission, he reported a sudden increase in pain, became short of breath and was severely

hypotensive. ECG showed *ventricular fibrillation* (a chaotic and disorganised rhythm of the ventricles) and *ventricular flutter* (a regular, rapid and organised contraction of the ventricles). Defibrillation was unsuccessful and the patient was pronounced dead at 11.35 p.m. A post-mortem examination showed the left ventricle of the heart was thicker than normal. The pericardial sac was filled with blood (Fig. CS34.1). There was a tear in the wall of the ascending aorta measuring 2.5 cm in length. The rest of the aorta was normal except for mildly atherosclerotic plaques and fatty streaks located in both the ascending and the descending aorta. There were no other abnormal findings. Histological examination of the aortic tissues showed focal disappearance of elastic fibres and accumulation of glycosaminoglycans characteristic of cystic medial necrosis.

Diagnosis

Diagnosis: Cystic medial necrosis of the aorta with aortic dissection and intrapericardial aortic rupture and cardiac tamponade.

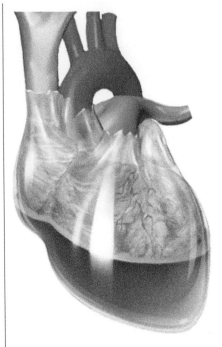

Figure CS34.1 Cardiac tamponade. This is a condition in which the space between the heart and the pericardium becomes filled with an abnormal accumulation of fluid, for example, blood, as in this case. This accumulation of fluid exerts pressure on the heart, compressing it and reducing its pumping capacity. *Source:* Blausen. com staff (2014) / WikiJournal of Medicine / licensed under CC BY 4.0.

Discussion

Aortic dissection (see also Chapter 9) is caused by a tear in the tunica intima (inner layer) of the aorta, which exposes the tunica media (middle layer) to blood from the lumen of the aorta. This blood is at high pressure and the medial layer splits into two (*dissection*), causing a false lumen along which blood can travel. The dissection may cause obstruction to branches of the aorta as it spreads along its length, which may in turn lead to cerebral or renal ischaemia. The false lumen may go on to rupture externally into the left pleural space or pericardium (as in this case) or internally into the true lumen of the aorta. The sudden increase in pressure on the heart as a result of the presence of blood in the pericardial sac often leads to heart failure (this type of heart failure is often referred to as cardiac tamponade).

Aortic dissection develops mostly in people aged between 40 and 60 years old with hypertension and results from the damaging effect of high blood pressure on the aortic wall. In younger patients, aortic rupture sometimes develops as a complication of Marfan syndrome (see also Case Study 50). This autosomal dominant condition is characterised by skeletal, ocular and cardiovascular manifestations and is caused by a defect in *fibrillin,* a connective tissue protein that participates in the formation of elastic fibres.

In the case patient, the histology of the aortic tissues was characteristic of a degeneration of the tunica media of the aorta, known as cystic medial necrosis. Cystic medial necrosis has been found in some cases of aortic rupture and aortic dissection, but its role in the pathogenesis of the lesion remains uncertain, since it can sometimes be seen in elderly patients as an incidental finding without aortic rupture. The patient also had an enlarged left ventricle, which is a sign of arterial hypertension.

Treatment and Prognosis

Aortic dissection and aortic rupture are serious conditions that frequently lead to sudden death due to massive bleeding. Lesions in the descending aorta can be corrected by surgery. Intrapericardial rupture of the ascending aorta is almost invariably lethal due to cardiac tamponade.

Questions

1   What is the definition of 'sudden death'?
2   What are the main causes of sudden death?

Answers on p. 544.

Case Study 35   A 2-day History of Severe Chest Pain in a Care Home Worker

Clinical Presentation

In September 2020, a 54-year-old male care home worker was admitted to hospital following a 2-day history of severe chest pain, which was made substantially worse when lying flat but was partially relieved when he sat up and leaned forward. He had a 30-pack-year smoking history. On examination, his temperature was 36.9°C, blood pressure 141/87 mmHg, heart rate 108 beats/min, respiratory rate 16 breaths per minute and oxygen saturation 98.5%. Recently, several of the residents of the care home in which he worked had tested positive for SARS-CoV-2. There was a detectable pericardial 'friction rub' (which sounded like 'sandpaper rubbed on wood') produced by friction of the heart against the pericardium, a finding that is typically seen in pericarditis; the intensity of the rub varying with the phase of the cardiac cycle rather than the respiratory cycle. ECG showed ST segment elevation in most leads with PR interval depression.

Investigations

Given the patient's exposure history, a swab was sent for a PCR test for SARS-CoV-2. In the meantime, laboratory investigations revealed a normal blood work-up. Transthoracic echocardiogram (TTE) showed normal systolic and diastolic function and normal valves. Cardiac MRI revealed a moderately thickened pericardium, bright signal on T2-weighted images and gadolinium uptake in the pericardium, suggesting pericardial inflammation. There was no evidence of a pericardial effusion.

Diagnosis

The PCR test for SARS CoV-2 RNA was positive. This patient likely has cardiac disease that may or not be related to coronavirus infection. Pericarditis is the most likely diagnosis. In this case, three of the four diagnostic signs of pericarditis were present: (i) sharp pleuritic chest pain improved by sitting up and leaning forward; (ii) friction rub; (iii) widespread ST elevation with PR depression on ECG. The fourth, pericardial effusion, was not observed in this case (if there was a pericardial effusion, there would usually be no pericardial rub, since the rub is generated from the friction of the two layers of pericardium).

Diagnosis: Acute pericarditis associated with SARS-CoV-2 infection.

Discussion

Pericarditis can be classified based on when symptoms occur. If a person experiences pericarditis symptoms for the first time, it is called *acute pericarditis*. Acute pericarditis is of sudden onset and is only rarely a cause of hospital admissions for chest pain. If the symptoms last for more than 4–6 weeks but less than 3 months, it is known as *incessant pericarditis*. *Chronic pericarditis* is when the symptoms last for more than 3 months. *Recurrent pericarditis* happens when a person experiences acute pericarditis symptoms again, at least 4–6 weeks after the symptoms of a previous episode have resolved. These recurrences can occur months or even years after the initial episode.

Acute pericarditis can be caused by different factors (Table CS35.1). Viruses are a common cause of pericarditis in resource-rich countries. Moreover, evidence has emerged recently linking SARS-CoV-2 infection with the development of pericarditis, though it is not clear if the virus infects the pericardium directly, or if the effects are indirect. A significant number (40–85%) of cases of acute pericarditis are of unknown aetiology. Establishing an underlying cause is often not possible because of the risks associated with obtaining diagnostic pericardial material. In most

Table CS35.1 Major groups of aetiological agents responsible for pericarditis.

Aetiological factor	Examples
Viral infections	Enteroviruses (e.g. Coxsackievirus B and Echoviruses), herpesviruses (e.g. Epstein-Barr virus and Cytomegalovirus), adenoviruses (especially in children), SARS-CoV-2
Bacterial infections	*Mycobacterium tuberculosis, Coxiella burnetii, Borrelia burgdorferi*
Autoimmune disorders	Systemic lupus erythematosus, Sjögren syndrome, rheumatoid arthritis, scleroderma
Vasculitides	Eosinophilic granulomatosis with polyangiitis, Takayasu disease
Other inflammatory	Sarcoidosis
Neoplastic	Secondary metastatic tumours, e.g. from lung, breast cancer, primary tumours – very rare
Metabolic	Uraemia
Pericardial/ myocardial injury	Early onset: trauma, transmural MI; delayed onset: Dressler syndrome, post-pericardiotomy syndrome

patients, acute pericarditis is mild and resolves spontaneously, though treatment with a nonsteroidal anti-inflammatory drug or a short course of a corticosteroid may be helpful. When a large pericardial effusion is produced, cardiac function may be compromised and *cardiac tamponade* can occur (Fig. CS35.1). Cardiac tamponade refers to the excessive accumulation of fluid, blood or other material in the pericardial sac surrounding the heart. *Beck's triad* is a classic set of three signs that indicate cardiac tamponade: (i) hypotension owing to increased external pressure on the heart, which reduces

the amount of diastolic flow back to the heart and so reduces the amount of blood available to pump out; (ii) jugular venous distention (JVD) – bulging of the jugular veins in the neck, which occurs due to the increased pressure in the superior vena cava caused by the fluid build-up. (iii) Muffled or distant heart sounds, due to the accumulation of fluid in the pericardial sac, which dampens the sound of the heart beating, making it difficult to hear clearly with a stethoscope.

Pericardial inflammation can also occur after a myocardial infarction and is classified into two types based on the timing and underlying mechanism of the inflammation. The first type is called *peri-infarction pericarditis* and is the most common form, occurring within the first few days after a large transmural myocardial infarction. This type of pericarditis is thought to be caused by an inflammatory response triggered by the injury to the heart muscle. Some studies report the incidence of pericarditis in transmural infarctions to be around 20–30%. The second type is called *post-myocardial infarction syndrome*, also known as *Dressler syndrome*, which typically occurs a few weeks after the myocardial infarction. This type of pericarditis is caused by an immune-mediated response to the damaged heart tissue and is very rare, with some estimates as low as 0.1–0.5%.

Treatment and Prognosis

Acute pericarditis is usually a self-limiting condition. However, it can recur in around one-third of patients. The patient was administered ibuprofen, a non-steroidal anti-inflammatory drug (NSAID), to reduce inflammation and ease the pain. Prescription-strength pain relievers may be used if the pain is severe. Oral colchicine was also prescribed 2 times per day for 2 weeks. Colchicine is a microtubule inhibitor and is useful in preventing relapse in pericarditis after the first presentation. Patients who respond poorly to

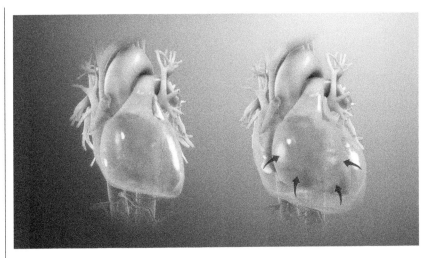

Figure CS35.1 Pericarditis causing pericardial effusion. Pericarditis can progress to pericardial effusion and eventually cardiac tamponade as shown on the right. A normal heart is shown on the left. *Source:* Scientific Animations / Wikimedia Commons / CC BY-SA 4.0.

these drugs may be treated with a short course of corticosteroids. If the pericarditis leads to significant pericardial effusion, then pericardiocentesis can be performed in which a sterile needle or a small catheter is used to drain the excess fluid from the pericardial cavity. Removal of the pericardium (pericardiectomy) may be necessary if the heart is permanently rigid due to constrictive pericarditis; in these cases, the pericardium heals with fibrosis and becomes permanently stiff. Most patients recover in 2–4 weeks with supportive care. This patient was discharged after 5 days and made a full recovery.

Questions

1 Dressler syndrome, also known as post-myocardial infarction syndrome, is a form of secondary pericarditis resulting from immune injury to the heart. The incidence of Dressler syndrome has decreased in recent years; why is this?

2 Pericarditis is reported as a complication of mRNA vaccines against SARS-CoV-2, especially in men under 40 years of age. Should men under 40 years of age be worried?

Answers on p. 544.

Case Study 36 Breathlessness, Cough and Dizziness after Mild Exercise

Clinical Presentation

A 69-year-old clinically obese male non-smoker, who was a known case of uncontrolled diabetes, presented to the Emergency Room after an episode of fainting. On taking the history, he was noted to have experienced being unusually breathless and fatigued after mild exercise for the past 4 months, for example, after walking his dog to the bottom of the street and back. This had become progressively worse in the last 3 weeks. It was also associated with a dry cough, dizziness and chest pain that was relieved on rest. He also mentioned that he fainted last month but did not worry about it. Family history revealed that his father had died aged 62 from 'heart disease' and his mother died of a 'stroke' aged 69 years.

Investigations

On examination, the patient's blood pressure was 145/105 mmHg. His pulse was regular, pulse volume was low and pulse character was described as 'slow-rising'. On auscultation of the heart, an ejection systolic murmur 4/6 with a crescendo-decrescendo ('diamond shaped') pattern, peaking in mid-systole, was heard on the 2nd intercostal space. The Valsalva manoeuvre strain reduced the intensity of the murmur and caused a decline in the carotid pulse volume; releasing the Valsalva manoeuvre increased the intensity of the murmur after several beats. The patient had diffuse fine crackles in both lung fields. There was moderate pitting oedema of both legs. A chest X-ray showed left ventricular enlargement and pulmonary oedema with bilateral pleural effusion. An ECG showed evidence of left ventricular hypertrophy (LVH) with possible left-axis deviation.

Diagnosis

The patient's signs and symptoms were consistent with moderate left-sided heart failure. The findings on physical examination suggested that valvular disease was probably responsible. The pulse character was described as 'slow-rising', a feature of significant aortic stenosis and often referred to as *'pulsus parvus et tardus'*, indicating a weak and delayed pulse upstroke in the arterial pulse waveform. The regular pulse helps to rule out atrial fibrillation. The diagnosis of aortic valve stenosis in a congenitally bicuspid aortic valve, with high valve gradient, was confirmed by imaging using transthoracic echocardiography (TTE). This high gradient signifies significant obstruction to blood flow across the valve and is a key diagnostic criterion for severe aortic stenosis. The TTE confirmed LVH, a common consequence of chronic pressure overload associated with severe aortic stenosis.

Given the patient's male gender, age, obesity and family history of atherosclerotic disease, his diagnostic evaluation included a CT scan of the coronary arteries. This revealed focal narrowing in some coronary vessels but without significant stenosis or flow-limiting lesions. The patient was referred for surgical replacement of the aortic valve, but he wanted to seek a different opinion. However, 3 weeks after his last cardiology appointment, he was found dead at home. An autopsy revealed an enlarged heart weighing 455 g (a normal heart usually weighs between 250 and 350 g) and an aortic valve with extensive calcification and stenosis. There was notable LVH (as described above, due to the increased workload of pumping blood against the narrowed aortic valve) and pulmonary oedema (resulting from backpressure in the pulmonary circulation due to reduced blood flow through the aortic valve).

Diagnosis: Calcific aortic valve stenosis.

Discussion

Valvular heart disease is due to *stenosis* (failure of the valve to open completely, thereby obstructing flow forward) or *insufficiency* (failure to close properly, thereby allowing blood to flow back, often referred to as *regurgitation*), or both. Calcific aortic valve stenosis is the most common valvular abnormality and the third most common cardiovascular disease in the Western world after hypertension and coronary artery disease. It is due to the calcification and fibrosis of either a previously normal valve or a congenitally bicuspid valve (Fig. CS36.1). Usually, the aortic valve has three cusps. A congenitally bicuspid aortic valve (BAV) has only two cusps and occurs in approximately 1% of the population. The prevalence of aortic valve stenosis increases from 2% in people over 65 years, to 4% in those >85 years of age, although aortic valve stenosis occurring in BAV presents earlier, as in this case.

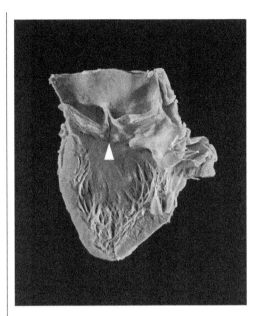

Figure CS36.1 Bicuspid aortic valve (white arrowhead) is associated with an increased risk of aortic valve stenosis. Licence: Attribution 4.0 International (CC BY 4.0). *Source:* Michael Frank / Wellcome Collection / CC BY 4.0.

Aortic valve stenosis is a progressive condition and once heart failure develops, survival is usually less than 2 years, unless the valve is surgically replaced. The damaged valve can be surgically replaced with either a mechanical valve, which will need lifelong anticoagulation, a bioprosthetic valve, or by transcatheter aortic valve implantation (TAVI). TAVI is a minimally invasive procedure in which a bioprosthetic valve is delivered to the aortic valve site via small incisions, typically in the groin or chest. TAVI is often considered for patients who are unsuitable for surgery, but increasingly is also being performed in younger, intermediate-risk, patients. It also offers the advantage of shorter recovery times and reduced postoperative complications.

The pathogenesis of aortic valve stenosis is inflammation, leading to the accumulation of lipids and calcification that has many similarities to atherosclerosis (Chapter 9). The aortic valve is subject to considerable mechanical stresses during the lifetime of an individual. These mechanical stresses can damage the endothelial cells of the valve with loss of barrier function against metabolic, mechanical and inflammatory insults. Endothelial injury represents a key initiating step in the pathogenesis of aortic valve stenosis (as it does in atherosclerosis) with subsequent infiltration of inflammatory cells into subendothelial valvular tissues. Low-density lipoprotein (LDLs) and lipoprotein A are deposited in the early lesions of aortic valve stenosis. These lipids undergo oxidative modifications leading to the generation of cytotoxic free radicals, including superoxide and oxygen peroxide and the secretion of pro-inflammatory and pro-fibrotic cytokines, pro-osteogenic signals with activation of fibroblasts. In turn, there is progressive fibrosis and calcification that is characteristic of aortic valve stenosis. Osteogenic factors in aortic valve stenosis include RANK (Receptor Activator of Nuclear Factor Kappa-B), which binds to RANK ligand (RANKL). In bone, RANKL-RANK interactions induce the activation of osteoclasts, leading to bone resorption. However, in valvular tissues, the binding of RANKL to RANK promotes osteoblast differentiation, leading to calcification and bone deposition.

The Aortic Stenosis Progression Observation Measuring Effects of Rosuvastatin (ASTRONOMER) study has identified an association between *metabolic syndrome* and the development and progression of aortic valve disease. Metabolic syndrome is a cluster of factors, which together increase the risk of serious cardiovascular disease (Chapter 9). Excess visceral adipose tissue could play an important role in the pathogenesis of aortic valve stenosis. In particular, *epicardial adipose tissue* (EAT) is implicated in promoting inflammation in aortic valve stenosis.

Treatment and Prognosis

The only effective treatment for haemodynamically high-gradient (defined as aortic transvalvular velocity ≥4 m/s and/or mean transvalvular pressure gradient ≥40 mmHg;

the aortic valve area [AVA] is typically ≤1 cm^2; with AVA indexed to body surface area ≤0.6 cm^2/m^2) severe aortic valve stenosis is valve replacement. However, this type of surgery has a perioperative mortality rate of around 4%, with a risk of prosthetic valve failure of around 1% per year. Watchful waiting is recommended for most people with asymptomatic aortic valve stenosis; mortality in patients with aortic valve stenosis managed by watchful waiting is not significantly different from that of age-matched patients without aortic stenosis. Indeed, the risks associated with surgery may outweigh the 1%

annual risk of sudden death associated with asymptomatic aortic stenosis. However, mortality increases after aortic stenosis becomes symptomatic and, in this group, without surgical intervention, the average overall survival rate is only 2–3 years. Mortality rates are kept under scrutiny in randomised controlled trials and may change in the fut*ure*.

Questions

1 What is the Valsalva manoeuvre?
2 What is epicardial adipose tissue?

Answers on p. 544.

Case Study 37   Worsening Breathlessness, Fatigue and Chest Discomfort

Clinical Presentation

A 40-year-old man presented to the cardiology clinic with worsening breathlessness, fatigue and chest discomfort. His symptoms had been present for the last 6 months, but recently they had limited his ability to perform routine activities. He reported a history of rheumatic fever at 12 years of age. He was a non-smoker and had no significant family history of cardiac disease. On examination, he was pale and fatigued. His blood pressure was 140/90 mmHg and his heart rate was regular at 80 beats per minute. There was no evidence of jugular venous distension, peripheral oedema or cyanosis. Cardiovascular examination revealed a diastolic murmur at the apex, of low pitch and rumbling in character. Respiratory examination revealed bilateral crepitations in the lower lung fields.

Investigations

Electrocardiography (ECG) revealed sinus rhythm with left atrial enlargement. Chest X-ray revealed an enlarged left atrium

and prominent pulmonary artery markings, particularly in the upper lung fields, with straightening of the left heart border. Transthoracic echocardiography (TTE) revealed severe mitral valve stenosis with restricted valve leaflet motion and a reduced valve area, consistent with the clinical and radiographic findings. Additionally, TTE confirmed left atrial enlargement, further supporting the diagnosis of mitral valve stenosis as the underlying cardiac condition.

Diagnosis

Based on the clinical presentation, history of rheumatic fever and imaging findings, a diagnosis of chronic rheumatic heart disease (RHD) with mitral stenosis was made. In mitral valve stenosis, the left atrium undergoes enlargement as a result of increased pressure within the atrium due to the restricted flow of blood from the left atrium to the left ventricle. This enlargement is visualised on the chest X-ray as a straight left heart border. Mitral valve stenosis can also lead to increased pressure in the pulmonary circulation, causing the pulmonary arteries to become more prominent on imaging, further altering the chest X-ray left heart border.

Diagnosis: Chronic rheumatic heart valve disease.

Discussion

Heart valve disease is a broad term encompassing conditions that cause structural damage to the valves and result in functional abnormalities that can lead to heart failure. Heart valve disease primarily affects the elderly. Calcific aortic valve disease (see Case Study CS36) and degenerative mitral valve disease are the two most prevalent non-rheumatic valve diseases. However, rheumatic heart disease is the leading cause of cardiovascular death among children and young adults in low and middle-income countries.

Rheumatic fever is an autoimmune response to group A streptococcal infection, usually appearing several weeks after the initial throat infection and characterised by the appearance of antibodies against the streptococci that cross-react with host tissues (Fig. CS37.1). The most common presentations of acute rheumatic fever are:

- *pan-carditis* with *migratory polyarthritis* (joint inflammation and pain that affects multiple joints but which can appear to move from one joint to another)
- the presence of subcutaneous nodules
- *erythema marginatum* (a rash that appears as pink, non-itchy, raised and irregularly shaped patches on the trunk and limbs)
- *Sydenham chorea* (a neurological disorder that results in involuntary movements, especially affecting the limbs and facial muscles)

The autoimmune response also causes long-term valve damage and dysfunction – most commonly stenosis and regurgitation of the mitral valve. In patients with severe chronic mitral valve disease, surgical or percutaneous intervention is usually required to prevent complications such as atrial fibrillation, pulmonary hypertension and heart failure.

Treatment and Prognosis

Percutaneous mitral balloon valvotomy (PMBV) is a minimally invasive procedure used to treat mitral stenosis. During PMBV, a catheter is inserted into the femoral vein and guided to the left atrium through an atrial septal puncture. The catheter carries a deflated balloon at its tip. When the catheter reaches the narrowed mitral valve, the balloon is inflated. This inflation pushes apart the stiff, fused valve leaflets, creating a larger opening for blood to flow through. This results in improved cardiac output and relief of symptoms associated with mitral stenosis, such as shortness of breath, fatigue and chest discomfort. Before performing PMBV, cardiologists assess the suitability of the mitral valve using echocardiography (echo). The *Wilkins score* is a widely used echocardiographic scoring system that helps determine the suitability of the valve for PMBV. It assesses various structural aspects of the valve, such as the thickness and mobility of the valve leaflets and the presence of calcium deposits. A lower Wilkins score indicates a more suitable valve for PMBV. Treating patients with PMBV can delay or even obviate the need for invasive surgical mitral valve replacement. Our patient was referred for potential PMBV and in the meantime, was advised to limit physical activity and avoid strenuous exercise.

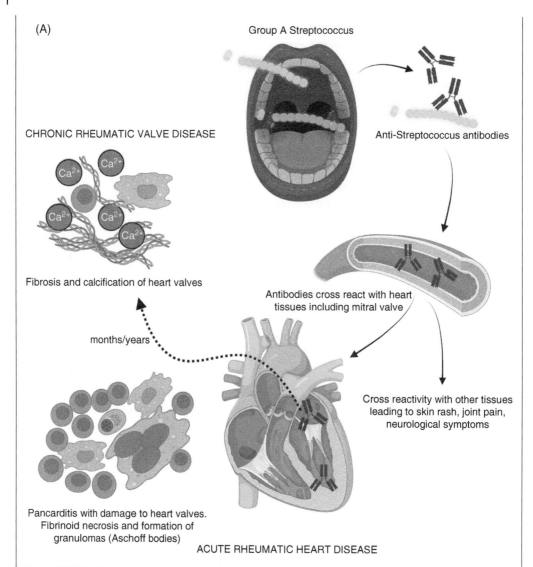

(A)

Group A Streptococcus

Anti-Streptococcus antibodies

CHRONIC RHEUMATIC VALVE DISEASE

Fibrosis and calcification of heart valves

Antibodies cross react with heart tissues including mitral valve

months/years

Cross reactivity with other tissues leading to skin rash, joint pain, neurological symptoms

Pancarditis with damage to heart valves. Fibrinoid necrosis and formation of granulomas (Aschoff bodies)

ACUTE RHEUMATIC HEART DISEASE

Figure CS37.1 Rheumatic heart disease. (A) Rheumatic heart disease is caused by throat infections with β-haemolytic (group A) streptococci. Anti-streptococcal antibodies are formed, which cross-react with endogenous tissue antigens in the heart, joints and other tissues. About 2 weeks after the initial infection, patients experience acute febrile polyarthritis, pan-carditis, the appearance of subcutaneous nodules, with rash and neurological problems. Chronic rheumatic heart disease is characterised by organisation of the acute inflammation with subsequent fibrosis and calcification of the heart valves. This results in valve leaflet thickening and commissural fusion and shortening, giving the valve a 'fish mouth' or 'buttonhole' appearance. Heart valves normally have two or more leaflets that open and close to regulate blood flow through the heart. Commissures are the points where the leaflets come together and form a seal to prevent the backflow of blood. In rheumatic heart disease, chronic inflammation and scarring cause the valve leaflets to thicken and become fused together at the commissures, making it difficult for them to separate and open properly. It also results in stiff valves that cannot close properly, resulting in varying degrees of valvular regurgitation. *Source:* Created with BioRender.com.

(B)

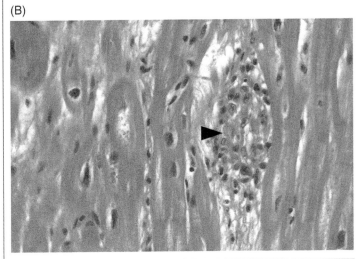

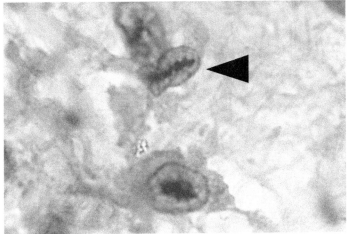

Figure CS37.1 (Cont'd) (B) Upper panel; the morphological features of acute rheumatic heart disease include pan-carditis with *Aschoff bodies* (rheumatic granulomas; black arrowhead) and fibrinoid necrosis affecting the endocardium (including the valves), pericardium and myocardium. Aschoff bodies are focal areas of inflammation in the connective tissue of the heart that are a hallmark of acute rheumatic fever. Fully developed Aschoff bodies are granulomas consisting of fibrin with infiltration by lymphocytes, occasional plasma cells and macrophages. *Source:* Nephron / Wikimedia Commons / CC BY-SA 3.0. Lower panel: Some of the macrophages fuse to form multinucleated giant cells, while others may become *Anitschkow* cells or 'caterpillar cells', which are characterised by their elongated, wavy nuclei that resemble the body of a caterpillar (black arrowhead). *Source:* Ed Uthman. / Wikimedia Commons / CC BY-SA 2.0.

Questions

1 What is transthoracic echocardiography (TTE)?

2 What are the major characteristics of beta-haemolytic streptococci?

Answers on p. 545.

Case Study 38   Sudden Onset of Severe Pain and Numbness in the Left Leg

Clinical Presentation

A 65-year-old man presented to the Emergency Department with severe pain in his left foot and leg that had started suddenly about 2 hours earlier. He reported no recent trauma or injury to the leg. The pain was worse when he tried to put his foot on the ground and was not relieved by rest or medication. He also reported a sensation of coldness in the affected foot. The patient had a history of hypertension, for which he was taking lisinopril (an ACE inhibitor), and a 40-pack-year history of smoking. He denied any history of peripheral arterial disease or blood clotting disorders.

On examination, the patient appeared in severe distress, with a heart rate of 110 beats per minute and a blood pressure of 160/90 mmHg. His left leg was cold to the touch. There was no palpable pulse in the left femoral, popliteal, dorsalis pedis or posterior tibial arteries. There was no swelling or erythema, but the toes were cyanotic and blanched on pressure with delayed capillary refill. Arterial Doppler (a non-invasive ultrasound technique used to assess blood flow and detect abnormalities or blockages in the arteries) signals were absent, but venous signals were present in the foot. The right leg had normal pulses, sensation and motor function.

Investigations

CT angiography of the aorta and lower extremities showed complete occlusion of the common femoral artery at the superior femoral and profunda femoral bifurcation with diffuse non-significant atherosclerotic changes and calcifications along the arterial tree. An ECG revealed atrial fibrillation. Blood tests showed elevated levels of creatine kinase and lactate dehydrogenase, indicating muscle damage. Laboratory values are summarised in Table CS38.1.

Table CS38.1   Selected laboratory results.

Test	Result	Reference range
Haemoglobin	**12.2 g/dL**	13.0–18.0
White blood cell count	9.5×10^9/L	4.0–11.0
Platelet count	250×10^9/L	150–400
Urea	3.2 mmol/L	2.5–6.7
Creatinine	72 μmol/L	59–104
Random blood glucose	**11.2 mmol/L**	3.5–8.9
Sodium	140 mmol/L	135–146
Potassium	4.0 mmol/L	3.5–5.0
Calcium	2.40 mmol/L	2.20–2.67
Prothrombin time	14 seconds	11–16
Activated partial thromboplastin time	32 seconds	26–35
Cholesterol	6.5 mmol/L	3.5–6.5
LDL	4.3 mmol/L	1.55–4.4
HDL	2.0 mmol/L	0.7–2.1
Triglycerides	1.9 mmol/L	0.7–2.1
Creatine kinase	**241 U/L**	24–195
Lactate dehydrogenase (LDH)	**284 U/L**	135–225

Values outside the reference range are in bold.

Diagnosis

Based on the patient's clinical presentation and diagnostic tests, he was diagnosed with acute limb ischaemia (a sudden decrease in blood flow to a limb that threatens the viability of the affected tissue) due to a complete occlusion of the left femoral artery.

Diagnosis: Acute lower limb ischaemia.

Discussion

Peripheral arterial disease refers to a group of conditions that affect the blood vessels outside of the heart and brain, primarily in the legs and feet. It is often caused by atherosclerosis, a build-up of plaque in the arteries that narrows and stiffens the vessels, reducing blood flow and oxygen supply to the affected tissues (see also Chapter 9). Acute limb ischaemia occurs due to a sudden reduction in blood flow to the affected limb. The most common cause is a thrombus that obstructs the arterial blood supply to the limb, but other causes such as emboli, trauma or arterial dissection can also result in acute limb ischaemia. The ischaemia can cause death of the tissue (infarction) supplied by the blocked vessel.

The '6 P's' of acute limb ischaemia refer to a group of symptoms that may indicate a severe and urgent lack of blood flow to a limb;

Pain: Severe, unrelenting pain in the affected limb, which may be described as burning, throbbing or cramping.

Pallor: Paleness of the skin in the affected limb, which may appear white or bluish in colour, particularly at the toes and distal foot.

Pulselessness: Absence of a palpable pulse in the affected limb, indicating reduced or absent blood flow.

Paraesthesia: Numbness, tingling, or pins-and-needles sensations in the affected limb, indicating nerve damage due to lack of oxygen.

Paralysis: In extreme cases, muscle weakness or paralysis may occur due to lack of blood flow and oxygen to the muscles.

Perishing cold: The affected limb feels cold with often a clear demarcation between ischaemic/threatened and non-ischaemic tissues.

Treatment and Prognosis

The patient was taken urgently to the operating room for embolectomy. An embolus was successfully removed from the left femoral artery and blood flow restored to the left leg. The patient was started on anticoagulation therapy with heparin and subsequently transitioned to warfarin following cardiology consultation. The patient had an uneventful recovery from surgery and was discharged to home on postoperative day 3. He was instructed to continue taking his anticoagulation medication and was referred to a vascular specialist for ongoing management of his peripheral vascular disease.

Questions

1 What are the initial steps in the management of acute limb ischaemia?
2 What are the potential complications of delayed treatment of acute limb ischaemia?

Answers on p. 545.

Case Study 39   Coughing Up Blood and Shortness of Breath in a 29-year-old Man

Clinical Presentation

A 29-year-old male schoolteacher presented to the Emergency Department with a 2-week history of haemoptysis (coughing up blood), along with increasing shortness of breath over the past month. In particular, he noticed that he was unable to climb three flights of stairs without significant discomfort. He smoked 10 cigarettes per day for the last 10 years and did not take alcohol. He reported feeling fatigued and having a low-grade fever. On examination, he had a respiratory rate of 24 breaths per minute and had scattered lung crepitations. His blood pressure was 155/80 Hg mm.

Investigations

A chest X-ray showed right basal consolidation with possible necrosis. A computed tomography (CT) scan of the chest revealed multiple pulmonary nodules and consolidations. An initial diagnosis of pneumonia was made and he was started on augmentin and salbutamol. However, there was no improvement and 5 days later, he visited his GP, reporting right-side chest pain, which was worse on taking a deep breath. Physical examination showed more marked crepitations in the right lower lobe and a pericardial rub. A repeat chest X-ray showed that the lung consolidation was more extensive.

Laboratory investigations revealed raised blood urea and creatinine (Table CS39.1). Urinalysis revealed protein and blood in the urine. He was anaemic (normochromic/normocytic; results not shown) with a raised

Table CS39.1 Selected laboratory investigations.

Test	Result	Reference range
Blood		
White blood cell count	13.1×10^9/L	4.0–11.0
Haemoglobin	11.0 g/dL	13.0–18.0
Creatinine	179 µmol/L	59–104
Erythrocyte sedimentation rate	22 mmHg	<15
C-reactive protein	25.2 mg/L	<10
Urea	8.2 mmol/L	2.5–6.7
Urine		
Protein	**Positive**	Negative
Blood	**Positive**	Negative

Values outside the reference range are in bold.

white cell count, ESR and CRP. Sputum cytology was negative for malignant cells and infection.

Diagnosis

Further investigations revealed a positive test for anti-neutrophil cytoplasmic antibodies (ANCA) with cytoplasmic staining (c-ANCA). Based on these findings, a diagnosis of granulomatosis with polyangiitis (GPA) was suspected and the patient was started on high-dose intravenous corticosteroids. He was subsequently referred to a rheumatologist for further management. During his treatment, the patient developed joint pain and swelling in his knees, and a skin rash on his lower legs consistent with cutaneous vasculitis. He underwent a kidney biopsy, which revealed pauci-immune necrotising glomerulonephritis, indicative of rapidly progressive glomerulonephritis (RPGN) confirming the diagnosis of GPA with renal involvement.

Diagnosis: Granulomatosis with polyangiitis (GPA) with renal involvement.

Discussion

ANCA (anti-neutrophil cytoplasmic antibody) systemic vasculitis is a group of rare autoimmune diseases that cause inflammation and damage to blood vessels throughout the body (see also Chapter 9). ANCA is a type of autoantibody that targets neutrophils. There are several types of ANCA systemic vasculitis:

1) Granulomatosis with polyangiitis (GPA), formerly known as Wegener granulomatosis, which mainly affects the respiratory tract and kidneys.
2) Microscopic polyangiitis (MPA), which typically affects the kidneys, lungs and skin.
3) Eosinophilic granulomatosis with polyangiitis (EGPA), formerly known as Churg-Strauss syndrome, which often involves the lungs, skin and nerves, and is associated with an elevated eosinophil count.

The symptoms of ANCA systemic vasculitis vary depending on the type of vasculitis and the organs involved, but include fever, fatigue, weight loss, joint and muscle pain, together with organ-specific symptoms such as cough, shortness of breath, skin rash and renal impairment.

Patients with ANCA-associated vasculitis have two main target antigens for ANCA: myeloperoxidase (MPO) and proteinase 3 (PR3), which are found in the granules of neutrophils and/or monocytes. Other 'minor' ANCAs have also been identified, including those targeting α-enolase, azurocidin, bactericidal permeability-increasing protein (BPI), cathepsin G, elastase, defensin, lactoferrin, lysosome-associated membrane glycoprotein 2 (LAMP2) and moesin.

MPO-ANCA patients typically show more features of MPA, while PR3-ANCA patients show more features of GPA.

Perinuclear ANCAs (p-ANCAs) stain around the nucleus and are primarily composed of MPO targeting antibodies. ANCAs that stain diffusely in the cytoplasm, are referred to as *cytoplasmic ANCAs* (c-ANCAs) and are mostly antibodies against PR3.

Rapidly progressive glomerulonephritis (RPGN) is a condition characterised by a rapid loss of kidney function, which typically presents with features of glomerulonephritis, including haematuria and proteinuria. RPGN is often associated with extensive 'crescent' formation, and therefore, is also known as *crescentic glomerulonephritis* (CGN; see also Case Study 32). From an immunological perspective, primary RPGN can be classified into three categories: *pauci-immune GN* (PICG), *anti-GBM GN* (Case Study 32) and *immune complex GN*. PICG, which is the most common cause of primary RPGN, is a type of glomerulonephritis in which there are few or no immune deposits observed by immunofluorescence or electron microscopy. In most cases, pauci-immune CGN is a component of a systemic small vessel vasculitis, such as GPA, as observed in this patient.

Clinical evidence and laboratory studies suggest that neutrophils play a critical role in the development of GPA. In GPA, activated neutrophils are present in the affected glomeruli of the kidneys and the severity of renal injury, as measured by serum creatinine levels, correlates with the number of activated intra-glomerular neutrophils. It has been demonstrated that ANCA activates cytokine-primed neutrophils, leading to an oxidative burst, degranulation and the release of inflammatory cytokines with associated damage to endothelial cells.

Autoimmune vasculitis has a multifactorial aetiology. Several genes related to susceptibility or resistance have been identified through genome-wide association studies (GWAS; Chapter 10). Major histocompatibility complex (MHC)-class II genes have demonstrated the strongest associations. However, environmental factors, including infectious agents, drugs such as propylthiouracil, hydralazine and cocaine, and airborne particulates like silica dust, can trigger the disease process. Toxic shock syndrome toxin-1 produced by *Staphylococcus aureus* is a well-described risk factor for relapse in patients with GPA.

The development of necrotising granulomas in GPA is believed to be triggered by infection, possibly by *S. aureus*, which can activate tissue-resident macrophages in the bronchial epithelium via Toll-like receptors (TLR; Chapter 6). The activated macrophages release proinflammatory cytokines such as TNF-α and IL-1β, which attract neutrophils and monocytes from the blood into the developing lesion. The recruited neutrophils release reactive oxygen species (ROS; Chapter 3) and lytic enzymes, which contribute to the formation of a necrotic core. The recruited monocytes differentiate into macrophages that secrete IL-23, causing the differentiation of T cells towards a Th17 phenotype. IL-17 released from Th17 cells plays a crucial role in the formation of the granuloma that surrounds the necrotic region.

Treatment and Prognosis

Treatment of GPA usually involves a combination of immunosuppressive medications, such as corticosteroids and cytotoxic drugs, to reduce inflammation and prevent damage to organs. In some cases, *plasmapheresis* (which removes the ANCA antibodies from the blood) may also be useful; patients who have had circulating ANCA removed by plasma exchange have a lower risk for progression to end-stage renal disease at 1 year than patients who did not receive plasma exchange. Targeted therapies that deplete B cells are also effective.

Remission rates for patients with GPA range from 30 to 93%. More than 50% of GPA patients who receive aggressive therapy for active disease are able to recover renal function and become dialysis-independent. However, relapse is common in GPA, with up to half of patients experiencing relapse within 5 years. ANCA status, treatment and target organ involvement are factors associated with relapse. The absence of kidney involvement is associated with a 100% 5-year survival rate, compared with around only 70% in those with kidney disease.

Permanent damage from GPA includes hearing loss, respiratory problems and end-stage kidney disease. Additionally, patients with GPA have an increased risk of cardiovascular events, urotoxic adverse events and the development of cancers associated with the immunosuppressive drugs. Severe, untreated GPA has a mortality rate of over 90% and even with current treatments, the 1-year mortality rate is around 11%.

The patient responded well to treatment and was eventually discharged from the hospital with a tapering course of oral corticosteroids and immunosuppressive therapy. Follow-up examinations revealed a resolution of his respiratory symptoms and improvement in his kidney function and skin rash. The patient continues to receive ongoing treatment and monitoring under the care of his rheumatologist.

Questions

1 What is pericardial rub?
2 The patient had evidence of necrosis in the inflamed lungs. Apart from the formation of necrosis within granulomas described earlier, what other pathological process might also be responsible for the development of necrosis in GPA?

Answers on p. 546.

Case Study 40 Chest Pain Radiating to the Left Arm

Clinical Presentation

A 53-year-old man was admitted to hospital with severe crushing central chest pain that radiated to the left arm, which started 1 hour ago while he was reading a book in the library. The pain was associated with sweating, nausea, one episode of vomiting and breathlessness. On examination, he was obese, pale and sweaty. His heart rate was 100 bpm and he was hypertensive (blood pressure: 170/100 mmHg).

Investigations

An electrocardiogram (ECG) showed right bundle branch block (RBBB) and raised ST segments in anterior chest leads. RBBB is when there is a delay or block in the pathway along which electrical impulses travel to make the right ventricle of the heart contract and thus to ensure that the right and left ventricles contract in synchrony. On an ECG, RBBB produces a specific pattern, indicating that the electrical signal is not being

Table CS40.1   Results of laboratory investigations.

Test	Result	Reference range
Blood		
Sodium	144 mmol/L	135–146
Potassium	4.3 mmol/L	3.5–5.0
Urea	5.7 mmol/L	2.5–6.7
Cholesterol	**8.2 mmol/L**	3.5–6.5
Random blood glucose	**11.6 mmol/L**	3.5–8.9
Urine (dipstick tests)		
Glucose	**2⁺***	0–trace
Protein	0	0–trace

Blood lipid profile		Optimal value
Triglycerides	1.4 mmol/L	<2.1
Cholesterol	**7.7 mmol/L**	<5.2
HDL	**1.2 mmol/L**	>1.0
Lipoprotein A	**811 mmol/L**	<300

* 2^+ glucose reading approximates to a urinary glucose of at least 28 mmol/L.
Values outside the reference range are in bold.

transmitted efficiently through the right side of the heart. It can be associated with anterior wall MI. The results of laboratory investigations are shown in Table CS40.1.

High-sensitivity cardiac troponin (hs-Tn), based on the specific troponin protein being measured (TnT or TnI) tests are now being used. This patient had a hs-TnI value of 141 pg/mL. The reference range is ≤15 pg/mL for females and ≤20 pg/mL for males; results above these values are indicative of myocardiocyte necrosis and suggest the possibility of myocardial infarction, requiring additional patient evaluation. An elevated and actionable value of hs-TnI is >100 pg/mL for males and >75 pg/mL for females, and indicates myocardial damage.

Diagnosis

The clinical history and features are suggestive of myocardial infarction. In this patient,

cardiac troponin levels were elevated above the normal reference range. Cardiac troponins are widely accepted as the best markers for measuring cardiac muscle damage. They enter the bloodstream several hours after the onset of injury and can remain raised for several days. Cardiac troponins are specific for heart muscle. Some studies show a correlation between cardiac troponin levels and long-term outcomes following acute myocardial episodes. High-sensitivity troponin-I tests can detect even minute amounts of cardiac troponins in the blood, allowing for earlier and more accurate diagnosis of cardiac events, especially in cases where damage might be minimal or in the early stages.

Diagnosis: Acute myocardial infarction.

Discussion

Myocardial infarction refers to the death of myocardial tissue usually as a result of ischaemia (inadequate blood flow) sufficient to produce lethal cell injury. Myocardial infarction is most often due to thrombosis in a coronary artery, usually at the site of a previous atheromatous plaque. Following MI, electrical impulses due to depolarization and repolarization of the myocardial cells are altered and these changes can be picked up on the surface ECG and used to determine whether an infarct has occurred and its likely location. Resulting abnormal heart rhythms may be life-threatening.

In myocardial infarction, the affected tissue goes through several histological changes as illustrated in Fig. CS40.1 and described below:

1) *Myofibre 'waviness':* Shortly after MI onset, myofibres in the affected area exhibit 'waviness' due to loss of alignment caused by disruption of structural integrity.
2) *Interstitial oedema:* A few hours after onset, there is an accumulation of fluid in the interstitial spaces between cells due to increased vascular permeability, contributing to tissue swelling.

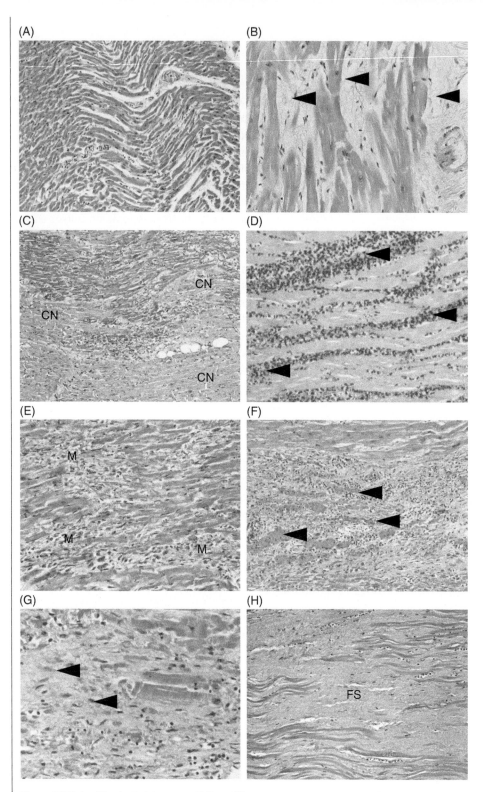

Figure CS40.1 Histological features of MI at different stages (without reperfusion). (A) Myofibre waviness. (B) Interstitial oedema (black arrowheads). (C) Hypereosinophilia and coagulative necrosis of cardiomyocytes (CN). (D) Heavy granulocyte infiltration with karyorrhexis (black arrowheads). (E) Macrophages (M) and lymphocyte infiltration with early removal of necrotic debris. (F) Granulation tissue with formation of microvessels (black arrowheads). (G) Fibroblast proliferation (black arrowheads) and early collagen deposition. (H) Dense fibrous scar (FS) replacing myocyte loss. All sections are stained with haematoxylin and eosin. *Source:* Katarzyna Michaud / Wikimedia Commons / CC BY 3.0. Katarzyna Michaud, Cristina Basso, Giulia d'Amati, Carla Giordano, Ivana Kholová, Stephen D. Preston, Stefania Rizzo, Sara Sabatasso, Mary N. Sheppard, Aryan Vink, Allard C. van der Wal & on behalf of the Association for European Cardiovascular Pathology (AECVP), CC BY 3.0 <https://creativecommons.org/licenses/by/3.0>, via Wikimedia Commons.

3) *Hyper-eosinophilia and coagulative necrosis of cardiomyocytes:* In the first 12–24 hours, eosinophilic staining of cardiomyocytes becomes evident. Coagulative necrosis occurs marking the presence of irreversible cellular damage.

4) *Neutrophil infiltration with karyorrhexis:* Around 1–3 days, there is a significant influx of neutrophils. Karyorrhexis, which is the fragmentation of cell nuclei, becomes visible.

5) *Macrophage and lymphocyte infiltration:* Within 3–7 days, macrophages and lymphocytes begin to infiltrate the necrotic tissue. Macrophages play a crucial role in phagocytosing necrotic debris and initiating tissue repair.

6) *Granulation tissue with formation of microvessels:* Around 1–2 weeks, granulation tissue forms, characterised by the presence of fibroblasts, myofibroblasts and capillaries.

7) *Fibroblast proliferation:* Within 1–3 weeks, fibroblasts proliferate and secrete collagen, leading to the initial formation of a collagen-based scar. The scar tissue gradually replaces the necrotic myocardium.

8) *Dense fibrous scar:* Over several weeks to months, the collagen scar becomes denser and matures. It replaces the lost cardiomyocytes and provides structural support to the healing area, contributing to the long-term stabilisation of the infarcted region.

Treatment and Prognosis

The management of a patient with an MI is in three stages:

First, as this is an emergency scenario, the patient is approached taking into consideration his airway, breathing and circulation. Accordingly, oxygen is administered along with analgesia in the form of morphine and sublingual nitrates. Routine MI medical therapy is given, taking into account co-morbidities and presentation. The ultimate goal is reperfusion (restoration of blood flow to the area of the heart muscle deprived of oxygen). *Percutaneous coronary intervention*

(PCI) is the modality of choice provided the 'door-to-balloon' time is no more than 90 minutes. PCI is a non-surgical procedure used to treat the narrowed coronary arteries of the heart. Essentially, it involves using a catheter, wire and balloon to open the coronary artery and then placing a stent to maintain patency. 'Door-to-balloon time' is the time from a patient's arrival at the hospital (the 'door') to the time when blood flow is restored to the blocked artery (by inflating the 'balloon', which pushes the thrombus aside). An earlier patient presentation after the onset of chest pain and a correspondingly shorter door-to-balloon time is associated with better patient outcomes. Otherwise, fibrinolysis is used when timely PCI is not available. Agents of choice are tissue plasminogen activators (tPA), specifically the recombinant form known as alteplase, or third-generation, reteplase. Patients are managed according to the *advanced cardiovascular life support* (ACLS) protocol; this is a set of clinical pathways for the urgent treatment of ACS (acute cardiac syndrome), cardiac arrest, stroke and other life-threatening medical emergencies.

Second, it is important to identify any risk factors for MI. The patient was a smoker, had untreated hypercholesterolaemia, high blood pressure associated with left ventricular hypertrophy (on subsequent echocardiography), suggesting longstanding previously undiagnosed hypertension. The elevated blood glucose level suggests type 2 diabetes (Chapter 13). Type 2 diabetes is frequently asymptomatic and may be present for several years before it is detected. The normal blood urea, sodium and potassium, and absence of proteinuria suggest that the patient has not yet developed renal injury. The early death of his father from heart disease suggests an inherited risk of MI which may be the result of genetically determined hypercholesterolaemia, hypertension, or type 2 diabetes. In the United Kingdom, Asian patients have an increased risk of death from MI, which is 40% above that of the British Caucasian population. They also have a four times greater risk of developing type 2 diabetes. Additional investigations and treatment (mainly addressing his risk factors with appropriate management, where possible) are necessary to prevent further MI or cardiac complications. Therefore, the patient must stop smoking as this is an important risk factor. Anti-hypertensives would be commenced as per guidelines, diabetic medications would be commenced (or re-adjusted) and his hyperlipidaemia managed (as described below).

Third, initiation of medical therapy which has been proven in statistical studies to influence ventricular remodelling and improve long term prognosis e.g. treatment with angiotensin-converting enzyme (ACE) inhibitors or angiotensin receptor blockers (ARBs).

The progress of patients who have suffered an MI is monitored in an outpatient clinic. In this case, the patient was clearly overweight at the time of his MI. Since then, he had seen a dietitian who had given him advice on healthy eating and suggested a reduction of his fat intake to less than 30% of his total energy intake.

To prevent further accumulation of cholesterol in his arteries, the patient was prescribed statins to lower his cholesterol levels. Statins are inhibitors of HMG-CoA reductase, a key enzyme involved in the synthesis of cholesterol by the liver. The patient was again advised to stop smoking and to take more exercise. His diabetes was successfully controlled by oral hypoglycaemic agents. Low-dose aspirin, as an antiplatelet agent, and β-blockers, were also prescribed. Both of these agents have been shown to reduce the incidence of re-infarction. Complications of myocardial infarction include:

- *Arrhythmias:* Disruption of the heart's normal electrical conduction can lead to irregular heart rhythms. Ventricular fibrillation, a life-threatening arrhythmia, is especially concerning and will lead to sudden cardiac death if not treated promptly.

- *Heart failure:* The heart's pumping ability can be compromised after an MI due to the loss of viable heart muscle. This can result in heart failure, defined as a condition in which the heart is unable to pump sufficient blood to meet the body's needs.
- *Cardiogenic shock:* A severe form of heart failure that occurs when a large portion of the heart muscle is damaged, leading to a significant decrease in blood pressure and inadequate perfusion of vital organs.
- *Structural complications:* MI can lead to structural abnormalities, including:
  - *Ventricular septal rupture (VSR):* A defect between the left and right ventricles due to necrosis and rupture of the interventricular septum.
  - *Free wall rupture:* Rupture of the heart muscle of the free wall, either leading to sudden cardiac death or to bleeding into the pericardial cavity resulting in cardiac tamponade.
  - *Papillary muscle rupture:* This can lead to severe mitral regurgitation and heart failure.
- *Pericarditis:* Inflammation of the pericardium, the sac-like structure around the heart. Immune–related post-MI pericarditis, also known as Dressler syndrome, is a rare form of post-MI pericarditis that arises weeks or months later.
- *Thromboembolism:* Blood clots can form within the left ventricle, especially if an area of the heart wall is not moving properly. These clots can dislodge and travel to other parts of the body as emboli and may result in cerebral embolisation and stroke or peripheral embolisation and its sequelae.
- *Ventricular aneurysm:* A weakened and protruding segment of heart muscle, potentially causing irregular heart rhythms or even cardiac rupture. Additionally, a thrombus may develop within the aneurysm.
- *Cardiac tamponade:* This is when fluid accumulates in the pericardial space, which can compress the heart and impair its ability to pump.

Questions

1 What is the physiological basis of left ventricular hypertrophy in long-standing hypertension?

2 How do you account for the pulmonary oedema observed in this patient?

Answers on p. 546.

Case Study 41   Rectal Bleeding with Diarrhoea

Clinical Presentation

A 17-year-old girl was referred to the gastroenterology department with a 6-month history of rectal bleeding and diarrhoea. When asked about the delay in seeing a doctor, she replied that her father had died when she was 6 years old from what her mother called 'bowel problems' and that she had been afraid she might have the same problem. Her family's medical history showed that her father died from a metastatic colonic adenocarcinoma when he was only 35 years old. The postmortem report revealed that he also had multiple polypoid adenomas situated throughout the entire length of his colon. The patient's brother died from hepatoblastoma, a rare form of liver cell cancer, when he was only 5 years old.

Investigations

The results of laboratory investigations are shown in Table CS41.1. The total white cell count was towards the upper limit of normal. The differential white cell count was normal. Red cells were microcytic and hypochromic.

The blood results indicated iron deficiency anaemia. Iron deficiency can arise as a result of dietary deficiency, malabsorption or chronic blood loss. In this case, in view of the clinical history and presentation, the most likely cause is chronic rectal blood loss. Chronic rectal bleeding and diarrhoea may be caused by cancer, inflammatory conditions of the colon, including ulcerative colitis and Crohn's disease (see Case Study 22), or may have an infectious cause. Examination of the colon and rectum by

Table CS41.1 Results of laboratory investigations.

Test	Result	Reference range
Total white cell count	10.6×10^9/L	4.0–11.0
Red cell count	**3.46×10^{12}/L**	3.8–5.8
Haemoglobin	**7.6 g/dL**	11.5–16.5
Haematocrit	**0.24 L/L**	0.37–0.47
Mean cell volume	**70.5 fL**	82–92
Mean cell haemoglobin	**21.96 pg**	27.0–32.0
Platelet count	**460×10^9/L**	150–400
White blood cell differential		
Neutrophils	70%	40–75
Lymphocytes	22%	20–45
Monocytes	6%	2–10
Eosinophils	2%	1–6
Basophils	0%	1

Values outside the reference range are in bold.

sigmoidoscopy is therefore indicated. On sigmoidoscopy, multiple polyps were observed throughout the patient's colon and rectum, some of which were biopsied. Histological examination of these polyps revealed that they were benign tumours of colonic epithelial cells (*adenomas*). Microbiological culture of the stools was negative.

Diagnosis

The presence of multiple polypoid adenomas in the colon indicated a diagnosis of familial adenomatous polyposis (FAP). This is consistent with the early onset of colonic adenocarcinoma in the patient's father, which had presumably arisen in at least one of the multiple adenomas situated throughout his colon. A mutation in exon 15 of the APC gene at codon 1309 was identified in our patient. This mutation is the most common in FAP families, occurring in about 10% of families worldwide.

Diagnosis: Familial adenomatous polyposis.

Discussion

FAP is an autosomal dominant condition, therefore, any offspring of an affected parent have a 50% risk of developing the disease. FAP is caused by germline mutations in the *APC* gene (see also Chapter 11). Hepatoblastoma is a relatively rare childhood tumour but is often found in association with FAP. A diagnosis of FAP in the deceased boy is therefore highly likely. People suspected of having FAP can have a

DNA test (usually from blood or saliva) to look for genetic alterations in the APC gene. This blood test can also be used to evaluate alterations in other genes, which might be less-common causes of polyposis. A pedigree of the case family revealed that there were two younger male siblings, aged 14 and 11, both of whom are at a 50% risk of developing the disease. The mutation was not found in the 14-year-old sibling, so further screening was not necessary in his case. However, the 11-year-old brother was found to have the mutation.

Treatment and Prognosis

Regular screening is recommended for people with FAP, including sigmoidoscopy or colonoscopy every 1–2 years starting at age 10–12, with yearly colonoscopy once polyps are found until a colectomy is planned. People with FAP may require a total colectomy, due to the high number of polyps and the associated inevitable development of colorectal cancer (Fig. CS41.1).

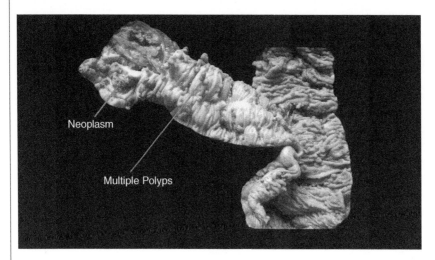

Figure CS41.1 Gross specimen of familial adenomatous polyposis. Note the multiple polyps and the presence of a cancer (neoplasm). *Source:* Dr. Roshan Nasimudeen / Wikimedia Commons / CC BY-SA 3.0.

Questions

1 FAP predisposes sufferers to develop colorectal cancer, but are there other genetic conditions associated with an increased risk of colorectal cancer?

2 Do different mutations in the APC gene confer variation in the phenotype of affected individuals?

Answers on p. 546.

Case Study 42   Heart Murmur and Cleft Palate

Clinical Presentation

A 10-day-old baby boy with cleft palate and a cardiac murmur required assisted ventilation. Several hours later, he developed twitching, neuromuscular irritability and convulsions.

Investigations

Laboratory investigations revealed a plasma calcium level of 1.6 mmol/L (reference range: 2.2–2.67). Values for all other analytes measured were within normal reference ranges. Investigation at a regional paediatric cardiothoracic centre by echocardiography revealed obstructive lesions of the left (pulmonary) outflow tract together with a ventral septal defect (VSD). A clinical geneticist reviewing the case noted that the facial features were slightly dysmorphic. Inner canthi were slightly displaced and the palpebral fissures were short. The root and bridge of the nose were noted to be wide and prominent. The mouth was relatively small. The ears were low set and posteriorly rotated. Deficient upper helices together with an increase in anteroposterior diameter gave a relatively circular shape to the ear. Previous family history revealed that a female half-sibling from the father's previous marriage had died from cardiac failure within 5 weeks of birth. A postmortem examination had revealed the presence of congenital heart defects together with hypoplastic thymus and parathyroid glands. Further questioning of the family also revealed that the patient's father had learning problems at school. He also had a cleft palate surgically repaired.

Blood samples were taken from the patient and his parents for chromosome analysis. Fluorescence *in situ* hybridisation (FISH) analysis of the patient's chromosomes using a probe specific to certain regions of chromosome 22 revealed a submicroscopic deletion on one copy of chromosome 22 at band q11 (Fig. CS42.1). This region encompasses the so-called DiGeorge commonly deleted region (DGCR; Fig. CS42.2). The same abnormality was also present in the father's case, but the mother's chromosomes were normal. The family pedigree is shown in Fig. CS42.3. Additional laboratory tests revealed reduced numbers of circulating T lymphocytes in the index case. A chest radiograph failed to show any conclusive evidence of thymus gland hypoplasia.

Diagnosis

Congenital cardiac defect, dysmorphic facies, reduced numbers of T-cells, indicating thymus gland hypoplasia, cleft palate and hypocalcaemia, together with a microdeletion of chromosome 22 at q11 suggests a diagnosis of DiGeorge syndrome (see also Chapter 6). The patient has inherited the chromosome abnormality from his father.

Diagnosis: DiGeorge syndrome.

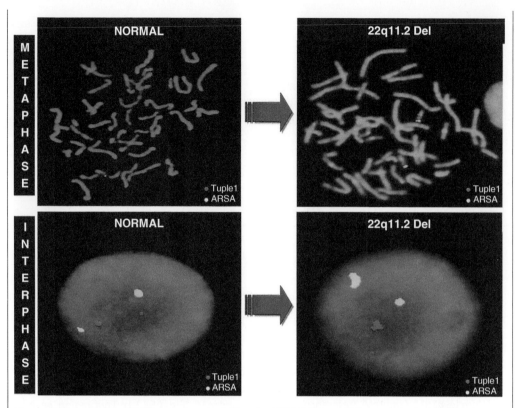

Figure CS42.1 Result of FISH analysis using LSI probe (TUPLE 1) from DiGeorge/velocardiofacial syndrome critical region. TUPLE 1 (HIRA) probe was labelled in Spectrum Orange and Arylsulfatase A (ARSA) in Spectrum Green as control. Absence of one of the orange signals indicates deletion of the TUPLE 1 locus at 22q11.2. Tonelli *et al.* Journal of Medical Case Reports 2007 1:167. *Source:* Adriano R Tonelli1 / Wikimedia Commons / CC BY 2.0.

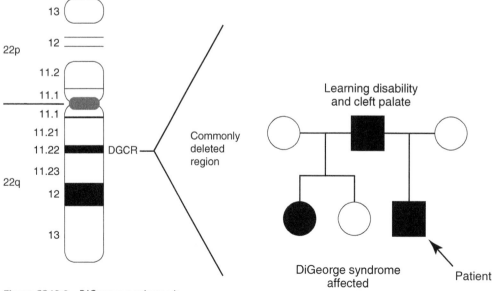

Figure CS42.2 DiGeorge syndrome is characterised by deletion of chromosome 22 within sub-band 22q11. The DiGeorge commonly deleted region (DGCR) spans approximately 2 megabases of DNA.

Figure CS42.3 Family pedigree of case family. Square symbols represent male members, and circles represent female members. Solid symbols represent affected individuals.

Discussion

DiGeorge syndrome comprises thymus and parathyroid gland hypoplasia, hypocalcaemia, cardiac outflow tract defects and dysmorphic facies. It is almost invariably associated with an interstitial microdeletion of chromosome 22 encompassing the DGCR, which comprises about 2 megabases of DNA.

DiGeorge syndrome is at the severe end of a spectrum of overlapping clinical disorders associated with the same deletion. *Conotruncal anomaly face syndrome* (dysmorphic facial appearance and cardiac outflow defects) and *velocardiofacial syndrome* (VCFS) (cleft palate, cardiac anomalies and characteristic facial appearance) show a deletion within 22q11 in most cases. All three syndromes show considerable phenotypic overlap and variability, even within a single family. The acronym CATCH-22 has been coined as an umbrella term for this overlapping group of syndromes.

Mnemonic

C: congenital heart disease (particularly conotruncal anomalies)
A: abnormal facies (hypertelorism, low set ears, short philtrum, among others)
T: thymic hypoplasia
C: cleft palate/cellular immune deficiency
H: hypoparathyroidism with hypocalcaemia
22: deletion located on chromosome 22

Of the 30–50 genes in the deleted region, a number have been identified as potentially pathogenic. Among these is the T-box transcription factor-1 (TBX1) gene. TBX1 has an important function in embryonic development, especially in the differentiation and migration of neural crest cells. Point mutations in TBX1 have also been observed in individuals with DiGeorge syndrome.

Treatment and Prognosis

Convulsions were controlled by the slow administration of calcium gluconate in dextrose with cardiac monitoring. It is important to monitor heart rate during treatment since bradycardia may develop. Any congenital heart defects will require surgery in an appropriate unit. It should be noted that thymus gland hypoplasia in DiGeorge syndrome, whilst resulting in the underproduction of T lymphocytes, does not always produce a life-threatening immunodeficiency. However, common infections may occur frequently up to 2–3 years of age, which is then often followed by a spontaneous improvement. Moderate to severe developmental delay may be expected to occur in approximately half of all patients with DiGeorge syndrome who survive beyond infancy. Symptoms of deafness may be present in older children and this should be confirmed by formal testing. Genetic counselling of parents of children with DiGeorge syndrome depends on whether a parent has the deletion. When one parent has the deletion, the risk of any offspring being affected is 50%. The value of FISH in the diagnosis of chromosome abnormalities is well illustrated in this case. FISH may also be useful in the prenatal diagnosis of DiGeorge syndrome. Screening for microdeletions of 22q11 using FISH has been successfully accomplished on cells from chorionic villus samples at 11–12 weeks of pregnancy. Counselling for this approach may still present clinical and ethical dilemmas because of the phenotypic variability that is seen in affected individuals.

Questions

1 Does the risk of having affected offspring increase or decrease in a man or woman who is a known deletion carrier who has two 'normal' (i.e. deletion-negative) offspring?
2 What causes hypocalcaemia in DiGeorge syndrome?

Answers on p. 547.

Case Study 43   Meconium Ileus in a Newborn

Clinical Presentation

A baby girl was born at term via normal vaginal delivery. The parents had noticed the passing of small volumes of very sticky greenish stools since birth. They also thought that her 'tummy was swollen'. On examination, the baby's abdomen was indeed distended, but otherwise, she appeared well, with stable vital signs. Auscultation revealed decreased bowel sounds. Meconium ileus – obstruction of the intestines of newborns with thick meconium – was suspected. Meconium ileus occurs in 15–20% of cystic fibrosis (CF) patients.

Investigations

An X-ray revealed dilated intestinal loops with a 'soap bubble' appearance (due to small air bubbles mixed with the meconium), confirming meconium ileus. The UK National Health Service (NHS) offers newborn screening for CF, typically performed by heel prick blood test done within the first few days of life. The initial step of CF screening involves measuring immunoreactive trypsinogen (IRT) levels. In this patient, elevated IRT levels were detected, indicating a potential CF risk.

Diagnosis

As the initial screening test suggested the child may have cystic fibrosis, additional tests were undertaken:

1) A sweat test – to measure the amount of chloride in sweat, which will be abnormally high in someone with cystic fibrosis. In a sweat test, there are three categories of result:
   - Chloride level less than or equal to 29 mmol/L = CF is unlikely.
   - Between 30 and 59 mmol/L = CF is possible and additional testing is needed.
   - Greater than or equal to 60 mmol/L = CF is likely.

In this baby, the chloride level in sweat was 65 mmol/L.

2) A genetic test to look for a CF mutation.
   Testing for specific CF-causing gene mutations revealed that this child was a compound heterozygote for the DeltaF508 and 621+1G>T mutations.

Diagnosis: Cystic fibrosis.

Discussion

CF is one of the most common autosomal recessive conditions, with an average carrier frequency of approximately 1 in 20 and an incidence of 1 in 2500 in North European Caucasian populations. The disease is much less common in some ethnic populations. Confirmation of the condition can be provided by genetic tests. Mutations in the cystic fibrosis transmembrane conductance regulator (CFTR) gene cause cystic fibrosis (see also Chapter 10). The DeltaF508 mutation accounts for 75% of all CF mutations seen in the United Kingdom. Genetics laboratories routinely test for the presence of the most common mutations. As of the most recent date (29 April 2023), a total of 804 variants are annotated on the CFTR2 website (https://cftr2.org/mutations_history); 719 of these are listed as CF-causing and 49 are variants of 'varying clinical consequence'. This baby's parents were tested to confirm their CF carrier status (Fig. CS43.1). The mother was heterozygous for the DeltaF508 cystic fibrosis mutation. The father was confirmed as the carrier of the 621+1G>T mutation. The father of the case patient has a brother (individual number 4) who did not have a mutation. However, the brother's pregnant wife (5) was found to be a carrier of the DeltaF508 mutation, despite having no family history of the condition.

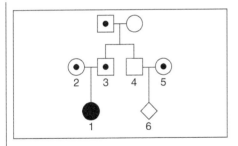

Figure CS43.1 Pedigree analysis of the CF patient and her family. The affected female is represented as a filled circle (1). The child's mother (2) was a carrier of the deltaF508 mutation and the child's father (3) was a carrier of the 621+1 G>T mutation. Analysis of the father's brother (4) and his female partner (5) revealed that only the female partner was a carrier. Their unborn child (6) has a 50% risk of being a carrier.

Treatment and Prognosis

In this patient, ileal obstruction was resolved by rectal infusion of gastrografin and acetylcysteine enemas. In some cases of CF-associated meconium ileus, surgical intervention is necessary to alleviate the blockage, or address complications such as intestinal atresia or perforation.

CF is characterised by abnormally thick mucus lining the epithelial tracts. Bronchioles often become obstructed. Patients are prone to recurring bacterial infections and hence long-term damage, which can eventually lead to bronchiectasis (see also Case Study 18). Pancreatic ducts that deliver digestive enzymes to the intestine can also become occluded and this causes malabsorption of fats and fat-soluble vitamins, but can be treated by oral ingestion of supplementary digestive enzymes taken with food, together with a high-calorie diet and extra vitamins.

Management of lung disease involves the use of bronchodilators, mucolytics (e.g. Pulmozyme) airway clearance techniques, antibiotics, and in some cases, anti-inflammatory medications. Chronic cystic fibrosis lung disease is characterised by bacterial colonisation, with for example, *Staphylococcus aureus*, *Pseudomonas aeruginosa*, or *Burkholderia*

cepacia, as well as fungal colonisation, primarily by *Aspergillus fumigatus*, with colonisation patterns changing over time with disease progression and patient age. Infertility is observed in approximately 95% of males due to blockage of the vas deferens. Some males with CF may be only mildly affected and their only physical symptom is congenital bilateral absence of the vas deferens (CBAVD). Subfertility is often seen in females.

The prognosis for individuals born with CF has improved greatly over the last 30 years. Based on 2019 data, the US Cystic Fibrosis Foundation Registry Report calculated the predicted median survival of a child born with CF that year to be 48.4 years. The greater longevity for CF patients is the result of a comprehensive team approach to care, the use of bronchodilators and anti-inflammatory drugs, exercise to treat pulmonary disease, improved nutrition and better management of complications, including respiratory failure. Moreover, newer drugs have been approved in the last few years that target specific CFTR abnormalities (see also Chapter 10). They include a CFTR modulating triple therapy known as Trikafta, combining three drugs – elexacaftor and tezacaftor, which are 'correctors' (they bind to mutant CFTR protein to allow its proper folding and presentation at the cell surface), and ivacaftor, a 'potentiator' (it allows for an increase in chloride ion flow). In some CF sufferers, CFTR drugs can reverse CF-associated pathology, including bronchiectasis (Case Study 18)

Questions

1 Why are elevated levels of chloride ions helpful in the diagnosis of cystic fibrosis?
2 What is the immunoreactive trypsinogen test?
3 What happens if the IRT levels are elevated, but no CFTR variant is detected by genetic testing?

Answers on p. 547.

Case Study 44   Severe Vaginal Bleeding in Pregnancy

Clinical Presentation

A 42-year-old woman, pregnant at 14 weeks of gestation, was admitted to hospital following several episodes of heavy vaginal bleeding with the passage of clots. She had two normal births previously. Otherwise, her medical history was unremarkable. On examination, she was pale, in pain (7/10 intensity), but her temperature was normal with a pulse rate of 110 beats per minute and a blood pressure of 135/75 mmHg. Examination revealed a tender abdomen.

Investigations

Ultrasound examination revealed embryonic tissues and a placental mass with a diffuse multi-cystic appearance. Laboratory tests showed an elevated serum beta-human chorionic gonadotropin (β-hCG) level of 13 420 mIU/mL (reference range <5). Vacuum aspiration and ultrasound-guided curettage were performed in the operating theatre. Vacuum aspiration is a procedure that uses suction to remove tissue from the uterus, while ultrasound-guided curettage involves the scraping away of cells from the uterine lining under ultrasound imaging. Microscopic examination of the removed tissue by a pathologist showed the typical histological appearance of a so-called partial hydatidiform mole characterised by an overgrowth of villous trophoblasts with cystic 'swollen' villi.

Diagnosis

The clinical picture together with the raised hCG indicated *gestational trophoblastic disease (GTD)*. GTD is a spectrum of disorders (see below). Partial mole differs from a complete hydatidiform mole, both in terms of cytogenetics and microscopic appearance. In particular, embryonic tissues are present in partial moles but are absent in complete moles.

Immunostaining for p57 can also help distinguish partial mole (p57+) from complete mole (p57–) in difficult cases. The $p57^{KIP2}$ gene encodes the p57 protein, an inhibitor of the cell cycle, which is paternally imprinted and expressed from the maternal allele. P57 immunostaining was positive in this case, meaning that the mole contained maternally derived tissue (and hence was a partial mole, see below). A negative p57 immunostain would indicate that only paternally derived tissues were present (and so this would be a complete mole). Taken together, the imaging and histology, together with the presence of embryonic tissues, are highly suggestive of a partial hydatidiform mole.

Diagnosis: Partial hydatidiform mole.

Discussion

Gestational trophoblastic disease (GTD) arises from the abnormal proliferation of placental trophoblastic tissue and constitutes a spectrum of disease that includes: hydatidiform mole (complete or partial), invasive mole, choriocarcinoma, placental site trophoblastic tumour and epithelioid trophoblastic tumour (Fig. CS44.1). The latter four entities being referred to as gestational trophoblastic neoplasia (GTN). GTN are aggressive conditions that can metastasize and require treatments that include chemotherapy. Early diagnosis of GTD is important to allow the appropriate treatment of choice as well as preserving fertility for subsequent pregnancies when possible.

Molar pregnancy most often presents with vaginal bleeding usually between 6 and 16 weeks of gestation. hCG is produced by the trophoblast cells and therefore quantitative measurement of hCG is useful; complete moles are usually associated with much higher levels of hCG compared with partial moles. Ultrasonography is the imaging

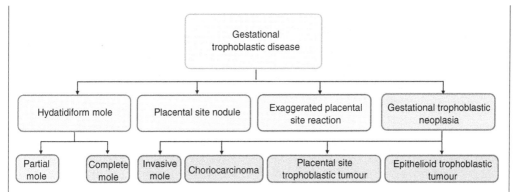

Figure CS44.1 Classification of gestational trophoblastic disease (GTD) by the WHO 2020 Classification of Female Genital Tumours. *Source:* Created with BioRender.com.

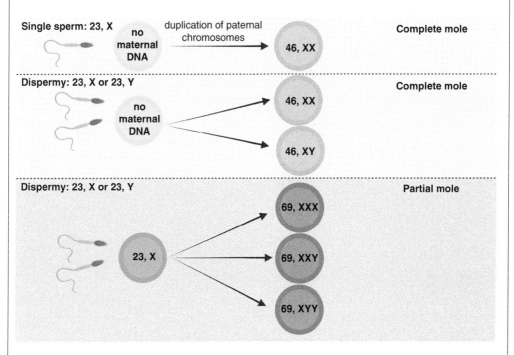

Figure CS44.2 Development of hydatidiform moles. Complete hydatidiform moles can occur when an empty ovum (which has lost maternal chromosomes) is fertilised either by a single sperm (in which case the sperm DNA is duplicated) or by two sperms. Thus, complete hydatidiform moles contain 46 paternally derived chromosomes. Partial hydatidiform moles contain both male and female DNA and usually have a triploid karyotype, for example, as a result of dispermic fertilisation of a normal ovum. *Source:* Created with BioRender.com.

modality of choice for the diagnosis of hydatidiform mole with confirmation by histopathology. A complete mole is the result of fertilisation of an ovum lacking maternal chromosomes either by two sperm or by a single sperm (the chromosomes of which are duplicated) (Fig. CS44.2). Complete moles are thus typically diploid (either 46, XX, or 46, XY), but only paternal chromosomes are present. In contrast, partial moles arise when a normal sperm fertilises a normal ovum and then duplicates its chromosomes or when two sperm fertilise a normal ovum. Thus, the karyotype is usually triploid, either 69XXX, 69XXY or 69XYY. In partial moles, both maternal and paternal chromosomes are present. Complete molar pregnancies are also classed as germ cell tumours (see Case Study 61).

Treatment and Prognosis

After termination of the molar pregnancy, the patient was followed up by hCG testing. This is required to detect any residual disease or progression to more serious forms of GTD. In this patient, hCG levels returned to normal 3 months after surgery and remained so for the following 6 months with no subsequent adverse events. During this period, the patient was advised to take precautions to prevent another pregnancy. The patient gave birth to a healthy baby boy 12 months after screening was complete.

Questions

1 What are the characteristics of the more serious forms of GTD?
2 Can choriocarcinoma occur in the absence of pregnancy?
3 Why is there no embryo in a complete molar pregnancy?

Answers on p. 547.

Case Study 45   Recurrent Miscarriage

Clinical Presentation

A 36-year-old woman and her 28-year-old husband had been attempting to conceive for 5 years. She had experienced four miscarriages, three of which were in the first trimester and the longest gestational period was 14 weeks.

Investigations

Recurrent miscarriage is often defined as the loss of three or more pregnancies prior to 20 weeks of gestation. The cause of recurrent miscarriage can be very difficult to determine. The first step is a detailed personal and family history. Exclusion of anatomical disorders or uterine abnormalities (e.g. cervical incompetence and fibroids) is important. Endocrine factors such as inadequate luteal function and polycystic ovary disease should also be considered. Immunological causes are autoimmune disease (e.g. systemic lupus erythematosus and rheumatoid arthritis, see Chapter 6) and the autoimmune-related condition, antiphospholipid syndrome (see also Chapter 8). A range of obstetric investigations at local hospitals revealed no problems in this couple; a visit to a specialist infertility centre resulted in blood samples being sent for chromosome analysis. The male partner's karyotype was normal, but the female was shown to be carrying a balanced reciprocal translocation of material between the long arm of chromosome 1 and the long arm of chromosome 11.

Diagnosis

Cytogenetic analysis revealed the karyotype: 46, XX t(1,11)(q44;q21). This indicated a female karyotype with an apparently balanced reciprocal translocation between the distal long arm of chromosome 1 and the long arm of chromosome 11. This is consistent with recurrent miscarriage due to the formation of unbalanced products of the translocation. Although it is likely that the unbalanced products will lead to foetal loss, there is an associated risk of an abnormal live-born child with mental or physical abnormality coming to term. An empirical estimate of this risk is between 1 in 20 and 1 in 100; a family study may help to define the figure more accurately and to identify other family members who may carry the translocation. However, if the translocation is the only factor affecting the foetal loss, then this couple does have the potential to establish a successful pregnancy with a normal or balanced chromosome complement. Any established pregnancy could be monitored by prenatal diagnosis. The couple could benefit from referral for genetic counselling, when the risks associated with this translocation could be discussed and a family study instigated.

Diagnosis: Balanced translocation involving chromosomes 1 and 11.

Discussion

Chromosome abnormality is a very common reason for foetal loss; it is thought that 50–60% of early miscarriage is due to this cause. However, many are sporadic events associated with errors in gametogenesis and do not recur. For a small number of couples with recurrent miscarriages, a balanced chromosome rearrangement can be responsible for the high frequency of pregnancy loss. These rearrangements, which include the translocation of material between two or more chromosomes or inversion of material within a chromosome, do not affect the health or phenotype of the carrier in any way other than their reproductive fitness, since the number of genes present is not affected. When the meiotic reduction division occurs to produce sperm or ova with a haploid karyotype in a translocation carrier, some gametes will inherit only one of the translocated chromosomes without its reciprocal partner, so resulting in chromosome imbalance in

those gametes. Depending on the type and degree of imbalance, the resulting conception may abort or develop into a live-born child with physical and mental abnormalities. Less commonly, a balanced familial translocation is detected by the birth of an abnormal child, rather than by recurrent miscarriage.

Treatment and Prognosis

To some extent, the likelihood of lethality of the imbalance or the expected prognosis for a live-born child can be predicted by knowledge of the chromosomal regions involved and by a family study of carriers. This is why referral to genetic counselling for the couple is important following diagnosis; prenatal foetal chromosome analysis should only be considered if there is a significant risk of a live-born abnormal child as no established pregnancy should be put at risk by that test unnecessarily. For most couples, an explanation for their recurrent miscarriages is a positive event, particularly if the diagnosis does not exclude the possibility of a successful pregnancy. Identification of carrier status can also help a couple decide on whether assisted reproduction techniques (ART) might be appropriate.

Questions

1 Fertility problems affect around 1 in 10 couples attempting to conceive. At what stage should chromosome analysis be considered?
2 For couples, in which one partner carries a balanced translocation and is experiencing fertility issues or recurrent miscarriages, assisted reproductive technologies (ART) may offer a viable path to a successful pregnancy. What kinds of ART approaches could be considered?

Answers on p. 548.

Case Study 46   Seizures, Aggressive Behaviour; Rett Syndrome-Affected Sister

Clinical Presentation

The index patient was an 8-year-old girl with non-specific developmental delay, seizures and aggressive behaviour. She was small for her age. She was referred for cytogenetic analysis coincidentally as part of a study of Rett syndrome patients. Rett syndrome is a progressive neurological disorder, limited to females (boys who have a similar mutation typically die shortly after birth), characterised by psychomotor deterioration and specific behavioural traits, notably stereotypical hand movements, for example hand-wringing. Rett syndrome is due to a mutation in the MECP2 gene on the X chromosome. It almost always occurs as a new (*de novo*) mutation, with less than 1% of cases being inherited. Diagnosis is based on the symptoms and can be confirmed with genetic testing. Her older sister had been diagnosed with Rett syndrome and a cytogenetic survey was being conducted on all such patients and their first-degree relatives (parents and siblings).

Investigations

Examination of the index patient revealed a subtle deletion of the tip of the short arm of one copy of chromosome 1. Although the loss of material appeared to be consistent, the small size of the deletion, at the limit of resolution of light microscopy, did not allow exclusion of the possibility that the patient had a balanced rearrangement involving this region. Fluorescence in situ hybridisation (FISH) studies were performed using a probe that maps to the tip of the short arm of chromosome 1 at 1p36.3. This confirmed that the index patient had a terminal deletion of the tip of the short arm of chromosome 1.

Diagnosis

FISH analysis showed that the index patient had a deletion at the tip of the short arm of chromosome 1. Her karyotype may be written as 46,XX,del(1)(p36.3) *de novo*. Clinical features associated with this particular deletion include (percentage of patients affected in brackets): large anterior fontanelle (100%), moderate to severe developmental delay (92%), growth delay (85%), eye/vision problems (75%), hearing deficits (56%), abusive/aggressive behaviour (56%) and deep-set eyes (50%).

Diagnosis: 1p36 deletion syndrome.

Discussion

1p36 deletion syndrome is a rare genetic disorder caused by a deletion of a segment of DNA on the outermost band of the short arm of chromosome 1. It is one of the most common deletion syndromes and affects approximately one in every 5000–10 000 births. This case illustrates the power of FISH to detect the presence of a subtle chromosome abnormality and to distinguish this patient from her chromosomally normal sister diagnosed with Rett syndrome. FISH confirmed the presence of a suspected deletion, but more importantly it:

- excluded the possibility that the deleted material had been translocated elsewhere in the karyotype, thus confirming that there was a chromosomal imbalance.
- confirmed that the abnormality was not present in either parent or her sister in balanced or unbalanced form.

These two observations clearly have clinical consequences. The most important of which is that the clinician can assign some, if not all, of the child's clinical problems to the presence of the deletion. Many parents in such a situation find it helpful to have a reason for the problem.

Questions

1 If the deleted 1p had been found in apparently unbalanced form in one or other parent, what further tests might have been performed?

2 From which parent was the *de novo* deleted chromosome 1 most likely to have been inherited?

Answers on p. 548.

Case Study 47 Sudden Onset of Severe Pain in the Leg

Clinical Presentation

A 20-year-old man with known sickle cell disease attended casualty with sudden onset of pain in his right thigh of 6 hours duration. He had initially tried to control the pain with oral paracetamol, but this had little effect. On examination, the patient was in obvious discomfort and reluctant to move his right leg. The pain was localised to the right thigh, but there was nothing to see on the examination. He was febrile (temperature 37.8°C) and the sclerae were yellow, indicating mild jaundice.

Investigations

The results of laboratory investigations are shown in Table CS47.1. The patient was anaemic with a high reticulocyte count, indicating haemolysis. The blood film showed sickled red blood cells consistent with sickle cell anaemia. There was neutrophilia and mild thrombocytosis, which could indicate acute infection or inflammation. However, levels of the C-reactive protein (CRP) were not dramatically raised indicating that an infective cause of the bone pain was unlikely. The elevated bilirubin, in the absence of abnormal liver enzymes, suggested a haemolytic disorder.

Diagnosis

The patient was known to have sickle cell disease from childhood, confirmed by

haemoglobin electrophoresis showing 98% haemoglobin S (HbS) and 2% haemoglobin F (HbF). He had been relatively well during childhood apart from some swelling and inflammation of the hands at the age of 5 years, and an episode of pneumonia at the age of 16. The current acute onset of pain and splinting (holding the limb rigid) with no evidence of bony abnormality on X-ray is typical of an acute sickle cell crisis.

Diagnosis: Acute sickle cell crisis involving the right femur.

Discussion

Sickle cell disease is one of a group of disorders classified as the haemoglobinopathies (see also Chapter 7). Haemoglobin is present in solution in the red cell as a tetramer composed of two α-chains and two non-α-chains. The majority of adult haemoglobin is composed of two α- and two β-chains (HbA) with small quantities of haemoglobin composed of two α- and two δ-chains (HbA$_2$), and haemoglobin made up of two α- and two γ-chains (HbF).

Haemoglobinopathies are of two types: (i) quantitative disorders; the thalassaemias, in which there is deficient production of normal haemoglobin chains, usually due to gene deletion, and; (ii) qualitative disorders, such as sickle cell disease where there is

Table CS47.1 Results of laboratory and clinical investigations.

Test	Result	Reference range
Blood		
Total white cell count	**20.0×10^9/L**	4.0–11.0
Red cell count	**2.2×10^{12}/L**	4.5–6.5
Haemoglobin	**7.1 g/dL**	13.0–18.0
Haematocrit	**0.26**	0.40–0.54
Mean cell volume	**98.7 fL**	82–92
Mean cell haemoglobin	32.3 pg	27–32
Platelets	**460×10^9/L**	150–400
Absolute reticulocyte count	**279×10^9/L**	50–100
Bilirubin	**42 µmol/L**	<21
Alanine transaminase	25 U/L	5–40
Alkaline phosphatase	70 U/L	25–115
Urea	**2.0 mmol/L**	2.5–6.7
Creatinine	71 µmol/L	59–104
C-reactive protein (CRP)	**20 mg/L**	<10
Blood culture White blood cell differential	No growth	
Neutrophils	**86%**	40–75
Lymphocytes	**4%**	20–45
Monocytes	8%	2–10
Eosinophils	2%	1–6
Red blood cell morphology	**Sickled red cells +**	—
Radiology		
X-ray lumbar spine, right femur	No abnormality seen	—

Values outside the reference range are in bold.

production of an abnormal haemoglobin. The gene defect in sickle cell disease is a point mutation (GAG codon changing to GTG) in the β-globin gene, which results in glutamate (E/Glu) being substituted by valine (V/Val) at position 6 (E6V substitution). Sickle cell disease is found with the highest frequency in sub-Saharan Africa, parts of India, and the Middle East. However, migration of affected individuals/carriers to lower-prevalence countries means that some resource-rich nations now have high rates of sickle cell disease.

Under conditions of reduced oxygenation, HbS precipitates into elongated rigid polymers forming characteristic sickle cells (see also Chapter 7), that do not flow well through small vessels and are more adherent than normal to vascular endothelium, leading to vascular occlusion and sickle cell crises, as well as chronic red cell haemolysis. This process occurs if the percentage of HbS is high in the red cells as in homozygous sickle cell anaemia (HbSS – when the patient has inherited HbS from both parents), or in sickle cell β-thalassaemia (when the patient has inherited HbS from one parent and a non-producing haemoglobin gene from the other parent). Patients who have inherited only one HbS gene, while the other is normal, have sickle cell trait (the carrier state) and are normally asymptomatic with normal blood parameters. Rarely, people with sickle cell trait may undergo pain crises if, for example, there is increased pressure, for example when scuba diving, or when atmospheric oxygen levels are low, for example at high altitudes.

Treatment and Prognosis

As the vaso-occlusive event cannot be reversed, treatment of the crisis in sickle cell disease is purely symptomatic and centred on pain relief. Hydroxyurea and L-glutamine reduce the frequency of painful crises. Crizanlizumab, given by injection, can also help reduce the frequency of pain crises. Crizanlizumab is an antibody directed against the adhesion molecule, P-selectin that is found on endothelial cells and platelets and which contributes to the cell–cell interactions that are important in the vaso-occlusion and sickle cell-related pain crises. Voxelotor can lower the risk of anaemia and improve blood flow. Voxelotor is an HbS polymerisation inhibitor. As sickle cell patients can become dehydrated due to poor renal concentrating ability, intravenous fluid replacement is also given. Sickle cell crises can occur spontaneously or may be precipitated by dehydration, cold, infection or stress.

It is important to exclude a treatable infection as the cause of the pain in sickle cell patients. The CRP result is against an infection, but a chest X-ray, blood and urine culture are usually taken and the patient started on a penicillin antibiotic, such as amoxycillin. The painful crisis subsides over 4–6 days, though the time course can be variable.

Sickle cell disease is a chronic debilitating condition with a reduced life expectancy, though the clinical variability of the condition means that some individuals may never experience any symptoms. Usually, sickle cell disease presents in early childhood with vaso-occlusion within the bones of the hands and feet, causing pain and swelling of the digits (*dactylitis*). Patients are also prone to acute splenic sequestration causing sudden severe anaemia as their blood volume pools and sequesters within a rapidly enlarging spleen. The spleen eventually infarcts leading to hyposplenism. Repeated infections are also common in childhood. In adulthood, painful crises in long bones predominate. Other complications are vaso-occlusion of cerebral vessels leading to cerebrovascular accident (stroke), while sickling within the lung (chest syndrome) causes increasing hypoxia, which has a poor prognosis if not dealt with urgently. Later, repeated bone infarctions result in joint problems, while chronic organ damage can occur in the lungs and kidneys. Stem cell transplantation is the only cure for sickle cell disease, but because of the high mortality rate, it is usually recommended only for people, normally children, with severe symptoms.

Questions

1 Why has the sickle gene predominated in equatorial Africa?
2 Why is the blood urea concentration low in patients with sickle cell anaemia?
3 Why may the patient not have the severity of symptoms which would be typical for the degree of anaemia?
4 What new treatments are on the horizon for people with sickle cell disease?

Answers on p. 548.

Case Study 48   Easy Bruising and Recurrent Nose Bleeds

Clinical Presentation

A 5-year-old boy presented with a 12-month history of easy bruising and recurrent nose-bleeds. His parents reported an episode of prolonged bleeding following a tonsillec-tomy at the age of 3 years. On examination, he looked well but had several large bruises on his upper and lower limbs. His mother also had a life-long history of easy bruising and suffered with menorrhagia (excessive menstrual bleeding). The child's 10-year-old sister was well with no history of bleeding problems.

Investigations

The results of laboratory investigations are shown in Table CS48.1. The platelet count and the prothrombin time (PT) were normal, but the activated partial thromboplastin time (APTT) was prolonged. Factor VIII and von Willebrand factor (VWF) antigen levels were both reduced to 30% (not shown in the table). VWF activity was reduced to 2% (not shown). The VWF antigen test measures the total amount of VWF present in the blood. It indicates the quantity of VWF protein, regardless of its functional activity. The VWF activity assay assesses the functional activity of VWF by measuring its ability to bind to platelets in the presence of ristocetin, which promotes the platelet–VWF interaction. This test specifically evaluates how well VWF can facilitate platelet adhesion and clot forma-tion. VWF multimer analysis showed loss of large molecular weight multimers. This test typically uses gel electrophoresis to sepa-rate and visualise the different-sized VWF multimers.

Diagnosis

Reduced levels of Factor VIII and VWF antigen are consistent with a diagnosis of von Willebrand disease. The VWF activity levels were lower than the VWF antigen levels. These findings together with the absence of high molecular weight multim-ers confirmed the diagnosis of type 2A von Willebrand disease.

Diagnosis: Type 2A von Willebrand disease (VWD).

Discussion

VWD is the commonest inherited bleeding disorder and is characterised by either a quantitative or a qualitative abnormality of VWF, a large multimeric glycoprotein that is present in the blood plasma, the sub-endothelial matrix, as well as in storage granules in endothelial cells (Weibel-Palade bodies) and in the α-granules of platelets. It may be measured as von Willebrand antigen or using a biological assay to determine its functional activity. VWF has several impor-tant functions in haemostasis (Chapter 8). For example, it mediates platelet adhesion to the sub-endothelium at sites of vascular injury, resulting in the formation of platelet

Table CS48.1   Results of laboratory investigations.

Test	Result	Reference range
Total white cell count	8×10^9/L	5.0–15.0
Haemoglobin	12 g/dL	12.0–14.0
Platelet count	210×10^9/L	150–400
Prothrombin time	11 seconds	11–16
Activated partial thromboplastin time (APTT)	**50 seconds**	26–35
APTT 50:50 mix	35 seconds	26–35

Values outside the reference range are in bold.

thrombi, activates bound platelets causing platelet aggregation and serves as a carrier protein for Factor VIII, preventing its degradation.

VWD is a heterogeneous disorder that results from a variety of genetic defects. Although estimated to affect 1% of the population, its true incidence is not well defined. This is because it often presents only as a mild disease. VWD usually shows an autosomal dominant pattern of inheritance, though occasionally, autosomal recessive and sporadic cases can be observed. There is a mild to moderate bleeding tendency, typified by mucocutaneous bleeding, including epistaxis (nose bleeds), easy bruising and menorrhagia (which can result in iron deficiency anaemia). Patients with severe disease may have a clinical picture that is similar to haemophilia A or B with formation of large haematomas and haemarthroses (joint bleeds). Clinical history may reveal previous episodes of bleeding, especially following minor surgery, such as dental extraction, and can also be useful in identifying symptoms in parents or siblings, as in this case. VWD is classified into types 1–3:

- **Type 1: Partial quantitative VWF deficiency**. Parallel reduction of VWF antigen and activity
- **Type 2: Qualitative VWF deficiency**. Subdivided into:
  - Type 2A-absence of high molecular weight multimers, more marked reduction in VWF activity compared to antigen levels.
  - Type 2B-loss of high molecular weight multimers with increased binding of platelets, which may lead to thrombocytopenia.
  - Type 2M-qualitative deficiency, as in 2A, but with a normal multimer pattern.
  - Type 2N-decreased affinity for Factor VIII resulting in accelerated clearance of unbound Factor VIII and markedly reduced

Factor VIII levels. This subtype is often misdiagnosed as mild haemophilia A.
- **Type 3: Virtual complete deficiency of VWF.** Antigen and activity levels <5% of normal.

A routine full blood count should be performed to check the platelet count and look for iron deficiency anaemia due to chronic blood loss. Coagulation screening tests (prothrombin time, PT; activated partial thromboplastin time, APTT) may be normal, especially in type 1 patients, and therefore a normal PT and APTT should not exclude the diagnosis. If the plasma VWF level is significantly reduced, however, the associated low Factor VIII level may result in a prolonged APTT (as in this patient). A bleeding time test (which assesses the ability of blood vessels and platelets to form a clot and stop bleeding, primarily evaluating platelet function and the interaction between platelets and blood vessel walls) is not essential for diagnosis and may be difficult to perform in young children. The bleeding time is usually prolonged, but as with coagulation tests, may also be normal.

Levels of VWF may be altered by other genetic and environmental influences. For example, patients with blood group O have lower levels than those with other blood groups. Infection may increase levels, as does stress (e.g. if there was difficulty in obtaining a blood sample). Levels also increase during pregnancy and in patients receiving oestrogen therapy.

In type 1 VWD, which accounts for around 80% of cases, a single missense mutation leads to abnormal VWF subunits that hinder proper dimerisation. Type 2 VWD, accounting for approximately 20% of cases, is characterised by mutations that affect specific functional domains or multimer assembly of VWF. In type 2A, impaired multimerisation or increased susceptibility to ADAMTS13 (a protease enzyme) causes a selective loss of intermediate and large multimers, resulting in diminished activity of the binding domains for platelet receptors. Type 2B is characterised by gain-of-function mutations in the A2 domain, leading to an increased affinity of VWF for platelet receptors and a lack of high-molecular-weight multimers. Mutations that selectively impair VWF activity are found in type 2M patients and mutations that reduce the carrier activity of VWF for Factor VIII are found in type 2N patients. Type 3 VWD is caused by compound heterozygous or homozygous mutations or deletions that result in a nonfunctional VWF allele. Type 3 VWD is a rare recessive form. Carriers of a single non-functional allele, such as the parents of type 3 patients, have VWF levels that are usually not low enough to cause significant bleeding.

Treatment and Prognosis

Some patients may not need treatment. However, if it is required, it is important to consider the type of VWD, the nature of the haemostatic challenge and the patient's bleeding history. Desmopressin/DDAVP, is a synthetic vasopressin analogue that stimulates release of endogenous VWF from endothelial cell stores resulting in a two- to fourfold increase in plasma levels of both VWF and Factor VIII within around 30 minutes of an intravenous injection. It can also be given subcutaneously or intra-nasally. It is usually used in patients with type 1 disease, although some do not respond. In cases where desmopressin is not effective or for type 3 VWD, replacement therapy with VWF-containing products may be required; they provide the missing or deficient VWF and FVIII. Replacement therapy may be given on-demand for bleeding episodes or as prophylaxis to prevent bleeding. Anti-fibrinolytic medications such as tranexamic acid or aminocaproic acid can be used in combination with desmopressin or as replacement therapy to control bleeding. Patients with VWD should not use aspirin or other non-steroidal anti-inflammatory drugs that interfere with platelet function. As VWD is an inherited disorder, all affected families should receive genetic counselling.

Questions

1 To determine the cause of a prolonged APTT, a mixing study is often performed. This involves mixing the patient's plasma with an equal amount of normal plasma and then re-testing the APTT. Why is the mixing test performed?
2 Does the severity of VWD alter during pregnancy?
3 Can VWD be acquired?

Answers on p. 549.

Case Study 49   Neonate with Dysmorphic Features and High-Pitched Cry

Clinical Presentation

A female baby who was not as big as would be expected due to intrauterine growth restriction (IUGR), was observed at birth to have dysmorphic features comprising downward slanting palpebral fissures, a wide and depressed nasal bridge, *hypertelorism* (wide-set eyes), slight *micrognathia* (receding chin), low-set ears, short neck and high-arched palate. The birth was normal and Apgar scores were 8 and 10 at 1 and 5 minutes. The only other physical features were mild jaundice and anteriorly placed anus, but the child also had a very distinctive high-pitched, piercing cry, reminiscent of a mewling kitten. The baby was discharged and at a review appointment 3 weeks later, was feeding well. The jaundice had resolved and she could fix objects and follow movement. The high-pitched cry was still pronounced. Cardiovascular, respiratory and abdominal examinations were unremarkable. There was a patent soft fontanelle and normal Moro and palmar grasp reflexes.

Investigations

In view of the dysmorphic features and unusual cry, blood samples were taken for chromosome analysis. This showed a deletion of part of the short arm of one copy of chromosome 5.

Diagnosis

The cytogenetic analysis showed the karyotype: 46, XX, del(5)(p14). This indicated a female karyotype with a deletion of the distal region of the short arm of chromosome 5, between band p14 and the telomere. *In situ* hybridisation with a gene probe specific for the locus D5S23, critical for the Cri du chat syndrome, showed deletion of the p15.2 region (see also Chapter 10).

Diagnosis: Cri du chat syndrome.

Discussion

As the name implies, Cri du chat syndrome is characterised by a high-pitched, shrill cry in infancy, caused by hypertrophy of the larynx. The cry is monotone and almost an octave higher than normal; its resemblance to a mewling kitten is striking and often suggests the diagnosis. The syndrome was first described in 1963 and was one of the earliest clinical syndromes to be associated with a structural chromosome abnormality. It is one of the more common chromosome deletion syndromes with an estimated incidence of 1 in 50 000 births. As well as the cat-like cry, the infantile cranio-facial dysmorphism is also very suggestive of the syndrome. The small head circumference, the wide apart and deep-set eyes, broad flat nasal bridge and slightly receding chin give a characteristic moon-like face.

Not all deletions of the short arm of chromosome 5 give rise to Cri du chat syndrome. Deletions in this region have been associated with varying clinical phenotypes, from severe developmental delay and dysmorphism to clinically normal phenotypes, suggesting a critical region specific to the syndrome. Molecular and phenotypic mapping of short arm 5 deletions has allowed delineation of this critical region to within band 5p15.2. The size of this region is around 2 Mb of DNA, which could code for between 2 and 40 genes. A gene probe for the D5S23 locus can be used with the in-situ hybridisation technique to assess likely prognosis in short arm 5 deletion cases.

Treatment and Prognosis

Malformations are relatively rare in this syndrome and life expectancy is generally not reduced. However, the prognosis for affected individuals is poor, as developmental delay is mostly severe, with some patients having an IQ less than 25 and not attaining language or motor skills. The characteristic dysmorphism of infancy is lost with age and the face becomes thin. The cat-cry disappears quite rapidly, sometimes within weeks of birth. Parental chromosomes should be investigated as a small proportion of cases are due to unbalanced forms of balanced inherited rearrangements. Families are usually offered genetic counselling as the diagnosis can be a traumatic event and they have many questions to be answered. Affected children are monitored closely and regularly by paediatricians for developmental assessment.

Questions

1 What is the Apgar score?
2 Would the expected prognosis be the same for all cases of Cri du chat syndrome?

Answers on p. 549.

Case Study 50   Chest Pain and Shortness of Breath

Clinical Presentation

A 24-year-old man was referred to the cardiology department by his GP after he reported experiencing chest pain and shortness of breath. Upon physical examination, the patient's blood pressure was found to be slightly elevated at 140/90 mmHg and his heart rate was 88 beats per minute. It was notable that the patient had experienced blurred and occasional double vision, which was diagnosed 6 months ago as *ectopia lentis* (displacement of the natural crystalline lens from the patellar fossa). It was also noted that the patient was very tall and thin with long arms, legs, fingers and toes, and that he had an arched palate with crowded teeth and a small lower jaw. He also had numerous stretch marks on his trunk. The patient had an inguinal hernia that was repaired 13 years ago.

Investigations

Further examination revealed a systolic murmur on the left side of his chest. An echocardiogram was performed, which showed a thoracic aortic dissection measuring 5.6 cm in diameter. A CT angiogram confirmed the diagnosis. Laboratory results were generally unremarkable, although haemoglobin levels were below the normal reference range.

Diagnosis

Given the patient's young age at presentation and the pre-existing ectopia lentis, a diagnosis of Marfan syndrome was suspected. Marfan syndrome can be diagnosed using the Ghent nosology; a set of clinical criteria outlined by international expert opinion to facilitate an accurate recognition of this genetic syndrome and improve patient management and care. In 2010, an international panel of experts revised these criteria to reduce the likelihood of missed or premature diagnosis. The updated criteria give more weight to cardiovascular symptoms, with aortic root aneurysm and ectopia lentis considered as cardinal features. If these two features are present in the absence of any family history, a diagnosis of Marfan syndrome can be made. If only one cardinal feature is present, a positive *systemic score* or the detection of an FBN1 mutation (see below) is required for diagnosis. The systemic score is a quantitative assessment of the involvement of various systems in the body, including the skeletal, ocular, cardiovascular and pulmonary systems. Each system is scored based on the presence and severity of certain clinical criteria associated with Marfan syndrome. A positive systemic score indicates that the individual has a considerable number of clinical features associated with Marfan syndrome, even if they do not meet the strict diagnostic criteria based solely on the cardinal features of the disorder.

Aortic enlargement with aneurysm formation and consequent aortic dissection is a classic feature of Marfan syndrome. The weakened aortic wall is disrupted, and the pulsatile blood can fissure and travel within the aortic wall, separating its layers – this is *aortic dissection*. Sometimes, the artificial lumen that is created compresses branch arteries. The blood can re-enter the true lumen below by breaking back into the aorta. This can result in a relatively stable dissection. Of course, the damaged weakened wall may fail with resultant rupture and death. In some cases, the dissection can travel back to the aortic root, widening the aortic valve ring with resultant aortic incompetence and may compress the coronary artery origins with resultant cardiac muscle infarction. There are two major classification

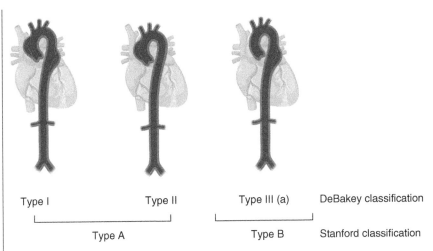

Type I   Type II   Type III (a)   DeBakey classification

Type A   Type B   Stanford classification

Figure CS50.1   Classification of aortic dissection. In the DeBakey classification: Type I involves the ascending aorta, arch and descending thoracic aorta and may progress to involve the abdominal aorta. Type II is confined to the ascending aorta. Type IIIa involves the descending thoracic aorta distal to the left subclavian artery and proximal to the celiac artery. Type IIIb dissection (not shown) involves the thoracic and abdominal aorta distal to the left subclavian artery and distal to the celiac artery. In the Stanford classification: Type A involves the ascending aorta and may progress to involve the arch and thoracoabdominal aorta. Type B involves the descending thoracic or thoracoabdominal aorta distal to the left subclavian artery without involvement of ascending aorta. *Source:* Created with BioRender.com.

systems (DeBakey and Stanford) for aortic dissection (Fig. CS50.1). Marfan patients are also prone to develop peripheral aneurysms.

Diagnosis: Thoracic aortic dissection in a patient with Marfan syndrome.

Discussion

Marfan syndrome is an autosomal dominant genetic disorder that affects approximately 1 in 5000 to 1 in 10 000 people worldwide. It is caused by a mutation in the FBN1 gene, which encodes fibrillin, a protein important for connective tissue formation. Around 75% of cases are inherited from one parent, the remaining cases arising as *de novo* mutations. Affected individuals are often tall and thin, with long extremities (Fig. CS50.2) as well as unusually flexible joints and a curved spine. The damage caused by the syndrome can be either mild or severe. If the aorta is affected, then this is potentially life-threatening. Marfan syndrome sufferers have

an increased risk of developing mitral valve prolapse.

The FBN1 gene is one of three genes (FBN1, FBN2 and FBN3) that encode fibrillins, structural extensible macromolecules that contribute to the integrity, elasticity and function of connective tissues. Fibrillins can exist as large bundles or as short individual microfibrils, depending on the connective tissue type. For example, in skin, they form a loose network; in tendons and periosteum, they run parallel to the long axis; and in muscular arteries, they encircle the lumen. Patients with other genetic disorders that affect the integrity or function of connective tissues, including Ehlers-Danlos syndrome and Loeys-Dietz syndrome, also have a higher risk of developing aortic aneurysm.

Treatment and Prognosis

The patient was advised to undergo surgical intervention. The procedure was performed under general anaesthesia and aortic root

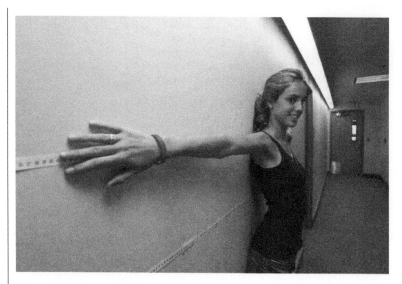

Figure CS50.2 Characteristic body shape of people affected by Marfan syndrome. Individuals with Marfan syndrome often exhibit a tall and slender physique, with disproportionately long arms, legs, fingers and toes. Other typical external features include a curved spine, a chest that sinks in or sticks out and a high-arched palate with crowded teeth. *Source:* Deborah Levenson. 2010 / Reproduced from John Wiley & Sons.

replacement was done with a mechanical valve. These procedures can be done via conventional open surgery and if suitable, sometimes via the endovascular route. Valve sparing or valve replacing techniques can be used; the latter is combined with coronary re-implantation. These techniques will require cardiac bypass with additional innominate and left carotid perfusion to maintain cerebral circulation. The patient tolerated the surgery well and was discharged after several days of recovery with regular check-ups. Medical management involves the use of ß-blockers and angiotensin receptor blockers, together with advice to avoid extreme stressful exercise (which causes increased blood pressure), which will reduce the risk of dissection and rupture.

Questions

1 What is Ehlers-Danlos syndrome?
2 What is the risk of a Marfan syndrome-affected father bearing an affected child?

Answers on p. 550.

Case Study 51   Back Pain, Lethargy and Weight Loss

Clinical Presentation

A 56-year-old woman attended her routine clinic appointment at the breast unit of her local hospital, reporting increasing lower back pain associated with general lethargy. Three years ago, following a mammogram, performed as part of the national screening programme, she had undergone surgery to remove primary breast cancer from the left breast. The doctor examining her on this occasion found tenderness over the lumbar spine and an enlarged lymph node in the axilla on the left side.

Investigations

The laboratory investigations shown in Table CS51.1 revealed a mild normocytic normochromic anaemia. The elevated alkaline phosphatase and calcium reflect probable bone resorption. The abnormal liver blood tests raised the suspicion of liver metastases. A fine needle aspirate of the axillary lymph gland was undertaken immediately in clinic and cytological examination showed the presence of poorly differentiated adenocarcinoma cells. X-ray of the lumbar spine showed a crush fracture of the 4th lumbar vertebra. An isotope bone scan revealed multiple 'hot' spots scattered throughout the skeleton. A chest x-ray was clear, but the liver ultrasound showed several small hypoechoic lesions (tissue does not reflect the ultrasound waves as well as other structures around it, appearing darker) suspicious of metastases. Mammography was performed on both breasts and revealed a spiculated mass 1.5 cm in diameter in the left breast. The spiculated or 'starburst' appearance is often associated with malignant tumours, especially invasive ductal carcinoma.

Diagnosis

The diagnosis of breast cancer generally involves a *triple assessment* approach that combines clinical evaluation, imaging tests and needle biopsy for a comprehensive analysis:

- *Clinical examination:* The first step is a thorough physical examination, including palpation of the breast tissue and the underarm area for lumps, skin changes or other abnormalities.
- *Imaging tests:* Mammography is the most commonly used imaging technique for breast cancer screening. It can detect changes in the breast up to 2 years before a patient or physician can feel them. Ultrasonography may also be used, especially for younger women with denser breast tissue, when mammography is less effective. Sometimes, other imaging techniques, such as MRI, are used for high-risk patients, or to better assess the extent of the disease.
- *Needle biopsy:* If abnormalities are detected during the clinical examination

Table CS51.1   Results of laboratory investigations.

Test	Result	Reference range
Haemoglobin	**10.0 g/dL**	11.5–16.5
Haematocrit	0.41 L/L	0.37–0.47
Mean cell volume	88.4 fL	82–92
Mean cell haemoglobin	27.2 pg	27–32
White cell count	7.3×10^9/L	4.0–11.0
Platelet count	163×10^9/L	150–400
Albumin	**32 g/L**	34–48
Calcium	**2.74 mmol/L**	2.2–2.67
Phosphate	1.25 mmol/L	0.8–1.5
Alkaline phosphatase	**182 U/L**	25–115
Bilirubin	15.4 µmol/L	<21
Alanine transaminase	**48 U/L**	5–40

Values outside the reference range are in bold.

or imaging, a needle biopsy is usually performed to take tissue samples from the suspicious area. These samples are then microscopically examined to check for cancer cells. Following an initial diagnosis of cancer by the pathologist, the case is discussed at a pre-operative tumour board and different surgical options will be considered.

The cancer that is eventually removed by the surgeon (for example, a lumpectomy) will be examined by the pathologist. The pathologist will provide crucial information about the type, grade and extent of local spread of the cancer, including the potential involvement of resection margins. For this purpose, the specimen is coated in different coloured inks to mark the different margins (Fig. CS51.1). These inks can be seen under the microscope. In this way, the pathologist can correctly identify which, if any, margins are involved. The pathologist will also provide information on the expression of different markers that are crucial in defining subsequent management (see below).

The presenting symptoms and clinical features in this patient were suggestive of breast cancer recurrence. The laboratory data are in keeping with a chronic illness and specifically indicate hypercalcaemia because of bone infiltration by metastatic tumour. The fine-needle aspiration cytology confirmed the presence of tumour cells in a local lymph node that were likely due to recurrence of the previously resected primary cancer. A needle biopsy of the new breast lump showed invasive ductal carcinoma. Immunohistochemistry revealed that the tumour cells were oestrogen receptor (ER)-positive, progesterone (PR)-negative and human epidermal growth factor receptor-2 (HER2)-negative (see below). Ki67 staining revealed 35% positivity. The tumour was classified as luminal B type (see below).

Figure CS51.1 'Inking' of a breast biopsy. This excisional biopsy was done for a palpable mass of the left breast. The lesion was considered highly suspicious for malignancy on mammography. Several years previously, the patient had invasive carcinoma of the right breast, with two positive right axillary nodes. At that time, she had an extended simple mastectomy on the right side, followed by reconstruction of the right breast and a reduction mammoplasty on the left. She had normal mammograms through 2010 and the abnormality in the left breast suddenly appeared in her 2011 mammogram. Orientation was provided to the pathologist by the surgeon in the operating room. The margins are colour-coded: red-left (lateral); green-right (medial); yellow-superior; blue-inferior; orange-anterior; black (not shown on this view)-posterior. This photograph was taken immediately after inking the specimen. As it turned out, on microscopic examination, the mass was shown to be fat necrosis, perfectly benign. Nevertheless, this sample illustrates the process of inking specimens using different coloured inks on each margin. These inks can be seen under the microscope and are used by the pathologist to determine if the resection margins are involved or not. *Source:* Ed Uthman. / Wikimedia Commons / CC BY 2.0.

When denoting recurrence in the TNM staging system for cancer, the 'r' prefix is used, which stands for 'recurrent'. The TNM system, established by the American Joint Committee on Cancer (AJCC) and the Union for International Cancer Control (UICC), typically uses this notation to indicate when a cancer has returned, either locally or in a distant site, after a period of remission or after initial treatment. Thus, this patient's stage can be written as rT2, rN1, rM1.

Diagnosis: Recurrent intra-mammary breast cancer with local and distant metastases. Luminal B type (ER+, PR−, HER2−).

Discussion

Globally, over 2 million women are diagnosed with breast cancer each year, a number that is increasing. Incidence and mortality rates vary significantly based on region and income levels, with higher incidence rates in high-income countries, but often better survival rates due to earlier detection and access to treatment. In contrast, in low- and middle-income countries, patients present at a later-stage and have poorer survival. Ethnicity also plays a role in the type and severity of the disease, with African and African-American women having higher rates of aggressive subtypes and lower survival rates. Data also indicate that women in resource-poor countries are generally diagnosed at a younger age compared to those in resource-rich nations, indicating a need to lower the age of entry into screening programs in resource-poor nations.

Approximately 10% of breast cancers are inherited. Key genes associated with hereditary breast cancer risk are the DNA repair genes, BRCA1 and BRCA2, which have a high penetrance. Individuals with mutations in these genes have a significantly elevated lifetime risk of developing breast cancer.

Other factors contribute to the risk of developing breast cancer, including hormonal influences, lifestyle and genetics. Early pregnancy and high oestrogen levels during pregnancy are associated with reduced risk. However, early menarche, lack of breast-feeding and late-onset menopause are associated with an increased risk. Modifiable risk factors, accounting for about 20% of cases, include obesity, physical inactivity and alcohol use. Central obesity has a stronger adverse effect on risk in women of Asian ancestry compared to non-Hispanic white women. Elevated BMI and breast density are also significant risk factors, particularly in postmenopausal women, including those of Asian descent.

Breast screening by mammography is routinely offered to women between the ages of 50 and 70 in the UK on a 3 yearly basis. Mammography is sensitive enough to detect early breast cancers less than 1 cm in size, which are difficult to detect on palpation. Unfortunately, even with optimal detection and treatment of early disease, many women will develop recurrence.

Breast cancer spreads locally via lymphatic vessels and more widely to distant sites via the bloodstream. The prognosis of patients with advanced disease is variable and is dictated primarily by the extent and site of metastases. Local recurrence and skeletal involvement are common and are not immediately life-threatening, but spread to the liver and the lungs carries a much poorer prognosis. Systemic therapy currently impacts little on chance of survival and so treatment is offered primarily to palliate symptoms of disease.

Both molecular and histological features of breast cancer significantly influence treatment strategies. Consequently, various classification systems based on these features have been established. The most commonly encountered breast cancer histological subtypes are ductal carcinoma (now known as 'no special type') and lobular carcinoma. These are invasive forms. The precursor lesions for these cancers are ductal carcinoma *in situ* and lobular carcinoma *in situ*. With respect to molecular classification, the so-called *intrinsic* categories are derived from a gene-expression profile

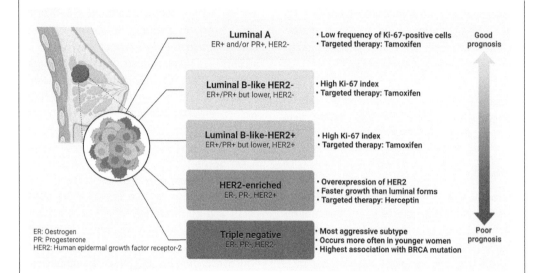

Figure CS51.2 The surrogate intrinsic classification categorises breast cancers into distinct subtypes. Luminal A tumours are slow-growing, ER and/or PR-positive, HER2-negative and primarily treated with hormone therapies like tamoxifen. Luminal B-like HER2-negative tumours grow faster, are ER-positive and often require a combination of hormone therapy and chemotherapy. Luminal B-like HER2-positive tumours, being HER2-positive, typically need hormone therapy, chemotherapy and HER2-targeted drugs. HER2-positive tumours, constituting 10–15% of breast cancers, are aggressive and treated with HER2-targeted treatments and chemotherapy. Triple-negative breast cancers are aggressive and mainly treated with chemotherapy. *Source:* Created with BioRender.com.

consisting of 50 genes (known as PAM50). In clinical settings, *surrogate intrinsic* types are commonly used (Fig. CS51.2). These are identified based on the presence of specific proteins assessed by immunohistochemistry, including ER, PR, HER2 and Ki67. There are five main surrogate intrinsic types:

- *Luminal A tumours:* These tumours express ER (Fig. CS51.3) and/or PR, but not HER2. They have a low expression of Ki67 (<20%). They are clinically low-grade, slow-growing and have a favourable prognosis. Hormone therapy with tamoxifen or aromatase inhibitors is the primary treatment, proving highly effective for people with this subtype.
- *Luminal B-like HER2− tumours:* These tumours are ER-positive and/or PR-positive, but expression is lower than in luminal

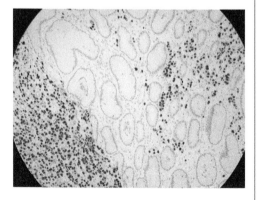

Figure CS51.3 Oestrogen receptor expression in breast cancer. Metastatic breast carcinoma in the stomach showing infiltration of oestrogen receptor-positive (brown stain) cancer cells. *Source:* Ed Uthman. / Wikimedia Commons / CC BY 2.0.

A. They display a higher expression of Ki67 (>20%) and carry a worse prognosis than Luminal A tumours. Treatment commonly involves a combination of hormone therapy and chemotherapy.

For both Luminal A and Luminal B-like tumours, additional novel treatment options have recently emerged. Patients with advanced ER-positive, HER2-negative tumours may benefit from treatment with a cyclin-dependent-kinase 4/6 inhibitor (e.g. abemaciclib) alongside hormonal therapy with the ER-antagonist, fulvestrant or an aromatase inhibitor, and also from the mTOR inhibitor, everolimus, alongside the aromatase inhibitor, exemestane. Abemaciclib also has a role in adjuvant therapy, alongside standard hormonal therapy for early-stage, high-risk ER-positive and HER2-negative disease.

- *Luminal B-like HER2+ tumours:* These tumours express lower levels of ER and/or PR, but also HER2. They have higher

Ki67 expression and typically require a combination of hormone therapy, chemotherapy and HER2-targeted drugs.

- *HER2+ tumours:* These tumours account for 10–15% of breast cancers and are characterised by high expression of HER2 (Fig. CS51.4) and a lack of ER and PR expression. Treatment usually includes drugs targeting HER2, combined with chemotherapy. HER2-targeting drugs are categorised by their mode of action and include monoclonal antibodies, cytotoxic drug-antibody conjugates and HER2 signalling inhibitors. Monoclonal antibodies include trastuzumab and the HER2 dimerisation inhibitor, pertuzumab. Drug conjugates include trastuzumab-emtansine and trastuzumab-deruxtecan. Small molecule HER2 signalling inhibitors include tucatinib, lapatinib and neratinib.

- *Triple-negative breast cancers (TNBC):* TNBC is marked by the absence of ER, PR and HER2 expression. These tumours are more

(A) (B) (C)

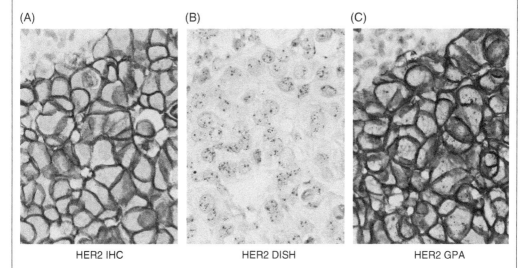

HER2 IHC   HER2 DISH   HER2 GPA

Figure CS51.4 Human epidermal growth factor receptor 2 (HER2) gene-protein assay (GPA). This assay combines HER2 immunohistochemistry (IHC) and HER2 dual in situ hybridization (DISH). HER2 IHC staining shows complete membrane staining (brown colour) of HER2 protein on a HER2 IHC 3+ case (A). HER2 DISH assay demonstrates amplified HER2 gene (black dots) and normal chromosome 17 centromere (CEN17, red dots) copy numbers (B). HER2 GPA that is a combination of HER2 IHC and HER2 DISH shows HER2 protein, HER2 gene and CEN17 (C). *Source:* Hiroaki Nitta et. 2016 / Reproduced from John Wiley & Sons.

common in younger and African-American women, are poorly differentiated, highly proliferative and are often detected at advanced stages. They are associated with mutations in the BRCA1 and BRCA2 genes.

Available treatment options for TNBC are limited, and chemotherapy is a primary mode of therapy. Immunotherapy with anti-PD-1 targeted monoclonal antibodies has emerged as an effective option, alongside chemotherapy. Chemotherapy-naïve patients with advanced TNBC may be offered atezolizumab with nanoparticle-albumen-bound paclitaxel, if the tumour PD-L1 level is greater than 1%, or alternatively, pembrolizumab with chemotherapy, if the combined PD-L1 positive score is greater than 10%. For locally advanced or high-risk early-stage TNBC, outcomes are improved with pembrolizumab in the adjuvant and neoadjuvant context compared to chemotherapy alone. For early-stage, high-risk TNBC, adjuvant treatment with the poly-ADP-ribose polymerase inhibitor Olaparib (see also Case Study 64) is also an option after surgery and conventional chemotherapy in the presence of germline BRCA1/2 mutation.

The intrinsic classification also dictates the pattern and timeline of metastatic progression. For instance, Luminal A tumours typically exhibit delayed recurrence, usually >5 years after initial diagnosis, and have a tendency to spread to the bone and lymph nodes. HER2-negative Luminal B-like tumours show a similar pattern of spread. TNBC predominantly metastasise to the brain and visceral organs, and relapses usually occur within 2–3 years after diagnosis.

Treatment and Prognosis

Despite optimal treatment, the patient had developed recurrent disease. Her main symptom was back pain associated with a vertebral crush fracture. The diagnosis was explained to her by the clinician, in the presence of the specialist nurse, offering additional support for the patient. She was immediately commenced on a non-steroidal analgesic and a single fraction of radiotherapy to the fractured vertebra was arranged for the following day. After discussion at the multidisciplinary tumour board, the patient was offered a CDK4/6 inhibitor, an aromatase inhibitor and Denosumab to reduce bone loss. Denosumab is a human IgG2 monoclonal antibody that binds to receptor activator of NF-κB ligand (RANKL) and competitively inhibits its binding to receptor activator of NF-κB (RANK).

Questions

1 What other complications of this woman's bone metastases should the clinician be alert to?

2 The patient attended her next appointment with her daughter, aged 30. Is the daughter's risk of developing breast cancer affected by her mother's illness?

3 Is it possible to predict prognosis in a breast cancer patient?

Answers on p. 550.

Case Study 52   Easy Bruising, Weight Loss and Abdominal Discomfort

Clinical Presentation

A 57-year-old woman presented to her family doctor with shortness of breath, weight loss and abdominal discomfort of 2 months duration. She had also noticed bruising after minor trauma. On examination, she was pale and had marked splenomegaly, which was the cause of her abdominal discomfort. Her heart rate was regular at 96 bpm and blood pressure was normal.

Investigations

The results of laboratory investigations are shown in Table CS52.1. There was an elevated white cell count, including a basophilia, and an increase in immature cells of the myeloid series (blast cells, myelocytes and metamyelocytes). These immature haemopoietic cells are normally found in the bone marrow, but not in peripheral blood. This blood picture is consistent with a diagnosis of either leukaemia or leukaemoid reaction.

Table CS52.1 Results of laboratory investigations.

Test	Result	Reference range
Blood		
White cell count	**98.6 × 10^9/L**	4.0–11.0
Red cell count	**3.4 × 10^{12}/L**	3.8–5.8
Haemoglobin	**11.0 g/dL**	11.5–16.5
Haematocrit	0.44 L/L	0.37–0.47
Mean cell volume	84 fL	82–92
Mean cell haemoglobin	31.30 pg	27.0–32.0
Platelets	**110 × 10^9/L**	150–400
Vitamin B12	**1150 ng/L**	160–760
Serum folate	3.0 µg/L	3–20
Urea	**7.5 mmol/L**	2.5–6.7
Urate	**0.52 mmol/L**	0.18–0.42
White blood cell differential		
Neutrophils	45%	40–75
Lymphocytes	**5%**	20–45
Monocytes	4%	2–10
Eosinophils	4%	1–6
Basophils	**8%**	1
Metamyelocytes	**10%**	—
Myelocytes	**21%**	—
Blast cells	**3%**	—

Values outside the reference range are in bold.

Cytogenetic analysis showed a reciprocal translocation involving chromosomes 9 and 22: t(9,22) (q34;q11).

Diagnosis

The patient was mildly anaemic with a very high white cell count, far in excess of that usually encountered as a result of infection or tissue damage. The differential leucocyte count, showing a range of cells from immature blast cells to relatively mature myelocytes and metamyelocytes, was consistent with a diagnosis of chronic myeloid leukaemia (CML). The serum levels of vitamin B12 were elevated which is observed in some CML patients. CML was confirmed by cytogenetic investigations showing the characteristic translocation involving chromosomes 9 and 22.

Diagnosis: Chronic myeloid leukaemia (CML).

Discussion

CML is a myeloid neoplasm. Three main categories of myeloid neoplasm exist; *acute myeloid leukaemias*, characterised by immature myeloid cells, known as blasts (Case Study 58); *myelodysplastic disorders* in which there is defective maturation of myeloid progenitors, and *myeloproliferative neoplasms*, characterised by over-production of one or more types of blood cells. CML is a myeloproliferative neoplasm.

All myeloproliferative neoplasms are characterised by the presence of constitutively activated tyrosine kinases (Table CS52.2), which promote increased growth and survival, but do not generally impede differentiation. The molecular hallmark of CML is the presence of the 9:22 translocation, giving rise to a small chromosome 22 (22q–) called the *Philadelphia chromosome* (Fig. CS52.1). As a result of this translocation, the ABL oncogene is transferred from chromosome 9 to

Table CS52.2 WHO 2022 classification of myeloproliferative neoplasms with associated genetic alterations.

Neoplasm	Genetic alteration
Chronic myeloid leukaemia	BCR-ABL fusion gene
Polycythaemia vera	JAK2-V617F
Essential thrombocythaemia	JAK2-V617F, MPL, calreticulin (CALR)
Primary myelofibrosis	JAK2-V617F, MPL, calreticulin (CALR)
Chronic neutrophilic leukaemia	CSF3R mutations
Chronic eosinophilic leukaemia	PCM1-JAK2, ETV6-JAK2 or BCR-JAK2 fusion genes
Juvenile myelomonocytic leukaemia	Ras pathway, commonly, PTPN11, NRAS, KRAS, NF1 or CBL
Myeloproliferative neoplasm NOS	Various

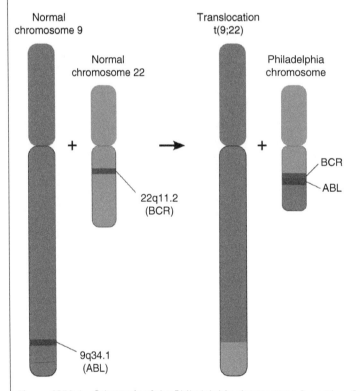

Figure CS52.1 Schematic of the Philadelphia chromosome formation. *Source:* unknown / Wikimedia Commons / licensed under CC BY 3.0.

combine with the BCR gene on chromosome 22. The result is a fusion gene (BCR-ABL)

Several critical tests are employed in the diagnosis and management of CML. A blood film examination is conducted to review the appearance and number of blood cells. A bone marrow aspirate is essential to determine the percentage of blasts present in the bone marrow, which helps differentiate between the chronic or accelerated phase of the disease. Additionally, a trephine biopsy is performed to assess the level of fibrosis in the bone marrow. Cytogenetic analysis identifies the genetic alteration present. The use of quantitative real-time PCR, not only confirms the specific transcript associated with the fusion gene, but also aids in subsequent response assessment; response to therapy is closely monitored by quantitative PCR. If patients do not achieve a deep molecular response as defined by the European Leukemia Net (ELN) group, it is an indicator that a change in therapy might be necessary.

The incidence of CML is approximately 1 per 100 000 of the population per year in the United Kingdom. The excessive *myelopoiesis* (white blood cell formation), which is a feature of CML, usually causes a reduction in the rate of erythropoiesis and thrombopoiesis, resulting in lower numbers of circulating erythrocytes and platelets, respectively. However, this is not always the case and red cell and platelet counts may be normal or raised in some patients with CML.

The precise function of the normal ABL protein is not clear, although recent evidence suggests it is a nuclear kinase that is activated by DNA-damaging agents and mediates growth arrest in a p53-dependent manner. However, the BCR-ABL fusion protein has a cytoplasmic location and acts as a tyrosine kinase that interacts with RAS and can activate MYC and cyclin D1 (see also Chapter 3) resulting in increased cell proliferation.

Treatment and Prognosis

Treatment of patients with CML aims to reduce the leucocyte count to normal and to prevent progression to its more aggressive terminal stage; left untreated most patients with CML eventually undergo 'blast transformation', in which a distinct change in the population of leukaemia cells occurs. In the chronic phase, most of the proliferating cells are relatively mature, whereas after transformation, the majority of the cells in the peripheral blood are myeloblasts or other primitive cells. At this stage, the disease has many more features of acute leukaemia and is much more resistant to treatment than when in the chronic phase. Survival after blast transformation is typically only 1–3 months.

Imatinib was the first tyrosine kinase inhibitor (TKI) used to treat CML. Since then, additional TKI have been developed, meaning that patients who do not respond, or cannot tolerate first-line therapy, now have options. Importantly, this has meant that the number of patients who develop accelerated or blast phase disease has decreased dramatically and the 10-year overall survival rate for CML is now 80–90%; the majority of CML patients can now expect a life expectancy close to normal.

Allogeneic stem cell transplantation (allo-SCT) remains a therapeutic option for a select group of CML patients. While TKIs have revolutionised CML therapy, there remains a small subset of patients who either do not respond to multiple TKIs (TKI-resistant) or cannot tolerate their side effects (TKI-intolerant). In such cases, the disease can progress and the therapeutic options are limited. Allo-SCT offers a potential cure for these patients. During allo-SCT, a patient receives healthy stem cells from a donor, which can help to rebuild healthy bone marrow and subsequently produce normal blood cells. The procedure also introduces a new donor immune system that can recognise and kill any residual leukaemia cells (graft-versus-leukaemia effect). However, allo-SCT comes with risks, including graft-versus-host disease (GvHD) in which the donated immune cells attack the recipient's healthy tissues (see Case Study 33), as well as infections due to immunosuppression. Owing to these risks, allo-SCT is generally reserved for those CML patients for whom multiple TKIs have failed or are not tolerated and in whom the potential benefits of the transplant outweigh the risks

Questions

1 Transformation of CML to 'blast crisis' involves the acquisition of new genetic abnormalities in the cancer cells. What are the most common genetic changes associated with blast crisis?

2 Do other types of leukaemia undergo blast crisis?

3 Why are serum vitamin B12 concentrations typically very high in CML?

Answers on p. 551.

Case Study 53   Change of Bowel Habit and Bleeding from the Rectum

Clinical Presentation

A 65-year-old man presented to his family doctor complaining of an alteration in his bowel habit over the past few months. This had taken the form of alternating constipation and diarrhoea, but in the last week, he had also noticed fresh blood in his motions. On examination, he was slightly pale, but there were no other external features of note. The doctor carried out a rectal examination and thought he could feel a mass in the rectum. The patient was therefore referred to a consultant surgeon for further investigation.

Investigations

On examination, the surgeon confirmed the presence of a lesion in the rectum at about 8 cm. from the anal margin and carried out a flexible sigmoidoscopy. This enabled him to view the tumour mass in the rectum and to take a biopsy. No other lesion was found in the sigmoid colon or the rectum. He also had an MRI to assess the extent of tumour and a CT TAP (scan of the thorax, abdomen and pelvis) to check for distant tumour spread.

Diagnosis

The histopathology of the rectal biopsy confirmed the diagnosis of an adenocarcinoma of the rectum. The carcinoma was described as well-differentiated, meaning that it closely resembled the normal glandular epithelium from which it had arisen (Fig. CS53.1). No other lesions were detected in the colon or the rectum. No metastases were found in the lungs or the liver. The surgeon assessed that the tumour had not spread into the surrounding pelvic tissues. The tumour and adjacent bowel were later removed by surgery.

Diagnosis: Adenocarcinoma of the rectum.

Discussion

Adenocarcinoma of the colon or rectum is a common cancer. A change of bowel habit and bleeding from the rectum are the most common presenting symptoms. The disease is more common in men than in women and increases in frequency with advancing age, especially after 55 years. Predisposing conditions include; inflammatory bowel disease (e.g. ulcerative colitis and Crohn's

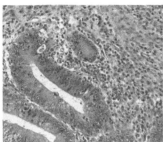 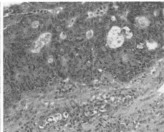 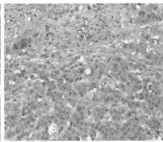

Figure CS53.1 Different histological grades of colorectal adenocarcinoma. From left to right: Well-differentiated, moderately differentiated and poorly differentiated colorectal adenocarcinoma, original. H&E stain. With increasing grade, the glandular architecture becomes increasingly less like that of normal colonic mucosa. *Source:* Panyun Zhou et al. 2022 / With permission of Springer Nature / CC BY 4.0.

disease; Case Study 22), familial adenomatous polyposis (FAP; Case Study 41), hereditary non-polyposis colorectal cancer (HNPPC) and pre-existing polyps of the large bowel. Many colorectal cancers (70–90%) arise from pre-existing benign polyps (known as adenomas). Some tumours may arise *de novo*, that is from non-polypoid colonic mucosa. Polyps that are greater than 2 cm. in diameter, have a villous pattern, or show dysplasia (pre-malignant histological changes) in the epithelium, are more likely to transform to a carcinoma than are other polyps.

Colorectal cancers, whether sporadic (non-inherited) or inherited, show multiple genetic alterations at the molecular level. Rather than being a single, uniform disease, colorectal cancers comprise a group of molecularly heterogeneous diseases that are characterised by a range of genomic and epigenomic defects. One classification system divides colorectal cancer into four consensus molecular subtypes (CMS):

- CMS1 (microsatellite instability immune, 14%), hypermutated, microsatellite unstable and strong immune activation.
- CMS2 (canonical, 37%), epithelial, marked WNT and MYC signalling activation.

- CMS3 (metabolic, 13%), epithelial and evident metabolic dysregulation.
- CMS4 (mesenchymal, 23%), prominent transforming growth factor-β activation, stromal invasion and angiogenesis.
- Samples with mixed features (13%), could represent a transitional phenotype or intra-tumoural heterogeneity.

Colorectal cancer spreads in four ways and the extent of spread is important in determining patient outcome. Local spread involves the pelvic organs, such as the bladder or the uterus. Lymphatic spread involves the local and regional lymph nodes and then the rest of the body. Blood spread results in metastases in the liver, lungs or other major organs and the tumour can also spread directly in the abdominal cavity, known as trans-coelomic spread.

Accurate assessment of the tumour and surrounding bowel by the pathologist is essential to determine the extent of tumour spread. Depth of invasion, rather than the size of the tumour, is the best measure of the degree of local spread. In rectal cancer, it is important to establish whether the peripheral or circumferential margin of the specimen that is removed surgically is involved by tumour. Lymph node involvement is assessed by examining all the lymph nodes from the mesorectal fat histologically for the presence of metastases. Blood vessels in the mesorectum are also examined for tumour involvement.

This information is used to classify colorectal cancer based on the TNM staging system. Tumour (T) describes the size of the tumour; Tis means carcinoma *in situ*, that is tumour is only present in the mucosa. T1 means tumour is within the submucosa, T2, if it has grown into the muscle layer and T3, if present in the subserosa. T4 is split into two stages, T4a and T4b: T4a means tumour has grown through the outer lining of the bowel wall and has spread into the visceral peritoneum. T4b means the tumour has grown through the bowel wall into nearby organs. Node (N) designates involvement of lymph nodes. If there is no involvement, then this is written as N0. N1a means cancer cells are present in one lymph node. N1b, if there are cancer cells in two or three nearby lymph nodes, and N1c when adjacent lymph nodes are free of cancer, but tumour is present in surrounding tissues. N2 is split into 2 stages – N2a means there are cancer cells in 4–6 nearby lymph nodes and N2b means there are cancer cells in 7 or more lymph nodes. M0 means the cancer has not spread to other organs, whereas M1 means the cancer has spread to other parts of the body; M1a means the cancer has spread to 1 distant site or organ. M1b indicates the cancer has spread to two or more distant sites or organs.

Treatment and Prognosis

The rectal carcinoma in this patient had invaded through to the muscle layer but no further. Four of the 15 lymph nodes recovered were involved by cancer. There was no evidence of liver metastases. The patient had several forms of treatment. First, because the tumour, on clinical assessment, was not thought to involve the surrounding tissues, the patient received short-term high-dose radiotherapy to the tumour pre-operatively. He then had an anterior resection of the rectum with total mesorectal excision and an end-to-end anastomosis of the proximal large bowel to the remaining rectal stump. Subsequently, following post-operative recovery, he was given a course of chemotherapy.

Epidermal growth factor receptor (EGFR) is a protein that helps the cancer cells to grow. Drugs that target EGFR can be used to treat some patients with advanced colorectal cancer. However, these drugs are not effective in colorectal cancers that have mutations in the KRAS, NRAS or BRAF gene. Doctors now commonly test tumours for these genetic changes because drugs are now available that can target the activity of the mutated genes.

Questions

1 Based on the information given above, what is the TNM stage for this patient?
2 Does the size of a rectal carcinoma have any bearing on the outcome of the disease?
3 Would the systematic removal of adenomas reduce the incidence of colorectal cancer to zero?

Answers on p. 551.

Case Study 54   Slurred Speech and Limb Weakness in a 6-year-old Girl

Clinical Presentation

A 6-year-old girl presented with a progressive 3-week history of slurred speech, trouble maintaining balance with limb weakness and a prominent drooping of the left side of the face. She had a *positive Babinski sign* (when the sole of the foot is brushed gently, the big toe bends up and back to the top of the foot and the other toes fan out). A positive Babinski sign is normal in infancy but not in children or adults. She also complained of severe headaches, which were worse in the morning and which were sometimes accompanied by nausea and vomiting.

Investigations

On magnetic resonance imaging (MRI), a tumour was found involving >50% of the pontine axial diameter, with invasion of the basilar artery. The tumour margins were poorly defined on imaging which is typical for *diffuse midline glioma* (DMG). The parents consented to a stereotactic biopsy, which confirmed the diagnosis. Molecular analysis of tumour tissue revealed a mutation at codon 27 (lysine to methionine; K27M) in a gene encoding Histone H3.

Diagnosis

A diagnosis of DMG was suggested based on the clinical history, physical examination and radiographic tests. Histopathology and molecular analysis confirmed the diagnosis.

Diagnosis: Diffuse midline glioma (DMG): H3 K27M-altered.

Discussion

DMGs are classified by the WHO as grade 4 paediatric type diffuse high-grade gliomas.

The term 'DMG' has now superseded the previous nomenclature used for these tumours-'diffuse intrinsic pontine glioma (DIPG)', to reflect the fact that they do not derive exclusively from the pons/brainstem, but may also originate in the other midline structures, including the thalami, the ganglio-capsular region, the cerebellum, cerebellar peduncles, the third ventricle, the hypothalamus, the pineal region and the spinal cord. DMGs are aggressive high-grade gliomas with predominantly astrocytic differentiation that occur in children, with a peak incidence between 6 and 9 years. They account for 10–15% of all paediatric brain cancers. Outlook is, unfortunately, bleak, with a median overall survival of only 8–12 months; fewer than 10% of children are alive at 2 years from diagnosis. Given the rapid progression of DMG, a short history is classical, as in this case. Although DMG originating in the brainstem has a heterogeneous clinical presentation, patients often present with the 'classic triad' of cranial nerve palsies, long tract signs and cerebellar signs:

Cranial nerve palsies

- *Abducens palsy or sixth nerve palsy:* one eye turned inwards.
- *Diplopia (double vision):* which may have contributed to the headaches in this patient.
- *Facial asymmetry:* muscles on one side of the face not working properly, so one side appears to droop.

Cerebellar signs (signs of pressure on the cerebellar area at the back and base of the brain)

- *Ataxia:* loss of control of body movements
- *Dysmetria:* lack of co-ordination, sometimes an inability to judge distance.
- *Dysarthria:* not speaking clearly. Manifest in this case as slurred speech.

Long tract signs (signs of damage to long tract nerves originating in the brain and spinal cord)

- Decreased strength
- *Hyperreflexia:* overactive reflexes
- *Upward Babinski:* big toe reflexes upward instead of downward

Until recently, surgical biopsy was not routinely used in the diagnosis of DMG of the brainstem due to concerns about the risks associated with obtaining tissue from this site. However, practice is changing and biopsies are beginning to be offered in light of the results of a large, international Phase II clinical trial, BIOMEDE, that recruited 279 patients and recorded no biopsy-related deaths. In biopsy samples of tumours with an H3K27M mutation, immunohistochemistry demonstrates strong nuclear expression of H3K27M and loss of H3K27me3 expression (Fig. CS54.1). The underlying molecular abnormality in this case was a mutation at codon 27 (lysine to methionine, K27M) in a gene encoding histone H3. These mutations and other alterations of H3K27me3 are diagnostic of the recently-defined entity from the 2021 WHO classification of tumours of the central nervous system – DMG, H3K27M-altered. These mutations interfere with the function of the EZH2 protein, which is responsible for adding methyl groups to H3K27. The mutant H3 protein's interaction with EZH2 results in a global reduction of methylation at H3K27, leading to dysregulation of various cellular processes.

Treatment and Prognosis

Because brain stem DMG grows diffusely and infiltrates critical brainstem structures, surgical resection is not possible. Radiotherapy remains the mainstay of treatment and this provides temporary improvement of the patient's symptoms. Radiotherapy increases overall survival on average by around 3 months. Chemotherapy is not effective in DMG. Though both clinical and radiographic responses were initially observed, local recurrence occurred 4 months after radiotherapy and the patient died three weeks later. The parents were hopeful of enrolling

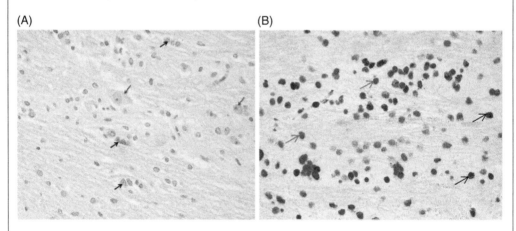

Figure CS54.1 Biopsy of a diffuse midline glioma. (A) Haematoxylin and eosin (H&E) stain showing neoplastic cells with nuclear atypia (black arrows) and intrinsic pontine neurons entrapped by neoplasm (red arrows). (B) H3K27me3 immunohistochemistry stain showing the presence of wild-type protein (black arrows) and mutant protein (red arrows) protein. *Source:* Sison J / Wikimedia Commons / CC BY 3.0.

their daughter in a Phase I trial to test a new immunotherapy drug, but regrettably she did not survive long enough. However, the parents were anxious to help research into this currently incurable form of cancer and consented to allow tumour tissue to be taken at autopsy. This was donated to a tissue repository for use in future research studies.

There are several clinical trials, both open and in development, for H3K27-altered DMGs. The recently opened CARMIGO trial is a non-randomised, open-label Phase I trial of GD-2 directed chimeric antigen receptor (CAR)-T-cells (GD-2 is a disialoganglioside, a ganglioside that contains two sialic acid residues, which is highly expressed on H3K27M-mutated glioma cells), delivered initially intravenously and then intra-ventricularly via an Ommaya reservoir (a ventricular access device for the purpose of repetitive access to the intrathecal space). CAR T-cells are a type of immunotherapy in which a patient's own T-cells are genetically engineered to produce chimeric antigen receptors (CARs) that target specific proteins on cancer cells. Once infused back into the patient, these modified T-cells seek out and destroy the cancer cells bearing the target protein (see also Case Study 55). The ACTION study is a Phase 3 randomised, placebo-controlled trial of ONC201, a small molecule, of the imipridone family, that is delivered orally. ONC201, also known as dordaviprone, is a unique therapeutic agent that operates by selectively interacting with two distinct targets. First, it binds to the dopamine receptor D2 (DRD2), which is a G-protein coupled receptor. By targeting DRD2, ONC201 influences DRD2-mediated signalling to disrupt tumour cell growth and survival. Second, ONC201 interacts with the mitochondrial protease ClpP, a component of the mitochondrial matrix. By binding to ClpP, ONC201 can affect mitochondrial function, leading to increased stress and promoting cancer cell apoptosis. These trials provide hope for patients and families with DMGs, where there has been none for several decades.

Questions

1 What is H3K27me3?
2 What is the blood–brain barrier and why is it important in the treatment of DMG?

Answers on p. 552.

Case Study 55   Fatigue, Loss of Appetite and Inguinal Lymph Node Enlargement

Clinical Presentation

A 67-year-old man presented to his General Practitioner with a painless lump in his left groin that was 'slowly getting bigger', fatigue and loss of appetite of 12 weeks duration. He had noticed that he had lost weight as his clothes felt looser. He had no significant medical history other than a left inguinal hernia repair aged 5 years. Upon physical examination, there was a 6 cm inguinal mass and evidence of enlarged axillary lymph nodes on both sides, with splenomegaly.

Investigations

The laboratory workup showed that the patient was anaemic and had a lower-than-normal platelet count and a reduced white blood cell count. He underwent an excisional lymph node biopsy and bone marrow trephine biopsy. The reporting pathologist described diffuse sheets of large lymphoid blasts (Fig. CS55.1A) replacing normal lymph node tissues with bone marrow involvement, which was consistent with the reduced platelet and white blood cell count.

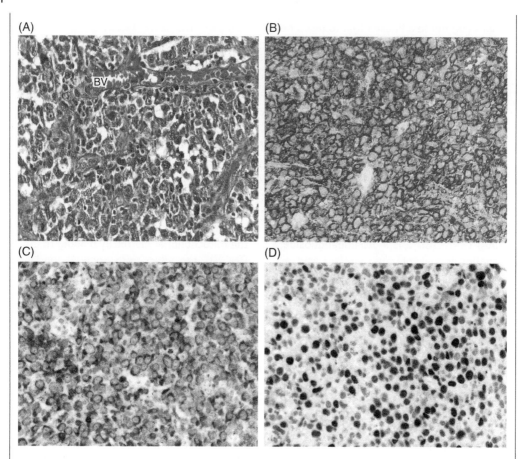

Figure CS55.1 Histopathology of diffuse large B-cell lymphoma. Left upper panel (A) shows an Haematoxylin and Eosin stained section displaying sheets of large lymphoma cells. A blood vessel is labelled 'BV'. *Source:* CoRus13 / Wikimedia Commons / CC BY SA 4.0. Right upper panel (B) shows expression of CD20 by lymphoma cells in a case of 'double-hit' lymphoma using immunohistochemistry (brown staining). Lower left panel (C) shows over-expression of BCL2 in the same case. Lower right panel (D) shows over-expression of MYC also in the same case.

Immunohistochemistry of tumour cells (Fig. CS55.1B-D) showed they were: BCL-2(+), BCL-6(+), MYC(+), CD30(+), CD10(+), CD20(+), CD3(−), EBER (Epstein-Barr virus encoded RNA) *in situ* hybridisation(−), with a Ki67 proliferation index of 95%. Almost all DLBCL express the B-cell marker, CD20, which is the target of the therapeutic monoclonal antibody, rituximab (Fig. CS55.1B), as well as CD19, which is a target of chimeric antigen receptor (CAR) T-cell therapy (not shown, see below). A fluorescence *in situ* hybridisation (FISH) test identified the presence of the t(14;18) translocation in 73% of cells scored

and the t(8;14) translocation in 87% of cells scored (Fig. CS55.2). These translocations are responsible for the over-expression of BCL2 and MYC, respectively. Additional tests showed a raised lactate dehydrogenase (LDH) level of 271 U/L (normal range: 135−225 U/L). A PET scan and a CT scan demonstrated bilateral axillary lymphadenopathy, mediastinal lymphadenopathy, splenomegaly and mild hepatomegaly. Based on these data, the patient was classified as having stage IVB DLBCL (see below for precise diagnostic categorisation).

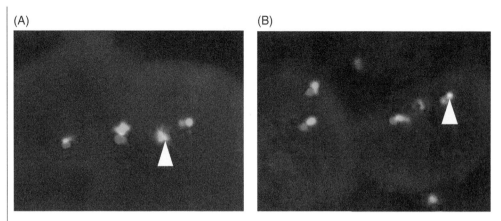

Figure CS55.2 Fluorescence in situ hybridisation (FISH) to detect BCL2 and MYC translocations in DLBCL. (A) Left panel shows the presence of a t(14;18) translocation indicated by the juxtaposition of two probes, one recognising a sequence on chromosome 14 (green) and the other a sequence on chromosome 18 (red). When they are in close proximity the fluorescence appears yellow. (B) Right panel shows the presence of a t(8;14) translocation using a similar methodology.

Diagnosis

The histological features and immunohistochemistry showed a high-grade B-cell lymphoma with large cell morphology. The presence of translocations involving MYC and BCL-2 is consistent with a so-called double hit molecular aberration.

Diagnosis: Diffuse large B-cell lymphoma/high-grade B-cell lymphoma with MYC and BCL-2 rearrangements (DLBCL/HGBL-MYC/BCL2)/High-grade B-cell lymphoma with MYC and BCL-2 rearrangements, (HGBL-MYC/BCL2) (ICC 2022). This is colloquially known as a 'double hit' lymphoma.

Discussion

Large B-cell lymphomas are the most common form of non-Hodgkin lymphoma, accounting for 25–35% of cases. DLBCL/HGBL-MYC/BCL-2 carries a particularly poor prognosis. Dysregulation of MYC, a transcription factor with paradoxical roles in increasing both cell proliferation and apoptosis, affects the transcription of up to 15% of the human genome and plays an important role in DLBCL development. Tumours harbouring MYC rearrangements are often found in combination with BCL-2 and/or BCL-6 rearrangements; conventionally these tumours are known as *double hit* or *triple hit* lymphomas (DHL/THL). These tumours have a 2-year survival rate of approximately 30–50%.

Tumours with overexpression of MYC with BCL-2, but which lack translocations are defined as *double expresser* B-cell lymphoma (DEL). Over-expressed BCL-2 protein protects MYC over-expressing cells from MYC-induced apoptosis without affecting MYC's ability to drive cell proliferation.

The updated WHO classification (2022) and International Consensus Classification of large B-cell lymphomas comprises a wide spectrum of tumours. DLBCL not otherwise specified (NOS) is the most common entity (Table CS55.1).

Table CS55.1 2022 WHO classification of large B-cell lymphomas.

Large B-cell lymphomas
Diffuse large B-cell lymphoma, NOS
T-cell/histiocyte-rich large B-cell lymphoma
Diffuse large B-cell lymphoma/high-grade B-cell lymphoma with *MYC* and *BCL-2* rearrangements
ALK-positive large B-cell lymphoma
Large B-cell lymphoma with *IRF4* rearrangement
High-grade B-cell lymphoma with 11q aberrations
Lymphomatoid granulomatosis
EBV-positive diffuse large B-cell lymphoma
Diffuse large B-cell lymphoma associated with chronic inflammation
Fibrin-associated large B-cell lymphoma
Fluid overload-associated large B-cell lymphoma
Plasmablastic lymphoma
Primary large B-cell lymphoma of immune-privileged sites
Primary cutaneous diffuse large B-cell lymphoma, leg type
Intravascular large B-cell lymphoma
Primary mediastinal large B-cell lymphoma
Mediastinal grey zone lymphoma
High-grade B-cell lymphoma, NOS

Treatment and Prognosis

The Lugano classification is used to stage lymphoma:

Limited stages:

Stage I: Involves a single lymph node or a group of adjacent nodes.

Stage IE: The cancer is present in one organ or site outside the lymph nodes.

Stage II: Involves two or more lymph node groups on the same side of the diaphragm.

Stage IIE: Cancer extends from a nodal site to a nearby organ or site, possibly involving other nearby nodes.

Advanced stages:

Stage III: Involvement of lymph node regions on both sides of the diaphragm.

Stage III(1): Also involves the spleen or nearby lymph nodes, such as the splenic, hilar, celiac or portal nodes.

Stage III(2): Extends to the para-aortic, iliac, inguinal or mesenteric nodes.

Stage IV: The cancer has spread widely into at least one organ or tissue beyond the lymph nodes.

Additional designations:

(A): No systemic symptoms are present.

(B): Presence of systemic symptoms such as fever, night sweats or unexplained weight loss.

(E): Indicates extranodal contiguous extension that can still be treated with a radiation field appropriate for nodal disease of similar extent.

(bulky): Indicates the presence of a large mass of lymph nodes.

The Lugano classification is applied based on the results of clinical examinations, imaging studies (including PET-CT or CT scans) and biopsy findings to understand the extent and progression of lymphoma. In everyday practice, this staging helps clinicians to determine the most appropriate treatment plan for patients, considering factors, such as the stage of lymphoma, the patient's overall health and the presence of systemic symptoms or extranodal disease. Each stage has associated treatment protocols that often involve combinations of chemotherapy, radiation therapy, immunotherapy, targeted therapy or stem cell transplant, depending on the specific type and stage of lymphoma.

Clinical scoring systems have been developed to risk-stratify DLBCL patients. The International Prognostic Index (IPI) scoring system was the first to be extensively applied and remains in use today. The IPI score assigns 1 point to each negative prognostic factor: age >60 years; raised LDH; stage III/IV disease; Eastern Cooperative Oncology Group (ECOG) performance status ≥2; and >1 site with extra-nodal involvement. Patients are then assigned to a risk group: 0/1 = low risk, 2 = low-intermediate risk, 3 = high-intermediate risk and 4/5 = high risk. High levels of LDH reflect high rates of cell turnover; when cells die, they release LDH into the bloodstream.

Standard treatment was given to this patient with six cycles of chemotherapy with the R-CHOP regimen (rituximab, cyclophosphamide doxorubicin (hydroxydaunomycin), vincristine (oncovin), prednisone). The patient achieved a complete remission. His first post-treatment PET/CT scan was unremarkable. However, one year later, he presented with recurrent axillary lymphadenopathy. Repeat biopsy confirmed relapse of his disease. He was deemed to be ineligible for high-dose chemotherapy and autologous

Table CS55.2 CAR-T cell therapies approved for DLBCL.

Drug	Target	Indication
Axicabtagene ciloleucel (Yescarta)	CD19	Diffuse large B-cell lymphoma (second line)
		Follicular lymphoma (third line)
Lisocabtagene maraleucel (Breyanzi)	CD19	Diffuse large B-cell lymphoma (third line)
Tisagenlecleucel (Kymriah)	CD19	Diffuse large B-cell lymphoma (third line)
		B-cell precursor ALL (third line)
		Follicular lymphoma (third line)

stem cell transplant (ASCT). The patient had a partial response to salvage chemotherapy with ICE (ifosfamide, carboplatin and etoposide), which was followed by disease progression 6 months later. He was further debulked with rituximab, gemcitabine, dexamethasone and cisplatin regimen (R-GDP) to a very good partial response and was consolidated with a one-time infusion of CD19-directed CAR-T-cells (axicabtagene ciloleucel; Table CS55.2). The patient achieved a complete remission following this treatment and remains in remission 1 year later.

Around one-third of patients with DLBCL are refractory to first-line treatment or will relapse, despite the use of the CD20-targeting monoclonal antibody, rituximab, which, when added to CHOP chemotherapy, dramatically improves survival. However, treatment options for DLBCL patients have now been expanded with the approval of CD19-directed CAR-T-cell therapies in the past 5 years. CAR-T therapy may be suitable

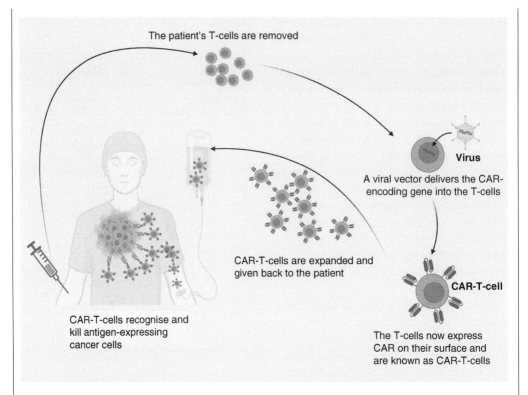

The patient's T-cells are removed

Virus

A viral vector delivers the CAR-encoding gene into the T-cells

CAR-T-cell

CAR-T-cells are expanded and given back to the patient

The T-cells now express CAR on their surface and are known as CAR-T-cells

CAR-T-cells recognise and kill antigen-expressing cancer cells

Figure CS55.3 CAR-T-cell therapy; basic principles. CAR-T cell therapy involves taking a patient's own T-cells and genetically modifying them to express chimeric antigen receptors (CARs) on their surface. These CARs are designed to recognise tumour-associated antigens. Once engineered, the CAR-T-cells are expanded in the laboratory and then infused back into the patient. In the patient's body, these CAR-T-cells recognise and bind to cancer cells expressing the target antigen, leading to T-cell activation, proliferation and destruction of the cancer cells. This approach harnesses the immune system's potent cytotoxic response to effectively target and eliminate cancer cells, offering a personalised immunotherapy for certain types of cancer. *Source:* Created with BioRender.com.

for older patients who are ineligible for autologous stem cell transplant, such as our patient described here. CAR-T-cell therapy uses T-cells taken either from the patient to be treated (autologous) or from a donor (allogeneic), which are then genetically engineered *ex vivo* to express CARs that recognise the tumour cells (Fig. CS55.3). The T-cells are then infused back into the patient. Post-relapse, patients receiving CAR T-cell therapy are reported to achieve long-term remission in 30–40% of cases. This constitutes a marked improvement in outcome for a previously poor prognosis cohort. At the time of writing, three CAR-T-cell therapies are FDA approved for the treatment of DLBCL (Table CS55.2). In all three cases, CD19 is the target of choice.

Questions

1 Calculate the IPI for this patient. His ECOG performance score was 1 (see below)

ECOG performance score
0=Asymptomatic
1=Symptomatic but completely ambulatory
2=Symptomatic, <50% in bed during the day, ambulatory and capable of all self care but unable to carry out any work activities.
3=Symptomatic, >50% in bed, but not bedbound, capable of only limited self-care
4=Bedbound
5=Dead

2 What is the significance of the Cell of Origin (COO) classification of DLBCL?

Answers on p. 552.

Case Study 56 Sudden Onset of Confusion, Twitching and Convulsions

Clinical Presentation

A 57-year-old man presented to the Emergency Department with a sudden onset of confusion and twitching in his left arm and leg, followed by generalised convulsions that lasted for about 2 minutes. He had no significant medical history and had never experienced anything like it before. He regained consciousness afterwards but was confused and disoriented. The patient was transferred to the neurology department for further evaluation.

Investigations

A brain CT scan was performed, which showed a large, heterogeneously enhancing mass in the right frontal lobe with surrounding oedema. The mass measured approximately 4.5 × 4 cm. On MRI with contrast enhancement (using a gadolinium-based compound that is sensitive to magnetic fields), the tumour appeared as a highly enhancing mass with central necrosis, surrounded by oedema. The enhancement was irregular and heterogeneous, which is typical of glioblastoma multiforme (GBM). The surrounding oedema extended up to 5 cm beyond the tumour borders. GBM typically has indistinct tumour borders as shown in Fig. CS56.1.

Diagnosis

The patient was prescribed the antiseizure medication, levetiracetam, to control his seizures. Due to the presence of a tumour, the patient was considered to be at high risk of future seizures. His condition was categorised as structural epilepsy. A biopsy of the mass confirmed a GBM, a highly malignant tumour of glial cells. The histology report described a highly infiltrative

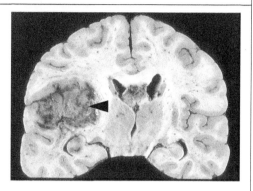

Figure CS56.1 Macroscopic pathology of glioblastoma multiforme (GBM). GBM is an aggressive malignant brain tumour that originates from glial cells and tends to have infiltrative growth with areas of necrosis and haemorrhage. Tumour is indicated by a black arrowhead. *Source:* Sbrandner / Wikimedia Commons / CC BY-SA 4.0.

and proliferative mass of neoplastic glial cells with necrosis (Fig. CS56.2) and microvascular proliferation. The tumour cells were described as anaplastic with large, hyperchromatic (darkly staining), nuclei. Immunohistochemistry showed that the tumour cells had a high proliferation rate (75%, as measured by Ki67 staining) and expressed glial fibrillary acidic protein (GFAP), a marker of astrocyte differentiation. Immunohistochemistry with the anti-IDH1 R132H antibody was used to determine if there was a mutation in the isocitrate dehydrogenase (NADP(+))-1 (IDH1) gene. This antibody specifically recognises the R132H mutation in IDH1, which is the most frequent IDH1 mutation in gliomas. However, the tumour cells did not stain with the anti-IDH1 R132H antibody. Sequencing confirmed that the IDH1 and IDH2 genes were wild-type (i.e. not mutated).

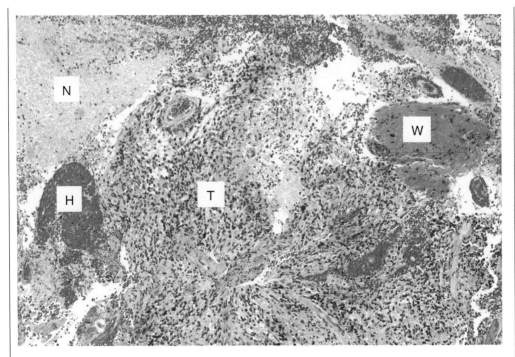

Figure CS56.2 Photomicrograph of glioblastoma multiforme. HPS (haematoxylin phloxine saffron) stain. The images show the characteristic features of glioblastoma: tumour (T) with necrosis (N) and areas of haemorrhage (H). A piece of remnant near-normal white matter (W) is present. *Source:* Nephron / Wikimedia Commons / CC BY-SA 3.0.

Diagnosis: Glioblastoma multiforme.

Discussion

The World Health Organization (WHO) grading system for CNS tumours uses a four-tiered scale:

Grade I: Tumours are typically benign and slow-growing with well-differentiated cells. They are generally curable with surgery and may not require additional treatment.

Grade II: Tumours are well-differentiated and considered to be low-grade. Typically, they are slow-growing tumours and have a better prognosis than high-grade tumours.

Grade III: These are high-grade tumours with more abnormal cell morphology, increased mitotic index (the number of cells in mitosis) and increased cellular proliferation (measured by Ki67 staining). They are more aggressive than low-grade tumours.

Grade IV: These tumours are the most aggressive. They have highly abnormal cell morphology, increased mitotic activity and higher rates of cell proliferation. They also often exhibit areas of necrosis and/or microvascular proliferation. Grade IV tumours include glioblastoma multiforme (GBM), which is the most common and aggressive type of primary brain tumour found in adults.

The presence of the tumour and cerebral oedema can lead to increased intracranial pressure (ICP), which can result in a range of neurological signs and symptoms which include: headaches, which are often worse in the morning or when lying down; nausea and vomiting, which may be severe and frequent; seizures; visual disturbances, including blurring or double vision; confusion or altered mental status; weakness or numbness; and swelling of the optic disc (papilledema). Cerebral herniation is a potentially life-threatening complication of raised intracranial pressure in which brain tissue is displaced from its normal position due to increased pressure within the skull. Cerebral herniation requires urgent intervention to reduce intracranial pressure and prevent further neurological damage. Treatment may include medications to reduce cerebral oedema and intracranial pressure, as well as surgical interventions such as craniotomy or decompressive hemicraniectomy.

Recent studies have revealed several molecular subtypes of GBM:

- *Classical subtype:* characterised by the amplification of the epidermal growth factor receptor (EGFR) gene, which leads to increased activity of the EGFR signalling pathway. In addition, there is often a loss of chromosome 10q and high expression of genes associated with cell cycle progression and DNA repair, including cyclin-dependent kinases (CDK, see Chapter 3) and checkpoint kinase 1 (CHK1).
- *Mesenchymal subtype:* associated with loss of the NF1 gene, which encodes a protein that negatively regulates RAS signalling (see also Chapter 11). Loss of NF1 leads to increased RAS activity and activation of the nuclear factor kappa B (NF-κB) pathway. Tumours in this category also have high expression of genes associated with inflammation and angiogenesis, such as vascular endothelial growth factor (VEGF) and interleukin-6 (IL-6).
- *Proneural subtype*: characterised by mutations in the IDH1 or IDH2 genes, resulting in the production of an oncometabolite called 2-hydroxyglutarate (2-HG), which can interfere with the normal epigenetic regulation of gene expression. Tumours often have a co-deletion of chromosome 1p and 19q, which is associated with a better prognosis, and high expression of genes associated with neural differentiation, such as oligodendrocyte lineage transcription factor 2 (OLIG2) and neural cell adhesion molecule (NCAM).
- *Neural subtype*: is typified by high expression of genes associated with neuronal function, such as neurofilament proteins and synaptic proteins. Tumours in this subtype also have low expression of genes associated with proliferation and angiogenesis (e.g. CDKs and VEGF). This subtype is relatively rare and has a better prognosis than the other subtypes.

The detection of additional molecular abnormalities can provide useful prognostic information. For example, methylation of the O6-methylguanine-DNA methyltransferase (MGMT) promoter. MGMT is a DNA repair enzyme that can remove alkyl adducts from DNA. MGMT expression confers resistance to chemotherapy with alkylating agents, such as temozolomide, which is a standard treatment for GBM patients. Thus, MGMT promoter methylation results in reduced MGMT expression and is associated with better outcomes in patients with GBM who receive temozolomide.

Treatment and Prognosis

GBM is a highly aggressive form of brain cancer with a poor prognosis. The presentation of new onset seizures in an adult patient should always raise suspicion of a possible underlying brain tumour. Despite the challenges associated with treating GBM, early diagnosis and treatment can improve the patient's quality of life and survival. Unfortunately, the patient's condition rapidly deteriorated with worsening neurological symptoms. Despite the best efforts of his medical team, the patient passed away 3 months after his initial diagnosis.

Questions

1 What is microvascular proliferation?
2 What is meant by heterogeneously enhancing areas on imaging?

Answers on p. 552.

Case Study 57   Night Sweats and Cervical Lymph Node Enlargement

Clinical Presentation

A 20-year-old woman presented to her General Practitioner with a 2-month history of drenching night sweats and pruritus (itching). Physical examination revealed a mobile and firm 2.5 cm-diameter lymph node on the left side of the neck. There was no evidence of other palpably enlarged nodes and the liver and spleen were not enlarged. An intermittent pyrexia (fever) of 39°C was noted.

Investigations

Investigations revealed a normal blood count with a raised erythrocyte sedimentation rate (ESR). A cervical lymph node biopsy was taken (Fig. CS57.1) and this showed loss of the normal lymph node architecture with nodules composed of lymphocytes, eosinophils, macrophages and Hodgkin/Reed-Sternberg (HRS) cells (Figs. CS57.2–CS57.4) surrounded by broad bands of fibrous tissue. A chest radiograph showed widening of the mediastinum. A positron emission tomography (PET) CT scan involves the injection of a radioactive chemical, fluorodeoxyglucose (FDG), that detects metabolically active lymphadenopathy or malignant organ infiltration. For this patient, the PET CT, in addition to the left cervical lymphadenopathy, showed gross enlargement of the mediastinal lymph nodes with an increase in FDG uptake, without any other lymphadenopathy or hepatosplenomegaly. As no marrow infiltration was noted on PET CT, an iliac crest bone marrow trephine biopsy was not required as part of staging. CT scans were performed on the abdomen and no other enlarged lymph nodes were found and the liver and spleen were normal. The elevated ESR likely reflects the inflammatory basis of Hodgkin lymphoma (see below).

Diagnosis

The histological appearances are characteristic of *classical Hodgkin lymphoma, nodular sclerosis* subtype *(cHL, NS)*. cHL, NS is the most common subtype of classical Hodgkin lymphoma and its histological features are distinct. Upon microscopic examination, the affected lymph nodes typically show a nodular growth pattern with broad fibrous bands (sclerosis) dividing the tissue into nodules. These nodules are usually composed of a mixture of cells, including small lymphocytes, eosinophils, histiocytes (tissue macrophages) and plasma cells. However, the hallmark of cHL is the presence of

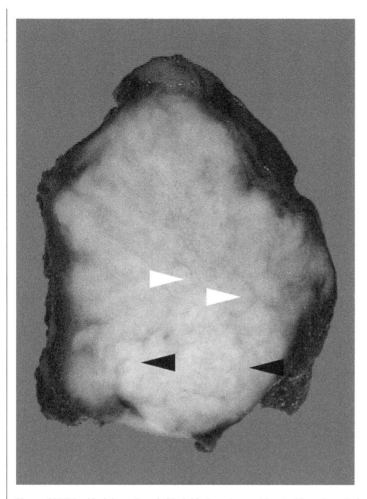

Figure CS57.1 Nodular sclerosis Hodgkin lymphoma; bisected lymph node from the mediastinum: The distinct nodules on the cut surface (white arrowheads) of this lymph node are separated by paler collagenous tissue (black arrowheads) strongly suggest the diagnosis. PEIR Digital Library (Pathology image database). Image# 415054. Image and description are from the AFIP Atlas of Tumor Pathology. *Source:* The Armed Forces Institute of Pathology (AFIP) / Wikimedia Commons / Public Domain.

bi-nucleated Reed-Sternberg (RS; Fig. CS57.4) cells or their mononuclear variants, known as 'Hodgkin' cells (collectively referred to as HRS cells). HRS cells have prominent nucleoli and a characteristic 'owl's eye' appearance. RS cells in the correct inflammatory milieu are necessary for a definitive diagnosis of Hodgkin lymphoma. Patients are staged using a combination of routine radiography and CT scanning. Disease stage is one of the main prognostic factors in HL. The patient was found to have stage IIB disease (see Table CS57.1).

Diagnosis: Stage IIB cHL, NS (WHO 2022/ ICC 2022).

Discussion

Hodgkin lymphoma is a B-cell lymphoma characterised by a minority of B-cell-derived large blasts surrounded by a mixture of

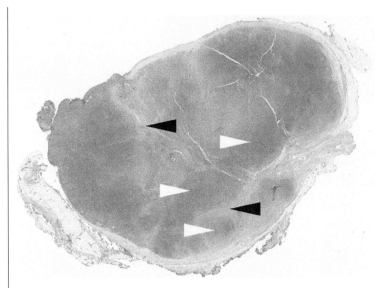

Figure CS57.2 Low power photomicrograph of nodular sclerosis Hodgkin lymphoma. Nodules are seen as purple-blue areas (white arrowheads) separated by pink fibrous tissue (black arrowheads).

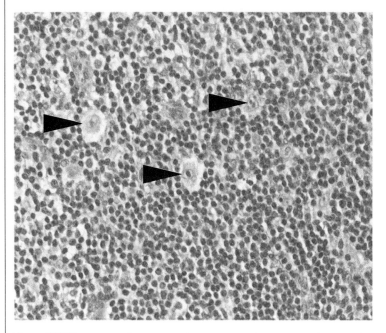

Figure CS57.3 Intermediate power photomicrograph of classical Hodgkin lymphoma. Shows a mixed inflammatory infiltrate comprising lymphocytes, eosinophils and some macrophages surrounding HRS cells (black arrowheads).

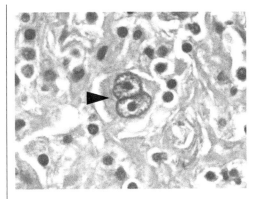

Figure CS57.4 High power photomicrograph showing a Reed-Sternberg cell (black arrowhead) with typical binucleated 'owl's eye' appearance. *Source:* Nva1991 / Wikimedia Commons / CC BY-SA 3.0.

Table CS57.1 Staging system for Hodgkin lymphoma.

Stage	Extent of spread
Stage I:	Involvement of a single node region
Stage II:	Involvement of two or more lymph node regions on the same side of the diaphragm
Stage III:	Involvement of lymph node regions on both sides of the diaphragm
Stage IV:	Extensive extra-lymphatic involvement and/or disseminated disease with involvement of liver, bone marrow, lung or skin

B Symptoms: Unexplained weight loss of more than 10% of the previous body weight during the past 6 months. Unexplained persistent or recurrent fever with temperatures above 38°C during the previous month. Recurrent drenching night sweats during the previous month.

non-malignant cells. There are two major forms; classic Hodgkin lymphoma (cHL) and nodular lymphocyte predominant Hodgkin lymphoma (NLPHL; WHO 2022; Table CS57.2)/Nodular lymphocyte predominant B-cell lymphoma (NLPBL; ICC 2022). cHL and NLPHL/NLPBL are diagnosed and sub-classified based on the appearance and

Table CS57.2 Classification of Hodgkin lymphoma (WHO classification).

Hodgkin lymphoma histological subtype	Features
Lymphocyte-predominant Hodgkin lymphoma	Few malignant cells, numerous non-malignant lymphocytes, tumour cells express B cell markers
Classic Hodgkin lymphoma (four subtypes)	HRS cells usually lack B cell gene expression and are CD30 and CD15 positive
Nodular sclerosis	Nodules of tumour separated by bands of fibrous tissue
Mixed cellularity	Moderate numbers of HRS cells with a mixture of different non-malignant cells
Lymphocyte depletion	Numerous HRS cells, few non-malignant cells
Lymphocyte-rich	Few HRS cells, numerous lymphocytes

immunophenotype of the characteristic HRS cells and the nature of the non-malignant cell population. The lesional HRS cells of cHL show loss of the B-cell programme on immunophenotyping. In contrast, the lesional cells of NLPHL/NLPBL are called 'lymphocyte-predominant' (LP) cells and show a retained B-cell programme. This retained B-cell programme has led to a recent renaming of this entity as nodular lymphocyte predominant B-cell lymphoma in the ICC classification.

The incidence of cHL in the UK is about 2/100 000. There is a male preponderance with a 3:2 male-to-female ratio and a bimodal age distribution with a distinct peak in the 15–34-year age group, with the incidence increasing into the sixth decade. In the younger patients, the male predominance is less marked and most cases are of the cHL, NS type.

Patients with cHL commonly present with lymphadenopathy. The mediastinum is frequently involved and most patients have

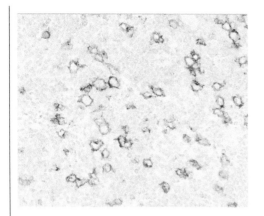

Figure CS57.5   CD30 expression in classical Hodgkin lymphoma. Image shows a case of cHL stained with antibodies recognising CD30. CD30 is strongly expressed by HRS cells (brown stain), the target for the antibody drug conjugate, Brentuximab vedotin.

stage I or II disease at presentation. The disease spreads contiguously to adjacent nodes with late involvement of the bone marrow and extranodal sites.

It is essential that the pattern of disease at presentation is considered together with the biopsy features when making a diagnosis. The accurate diagnosis of cHL involves demonstrating destruction of the normal nodal architecture by an infiltrate that includes HRS cells in the appropriate cellular background. The disease predominantly involves the lymph nodes and pure extranodal manifestations are rare, and should prompt consideration of an alternative diagnosis. Phenotyping of the HRS cells by immunohistochemistry is also essential for the accurate diagnosis of cHL. In every case of cHL, HRS cells express CD30 (Fig. CS57.5) CD30 is a growth factor receptor of the tumour necrosis factor receptor (TNFR) family. HRS cells of cHL also often express CD15. The lesional HRS cells show partial loss of the B-cell programme, including absent CD20 expression and weak PAX5 expression. In most cases of cHL, the HRS cells are clonal and of B-cell origin. Since B-lymphocytes rearrange their immunoglobulin genes early in their development and each rearrangement is unique to an individual B lymphocyte, analysis of rearrangements can establish whether a clonal population of B cells is present. Detecting B-cell clonality by conventional bulk methods, such as PCR and sequencing, can be problematic in HL, as the lesional cells are in the minority compared to the non-neoplastic inflammatory cells.

Approximately 40% of cases of cHL from Europe and the USA harbour the Epstein-Barr virus (EBV) within the HRS cells. There is now convincing evidence implicating EBV directly in cHL development. For example, EBV's oncogene, latent membrane protein (LMP1, see also Chapter 11), is consistently expressed in HRS cells. The life cycle of EBV in host B-cells is discussed in more detail in Case Study 67.

Some of the pathological and clinical features of cHL are not due to tumour invasion but are related to cytokine release by HRS cells. Cytokines are responsible for the fever, night sweats, weight loss and pruritus experienced by some patients; collectively known as 'B' symptoms. As well as mediating systemic effects, cytokine production may govern the interaction between the HRS cells and the non-malignant cell populations that make up most of the mass of the affected tissue. Thus, the non-malignant cells support the growth of HRS cells; they express many growth promoting ligands (such as CD40L and CD30L) for growth factor receptors (CD40, CD30) expressed on HRS cells. Other cytokines produced by HRS cells (e.g. IL-10) may inhibit cytotoxic T cell (CTL) responses to the tumour cells or produce the fibrosis characteristic of cHL, NS (e.g. TGFβ and IL-1).

Treatment and Prognosis

The treatment of patients with cHL is stage dependent:

Stages IA and IIA, Favourable

This group includes patients who do not have any unfavourable factors. Treatment is usually with chemotherapy (e.g. two to four cycles of the ABVD regimen-adriamycin [doxorubicin], bleomycin, vinblastine and dacarbazine), followed by radiation to the initial site of the disease (involved site radiation therapy, or ISRT). Another option is chemotherapy alone (usually three to six cycles) in selected patients. If a person cannot tolerate chemotherapy, radiation therapy alone is an option.

Stages I and II, Unfavourable

This group includes cHL patients who have one or more of the following risk factors:

- Bulky tumour
- Disease is in three or more different areas of lymph nodes.
- There is cancer outside the lymph nodes (i.e. extranodal involvement)
- 'B' symptoms are present
- ESR (erythrocyte sedimentation rate) is high.

Treatment for these patients is generally more intense, typically with chemotherapy (usually ABVD four to six cycles, or other regimens such as three cycles of Stanford V). Radiation therapy (involved field radiation therapy, or IFRT) is usually given to the sites of the tumour, especially if there is bulky disease.

Stages III and IV

More intense chemotherapy regimens are used for these patients, for example ABVD, Stanford V or BEACOPP regimen (bleomycin, etoposide, doxorubicin, cyclophosphamide, vincristine, procarbazine and prednisone). A relatively small number of patients have disease that is refractory to standard treatment or will relapse. Some of these patients will respond to second-line chemotherapy. High-dose chemotherapy used together with autologous haemopoietic stem cell transplant may cure patients with relapsed or refractory disease.

The last few years have seen the emergence of new drugs for cHL patients, which include the antibody-drug conjugate, brentuximab vedotin. This is an antibody to CD30 conjugated to the poison, monomethyl auristatin E. Drugs that block the programmed death-1 (PD-1) immune checkpoint pathway can also be used. Pembrolizumab and nivolumab are two anti-PD-1 antibodies that are approved for the treatment of cHL patients.

Questions

1 If cHL is a B-cell lymphoma, why is rituximab not used to treat patients?
2 What are the late effects of treatment in people who survive cHL?

Answers on p. 553.

Case Study 58   Bleeding from the Nose and Gums with Fever

Clinical Presentation

A 57-year-old man presented to his family doctor with bleeding from the nose and gums, red urine and faintness of 1-week duration. On examination, he was very pale and had a petechial rash on his arms and legs (Fig. CS58.1). The patient was febrile (temperature: 39°C). The heart rate was regular at 100 b.p.m. and his blood pressure was normal.

Investigations

The results of laboratory investigations are shown in Table CS58.1. The total and differential white blood cell counts are abnormal, with a significant increase in the white blood cell count due to the presence of promyelocytes (large cells with numerous primary granules that coalesce into needle-like aggregates, called *Auer rods*; Fig. CS58.2).

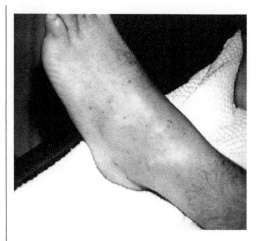

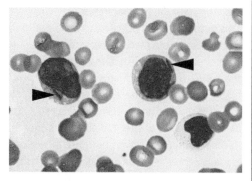

Figure CS58.2 Bone marrow: Auer rods. Two myeloblasts, each with a single prominent Auer rod (arrowheads). Wright-Giemsa stain. *Source:* The Armed Forces Institute of Pathology (AFIP) / Wikimedia Commons / Public Domain.

Figure CS58.1 Petechia: Small red or purple spots caused by bleeding into the skin. *Source:* Seasurfer / Wikimedia Coimmons / Public Domain.

Table CS58.1 Results of laboratory investigations.

Test	Result	Reference range
Blood		
Total white cell count	**48 × 10⁹/L**	4.0–11.0
Red cell count	**1.1 × 10¹²/L**	4.5–6.5
Haemoglobin	**6 g/dL**	13.0–18.0
Haematocrit	**0.26 L/L**	0.40–0.54
Mean cell volume	85 fL	82–92
Mean cell haemoglobin	28 pg	27–32
Platelet count	**8 × 10⁹/L**	150–400
Urea	**7.0 mmol/L**	2.5–6.7
Urate	**0.6 mmol/L**	0.18–0.42
Plasma fibrinogen	**1.1 g/L**	1.5–4.0
Prothrombin time (PT)	**21 seconds**	11–16
Activated partial thromboplastin time (APTT)	**41 seconds**	26–35
White cell differential		
Neutrophils	**10%**	40–75
Lymphocytes	**4%**	20–45
Monocytes	**1%**	2–10
Eosinophils	**0%**	1–6
Basophils	**0%**	1
Promyelocytes*	**85%**	—

* Most have Auer rods.
Values outside the reference range are in bold.

Promyelocytes are normally without Auer rods and are usually present only in the bone marrow and not in the peripheral blood. Auer rods are seen only in *acute myeloid leukaemia* (AML). Cytogenetic studies showed a reciprocal translocation between chromosomes 15 and 17; t(15;17) (q22;q12). This translocation involves two oncogenes; the retinoic acid receptor (*RAR*) gene and the *PML* gene.

Diagnosis

The patient was severely anaemic with a very high white cell count, far in excess of that usually encountered as a result of infection. The patient also had severe *thrombocytopenia* (very low platelet count), which caused the bleeding (indicated by the raised PT and APTT) and petechial rash and neutropenia (low neutrophil count), which predisposed the patient to infections and was probably responsible for the fever. The blood picture was consistent with a diagnosis of AML of acute promyelocytic leukaemia (APL) type.

Flow cytometry is widely used in the diagnosis of haematological malignancies. In flow cytometry, peripheral blood or bone marrow samples are collected from the patient and stained with fluorescently labelled antibodies that target specific cell surface markers, intracellular proteins or other cellular components. The labelled cells are passed through a flow cytometer, in which laser beams excite the fluorescent tags, allowing the identification and quantification of cell populations based on the emitted fluorescence signals. The analysis provides essential information on the presence, percentage and phenotype of blood cells (including abnormal promyelocytes in this case), which is crucial for diagnosis, risk stratification and treatment monitoring. Characteristic immunophenotypic markers such as CD33, CD13 and sometimes CD117

are expressed on APL cells, whereas markers such as CD34 and HLA-DR are usually absent. This information can help clinicians differentiate APL from other forms of leukaemia and guide therapeutic decision-making. Cytogenetic analysis is also important in the diagnosis of leukaemia. In this case, the diagnosis of APL was confirmed by the presence of the characteristic translocation involving chromosomes 15 and 17, t(15;17) (q22;q12).

Diagnosis: AML of APL type.

Discussion

AML is more common in adults than in children with a peak incidence in late adult life (>65 years old); 60% of AML patients are over the age of 60 years. AML is classified primarily based on the defining genetic abnormalities with some contribution from cell morphology:

- *Genetic classification:* A significant advancement in AML classification has been the identification of recurring genetic abnormalities that play a crucial role in disease prognosis and treatment. This approach categorises AML based on specific genetic mutations, chromosomal translocations and molecular markers. Some common examples of genetic abnormalities include mutations in genes, such as FLT3, NPM1, CEBPA and chromosomal translocations such as t(15;17), which is characteristic of APL. This genetic classification not only aids in diagnosing and predicting the outcome of AML, but also helps guide targeted therapies. Different subtypes of AML with specific genetic aberrations respond differently to various treatments, making this approach essential for personalised treatment strategies.
- *Morphology-based classification:* Historically, AML subtypes were primarily classified based on the morphology of the leukaemic

cells, including their appearance under a microscope and how they resemble normal bone marrow cells at various stages of development. This approach was crucial before the discovery of genetic abnormalities and continues to be relevant for cases where specific genetic markers are not present.

APL has a specific association with disseminated intravascular coagulation (DIC). Patients may present with severe bleeding and this may increase following treatment as the blast cells break down, leading to further consumption of clotting factors and platelets.

The translocation in APL links ligand and DNA binding sequences of the *RAR* gene on chromosome 17 to sequences of the *PML* gene on chromosome 15. The resultant fusion protein arrests differentiation in the promyelocyte stage. The chimeric PML-RARα protein retains the DNA and ligand-binding domains of RARα and is able to suppress PML/p53-driven transcription of retinoic acid target genes. These target genes are involved in self-renewal or differentiation following recruitment of transcription co-repressor complexes.

Treatment and Prognosis

Therapy of APL is completely different from the other types of AML and has been revolutionised by the introduction of all-trans retinoic acid (ATRA; tretinoin) and arsenic trioxide (ATO). Many clinical studies have demonstrated the efficacy of ATRA plus ATO, with or without chemotherapy. ATRA and ATO work by inducing differentiation of the abnormal promyelocytes so that differentiated myeloid cells are produced that have a limited lifespan in the circulation. ATRA and ATO both act in part through their ability to degrade the PML-RARα oncoprotein. Differentiation syndrome (DS), previously known as retinoic acid syndrome, is a common life-threatening complication in patients with APL treated with ATRA and/or ATO, and is characterised by hyper-leucocytosis associated with unexplained fever, respiratory distress, weight increase, lower-extremity oedema, dyspnoea, pleural or pericardial effusions, hypotension and/or acute renal failure. It is typically managed with steroid treatment and supportive care, including involvement of the intensive care team if significant respiratory distress occurs. Careful monitoring of coagulation profiles, and transfusion of platelets and plasma products to avoid bleeding, is critical during the induction phase of treatment.

Questions

1   What are Auer rods composed of?
2   What is the significance of the raised uric acid levels in a patient with leukaemia?

Answers on p. 553.

Case Study 59   'Moon' Face, Breathlessness and 'Stretch' Marks in a 65-year-old Woman

Clinical Presentation

A 65-year-old woman with a 2-month history of a new cough was taken to the Emergency Department because of a 3-week history of worsening symptoms, including dyspnoea at rest, chest tightness and swelling of her lower legs. She had smoked around 20 cigarettes a day from her early 20s but gave up when her husband died of oral cancer 18 months ago. The patient noted that several of her family members had said she looked 'plumper' than usual. She had also developed the appearance of a fatty 'hump' between the shoulders, a rounded 'moon' face and pinkish purple 'stretch' marks on her skin, the latter of which she attributed to changes in her weight. Symptomatic medications and a course of antibiotics had not improved her condition. Her performance score (Eastern Cooperative Oncology Group; ECOG criteria; see Case Study 55) was 2 (ambulatory and capable of all self-care but unable to carry out any work activities; up and about more than 50% of waking hours).

Investigations

On admission, the patient was hypertensive (blood pressure 160/90 mmHg) and she had oedema of both lower limbs. Chest auscultation showed diminished respiratory sounds on the right side, with soft wheezing. A chest X-ray revealed a mass at the hilum of the right lung with enlarged mediastinal lymph nodes. A subsequent CT thorax confirmed the presence of a 6 cm primary tumour in the right lung with ipsilateral nodal involvement (nodes on the same side of the body as the tumour are involved) and metastases in the left lung. Abdominal CT scan and CT brain were normal. Laboratory tests revealed that she had severe hypokalaemia with a potassium level of 1.8 mmol/L (reference range: 3.5–5.0), hyponatraemia with a sodium level of 128 mmol/L (reference range: 135–146) with metabolic alkalosis (bicarbonate 35 mmol/L, reference range: 22–30; increased arterial pH of 7.55, reference range: 7.35–7.45; arterial PCO_2 of 4.5 kPa, reference range: 4.8–6.1) and a lymphopenia of 1.18×10^9/L (reference range: 1.5–4.0).

An endobronchial ultrasound-guided fine needle aspirate (EBUS-FNA) was taken and cytological examination revealed the presence of small malignant cells clustered in groups with markedly increased nuclear to cytoplasmic ratio (Fig. CS59.1). A needle core biopsy was also taken and after 2 days processing in the histopathology department, the pathologist issued a preliminary report of a 'malignant tumour consistent with small cell carcinoma' (Fig. CS59.2). Subsequently, immunohistochemistry (IHC) tests confirmed this diagnosis with tumour cells showing positivity for synaptophysin, chromogranin A and

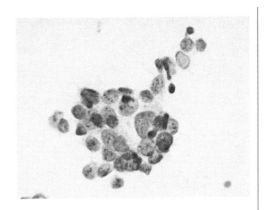

Figure CS59.1   Photomicrograph of the tumour cells of small cell lung cancer (SCLC). Fine needle aspiration specimen. 'Small cell' is so named as the tumour cells are small relative to other carcinomas, for example adenocarcinoma. SCLC cells are typically small, round to oval-shaped cells with scant cytoplasm and densely packed nuclei on cytology. SCLC cells also exhibit nuclei with finely dispersed chromatin (resembling 'salt') and darkly stained, punctate nucleoli (resembling 'pepper'). This unique nuclear appearance is a distinctive feature of SCLC and helps pathologists identify it microscopically. *Source:* William D Travis et al. 2012 / Reproduced from Springer Nature.

CD56. IHC also revealed that the tumour cells had a very high proliferative index (75%) as assessed by staining for the Ki67 antigen.

Given the clinical suspicion of Cushing syndrome, additional blood tests were ordered, including plasma corticotropin (adrenocorticotrophic hormone; ACTH), the levels of which were 121 ng/L (reference range 0–50 for samples collected at 09:00 hours). ACTH stimulates secretion of cortisol from the adrenal cortex and is subject to diurnal variation. A dexamethasone suppression test (DST) was also used to measure how cortisol levels change in response to dexamethasone and can be useful in distinguishing classical Cushing disease from Cushing syndrome. Dexamethasone binds to glucocorticoid receptors in the pituitary gland to suppress the secretion of ACTH. A low dose of dexamethasone suppresses cortisol in individuals with no pathology in endogenous cortisol production. High doses of dexamethasone negatively

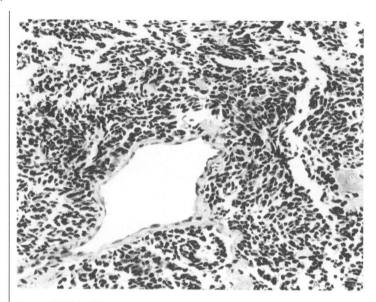

Figure CS59.2 Histopathological image of SCLC. CT-guided core needle biopsy. H&E stain. SCLC histology is characterised by the presence of small, round to oval-shaped, tumour cells with a high nuclear-to-cytoplasmic ratio and showing nuclear 'moulding' (nuclei of adjacent tumour cells appear compressed and elongated). *Source:* No machine-readable author provided / Wikimedia Commons / CC BY-SA 3.0.

feedback on pituitary neoplastic ACTH-producing cells (Cushing disease) but have no effect on cells ectopically producing ACTH (Cushing syndrome). In this patient, high-dose dexamethasone failed to suppress cortisol levels consistent with ectopic ACTH production.

Diagnosis

Cushing syndrome secondary to small cell lung cancer was suspected, given the clinical presentation and the laboratory findings, including the failure to suppress cortisol in the Dexamethasone suppression test.

Diagnosis: Small cell lung cancer with associated Cushing syndrome, stage T3N3M1a.

Discussion

Small cell lung cancer (SCLC) is believed to originate from neuroendocrine cells of the bronchus, which accounts for expression of the neuroendocrine markers, synaptophysin and chromogranin A, which can be helpful in distinguishing SCLC from other types of carcinomas, including non-small cell lung carcinomas (NSCLC), predominantly squamous cell carcinoma and adenocarcinomas (Case Study 60). Biallelic loss of TP53 and RB1 are the most common mutations observed in SLCLC.

SCLC may ectopically produce hormones, especially ACTH. Ectopic hormone production is an example of a *paraneoplastic syndrome* (Chapter 11). Paraneoplastic syndromes are defined as signs and symptoms of malignancy, usually systemic in nature, that are not a direct consequence of the effects of the tumour mass. Hypercalcaemia is another example of a paraneoplastic syndrome and is often due to ectopic production of parathyroid hormone-related peptide (PTHrP), for example by squamous carcinomas of the lung. Osteolytic bone lesion can also be responsible for hypercalcaemia in cancer patients.

The TNM classification system is favoured over the older Veterans Administration Lung Study Group (VALSG) system for staging SCLC. The TNM system offers more detailed anatomical information, which is useful for determining both prognosis and the most appropriate treatment options. It also provides more precise information on lymph node involvement. In contrast, the VALSG system only distinguishes between limited-stage and extensive-stage disease. For instance, VALSG does not differentiate between early-stage SCLC and locally advanced cases, whereas the TNM system does so by classifying tumours based on their size (T), lymph node involvement (N) and presence or absence of metastasis (M). Although the TNM is considered more precise, the VALSG system is still used in some settings.

Treatment and Prognosis

The management goals in this case were to treat the underlying SCLC, which should result in resolution of the Cushing syndrome. The patient commenced first-line chemo-immunotherapy with carboplatin etoposide and atezolizumab (an immune checkpoint inhibitor) and her condition noticeably improved (scored as ECOG 1; Case Study 55) with no dyspnoea at rest and only limited oedema of the lower ankles. Potassium, sodium and other blood parameters, including ACTH levels, returned to normal. The patient completed six cycles of chemotherapy with complete regression of the tumour masses and lymphadenopathy as assessed by CT scan. The chemotherapy was stopped, but she remained on maintenance atezolizumab and had regular follow-up CT scans and visits to the oncology

outpatient clinic. After 3 months of atezolizumab treatment, her disease returned with evidence of liver involvement on CT scan. The patient was referred for palliative care support and died 2 months later. Although the addition of immunotherapy to chemotherapy has improved survival and quality of life for patients, SCLC remains largely incurable. Notably, immune checkpoint inhibitors are well tolerated when used in combination with chemotherapy. Atezolizumab, used in this patient, is a type of immune checkpoint inhibitor, which recognises the programmed death ligand-1 (PDL-1) and blocks the interaction between PDL-1 (expressed on tumour cells, or on cells of the tumour microenvironment) and programmed cell death protein-1 (PD-1; expressed on T-cells). In some patients, this unleashes a pre-existing T-cell response against the tumour cells. Durvalumab is another monoclonal antibody that binds to PDL-1 and is also approved for the treatment of patients with SCLC.

Questions

1 Some patients with SCLC will present with a paraneoplastic syndrome known as Lambert–Eaton myasthenic syndrome (LEMS). What is LEMS?
2 Targeted therapies have revolutionised cancer therapies, but how are these new drugs named?

Answers on p. 554.

Case Study 60   Hoarseness and Weight Loss

Clinical Presentation

A 71-year-old retired plumber attended his family doctor's surgery regularly for monitoring of his diabetes (which was controlled by dietary restriction) and his hypertension. On one visit, his doctor noticed that he had lost about 10 pounds in weight and was about to commend his efforts at dieting

when she noticed how hoarse his voice sounded. The patient had noticed this symptom for approximately 6 weeks, but since there was no pain in his throat, he had thought little of it. He put it down to the strain of a smoker's cough. In fact, he had brought up flecks of blood. The patient had smoked at least 20 cigarettes daily since serving in the army. On questioning, he admitted to being rather short of breath for the last few months and to feeling rather wheezy at night. On examination, the patient was pale and was mildly breathless at rest. He was not blue (which would indicate hypoxia), but his fingers were club-shaped, which is often a sign of heart or lung disease. There was a pronounced wheeze localised to the upper lung, but no evidence of infection or fluid around the lung. There was no lymphadenopathy and the liver was not palpable. Urgent investigations were arranged.

Investigations

The results of laboratory investigations are shown in Table CS60.1. They show a mild normocytic normochromic anaemia and a raised erythrocyte sedimentation rate (ESR). The total leucocyte count was high, the majority of cells being neutrophils. A blood film showed the presence of immature leucocytes and nucleated red blood cells. This finding (known as a *leucoerythroblastic blood picture*) suggests bone marrow infiltration by tumour, which replaces normal haemopoietic cells. A chest

Table CS60.1 Results of selected laboratory investigations.

Test	Result	Reference range
Haemoglobin	**10.7 g/dL**	13.0–18.0
Red cell count	**3.84 × 10^{12}/L**	4.5–6.5
Haematocrit	**0.32 L/L**	0.40–0.54
Mean cell volume	84.6 fL	82–92
Mean cell haemoglobin	27.9 pg	27.0–32.0
Erythrocyte sedimentation rate	**84 mm/hour**	<20
Total white cell count	**13.6 × 10^9/L**	4.0–11.0
Neutrophils	**11.2 × 10^9/L (83%)**	2.0–7.5 (40–75%)
Lymphocytes	1.25 × 10^9/L **(9%)**	1.5–4.0 (20–45%)
Monocytes	**1.00 × 10^9/L** (7%)	0.2–0.8 (2–10%)
Eosinophils	0.10 × 10^9/L (1%)	0.04–0.4 (1–6%)
Basophils	0.05 × 10^9/L (<1%)	0.02–0.1 (1%)
Platelet count	258 × 10^9/L	150–400
Blood picture	Neutrophilia with some toxic granulation; occasional nucleated red blood cells and myelocytes present	
Urea	**8.4 mmol/L**	2.5–6.7
Creatinine	**134 µmol/L**	59–104
Urate	0.3 mmol/L	0.18–0.42
Phosphate	1.25 mmol/L	0.8–1.5
Total bilirubin	15.2 µmol/L	<21
Alkaline phosphatase	**135 U/L**	25–115
Alanine transaminase	7 U/L	5–40
Aspartate transaminase	18 U/L	10–40
Gamma glutamyltransferase	13 U/L	11–50

Values outside the reference range are in bold.

radiograph showed a large opacity in the left lung. The patient was referred urgently to the chest clinic and bronchoscopy was performed. He was found to have a tumour in the left main bronchus. Histological analysis of a biopsy specimen taken at bronchoscopy showed non-small cell lung cancer. Closer examination of his chest X-ray revealed the presence of several metastases in his ribs.

Diagnosis

The clinical features and laboratory data confirmed the presence of a tumour in the left lung. Histology showed this to be a non-small cell lung cancer. Tumour cells were large, undifferentiated and polygonal, lacking features of glandular or squamous differentiation. The WHO classification suggests that immunohistochemistry should be performed in any morphologically undifferentiated tumour, using antibodies against TTF-1 to exclude adenocarcinoma and p40 or p63 to exclude squamous cell carcinoma. Only tumours that are negative for these markers should be classified as large cell carcinoma. TTF1-negative adeno-carcinomas do occur and carry a poorer prognosis. Immunohistochemistry was per-formed in this case and revealed positive staining for TTF-1 and the tumour was therefore classified as an adenocarcinoma, despite the appearance of the large tumour cells. An example of a TTF-1-positive lung adenocarcinoma is shown in Fig. CS60.1. The presence of metastases in the ribs and the blood film results suggested significant involvement of bone marrow. The absence of any lymphadenopathy, hepatomegaly and the normal values for the enzymes, alanine transaminase and aspartate transaminase, suggested there was no metastatic spread to

Figure CS60.1 Micrograph showing a TTF-1 positive adenocarcinoma of the lung. *Source:* Nephron / Wikimedia Commons / CC BY-SA 4.0.

the lymph nodes or liver. A high level of alkaline phosphatase in the blood is common in bone metastases. The anaemia present in this case was almost certainly due to infiltration of the bone marrow by tumour, which had displaced normal haemopoietic tissue.

Diagnosis: Non-small cell lung carcinoma of the left lung; adenocarcinoma.

Discussion

From a clinical viewpoint, lung cancer can be divided into two main groups. The first comprises squamous carcinomas, adenocarcinomas and large cell carcinomas, collectively referred to as non-small cell lung cancers (NSCLC). These are clinically different from small cell lung cancers (SCLC; see Case Study 59). NSCLCs account for approximately 75% of all lung cancers. Adenocarcinomas are tumours of glandular lung epithelial cells, whereas squamous carcinomas are derived from squamous epithelium. Large cell lung carcinomas are NSLC that cannot be classified as adenocarcinomas or squamous carcinomas.

Over the last 80 years, the incidence of lung cancer has closely paralleled cigarette smoking patterns. In the mid-1970s, it was estimated that cigarette smoking was responsible for 95% of all lung neoplasms in men. Populations in which smoking has been reduced or low tar preparations favoured, have experienced a decline in the incidence of the disease. Women, who tended not to smoke before the 1940s, are increasingly victims of the disease. Other possible environmental carcinogens are asbestos, coal smoke, radon (in certain areas of the United Kingdom, such as Cornwall) and other atmospheric pollutants (see also Chapter 11). All the available evidence suggests that these other factors account for only a small minority of cases. While smoking remains a significant risk factor for all types of lung cancer, adenocarcinoma is the most common type seen in people who have never smoked or who are light smokers.

Adenocarcinomas tend to have a higher rate of specific mutations, such as EGFR, ALK and ROS1, among others. This higher mutational burden provides potential targets for precision medicine approaches, using targeted therapies designed specifically for tumours harbouring these mutations (see below).

Treatment and Prognosis

Molecular testing for EGFR mutations, BRAF V600E, as well as ALK, ROS1 and NTRK fusion genes, is now considered standard-of-care for patients with advanced NSCLC. The use of inhibitors targeting these molecular abnormalities has improved patient outcomes. The EGFR T790M mutation and KRAS mutations should also be tested for as they have been shown to confer resistance to EGFR inhibitor therapy. Immunotherapy with immune checkpoint inhibitors (ICI; see also Chapter 11), given as monotherapy or as a combination with other ICIs or with chemotherapy, has also improved survival for patients with NSCLC, even when used in heavily pre-treated patients.

Three weeks after commencing therapy, his wife rang the doctor for advice. The patient was confused and complaining of nausea, abdominal pain, constipation, thirst and constantly going to pass urine. The doctor suspected deterioration of his diabetes, but analysis of his urine and blood glucose screen were normal. However, his serum calcium was found to be high at 3.3 mmol/L (reference range 2.2–2.67). He was admitted to hospital and treated with rehydration and zoledronic acid to control the hypercalcaemia.

Questions

1 Was the patient's hoarseness caused by his tumour?
2 What were the possible causes of hypercalcaemia in this patient?

Answers on p. 554.

Case Study 61   Painless Lump in the Right Testicle

Clinical Presentation

A 33-year-old man presented to his primary care physician with a painless lump in his right testicle, which he noticed 2 weeks ago. He reported no other symptoms, apart from non-specific testicular discomfort. The patient had no significant past medical history and was not taking any medications. Upon physical examination, the physician identified a firm, non-tender mass in the right testis.

Investigations

The results of the patient's serum tumour markers were within normal limits for alpha-fetoprotein (AFP) and beta-human chorionic gonadotropin (β-hCG), but showed elevated levels of lactate dehydrogenase (LDH) (Table CS61.1). Scrotal ultrasound revealed a solitary 2.5 cm mass in the right testis. This was supported by the results of Doppler imaging, a specialised form of ultrasound technology that is particularly useful in the diagnosis and evaluation of testicular cancer. Unlike conventional ultrasound, which primarily provides structural details, Doppler imaging also allows for the assessment of blood flow within the testicular tissue. This is crucial because tumours often have a characteristic blood supply pattern that can be different from that of normal or benign tissue. By analysing the blood flow patterns, physicians can gain additional information to help distinguish between benign and malignant testicular conditions.

Diagnosis

A computed tomography (CT) scan of the chest, abdomen and pelvis showed no evidence of metastasis. A biopsy of the testicular mass showed the presence of a seminoma. Seminomas are known for their distinctive histological features, characterised by large, round tumour cells with abundant clear cytoplasm and distinct borders, surrounding lymphocyte-containing fibrovascular septa (Fig. CS61.1). The tumour cells are often described as having a 'fried egg' appearance.

Diagnosis: Seminoma of the testis.

Discussion

Seminoma is a type of germ cell tumour. Germ cell tumours (GCTs) can take on different histological forms (Fig. CS61.2) and are broadly divided into *seminomas* (also called 'dysgerminomas' in females and 'germinomas' when they occur in the CNS) that retain pluripotency and *non-seminomas* that take on a variety of differentiation states, and which include *embryonal carcinomas, teratomas* that differentiate into somatic cell lineages (endoderm, mesoderm and ectoderm), *yolk sac tumours* that resemble the foetal yolk sac, and *choriocarcinomas* that resemble the placenta.

The origin of GCTs is best explained if we consider the various potency states of cells in the early embryo, representing the different stages of increasing specialisation as described below and illustrated in Fig. CS61.2:

- *Omnipotent 2C state:* At the zygote and two-cell stage, cells are *omnipotent*, meaning

Table CS61.1   Results of selected laboratory investigations.

Test	Result	Reference range
Alpha-fetoprotein (AFP)	2.8kU/L	<5.8
Beta-human gonadotropin (β-hCG)	2.5 mU/mL	<5
Lactate dehydrogenase (LDH)	**420 U/L**	135–225

Values outside the reference range are in bold.

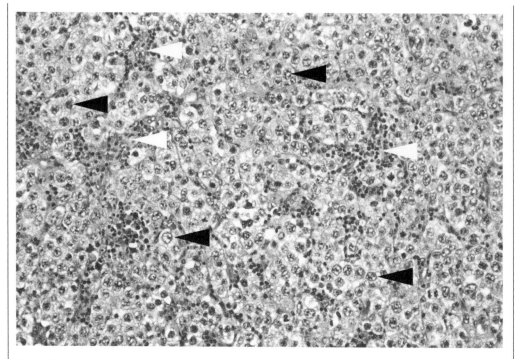

Figure CS61.1 Typical histological features of a seminoma. Seminomas are recognised for their distinctive histological characteristics, which include the presence of large, rounded tumour cells (black arrowheads) featuring abundant clear cytoplasm and well-defined borders, typically with a lymphocytic infiltrate (white arrowheads). *Source:* Nephron / Wikimedia Commons / CC BY-SA 3.0.

they have the potential to develop into a complete organism as well as into extra-embryonic tissues like the placenta.

- *Totipotent naive state:* Cells of the inner cell mass and pre-implantation epiblast are *totipotent*. They can give rise to all types of cells in the body but are not capable of developing into a complete organism on their own.
- *Pluripotent primed state:* Epiblast stem cells (EpiSCs) of the post-implantation epiblast are *pluripotent*, meaning they can differentiate into many, but not all, types of cells within the organism. They are more restricted in their developmental capacity compared to the naive state.

Primordial germ cells (PGCs) are derived from the pluripotent primed state of EpiSCs.

They represent a specialised committed lineage. PGCs migrate from the epiblast to the gonadal ridges, where they differentiate into either oocytes (in females) or spermatocytes (in males). Once PGCs reach the developing gonads, they are known as gonocytes. Gonocytes represent a transitional stage of germ cell development in the foetal gonads before they further mature into oogonia in females and spermatogonia in males. During migration and settling in the gonadal ridges, PGCs undergo extensive epigenetic reprogramming, including demethylation of their genome, resetting the epigenetic marks to a baseline state. This is crucial to allow the germ cells to give rise to a completely new organism upon fertilisation. Upon arrival in the gonadal ridge, PGCs undergo a series of mitotic divisions

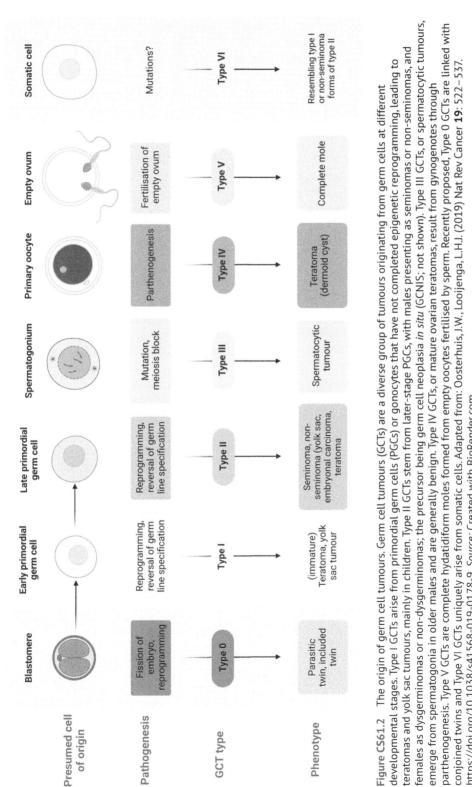

Figure CS61.2 The origin of germ cell tumours. Germ cell tumours (GCTs) are a diverse group of tumours originating from germ cells at different developmental stages. Type I GCTs arise from primordial germ cells (PGCs) or gonocytes that have not completed epigenetic reprogramming, leading to teratomas and yolk sac tumours, mainly in children. Type II GCTs stem from later-stage PGCs, with males presenting as seminomas or non-seminomas, and females as dysgerminomas or non-dysgerminomas; the precursor being germ cell neoplasia *in situ* (GCNIS; not shown). Type III GCTs, or spermatocytic tumours, emerge from spermatogonia in older males and are generally benign. Type IV GCTs, or mature ovarian teratomas, result from gynogenotes through parthenogenesis. Type V GCTs are complete hydatidiform moles formed from empty oocytes fertilised by sperm. Recently proposed, Type 0 GCTs are linked with conjoined twins and Type VI GCTs uniquely arise from somatic cells. Adapted from: Oosterhuis, J.W., Looijenga, L.H.J. (2019) Nat Rev Cancer **19**: 522–537. https://doi.org/10.1038/s41568-019-0178-9. *Source:* Created with BioRender.com.

to increase their number. These mitotic divisions occur during embryonic and foetal development. After these mitotic divisions, in females, germ cells will begin the process of meiosis during foetal development (but will be arrested in prophase I of meiosis until puberty). Each menstrual cycle after puberty, some of these primary oocytes resume meiosis as discussed below. In males, the process of meiosis does not begin until puberty and spermatogonia (derived from PGCs) undergo meiosis to form spermatocytes, which then develop into mature sperm, as described below.

Males:

- *Spermatogonia:* These are the precursor germ cells found in the testes. They undergo mitosis to produce more spermatogonia.
- *Primary spermatocytes:* Some spermatogonia differentiate into primary spermatocytes, which then enter meiosis I.
- *Secondary spermatocytes:* After meiosis I, each primary spermatocyte produces two secondary spermatocytes.
- *Spermatids:* Secondary spermatocytes undergo meiosis II to produce spermatids.
- *Spermatozoa:* Spermatids further differentiate into mature sperm cells, or spermatozoa.

Females:

- *Oogonia:* These are the precursor germ cells found in the ovaries. They divide mitotically during foetal development.
- *Primary oocytes:* Before or shortly after birth, oogonia differentiate into primary oocytes, which enter meiosis I but are arrested in prophase I until puberty.
- *Secondary oocytes:* Starting at puberty and with each menstrual cycle, a cohort of primary oocytes, around 10 or 20, enclosed within their follicles, are stimulated to resume meiosis. However, out of this cohort of primary oocytes, typically only one dominant follicle reaches full maturity and undergoes ovulation. The other follicles that began to mature will undergo atresia or degeneration.
- *Ovum and polar bodies:* The secondary oocyte begins meiosis II but is arrested at metaphase II. It remains in this state until fertilization. If fertilization by a sperm does not occur, the secondary oocyte will not complete meiosis II and will degrade. However, if a sperm penetrates the secondary oocyte, it triggers the completion of meiosis II. The result of this final division is the mature ovum (or egg) and a second polar body.

PGCs and germ cells exhibit properties of continuous proliferation, primarily due to high levels of telomerase activity (Chapter 3). This characteristic enables these cells to have seemingly unlimited replication potential, a trait of cancer cells. GCT can manifest in either the ovaries or testes, or in extragonadal areas, and are unique insofar as their potential for development is essentially dictated by the potency state of their originating cells, which can revert to omnipotent, totipotent or pluripotent stem cells.

There are five major types of GCTs (Fig. CS61.2), each characterised by unique, principally epigenomic, abnormalities. Recently, an additional two types have been included. Genetic predisposition factors impacting the originating cells and their microenvironments are likely responsible for the bilateral, multifocal and familial occurrences seen across the various GCT types. Unlike most other cancers, GCTs seldom arise due to somatic mutations that drive the disease. Instead, they develop when there is a failure to regulate the latent developmental capacity of their originating cells, leading to cellular reprogramming. Despite being generally driven by non-mutational origins, even malignant forms of GCTs are often characterised by a lack of TP53 mutations and are notably sensitive to DNA damage.

- *Type I GCT:* Extragonadal and gonadal Type I GCTs are believed to arise from PGCs or gonocytes that have not undergone complete epigenetic reprogramming. If these cells do not complete this reprogramming and somehow escape the normal developmental pathways, they can become cancerous. The main histological forms of Type I GCTs are teratomas and yolk sac tumours, and they are found more frequently in children than in adults.

- *Type II GCT:* Typically originate from later-stage PGCs. Type II tumours in males can further be classified into seminomas and non-seminomas, with their female equivalents being referred to as dysgerminomas and non-dysgerminomas respectively. Embryonal carcinomas serve as the stem cell component of non-seminomas and can further evolve into various specialised forms like yolk sac tumours, choriocarcinomas or teratomas. These tumours can appear individually or in a mixed form. The precursor lesion for these tumours is known as germ cell neoplasia *in situ* (GCNIS). Aberrant reprogramming of PGCs or gonocytes is also the driving mechanism in Type II GCTs, leading to these specialised cells reverting to a more primitive or stem-cell-like state.

- *Type III GCT:* Also known as spermatocytic tumours, they arise from spermatogonia, the earliest and most undifferentiated cells in the spermatogenic lineage and almost exclusively occur in the testis, primarily in older men. Histologically, they are composed of a mixture of small, medium and large-sized cells, reflecting different stages of spermatogenic differentiation. Spermatocytic tumours are typically benign and have a very low potential for metastasis.

- *Type IV GCT:* These are mature teratomas of the ovary, or so-called dermoid cysts. Dermoid cysts contain different tissues such as skin, hair and sometimes teeth. Rarely, a mature teratoma can contain structures that resemble a malformed foetus. This is termed *fetus in fetu* or *fetiform* teratoma. However, these structures are not viable foetuses. Instead, they are an unusual manifestation of the tumour's ability to generate various tissue types. In some cases, there can be malignant transformation within the cyst (Case Study 65). Previously, it was thought dermoid cysts developed from functional oocytes, but recent studies suggest they arise from *gynogenotes* through a process called *parthenogenesis*. Gynogenotes (or gynogenetic zygotes) are oocytes activated without sperm contribution and where all the genetic material is derived from the mother. In some species (but not in humans), gynogenotes can develop into viable organisms. Thus, dermoid cysts likely develop from primary oocytes that have bypassed the usual hormonal regulation that prevents meiosis. Since the halting of meiosis is controlled by hormones, this could explain why dermoid cysts typically form during a woman's reproductive years.

- *Type V GCT:* These are *complete hydatidiform moles* (complete moles) and are the result of an empty oocyte that is fertilised by sperm to form *androgenotes*. Androgenotes (or androgenetic zygotes) develop from an oocyte that has been activated, but in which all the genetic material is derived from the father (i.e. the sperm) and the maternal genetic content is either absent or inactive. Complete moles are discussed in more detail in Case Study 44.

Recently, two additional types of GCT have been proposed:

Type 0 GCT: These are a very rare subtype associated with the phenomenon of conjoined twins. Conjoined twins occur when, during the early stages of foetal development, the process of splitting of a single fertilised egg is not completed entirely, resulting in two foetuses that remain physically connected to each other. At the attachment site or the point of union between conjoined twins, tissues from both twins may interact and merge, creating a localised environment in which Type 0 GCTs can arise. The exact pathogenesis of Type 0 GCTs is poorly understood, but they are believed to result from aberrant germ cell development or differentiation at the conjoined twin attachment site.

Type VI GCT: This is a rare and unusual subgroup. Unlike the other types, Type VI GCTs are believed to originate from somatic cells, in other words, cells in the body that are not involved in reproduction. The transformation of a somatic cell into a germ cell tumour is an unusual event. This process entails a somatic cell acquiring mutations or undergoing certain changes that cause it to adopt characteristics of a germ cell and then form a tumour. The exact mechanisms and triggers for this transformation are only poorly understood.

Extragonadal GCTs are often located along the midline of the body. The most common midline locations include the mediastinum (area between the lungs), the retroperitoneum (area behind the abdominal cavity) and the pineal gland in the brain. During embryonic development, primordial germ cells originate in the yolk sac and follow a migratory pathway to the developing gonads (ovaries or testes). This migratory path often follows midline structures and it is believed that some germ cells may settle along this midline path. If these misplaced germ cells undergo malignant transformation, they can give rise to extragonadal GCTs in midline structures.

The histology of GCTs is similar in both males and females, whether occurring in the testis, ovary or extragonadal sites, further confirming their origin from common precursor cells. There are, however, epidemiological differences in the incidence of GCTs between males and females and in different age groups. In males, there are two peaks of testicular GCT incidence, one in early childhood at around age 3–4 years, and a second, much larger, peak that begins at puberty and is maximal at around age 30. In females, there is an early peak from age 0 to 2 representing the incidence in newborns and infants of sacrococcygeal teratoma, an extragonadal GCT. Beginning at age 5–6, the incidence of ovarian GCT increases with age, becoming maximal at age 20–25.

While the overall incidence of GCT (about 12 500 cases/year in the United States) is lower than that of common epithelial cancers, such as those of the lung, breast and prostate gland, testicular GCT is the most common cancer and the leading cause of cancer death in young men. Risk factors for testicular GCT include *disorders of sexual development* (DSD), *gonadal dysgenesis* (abnormal development of the gonads), *cryptorchidism* (undescended testis), environmental exposures and genetic associations, including family history; the risk of developing GCT is fourfold higher in a male with a father who had a GCT, and up to ninefold in a male whose brother had a GCT.

Testicular tumours can also develop from the supportive and hormone-producing tissues, or stroma, of the testicles and are known as gonadal stromal tumours. They constitute <5% of adult testicular tumours, but up to 20% of childhood testicular tumours. The two main types are *Leydig cell tumours* and *Sertoli cell tumours*. Only around 10% of each is malignant. Leydig cell tumours derive from the Leydig cells in the testicle that normally produce male sex hormones (e.g. testosterone). Most Leydig cell tumours do not spread beyond the testicle and are usually cured by orchiectomy (removal of the testicle). Sertoli cell tumours develop from Sertoli cells, which support and nourish the sperm-producing germ cells.

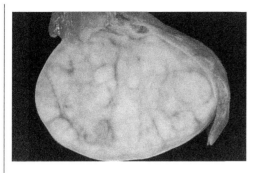

Figure CS61.3 Seminoma of the testis. Typically, seminoma appears as a well-defined, pale greyish-white, firm mass within the testicle. *Source:* Ed Uthman. / Wikimedia Commons / CC BY 2.0.

Treatment and Prognosis

The patient was referred to an oncologist, who recommended surgical removal of the affected testicle to confirm the diagnosis and stage the cancer. The pathologist described the gross appearance of the tumour as 'a well-defined ovoid mass within the testicular tissue, pale white in colour, lacking cystic or haemorrhagic areas (commonly seen in other testicular tumours). The tumour mass is well-demarcated from the surrounding testicular tissue' (Fig. CS61.3).

Following surgery, it is crucial to monitor the patient for any signs of disease recurrence or metastasis, especially given that the TNM staging was pT1 N0 M0 S0, indicating localised disease without spreading to lymph nodes or distant organs. All serum levels of tumour markers returned to normal. This is a good prognostic sign, but vigilance in follow-up is essential. Contrast-enhanced CT or MRI scans at 4–6 months and then at 12–18 months post-surgery should be deployed. The contrast agents used in these scans improve visualisation of tissue structures and can help detect any subtle changes that might suggest the presence of residual or recurrent cancer. These scheduled scans offer the advantage of tracking the stability of the patient's condition over time, making it easier to spot potential issues before they become more serious problems. Beyond these time-specific scans, the patient should follow regular guidelines for ongoing surveillance, which may include periodic imaging, physical examinations and serum tumour marker tests, to ensure that any signs of recurrence are caught and treated promptly.

The treatment plan for seminoma is largely determined by the stage at which the cancer is diagnosed, as well as other factors, such as patient age, overall health, patient compliance with surveillance and the presence of metastases. For patients with early-stage seminomas (Stage I) without high-risk features, active surveillance (following orchiectomy) with frequent imaging and serum tumour marker checks to monitor for any signs of disease progression is preferred. For non-compliant patients, or patients with high-risk features, a single cycle of chemotherapy or localised radiotherapy are the preferred options. For more advanced stages, or for cases that have spread to the lymph nodes or other organs (Stage II and Stage III), a combination of chemotherapy and possibly radiotherapy is typically employed. Chemotherapy regimens such as BEP (bleomycin, etoposide and cisplatin) or EP (etoposide and cisplatin) are commonly used in these scenarios. In certain instances, lymph node dissection may be performed to remove affected lymph nodes, especially if they remain enlarged after chemotherapy. Prognosis for seminoma, particularly when detected at an early stage, is generally very favourable. Five-year survival rates for early-stage seminoma exceed 95%, making it one of the most treatable forms of cancer.

Questions

1 What is the role of tumour markers in the management of patients with seminoma?
2 Testicular cancer is staged according to the TNM system, but what does 'S' mean?

Answers on p. 554.

Case Study 62 Sudden Onset of Lower Back Pain

Clinical Presentation

A 50-year-old man presented to his GP with lower back pain. He described the pain as coming on suddenly when he lifted some furniture for his son, who was moving house. Since then, he had a persistent dull backache and an occasional sharp pain when bending over. He reported no weakness or numbness and no shooting pain down either leg. There was no recent weight loss, fever or sweating. There was no point tenderness in the lower lumbar spine. Sensation and reflexes were normal.

Investigations

The differential diagnoses in this case include prolapsed intervertebral disc, muscular strain, vertebral fracture, discitis, spinal stenosis and occult cancer. Intra-abdominal or genitourinary pathology should also be considered, including abdominal aortic aneurysm, pancreatitis, renal colic, pyelonephritis and prostatic disease. The results of laboratory investigations are shown in Table CS62.1. The patient was anaemic with a raised ESR, CRP and hypercalcaemia. The raised creatine and urea indicate renal impairment. A lumbar spine X-ray revealed a fracture (L4) and multiple, small, well-circumscribed round, lytic, so-called punched-out, lesions. An example of these typical 'punched out' lesions is shown in Fig. CS62.1.

Diagnosis

Multiple myeloma was suspected based on the symptoms, anaemia, abnormal renal function and elevated calcium. The following mnemonic from the International Myeloma Working Group (IMWG) is helpful when remembering the most common presenting features of multiple myeloma:

Table CS62.1 Results of selected laboratory investigations.

Test	Result	Reference range
Blood		
Total white cell count	4.2×10^9/L	4.0–11.0
Red cell count	5.1×10^{12}/L	4.5–6.5
Haemoglobin	**11.2 g/dL**	13.0–18.0
Haematocrit	0.51 L/L	0.40–0.54
Platelet count	**141 × 10⁹/L**	150–400
Urea	**8.7 mmol/L**	2.5–6.7
Creatinine	**170 µmol/L**	59–104
Alkaline phosphatase	**251 U/L**	25–115
Calcium	**3.6 mmol/L**	2.20–2.67

Values outside the reference range are in bold.

CRAB

- Hyper**C**alcaemia: cytokines produced by cancer cells activate osteoclasts, which release calcium from bones
- **R**enal failure: light chains or intact immunoglobulins that deposit in renal tubules
- **A**naemia: bone marrow infiltrated by malignant plasma cells
- **B**one lesions: due to increased osteoclast activity

Multiple myeloma is a cancer of plasma cells. Plasma cells are B-cells that have terminally differentiated and produce antibodies (immunoglobulins, see Chapter 6). There are five subclasses of immunoglobulins – IgA, IgD, IgE, IgG and IgM. A normal population of B cells is polyclonal and thus produces multiple different types of antibodies. However, multiple myeloma is a tumour derived from a single B-cell, that is it is a monoclonal disease and only one type of abnormal immunoglobulin is produced in

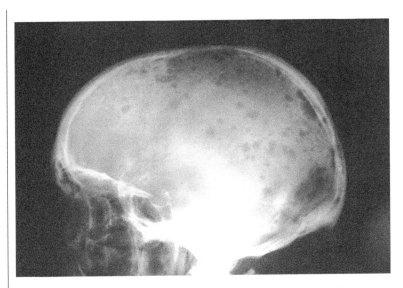

Figure CS62.1 Typical 'punched-out' bone lesion seen in myeloma. This X-ray image displays a striking example of punched-out lesions in the skull, a hallmark radiological feature of myeloma. The circular or ovoid lesions appear as well-defined radiolucent areas in the dense bone, resembling craters and result from the destructive activity of myeloma cells within the bone marrow.

increased amounts by the malignant plasma cells. For this reason, it is useful to measure the abnormal levels of immunoglobulin in the blood of people suspected of multiple myeloma. This patient had elevated blood levels of IgG (59 g/L, normal range: 7.0–18.0 g/L). IgG is the most common type of immunoglobulin produced by the malignant plasma cells in multiple myeloma patients. IgA is the next most common. Patients with IgA myeloma have a worse prognosis and survival than other Ig subtypes. IgM, IgE and IgD myeloma are far less common. Approximately 20% of multiple myeloma patients produce only an abnormal light chain (kappa or lambda) without an intact paraprotein. This is known as light chain multiple myeloma, and is associated with an inferior overall survival. Protein electrophoresis can separate proteins in the blood or the urine and is a useful tool to identify excess or abnormal proteins. Serum protein electrophoresis (SPEP) was performed on this patient and identified an abnormal peak (Fig. CS62.2, lower panel). This is often referred to as 'M' (monoclonal)

protein, paraprotein or 'M-Spike' protein and strongly suggests multiple myeloma. A bone marrow aspirate and trephine biopsy revealed 80% plasma cells. Immunohistochemistry showed that these plasma cells expressed lambda light chains, but not kappa light chains. A t(11;14) translocation was detected by fluorescence *in situ* hybridisation (FISH).

Each antibody molecule is composed of two heavy chains and two light chains. In the context of multiple myeloma, some cancerous plasma cells produce an excess of free light chains without the corresponding heavy chains. These excess light chains can be detected in the blood and urine and are sometimes referred to as 'free' light chains or Bence-Jones protein, named after the doctor who first found them in the urine of affected patients. The light chains can be immunoglobulin fragments or single homogeneous immunoglobulins. This is an important test in the diagnosis of multiple myeloma. A PET scan or whole-body MRI is also recommended in cases of suspected myeloma in order to

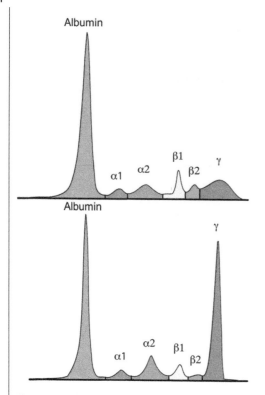

Figure CS62.2 Diagrammatic representation of the results of serum protein electrophoresis. Normal donor (upper panel) and myeloma patient (lower panel) are shown. Albumin (orange), α1 (pink), α-2 (blue), β-1 (yellow), β-2 (green) and γ (red) peaks are shown. Antibodies are mostly found in the γ region. The serum protein electrophoresis of the normal individual shows polyclonal antibodies (no peaks in the γ region). However, monoclonal antibody is present in the myeloma patient as indicated by the peak in the γ region. *Source:* Created with BioRender.com.

detect bone defects caused by osteoclast activation. This provides a more detailed view of the bone structures than plain radiographs and can detect impending bone/vertebral fractures or sites of disease that might require urgent radiotherapy. It is especially useful in cases where lesions are suspected but are not clearly visible on plain X-rays.

Diagnosis: Symptomatic multiple myeloma.

Discussion

Multiple myeloma is a tumour of B-cell origin characterised by hypercalcaemia, osteolytic bone lesions and pathological fractures. Back pain is one of the most common presenting symptoms with around 60% of patients having vertebral involvement at diagnosis. Some patients may have evidence of a monoclonal myelomatous process without symptoms- referred to as *asymptomatic or smouldering myeloma*. There are strict definitions that distinguish symptomatic and asymptomatic myeloma:

Symptomatic myeloma:

- Clonal plasma cells >10% on bone marrow biopsy or other affected tissues
- Presence of monoclonal immunoglobulin in either serum or urine
- Evidence of systemic organ damage caused by disease (presence of CRAB).

Asymptomatic myeloma (smouldering myeloma)

- Serum M paraprotein (IgG or IgA) >30 g/L or urinary M protein ⩾500 mg per 24 hours and/or clonal bone marrow plasma cells 10–60%.
- BUT absence of myeloma-related organ/tissue impairment (i.e. absence of CRAB) or amyloidosis (see also Case Study 16).

The IMWG, an assembly of global myeloma experts, continuously refines the diagnostic criteria for multiple myeloma in line with evolving scientific understanding and clinical observations. Historically, the diagnosis of multiple myeloma has hinged on the CRAB criteria. However, in a significant update, the IMWG recognised the need to encompass a broader range of biomarkers. These additional biomarkers serve as strong indicators of the potential progression to symptomatic stages and can help to identify patients who would benefit from early intervention even before conventional symptoms are observed. This expansion, now integrated into clinical practice guidelines since 2016, includes the following *myeloma-defining events* (MDE):

- *Bone marrow plasma cells:* A clonal bone marrow plasma cell percentage equal to or exceeding 60%.
- *Serum-free light chain ratio:* An involved-to-uninvolved serum-free light chain (FLC) ratio of 100 or more, provided the involved free light chain level is also 100 mg/L or higher.
- *MRI findings:* The presence of more than one focal lesion on MRI scans, with each lesion being at least 5 mm in size.

Patients within these categories have a high likelihood of disease progression within 2 years, reinforcing the necessity for prompt treatment.

An additional condition known as *monoclonal gammopathy of undetermined significance (MGUS)* is a clinically asymptomatic premalignant clonal plasma cell disorder with a propensity to develop into multiple myeloma.

Recurrent genetic alterations in multiple myeloma are commonly chromosome translocations involving the immunoglobulin heavy chain (IgH) gene locus on chromosome 14 (Table CS62.2) and hyperdiploidy (a gain in chromosome number). The translocations are caused by errors in VDJ recombination (Chapter 6) during early B cell development involving the breakage of chromosomes. These translocations juxtapose oncogenes (e.g. CCND1, CCND3, encoding cyclin D1 and cyclin D3, respectively, see also Chapter 3) close to strong enhancers within the IgH loci. These enhancers drive constitutive expression of the oncogene involved and contribute to transformation. Furthermore, mutations and deletions in chromosome 17p are associated with an inferior overall survival in myeloma patients. When the presence of del(17p) and/or translocation t(4;14) and/or translocation t(14;16) occurs alongside an elevated serum beta-2-microglobulin, an elevated lactate dehydrogenase and a low serum albumin, patients are classified as having high-risk myeloma which has a drastically reduced response to treatment and a poor prognosis (Revised Multiple Myeloma International Staging System).

The bone defects in myeloma patients are caused in part by the overproduction by myeloma cells of a factor known as receptor activator of NF-κB (RANK) ligand, which is a tumour necrosis factor receptor (TNFR) family member. Osteoclast maturation and activation are driven by the interaction of RANKL with RANK on the surface of osteoclasts. Under normal circumstances, RANKL-RANK interactions are negatively regulated by osteprotegerin (OPG), a soluble decoy receptor for RANKL. However, OPG levels are decreased in multiple myeloma. Loss of OPG increases osteoclast-mediated bone resorption.

Table CS62.2 Translocations involving chromosome 14 frequently observed in multiple myeloma.

Translocation	Translocation partner gene	Approximate frequency (%)
t(11;14)	Cyclin D1 (CCND1)	16
t(4;14)	FGFR3/ MMSET	15
t(6;14)	Cyclin D3 (CCND3)	6
t(14;20)	MAFB	2

Treatment and Prognosis

Myeloma is a chronic, relapsing/remitting condition that is essentially incurable. Therapy aims to control progression, prolong survival and maximise quality of life. The patient in this case had symptomatic myeloma and required active treatment. The nature of treatment offered depends upon the patient's age and general health. The patient was relatively young and fit and so received a combination therapy known as

'VRd', which includes bortezomib (brand name: Velcade), lenalidomide (brand name: Revlimid) and dexamethasone. Lenalidomide, is a derivative of thalidomide, which was used in the late 1950s and 1960s to treat pregnant women for 'morning sickness' and tragically caused severe birth defects in their children (Chapter 14). Lenalidomide has multiple mechanisms of action that are therapeutically beneficial in multiple myeloma; for example, it can increase IL-2 production by immune cells and hence is often referred to as an *immunomodulatory drug*. The patient also received the monoclonal antibody, daratumumab (which targets CD38 expressed on plasma cells; brand name: Darzalex). These treatments were followed by high-dose melphalan therapy with autologous stem cell transplantation (ASCT). After ASCT, the patient was maintained on lenalidomide, which is approved as a maintenance monotherapy for younger adult patients with newly diagnosed myeloma who have undergone ASCT. The patient was subsequently reviewed every 3 months on lenalidomide and on each occasion was tested for renal function, bone profile, serum immunoglobulins and serum protein electrophoresis. At the time of writing, 2 years after ASCT, the patient remains in clinical remission. The median survival is approximately 6–7 years. In patients eligible for ASCT, 4-year survival rates exceed 80%.

Questions

1 What are other possible causes of hypercalcaemia in this patient?

2 How would you treat the patient's hypercalcaemia?

Answers on p. 555.

Case Study 63   Facial Swelling with Discharge from the Nose

Clinical Presentation

A 65-year-old man presented to the Emergency Department after an acute episode of breathlessness. He admitted a 2-month history of facial swelling and intermittent purulent discharge from the right nostril with episodes of epistaxis (nose-bleeds), which he attributed to the side effects of COVID-19, which he had been ill with 1 month earlier. He had also unintentionally lost weight during the previous 8 months, which was associated with a loss of appetite in the last 2 months. He reported a single episode of drenching night sweats and occasional fever. Examination of the right nostril revealed a 1.5 cm ulcerated mass on the lateral wall, which had obliterated the nasal passage.

Investigations

Laboratory investigations revealed a decreased white blood cell count of 3.6 × 10^9/L (reference range 4–11) with anaemia (haemoglobin 10.5 g/dL; reference range 13.0–18.0), thrombocytopenia of 115 × 10^9/L (reference range 150–400). Lactate dehydrogenase (LDH) was elevated (828 U/L; reference range 135–225). Serum ferritin levels were also increased (321 nmol/L; reference range: 5.8–144).

A biopsy of the nasal mass was taken, and a haematoxylin and eosin-stained section revealed neoplastic lymphoid cells showing vascular invasion (also known as angio-invasion) and necrosis. Immunohistochemistry showed tumour cells had strong expression of CD45, CD2, cytoplasmic CD3ε, CD56 and granzyme B. CD4 and CD8 stains were negative. *In situ* hybridisation was also performed for detection of Epstein-Barr virus (EBV) and was positive. The proliferation index of tumour cells was 75%, as measured by Ki67 staining. The EBV load in the peripheral blood was 5060 IU/mL, which is elevated (in most cases, EBV is undetectable). Full-body PET-CT demonstrated disseminated disease with local invasion of the nasal turbinate and right paranasal sinuses, as well as bilateral axillary and hilar lymphadenopathy and pulmonary involvement. A bone marrow trephine biopsy showed 20–30% infiltration by neoplastic cells. Overall, this was in keeping with stage IV disease, as assessed by the Ann Arbor (AA) staging system.

Diagnosis

Histopathology showing neoplastic cells staining positively for NK cell markers, together with positivity for EBV, confirms the diagnosis of extranodal NK/T-cell lymphoma (ENKTL).

Diagnosis: Extranodal NK/T-cell lymphoma (ENKTL).

Discussion

Extranodal NK/T-cell lymphoma is a rare but highly aggressive type of non-Hodgkin lymphoma associated with EBV. Typically, ENKTL is characterised by extensive necrosis and angio-invasion and presents in extra-nodal sites, predominantly in the nasal cavity, nasopharynx, paranasal sinus or palate. ENKTL shows a high incidence in East Asian, Central and South American and Mexican populations, accounting for 7–10% of non-Hodgkin lymphoma in these populations. In contrast, ENKTL is much rarer in North America and Europe. ENKTL usually occurs in immunocompetent adults (median age of presentation 40–50 years) with a strong male predominance.

Most ENKTL are derived from cells of NK lineage, as evidenced in this case by positivity for CD56, CD2 and CD3ε (Fig. CS63.1A), lack of expression of CD4 and CD8, and the presence of a germline T-cell receptor (TCR). The remaining 15% or so of cases show cytotoxic T-cell differentiation with clonal TCR rearrangements. EBV is always present in the tumour cells in ENKTL and appears essential for the pathogenesis of the disease (Fig. CS63.1B). EBV is present in the memory B-cell pool of most healthy people (see also Case Study 67); but the presence of EBV in T or NK cells is generally seen only in lymphoproliferative disorders. The precise mechanisms by which a B-cell tropic virus causes NK cell or T-cell malignancies are not fully understood but are likely to involve virus-mediated activation of cell signalling pathways such as NF-kB, JAK-STAT and PI3K/AKT, as well as virus-induced epigenetic changes through alterations in methylation state and histone structure. Chromosome abnormalities are also common in ENKTL,

(A)

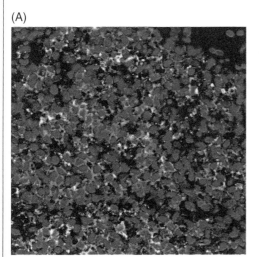

(B)

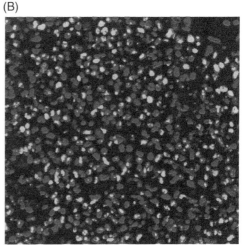

Figure CS63.1 Fluorescent immunohistochemistry and EBER-ISH on an ENKTL tumour. (A) Dual IHC for CD3e (red) and CD56 (yellow) highlights double-positivity of tumour cells for these markers. (B) The ENKTL tumour cells are also positive for EBER (green) indicating the presence of EBV.

with deletion of 6q21 the most frequently observed alteration. Mutations in MYC, PD-L1 and TP53 are also frequently observed.

Treatment and Prognosis

Prognosis is most commonly assessed using the prognostic index for natural killer cell lymphoma score (PINK-E) (Table CS63.1). Patients in low-risk (zero or one risk factor), intermediate-risk (two risk factors) and high-risk (three or more risk factors) groups have an estimated 3-year overall survival (OS) of 81% (95% CI 75–87%), 55% (95% CI 44–66%) and 28% (95% CI 18–40%), respectively. Detection of EBV in the blood by PCR is a poor prognostic marker at diagnosis and ongoing positivity after treatment is linked to a high rate of relapse.

The management of ENKTL differs greatly depending on whether the disease is localised (AA stage I/II) or disseminated (AA stage III/IV). Involved field radiotherapy is crucial for the management of localised disease and is generally combined with platinum-based chemotherapy. The timing of radiotherapy in relationship to chemotherapy, and if some patients can be treated with radiotherapy alone, is the subject of ongoing research. Historically, the outcomes for patients with stage III/IV disease treated with standard anthracycline-based regimens, used for

Table CS63.1 Prognostic index of natural killer lymphoma-EBV score (PINK-E). Each factor, when present, scores one point.

PINK-E
Age >60 years
Stage III/IV disease
Non-nasal primary localization
Distant lymph node involvement*
Detectable blood EBV DNA

* In primary nasal ENKTL, this includes axillary, infraclavicular and mediastinal nodes. This patient had four high-risk factors (age > 60 years, stage IV disease, detectable plasma EBV DNA and distant lymph node metastasis).
This is an update from the original PINK score that now includes EBV plasma load.

B-cell lymphomas, were very poor. However, patients treated with newer, more intensive, chemotherapy regimens, including aspargin-ase, have much better outcomes, but at the cost of significant toxicities. Asparginase can induce apoptosis of NK-cells *in vitro* and has single-agent activity in relapsed/refractory ENKTL. High-dose chemotherapy with autologous stem cell transplant is often used to consolidate remission and reduce the relapse risk in advanced-stage ENKTL; although, there are no prospective data to guide the use of this therapy. Allogeneic bone marrow transplant shows significant efficacy in preventing relapse but at the cost of a high rate of treatment-related mortality, making the overall survival benefit unclear and is therefore reserved for patients with relapsed disease or for those judged to be at significant risk of disease progression.

This patient was started on chemotherapy using the SMILE protocol (Dexamethasone [Steroid], Methotrexate, Ifosfamide, L-asparaginase and Etoposide). After three cycles of chemotherapy, the patient achieved a partial response by PET-CT, with significant disease regression at almost all sites combined with EBV viral loads dropping to near undetectable levels (<500 IU/mL). Historically, this would have been associated with a high risk of subsequent disease progression and poor overall survival compared to achieving a complete response (CR). The patient was therefore enrolled in a clinical trial investigating the efficacy of the immune checkpoint inhibitor, pembrolizumab. After seven cycles of treatment, he achieved a CR and, at the time of writing, remains disease-free.

Questions

1 What is the cellular target of pembrolizumab?
2 Calculate the expected survival for this patient based on the PINK-E prognosis classifier.

Answers on p. 555.

Case Study 64   Abdominal Bloating, Constipation and Frequent Urination

Clinical Presentation

A 57-year-old woman presented to her primary care physician with a history of abdominal bloating, constipation and frequent urination. She reported a gradual onset of these symptoms over the last few months. Her medical history was significant for hypertension and hyperlipidaemia. She had no history of cancer in her family. On physical examination, the patient's abdomen was distended and tender to palpation. Pelvic examination revealed a large, irregularly shaped, mass in the left adnexa, which was not mobile. There was no ascites or palpable lymphadenopathy.

Investigations

A pelvic ultrasound revealed a large, complex mass in the left ovary measuring approximately 10 cm in diameter. The right ovary appeared normal. A CT scan of the abdomen and pelvis confirmed the presence of a large left adnexal mass with multiple septations, suggestive of ovarian malignancy. The patient's laboratory results showed an elevated CA-125 level of 380 U/mL (normal range <35 U/mL) and an elevated white blood cell count with eosinophilia. According to the International Ovarian Tumor Analysis (IOTA) criteria, this was an ovarian 'M' (malignant) lesion, which should not be operated on by a gynaecologist. Instead, she was referred to a cancer centre because in 30% of cases, a multidisciplinary approach together with a surgical or urological oncologist will be needed as the surgical goal is to gain an R0 resection. A whole-body CT was performed to exclude liver or pulmonary metastases. Her case was presented and discussed at the preoperative interdisciplinary tumour conference.

The decision was taken to perform up-front surgery with frozen section (when the tissue from an operation is frozen in the histopathology lab' and a section cut from it, thus allowing a very rapid diagnosis, while the patient remains anaesthetised) but only if the patient's ECOG performance score allowed that. The cytoreductive surgery is a radical approach starting with peritonectomy, hysterectomy plus bilateral salpingo-oophorectomy, omentectomy and pelvic plus paraaortic lymphadenectomy.

Diagnosis

Histopathology of resected tumour revealed a high-grade serous carcinoma of the left ovary with involvement of the fallopian tube. The patient underwent a complete staging procedure, which included a total abdominal hysterectomy, bilateral salpingo-oophorectomy, omentectomy and lymph node dissection. The patient's tumour was staged as FIGO (International Federation of Gynecology and Obstetrics) stage III.

Diagnosis: High-grade serous carcinoma of the ovary, stage III.

Discussion

Ovarian epithelial cancers can be divided into two broad categories, type I and type II tumours. Type I tumours include low-grade serous, mucinous, endometrioid and clear cell carcinomas and are slow-growing tumours harbouring mutations in KRAS, BRAF, PTEN, PIK3CA, CTNNB1 and ARID1A. Type II tumours are the most common forms of epithelial ovarian cancers and include high-grade serous, endometrioid and undifferentiated carcinomas; they are aggressive, genetically highly unstable, cancers that are normally diagnosed only at advanced stages. High-grade serous ovarian carcinoma (HGSOC) is the most common and deadliest type of ovarian cancer, accounting for approximately 70% of all ovarian cancer cases. The exact cause of HGSOC is not yet fully understood, but several risk factors have been identified, including age, family history of ovarian and breast cancer, including the presence of germline mutations (such as BRCA1 and BRCA2), and environmental/lifestyle factors (such as tobacco use and obesity). Unlike low-grade epithelial cancers that develop from ovarian surface epithelial cells, HGSOC typically develops from precursor lesions called 'serous tubal intraepithelial carcinomas' (STICs) that originate in the fallopian tubes, rather than in the ovaries themselves. STICs are thought to arise from normal-appearing fallopian tube cells that undergo mutations or epigenetic changes, leading to the development of abnormal cells with the potential to become cancerous. The abnormal cells can detach from the fallopian tube and migrate to the surface of the ovary, where they form small cysts or nodules. Over time, these nodules grow and coalesce, forming larger tumours that invade surrounding tissues and organs. The cancer cells can also metastasise to other parts of the body through the lymphatic system or bloodstream.

At the molecular level, the most common alteration in HGSOC is a TP53 mutation, which is an early event, occurring in up to 96% of cases and resulting in the loss of function of the p53 tumour-suppressor protein. Other common mutations include alterations in genes involved in cell cycle regulation (such as RB1). These genetic alterations contribute to the hallmark features of HGSOC, including rapid growth, high-grade morphology and resistance to chemotherapy.

Treatment and Prognosis

The pathologist presented the case at the postoperative interdisciplinary tumour board. Nowadays, the therapy of ovarian cancer stands on three pillars: radical cytoreductive surgery with the goal of R0 or R1 resection, followed by chemotherapy and maintenance therapy. This patient's tumour was classified as FIGO stage III, indicating that it had spread beyond the ovaries but was still confined to the pelvis. Current first-line treatment for FIGO stages IIb-IV HGSOC includes debulking surgery followed by combination chemotherapy, usually with carboplatin and paclitaxel. Ovarian cancer is highly sensitive to chemotherapy, particularly to platinum. While many patients achieve remission following initial chemotherapy, many will also eventually relapse with the highest rate occurring during the first 2 years after primary treatment. The case patient underwent six cycles of adjuvant chemotherapy with carboplatin and paclitaxel.

Maintenance therapy has changed in recent years as many new biomarker-targeted molecular immunotherapies and immune checkpoint inhibitors have been added to the treatment guidelines. The current standard-of-care is bevacizumab, which is a therapeutic monoclonal antibody specific for the vascular endothelial growth factor (VEGF), a protein that promotes the growth of new blood vessels, and PARP inhibitors, such as Olaparib (Fig. CS64.1). Bevacizumab is used in patients irrespective of BRCA status, while Olaparib is primarily used in those who are BRCA mutation-positive. At the time of writing (18 months after completing treatment), our patient remains disease free.

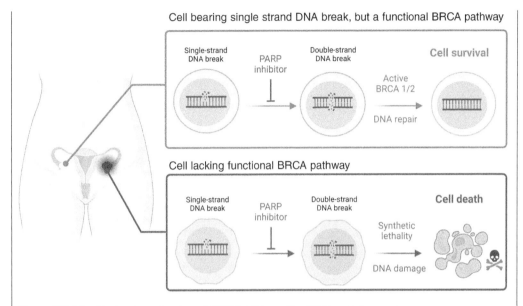

Figure CS64.1 Mechanism of action of PARP inhibitors. The BRCA proteins are tumour suppressor genes because they have a critical role in repairing double-stranded DNA damage. Therefore, BRCA proteins help maintain genomic integrity. Consequently, cells with mutated BRCA genes, often exhibiting loss of heterozygosity (LOH), experience DNA breaks and other genetic rearrangements. The enzyme, Poly (ADP-ribose) polymerase (PARP) is responsible for repairing single-stranded DNA breaks. PARP inhibitors prevent PARP from repairing single-stranded DNA breaks, leading to their accumulation and eventual conversion into double-stranded breaks. The accumulation of double-stranded breaks in cells with BRCA mutations ultimately results in cell death. Therefore, administering PARP inhibitors to cells with BRCA gene mutations triggers double-stranded DNA breaks and subsequent cell death. This targeted approach, known as synthetic lethality, is very effective in killing BRCA-mutant cancer cells while sparing normal cells that have functional BRCA proteins, making PARP inhibitors a potent targeted therapy for patients with BRCA-mutant cancers. *Source:* Created with BioRender.com.

Questions

1 Are there new targeted therapies for patients with HGSOC?

2 What is the role of tumour markers in the management of patients with ovarian cancer?

3 What is an R0 resection?

Answers on p. 555.

Case Study 65   2-month History of Abdominal Pain and Discomfort

Clinical Presentation

A 35-year-old woman presented to her GP with abdominal pain and discomfort for the past 2 months. She also reported a feeling of heaviness in the lower abdomen and increasing fatigue. She had no significant medical history and no family history of ovarian cancer. On physical examination, a palpable mass was found in the left lower abdomen.

Investigations

An ultrasound of the pelvis revealed a complex mass in the left ovary, measuring 8 cm in diameter. A CT scan confirmed the presence of a large left ovarian mass with areas of calcification and fat density. A CA-125 blood test was within the normal reference range. According to International Ovarian Tumor Analysis (IOTA) group criteria, this was originally classified as a 'B' (benign) lesion, which may be operated on by a general gynaecologist.

Diagnosis

Based on the imaging studies, the patient was thought to have an ovarian teratoma. The patient underwent a laparoscopic left salpingo-oophorectomy with the use of an 'endobag' to avoid spilling of contents which could induce a so-called *chemical peritonitis*, and the tumour was removed intact. On histopathological examination, a typical ovarian teratoma was found composed of multiple cysts lined by stratified squamous epithelium, containing keratinous material and hair (Fig. CS65.1). The cysts were filled with a yellowish, cheesy material (*sebum*). In addition, the cyst wall contained other tissue types; notably around two-thirds of the mass was composed of mature thyroid tissue. Bone, muscle, neural tissue and a fully developed tooth were also present. Further examination revealed a small focus of thyroid carcinoma within the thyroid tissue. This was identified by the pathologist as a papillary thyroid carcinoma, having a typical papillary growth pattern, nuclear grooves and intranuclear inclusions. Immunohistochemical staining for thyroglobulin and thyroid transcription factor-1 (TTF-1) were positive.

Diagnosis: Mature ovarian teratoma with a focus of thyroid cancer.

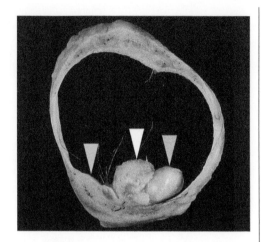

Figure CS65.1 Mature ovarian teratoma: The photograph shows a well-developed tooth (gold arrowhead) arising from the right side of the mural nodule ('Rokitansky nodule', white arrowhead) that contains most of the solid teratomatous elements. The central portion of the nodule contains mostly cutaneous tissues (skin, sweat glands and hair follicles), while the neural tissues extend into the wall toward the left (green arrowhead). *Source:* Ed Uthman / Wikimedia Commons / Public Domain.

Discussion

Mature ovarian teratomas are usually asymptomatic and are often discovered incidentally during routine pelvic examination or imaging studies. When they do cause symptoms, they can present with abdominal pain and bloating. The diagnosis of ovarian teratoma is usually made by imaging.

Teratomas arise from germ cells. The origins of germ cell tumours are discussed in detail in Case Study 61. Teratomas are unique in that they contain tissues from all three embryonic germ layers: ectoderm, mesoderm and endoderm, and can differentiate into various cell types. As a result, teratomas often contain hair, teeth, bone, muscle and neural tissue. Immunohistochemical staining

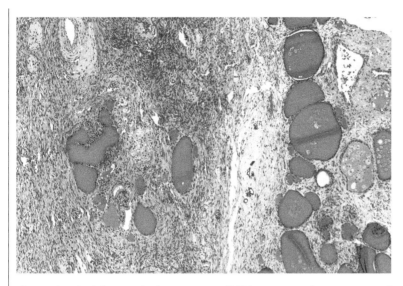

Figure CS65.2 Micrograph of a struma ovarii. This rare type of teratoma contains prominent thyroid follicles that stain pink on H&E. *Source:* Nephron / Wikimedia Commons / CC BY-SA 3.0.

can be used to identify the different tissue types within the teratoma. For example, neural tissue can be identified by the presence of neurons and glial cells, which stain positively in immunohistochemistry using markers such as synaptophysin and GFAP, respectively. Bone and cartilage can be identified by the presence of osteocytes and chondrocytes, which stain positively for osteocalcin and S100, respectively.

Thyroid cancer arising in a dermoid cyst is a rare, but well-documented, phenomenon. *Struma ovarii* is a type of ovarian teratoma that is composed primarily of thyroid tissue (Fig. CS65.2) and accounts for less than 3% of all ovarian teratomas. *Struma ovarii* can give rise to thyroid cancer, usually of the papillary type.

Treatment and Prognosis

In most centres, thyroidectomy is considered for diagnostic reasons in high grade malignant struma ovarii and metastatic disease. This is because detection of metastases in this case is only possible by radioactive iodine imaging; the presence of an intact thyroid gland poses a challenge because it would absorb almost all the radioactive iodine, thereby compromising the imaging. Ultimately, the decision on how to treat this patient must be made only after careful consideration at an interdisciplinary tumour board in a cancer centre and whole-body imaging performed to exclude metastases. The fact that the patient is of child-bearing age, and may wish to have more children, is an important consideration.

Questions

1 What is the difference between a mature and an immature teratoma?
2 What is an 'endobag'?

Answers on p. 556.

Case Study 66   Burning Sensation during Urination, Weight Loss and Skeletal Pain

Clinical Presentation

A 65-year-old man presented with both urgent and frequent urination and a urethral burning sensation during micturition, which had developed gradually during the course of the last 5 years. The intensity of the symptoms had fluctuated during this time, but had become significantly worse in the last 2 weeks before presentation.

Investigations

Laboratory investigations were generally unremarkable with the exception of a significantly raised serum prostate-specific antigen (PSA) of 12.3 ng/mL (reference value <4 ng/mL). Microbiological investigation of the urine was negative. Physical examination included digital rectal examination (DRE) of the prostate. The examining physician found a rough textured prostate with a lumpy surface. He underwent prostate parametric magnetic resonance imaging (MRI), which identified a PI-RADS 4 lesion in the right peripheral zone. PI-RADS, or Prostate Imaging Reporting and Data System, is a structured reporting scheme for evaluating the prostate gland for cancer. A PI-RADS 4 classification suggests that the lesion should be considered for biopsy or further clinical assessment due to the elevated risk of significant cancer. Histological examination of a subsequent prostatic biopsy revealed the presence of a poorly-differentiated adeno-carcinoma. The aggressiveness of prostate cancer is graded by the pathologist based on the Gleason system, which classifies the tumour tissue based on the degree of differentiation observed on histopathological examination. This patient had a Gleason score (GS) of 4 + 3. The GS is composed of two primary Gleason grades, each ranging from 1 to 5. The primary grade is the dominant pattern of the tumour (>50% of the

total pattern seen), in this case '4'. The next most frequent pattern is the secondary grade (<50%, >5%, of the total pattern), in this case '3'. These grades reflect how the cancer cells appear under a microscope:

- *Grade 1 and 2:* Discrete regular glands; circumscribed, rounded nodules. Not used for needle biopsy specimens. Very rarely used in grading radical prostatectomy specimens.
- *Grade 3:* While recognisable glands are still present, the cells are darker and some are beginning to invade the surrounding tissue, indicating an infiltrative pattern.
- *Grade 4:* This pattern corresponds to poorly differentiated carcinoma. Fewer recognisable glands are present and many cells are invading the surrounding tissue in neoplastic clumps.
- *Grade 5:* This pattern represents anaplastic carcinoma, the least differentiated form. The tissue lacks recognisable glands and there may be just sheets of cancer cells.

The GS is obtained by adding the primary and secondary grades. In this case, 4 (primary) + 3 (secondary) equals a Gleason score of 7. Cancers with a higher Gleason score are more aggressive and have a worse prognosis. Grade groups (GG) are now used to make the distinction between low-grade and high-grade tumours easier (GG1=GS up to 6; GG2= GS 3+4; GG3= GS 4+3; GG4= GS 8, GG5= GS 9-10).

DRE is an essential part of the clinical examination. A serum PSA is also important, especially in cases in which the cancer is clinically undetectable by DRE. In the case of an elevated PSA or suspicious DRE, the diagnostic process is followed by a prostate biopsy. Nowadays, MRI is standard before prostate biopsy for better targeting of pathological lesions (especially when prostate re-biopsy is indicated).

Diagnosis

The definitive diagnosis was established by biopsy and was supported by the high level of PSA in the serum.

Diagnosis: Poorly-differentiated adenocarcinoma of the prostate gland.

Discussion

Prostatic adenocarcinoma is one of the most common cancers affecting men. The incidence of this cancer increases dramatically in men over 40 years of age and is associated with changes in androgen levels. Despite an 80% prevalence at autopsy by the age of 80 years, the clinical incidence is much lower. This means that many adenocarcinomas will remain undiscovered during life and that their development is very slow, usually occurring over many years. Prostate cancer is often clinically silent and may have no specific presenting symptoms. On the other hand, it may cause lower urinary tract symptoms, mimicking benign nodular hyperplasia of the prostate, which itself is very common in men of older age. In fact, most prostatic cancers arise in patients with concomitant benign hyperplasia. Enlargement of the prostate gland, a common symptom of both benign prostatic hyperplasia and prostate cancer, causes obstruction of urine flow through the prostatic urethrae. This obstruction generates the sensation of urgency and difficulty in initiating urination/voiding, diminished stream size and force, increased frequency, incomplete bladder emptying and *nocturia* (urination during the night). Other signs and symptoms (e.g. urethral burning, presence of leucocytes and bacteria in the urine) may arise secondary to the prostate cancer and are due to inflammation or infection of the lower urinary tract (see also Case Study 13).

Approximately 9% of men diagnosed with prostate cancer report a familial history of prostate cancer. Key factors considered when evaluating familial risk include the number of affected members, their relationship with each other and the age when they were diagnosed. If a family has three or more members diagnosed with prostate cancer and at least two of them were diagnosed early (before the age of 55), it is described as a familial case. Men with immediate family members diagnosed with prostate cancer have double the risk of contracting the disease themselves. Additionally, certain inherited mutations in genes related to DNA repair, such as BRCA1, BRCA2, ATM, ATR and others, are associated with an increased risk of early-onset prostate cancer (diagnosis before 60 years of age). Men with these mutations also make up a significant portion of advanced or metastatic prostate cancer cases. Among these, mutations in BRCA2 and HOXB13 stand out as especially significant; mutations in BRCA2 increase the risk of prostate cancer by up to seven to eight times, while HOXB13 mutations triple it.

The outcome for someone diagnosed with prostate cancer varies greatly and is primarily based on the grade and stage of the tumour at the time of initial diagnosis. In resource-rich regions with advanced healthcare systems, like the United States and the United Kingdom, modern early detection methods, including PSA testing and DRE, allow for the diagnosis of most cases during the early stages of the disease. Current statistics suggest that roughly 80% of diagnoses occur when the disease is confined to the prostate, 15% with local metastases and 5% in cases where metastases have spread further. Men diagnosed early, when the disease is still localised, have 10-year survival rates reaching nearly 99%. On the other hand, for those diagnosed with advanced disease, particularly when there are distant metastases, the 5-year survival rate is only 30%.

The development of prostate cancer is closely linked to the accumulation over an individual's lifetime, of genetic mutations in prostate epithelial cells (in some cases on a background of inherited mutations that increase risk, as described above). High-grade prostatic intraepithelial neoplasia (PIN) is considered a precursor to carcinoma. Somatically acquired mutations include gene fusions involving the androgen receptor (AR)-regulated promoter regions and the ETS family of transcription factors. Notably, the TMPRSS2-ERG gene fusion is present in approximately 50% of biopsies from white men but is less prevalent in Black and Asian populations. Ethnic disparities in prostate cancer genetics exist, with Chinese patients showing different mutational patterns compared with Western patients. Crucially, the AR, which normally regulates prostate growth, becomes a key player in disease progression, with various alterations rendering it continuously active and thus able to drive tumour progression. Furthermore, genes associated with cell growth control and genetic stability, such as PTEN, TP53, RB1 and those involved in DNA damage repair, are more frequently mutated in the advanced stages of disease.

The primary method to detect asymptomatic prostate cancer is through screening. The main aim is to differentiate between tumours that require immediate treatment and those that can be managed with active surveillance. PSA is the main screening tool, but its efficacy is debated; some studies suggest its use can reduce mortality from prostate cancer, while other studies have found no significant reduction. There is a risk of overdiagnosis and therefore overtreatment. Moreover, elevated PSA levels can be observed in men without prostate cancer, which may cause unnecessary anxiety Current guidelines advise individualised decision-making regarding prostate cancer screening, taking into account factors such as age and family history.

Treatment and Prognosis

70–80% of prostate cancers are initially responsive to androgen deprivation/ withdrawal therapy (and are described as *hormone-sensitive*). However, after a period of time, in about 50% of cases, the cell clones that survive androgen ablation give rise to androgen-independent prostate carcinomas (referred to as *hormone-refractory*). Hormone-refractory tumours are clinically more aggressive and the prognosis of hormone refractory disease after relapse is poor.

For localised disease, the treatment of choice includes radical prostatectomy (robotic prostatectomy being the gold standard) or radiotherapy. Which modality is chosen depends on the patient's preference and co-morbidities. In cases of locally advanced disease, the preferred option is radiotherapy, usually combined with androgen deprivation therapy (e.g. with luteinizing hormone-releasing hormone [LHRH] analogues/antagonists, which suppress the secretion of LH from the anterior pituitary and indirectly reduce testosterone production). In low-risk prostate cancer patients (low PSA level, localised disease with Gleason ≤ 3 + 3), active surveillance is an option (monitoring with repeat PSA testing, MRI of the prostate and prostate re-biopsy). If there is PSA or Gleason score progression, radical treatment is started. Watchful waiting is an approach generally reserved for the elderly or for fragile patients with limited life expectancy. In these patients, only palliative treatment (e.g. with LHRH analogues/antagonists) is indicated if there is a rapid increase in PSA levels, or if performance status deteriorates, for example, due to skeletal metastasis.

In the present case, the patient was treated with localised radiotherapy and androgen deprivation for 3 years. After 1 year of good response (disappearance of subjective symptoms, decrease of PSA level), the patient experienced the same symptoms as before, together with additional symptoms, which included an aversion to food, vomiting, weight loss and fever. Laboratory examinations suggested an infection of the lower urinary tract. The patient received antimicrobial therapy. After a short period of time, he felt intense pain in the lumbar spine and in the pelvis. At this time, the serum PSA level was high (32 ng/mL) and the prostate gland was enlarged. Scintigraphy of the skeleton showed numerous metastases (Fig. CS66.1). CT TAP (thorax, abdomen and pelvis) confirmed bone metastases and pelvic lymphadenopathy. Lymph node biopsy revealed the presence of prostatic adenocarcinoma. The patient was commenced on abiraterone, an oral inhibitor of the cytochrome P450 (17alpha)-hydroxylase/17,20 lyase (CYP17) complex that is critical for androgen production, together with ongoing androgen deprivation therapy. Abiraterone, in combination with prednisone and androgen deprivation therapy is currently FDA-approved for use in men with metastatic castration-resistant prostate cancer, both in the pre- and post-chemotherapy settings. Patients with hormone-refractory prostate cancer have alternative options, including second-generation anti-androgens (e.g. enzalutamide, apalutamide). Other alternatives include autologous dendritic cell therapy (e.g., Sipuleucel-T/ Provenge). In this personalised treatment, dendritic cells are removed by leukapheresis, incubated with a fusion protein combining the antigen prostatic acid phosphatase (PAP) and the immune stimulator GM-CSF, and then reinfused into the patient to stimulate an immune response against prostate cancer cells. Another option is radium-223 (Xofigo), an alpha-emitting radiopharmaceutical used in castration-resistant prostate cancer patients

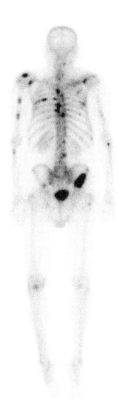

Figure CS66.1 Osseous metastases from prostate cancer; bone scan. *Source:* RadsWiki / Wikimedia Commons / CC BY-SA 3.0.

with symptomatic bone metastases and no known visceral metastatic disease. Radium-223 targets areas of increased bone turnover in bone metastases, delivering targeted radiation.

After this treatment, a decrease in the PSA level was noted and the patient felt better. A bone scan revealed regression of the metastatic deposits. However, 2 years later, there was a further relapse, accompanied by a raised PSA (123 ng/mL), skeletal pain, fever and weight loss. The patient was unfit for chemotherapy and died soon after this. An autopsy revealed metastatic deposits in abdominal lymph nodes, liver, adrenal glands and bone. Hypertrophy and chronic inflammation of the bladder were also found. The cause of death was bronchopneumonia.

The course of the disease in this patient is typical of hormone-refractory prostate cancer. After an initial very good response to treatment by anti-androgens this was followed by relapse. Despite subsequent androgen ablation, the tumour cells revealed only limited sensitivity and the tumour progressed. Infectious complications due to metastases are very often the cause of death.

Questions

1 Are there any other targeted molecular therapies available for patients with prostate cancer?
2 The BCL2 protein is found more commonly in androgen-independent prostate cancer when compared to androgen-sensitive tumours. What is the significance of this observation?

Answers on p. 556.

Case Study 67 Fever and Cervical Lymph Node Enlargement

Clinical Presentation

A 23-year-old woman with a history of renal transplantation 4 weeks ago for chronic renal failure secondary to hereditary thrombotic thrombocytopenic purpura, was admitted to the hospital with oliguria (reduced urine output), new onset cervical lymphadenopathy and renal failure. At the time of transplantation, the patient had been started on prednisone and azathioprine immunosuppression to prevent rejection of the allograft. However, 3 weeks later, the serum creatinine rose and a percutaneous renal biopsy yielded a diagnosis of acute rejection. This was treated by increasing the immunosuppression with an intravenous infusion of OKT3 (a mouse monoclonal anti-T-lymphocyte antibody). Eleven days after the monoclonal antibody infusion, the patient developed pharyngitis, cervical and submandibular lymphadenopathy and fever.

On examination, the patient was hypertensive (blood pressure: 170/95) and febrile (temperature: 38.5°C), confused and disoriented with a raised heart rate and respiratory rate. The tonsils were enlarged and there was

pharyngeal erythema (redness). There was bilateral submandibular and cervical lymphadenopathy, and hepatosplenomegaly.

Investigations

Laboratory investigations are shown in Table CS67.1. The clinical features and laboratory results suggested a systemic infection. Systemic infection was supported by pharyngitis, fever, lymphadenopathy and hepatosplenomegaly. This syndrome is suggestive of Epstein–Barr virus (EBV)-associated infectious mononucleosis (glandular fever) typically characterised by: pharyngitis, fever,

Table CS67.1 Results of selected investigations.

Test	Result	Reference range
Urea	**17 mmol/L**	2.5–6.7
Creatinine	**1112 μmol/L**	45–84
White cell count	**1.8 × 10⁹/L**	4.0–11.0
VCA-IgM titre	**1:128**	—

Values outside the reference range are in bold.

cervical lymphadenopathy, hepatospleno-megaly and a leucocytosis (increased white cell count) with atypical lymphocytes. Other infections, including Cytomegalovirus, Human herpesvirus 6, Human immunodefi-ciency virus and toxoplasmosis can produce similar or overlapping syndromes with lym-phadenopathy and systemic symptoms. However, pharyngitis is not part of the con-stellation of symptoms in these diseases and is typical of EBV-associated infectious mono-nucleosis. The depressed white blood cell count and absence of atypical lymphocytes in this patient are unusual in infectious mon-onucleosis but may reflect treatment with the *myelosuppressive* (bone marrow sup-pressing) and lympholytic agents employed (azathioprine, prednisone and OKT3). Detection of an immunoglobulin M (IgM) antibody titre to the EBV viral capsid antigen (VCA) is diagnostic of a recent EBV infection and is consistent with this interpretation.

In addition to infectious mononucleosis, a second diagnosis must be considered: that of post-transplant lymphoproliferative dis-ease (PTLD). Patients who are immunosup-pressed, particularly with respect to T-cell function, are at risk for developing uncon-trolled proliferation of EBV-infected B lym-phocytes. This disease process may occur with or without an infectious mononucleosis-like syndrome. Lymphadenopathy in the post-transplant setting must always raise this concern.

A submandibular lymph node was biop-sied and a diffuse proliferation of large, activated B lymphocytes (B-immunoblasts) was observed on histological assessment. Nephrectomy of the transplanted kidney was performed and showed focal proliferation comprising sheets of B-immunoblasts, with associated architectural disruption, as well as acute rejection. *In situ* hybridisation showed the presence of EBV within the immunoblasts in the submandibular lymph node and the kidney. PCR analysis revealed a clonal population of B-cells in both tissues. In this assay, the presence of a large number of identical B-cell receptor genes indicates a monoclonal expansion of B-cells. The tumour cells were also positive for the EBV-encoded latent membrane protein-1 (LMP1) and Epstein-Barr nuclear antigen-2 (EBNA2) (Fig. CS67.1A and B, respectively).

Diagnosis

The diagnosis of PTLD was confirmed by a study of the biopsied lymph node and the nephrectomy specimen, both of which showed

(A)

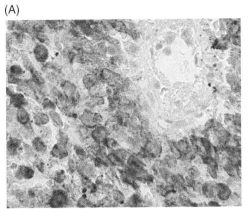

(B)

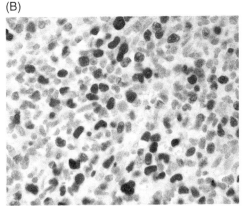

Figure CS67.1 Post-transplant lymphoproliferative disease. (A) This case was positive for the EBV-encoded latent membrane protein-1 (LMP1; brown staining in cytoplasm). (B) The same case was also positive for the Epstein-Barr nuclear antigen-2 (EBNA2; red nuclear staining).

proliferation of EBV-positive B-immunoblasts. Any potential confusion with benign lymphoid hyperplasia or a lymphoproliferation that may accompany infectious mononucleosis, was eliminated by the demonstration of diffuse sheets of blast cells distorting the native tissue architecture, monoclonal populations of cells and the identification of EBV within virtually all the blasts. The presence of one or more distinct clonal B-cell populations and the absence of chromosomal translocations is characteristic of PTLD. These features are consistent with a monomorphic type PTLD, specifically, EBV-positive diffuse large B cell lymphoma (DLBCL).

Diagnosis: DLBCL, EBV+, post-transplant lymphoproliferative disorder (WHO 2022)/ Monomorphic post-transplant lymphoproliferative disorder: EBV+ DLBCL (ICC 2022).

Note: Disease classification is ever evolving, incorporating the most recent clinical evidence and understanding of disease processes. Consequently, the classification of lymphomas has changed significantly in recent decades and will likely change in the coming years. At present, there are two classification systems of haematolymphoid malignancies, namely the World Health Organization, 2022 classification (WHO 2022) and the International Consensus Classification, 2022 (ICC 2022). For some entities, the names and definitions differ between these classifications. In such cases, both names for the entities are provided as indicated above.

Discussion

EBV-associated B-cell lymphoproliferative disease is a heterogeneous group of lymphoproliferative disorders with high incidence in patients with congenital immunodeficiencies (such as the X-linked lymphoproliferative syndrome; Chapter 6), HIV infection and in organ transplant recipients. Both solid organ transplantation (SOT) and allogeneic hematopoietic transplantation (HSCT) recipients are at risk. In many cases, the disease is due to the ability of EBV to immortalise human B lymphocytes. EBV-immortalised B lymphocytes will proliferate indefinitely *in vitro* and will grow as tumours in mice with *severe combined immunodeficiency* (SCID). In the presence of memory T-cells, *in vitro* or *in vivo*, the EBV-induced lymphoproliferation is suppressed. In immunosuppressed patients, the inadequate T-cell response allows the outgrowth of EBV-infected B-cells. Histologically, these tumours may have features of the spectrum of benign hyperplasia through to frank malignancy.

EBV transmission occurs primarily through oral contact. The virus attaches to B-cells through the interaction of the viral gp350 protein with the cellular CD21 receptor, followed by fusion, which is facilitated by binding of the virus gp42 protein to host MHC class II molecules. Infection can lead to infectious mononucleosis or be asymptomatic. Memory B-cells provide the virus with the means to persist long term in humans. In fact, once infected, the human host will carry EBV for their lifetime. Memory B-cells originate from the differentiation of infected naïve B-cells. Naïve B-cells become infected in the oropharynx or in the tonsils; the virus probably transiently infecting oropharyngeal epithelial cells (Fig. CS67.2). Once in B-cells, the virus expresses its full range of so-called *latency* genes. This form of latency is known as Latency III and requires the coordinated expression of 10 EBV latent genes; it is these genes that drive the proliferation and survival of infected cells. However, because of the high level of viral antigen expression, infected cells are targets for the immune system and so, to survive, the virus must downregulate viral gene expression. This appears to occur in a subset of infected B-cells through the normal process of B-cell

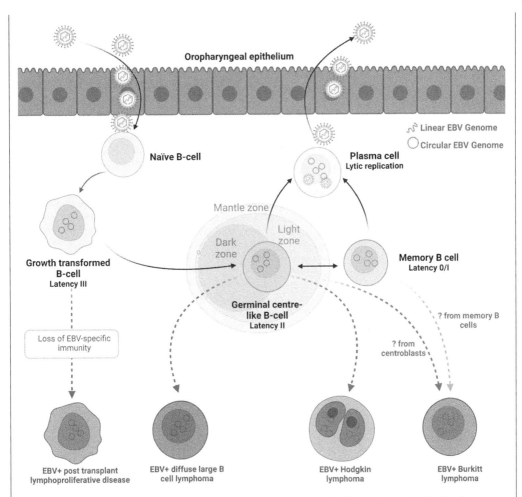

Figure CS67.2 The EBV life cycle and the origin of EBV-associated B-cell lymphomas. Naïve B-cells become infected with EBV in the oropharynx or tonsils, possibly involving a transient infection of oropharyngeal epithelial cells. Once inside B-cells, the virus expresses around ten or so latent genes that promote cell proliferation and survival (a stage known as Latency III). To evade the immune system, the virus down-regulates expression of most of these genes. This occurs when the infected B-cell differentiates, entering a germinal centre-like phase with expression of fewer genes, including the latent membrane proteins, LMP1 and LMP2 (Latency II). Infected germinal centre B-cells can differentiate into memory B-cells, in which there is an almost complete (latency I), or a complete (latency 0) shutdown of viral protein expression. Infected cells can also differentiate into plasma cells, in which the virus undergoes replication generating new virions that are released into the oropharynx. Loss of EBV-specific immunity allows growth-transformed EBV-infected cells to expand, eventually leading to the development of EBV-driven lymphoproliferative disease. Other EBV-associated B-cell lymphomas, including Hodgkin lymphoma, diffuse large B-cell lymphoma and Burkitt lymphoma, arise from distinct stages of the virus's life cycle in B-cells. The exact origin of EBV-positive Burkitt lymphoma is controversial. *Source:* Created with BioRender.com.

differentiation; the infected B-cells entering a germinal centre (GC)-like phase, in which they express a latency II program facilitated by expression of fewer viral proteins that include the latent membrane proteins, LMP1 (see also Chapter 11) and LMP2A; viral homologues of the cellular CD40 and B-cell receptor, respectively. Infected GC B-cells then differentiate into either memory B-cells with further down-regulation of virus gene expression (expressing either no viral proteins, known as Latency 0, or only the Epstein-Barr virus nuclear antigen-1 [EBNA1], known as Latency I) or they become plasma cells which replicate the virus close to the oropharynx, allowing virus release for transfer to other susceptible hosts. Loss of immunity, as in the case of this patient, can lead to the outgrowth, usually of Latency III cells and the development of EBV-driven lymphoproliferative disease. Other EBV-associated B-cell lymphomas arise from different stages of the EBV life cycle in B-cells, as shown in Fig. CS67.1. They include Hodgkin lymphoma (Case Study 57), diffuse large B-cell lymphoma (Case Study 55) and natural killer/T cell lymphomas (Case Study 63).

A tendency to involve extranodal sites is characteristic of these opportunistic EBV-associated lymphoproliferative diseases. Involvement of the allograft itself is also particularly common. Although the disease may mimic a viral illness or be entirely asymptomatic at presentation and appear histologically benign, the lymphoproliferation will result in the death of the patient in most instances, unless appropriate interventions are undertaken.

Treatment and Prognosis

Reduction of immunosuppression is the usual first step in the treatment of SOT-related PTLD. In HSCT-related PTLD, immunosuppression is mainly the consequence of the conditioning regimen, making reduction of immunosuppression less important. The monoclonal anti-CD20 antibody, rituximab, combined with reduced immunosuppression, is now the standard of care for the majority of CD20-positive PTLD. Patients who do not respond well to these approaches can be treated by combination immunochemotherapy. Adoptive immunotherapy approaches have been used with success and include using EBV-specific cytotoxic T cells which are either of autologous (i.e. derived from the recipient) or of allogeneic (derived from a donor or from a bank of partially HLA-matched donors) origin.

Questions

1 Apart from EBV, which other viruses are implicated in the development of lymphomas?
2 Is immunosuppression associated with a global increase in the risk of malignancy?
3 What is thrombotic thrombocytopenic purpura?

Answers on p. 557.

Case Study 68   Pigmented Skin Lesion on the Lower Back

Clinical Presentation

A 67-year-old male retired schoolteacher presented at his GP with a dark brown/black mottled irregular lesion approximately 1.5 cm in diameter on his lower back. He was hypertensive (160/90 mmHg) with no other significant medical history. The patient remembers that the lesion was originally a mole. He hadn't noticed that it had become bigger and less regular until his wife spotted it when he was getting undressed last week and remarked that it was crusty and appeared to have been bleeding. The lesion was also

significantly different in appearance compared to other moles and skin features on his body, so was exhibiting the so-called *ugly duckling* sign. This sign is often an indicator that a particular skin lesion is suspicious and merits further evaluation, as it stands out as distinctly different from neighbouring moles or skin markings. In terms of skin type, the patient was assessed to be Fitzpatrick skin type II-fair skin that burns easily and tans minimally, thereby placing him at a higher risk for skin cancers, including melanoma. Given the irregular, dark brown to black colour of the lesion, its mottled appearance and its recent changes, including crusting and bleeding, along with his skin type and the 'ugly duckling' sign, the clinical presentation was strongly suggestive of melanoma. Dermoscopy was performed. This is a non-invasive diagnostic tool that uses a specialised microscope to examine skin lesions in detail. This technique is particularly useful for identifying the structural characteristics within the lesion that might be indicative of melanoma, such as irregular pigmentation patterns, asymmetry and border irregularities. The dermoscopic findings were in line with the clinical suspicion of malignant melanoma, reinforcing the need for an urgent biopsy for definitive diagnosis. The patient's hypertension was also addressed as part of his overall medical evaluation.

One way to remember how melanoma presents is to use the mnemonic:

Asymmetry, **B**order, **C**olour, **D**iameter and **E**volution

- *Asymmetry:* Melanoma often appears asymmetrical.
- *Border:* Melanoma often has ill-defined borders.
- *Colour:* Melanoma is often more than one colour or shade.
- *Diameter:* Melanoma is usually larger than 6 mm in diameter.
- *Evolution:* Melanoma will often change over time.

Investigations

An excisional biopsy of the lesion was performed. Histological examination showed an ulcerated melanoma with tumour Breslow thickness of 4 mm. Breslow thickness is the measurement of the depth of the melanoma from the surface of the skin to the deepest point of invasion and is measured by the pathologist (Fig. CS68.1). Breslow thickness

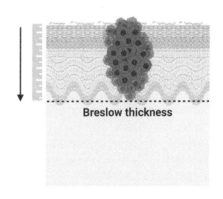

Figure CS68.1 Breslow thickness to assess depth of invasion of primary melanoma. *Source:* Created with BioRender.com.

is a key prognostic indicator and determines the 'T staging'. 'N staging' is determined by regional lymph node involvement that may be evident clinically or can be identified microscopically by sentinel lymph node biopsy (SLNB). SLNB is undertaken to provide additional prognostic information in thicker melanomas:

Melanomas 0.75–1.0 mm thick:

- Likelihood of microscopic involvement of a sentinel lymph node (SLN) at presentation is 5–10%.
- Risk is higher in the presence of histological features such as ulceration or an elevated tumour mitotic rate (>1 mitotic figures/mm^2).
- SLNB may be considered in this subgroup only if there is evidence of primary melanoma ulceration, lymphovascular invasion or mitotic index ⩾2.

Intermediate thickness melanomas (1.0–4.0 mm):

- Risk of metastatic spread to regional lymph nodes is 15–25%.
- Risk increases with increasing tumour thickness.
- SLNB is routinely offered, because positivity rates are 25–40%.

- Stage IIB (2–4 mm ulcerated) tumours are at high risk of recurrence and can be considered for adjuvant immunotherapy

Thick melanomas (>4.0 mm):

- SLNB is routinely offered because positivity rates are 25–40%.
- Stage IIB (>4 mm non-ulcerated) and IIC (>4 mm ulcerated) tumours are at high risk of recurrence and can be considered for adjuvant immunotherapy

Sentinel lymph node status

- A positive SLN designates stage III disease. Clinically evident regional lymph node spread is a poor prognostic indicator and patients are routinely offered adjuvant therapy, which has been shown to reduce recurrence by around 50%. On the other hand, while microscopic involvement of SLNs does alter staging, outcomes for patients with a primary melanoma <2 mm and positive SLN designated stage IIIA have outcomes similar to stage IIB patients and better outcomes compared with stage IIC melanoma. The role of adjuvant therapy in this patient group remains uncertain.

Diagnosis

Molecular analysis of tumour tissue revealed the presence of the V600E mutation in the BRAF oncogene (see also Chapter 11). Immunohistochemistry was performed on sections of the tumour using a monoclonal antibody specific to the mutant protein. The tumour cells were positive, confirming the presence of the mutation (Fig. CS68.2).

Diagnosis: BRAF V600E mutation-positive primary melanoma.

Discussion

Melanoma is a tumour that is usually derived from melanocytes in the skin. However, because melanocytes are derived from the neural crest, rarely, melanomas can develop at other

Figure CS68.2 Immunohistochemistry staining of BRAF V600E mutant tumour cells (brown) in a melanoma metastasis. *Source:* Jensflorian / Wikimedia Commons / CC BY-SA 4.0.

Table CS68.1 Risk factors for cutaneous melanoma.

Factor	Risk
Sun exposure	High UVB and UVA radiation exposure
Family history	In 5–10% of patients; 2.2-fold higher risk with at least one affected relative
Personal characteristics	Blue eyes, fair and/or red hair, pale complexion,
Prior lesions	e.g. benign and/or dysplastic melanocytic naevi
Immunosuppression	Transplantation, prior haematological cancer

locations, such as the mucosal membranes of the gastrointestinal and urogenital tracts, or the uveal tract of the eye. Some risk factors for melanoma are shown in Table CS68.1.

The Cancer Genome Atlas (TCGA) programme was the earliest to conduct large-scale molecular screens of melanoma. The TCGA identified major molecular subtypes of melanoma characterised by mutations in BRAF (accounting for just under 50% of cases), RAS (around 25%) and NF1 (15%). The remaining cases showed recurrent mutations in other genes, including BAP1, CDKN2A and in ocular melanomas, GNAQ and GNA11. The most frequent cutaneous melanoma mutation occurs in codon number 600 of BRAF, changing the amino acid, valine (V) to the amino acid, glutamic acid (E). Hence the terminology V600E. The V600E mutation confers constitutive activation upon the BRAF protein, meaning that the MAPK signalling pathway is not regulated by the availability of ligand at the cell surface (see also Chapter 11). Germline mutations in the CDKN2A gene, an important regulator of the cell cycle (Chapter 3), are present in some sporadic melanomas and in 20–40% of melanoma families. The presence of CDKN2A mutations is associated with a high number of affected family members, multiple primary melanomas as well as early onset of melanoma, and an increased risk of pancreatic cancer. Mutations in the other genes, for example CDK4, are less common. CDK4 is also an important protein involved in the regulation of the cell cycle (Chapter 3).

Treatment and Prognosis

The patient was referred for a wide local excision and sentinel lymph node biopsy. Histological examination showed no involvement of resection margins or sentinel nodes, staging the tumour as pT4bN0, overall stage IIC. Staging CT scans of head and body were clear and the patient declined adjuvant therapy. However, 2 months later, the patient developed small dark nodules in the skin surrounding the surgical scar and palpable lymph nodes in the left axillae and left cervical regions. The nodules and lymph nodes were removed surgically and the presence of metastatic deposits of melanoma confirmed in 3 of the resected nodes. Resected stage IIIC melanoma carries a high risk of recurrence, with 5-year and 10-year melanoma-specific survival rates around 70 and 60%, respectively. Owing to the presence of the BRAF V600E mutation, the patient was offered adjuvant therapy with the BRAF inhibitor, dabrafenib combined with the MEK inhibitor, trametinib for 1 year. Therapy was well tolerated except for intermittent episodes of hyperpyrexia, which was improved by temporary withdrawal of dabrafenib and trametinib.

Twelve months after treatment stopped, the patient was found to have multiple lung metastases on a routine surveillance CT scan. He was offered treatment with immune checkpoint inhibitors. Several different immune checkpoint inhibitor regimens are now available for treating metastatic melanoma, each offering patients a significant chance of long-term remission and possible cure. Immune checkpoint inhibitors used in melanoma include pembrolizumab and nivolumab (both PD-1 inhibitors) monotherapy, ipilimumab (CTLA-4 inhibitor) combined with nivolumab, or nivolumab combined with relatlimab (inhibitor of lymphocyte activation gene-3; LAG-3). Combinations of checkpoint inhibitors might be more effective compared with anti-PD1 monotherapy, but they are also more likely to result in severe side effects that can be both life-changing and life-threatening. Choice of treatment requires consideration of patient factors such as age, fitness and co-morbidities, tumour factors such as site and volume of disease, rate of disease growth, as well as patient preference. Tumour mutational burden (TMB), defined, as the number of somatic mutations per megabase of DNA, is very high in melanoma. In general, tumours with a high TMB respond better to immune checkpoint blockade. This is thought to be because tumours with high TMB generate more new (neo)-epitopes that can be recognised by the immune system (see also Chapter 6) and may explain why immunotherapy may be a much more effective treatment for melanoma compared with some other types of cancer.

Questions

1 How do immune checkpoint inhibitors work?
2 What are the side effects of immune checkpoint inhibitors?

Answers on p. 557.

Case Study 69   Shortness of Breath, Chest Pain and Fatigue in a Retired Coal Miner

Clinical Presentation

A 75-year-old man presented to his GP with a persistent cough, shortness of breath, chest pain and fatigue for the past 6 months. He was a retired coal miner with a history of heavy smoking, averaging one pack per day for the past 25 years. He also reported having worked in an asbestos factory for around 3 years when a young man. He reported a history of hypertension, for which he was taking lisinopril. He had no known allergies and had never been hospitalised for any respiratory illness. On physical examination, the patient was thin, appeared fatigued and was breathless. He had a barrel-shaped chest, diminished breath sounds on auscultation and a right ventricular heave. He also had clubbing of his fingers and swelling of both legs below the knees.

Investigations

A chest X-ray revealed bilateral nodular opacities and a diffuse reticular pattern, consistent with pneumoconiosis, confirmed by high-resolution CT scan. Pulmonary function tests showed a restrictive pattern characterised by reduced lung compliance and a reduced FVC. The ratio of FEV1 to FVC was normal (Table CS69.1).

Table CS69.1   Results of blood gas analysis and spirometry.

Test	Result	Reference range
Arterial pH	**7.2**	7.35–7.45
Partial pressure of oxygen (PaO_2)	**69 mmHg**	75–100
Partial pressure of carbon dioxide ($PaCO_2$)	**49 mmHg**	36–46
Oxygen saturation (SaO_2)	**89%**	>95
FEV1/FVC ratio	73%	>70

Values outside the reference range are in bold.

Diagnosis

In a patient with pneumoconiosis, the results of lung function tests would be expected to show a restrictive pattern with lower levels of oxygen in their blood (low PaO_2), higher levels of carbon dioxide (high $PaCO_2$) and lower oxygen saturation levels (low SaO_2). The ratio of FEV1 to FVC can help distinguish obstructive and restrictive lung diseases. In obstructive diseases, FEV1 is reduced owing to increased airway resistance to expiratory flow. A reduction in the FVC may also be present (because of premature airway closure in expiration) but is not affected as much as FEV1 and, thus, is not decreased in the same proportion as FEV1. In restrictive lung diseases, however, the FVC may be decreased more compared to the FEV1, thus giving an FEV1/FVC ratio of >70%.

One of the potential complications of advanced pneumoconiosis is right heart failure, also known as *cor pulmonale*. The chronic hypoxic state of these patients results in hypoxic pulmonary vasoconstriction with remodelling of the pulmonary vascular bed and pulmonary hypertension. Cor pulmonale occurs when the right ventricle must work harder to pump blood to the high-pressure pulmonary circulation, leading to right-sided heart enlargement and eventual heart failure.

Diagnosis: Pneumoconiosis with right heart failure.

Discussion

Pneumoconiosis is a category of interstitial lung disease caused by inhalation of dust particles. The disease is named according to the type of exposure. The most common types of pneumoconiosis are coal miner's lung, silicosis and asbestosis. Pneumoconiosis is often an occupational disease resulting

Table CS69.2   Examples of types of pneumoconiosis.

Type of pneumoconiosis	Causative dust	Industry/occupation
Coal worker's pneumoconiosis	Coal, carbon	Mining
Aluminosis	Aluminium	Mining and welding
Asbestosis	Asbestos	Shipbuilding, construction, insulation
Silicosis	Crystalline silica dust	Mining, sandblasting, construction
Bauxite fibrosis	Bauxite	Bauxite mining
Berylliosis	Beryllium	Aerospace and nuclear industries
Siderosis	Iron	Welding and iron foundries
Byssinosis	Cotton dust	Textile industry
Chalicosis	Fine dust from stonecutting	Stonecutting industry
Silicosiderosis	Mixed dust containing silica and iron	Mining
Labrador lung	Mixed dust containing iron, silica and anthophyllite (a type of asbestos)	Mining
Stannosis	Tin oxide	Tin mining
Talcosis	Talc	Mining
Baritosis	Barium	Mining

from prolonged exposure in industries such as mining, textile milling, shipbuilding, sandblasting and agriculture (Table CS69.2). Severity of symptoms varies depending on the intensity and duration of exposure. The classification of pneumoconiosis has been extended to include diseases caused by chemical fumes and vapours, such as nitrous oxide, benzene or insecticides.

This patient had occupational exposure to coal dust and asbestos; both may have contributed to his current condition. The reaction of the lung to mineral dusts is complex and depends on many variables, including the size, shape, solubility and reactivity of the particles. Particles that are more reactive trigger a more vigorous inflammatory response and initiate more intense fibroblast proliferation and collagen deposition. Tobacco smoking worsens the effects of all inhaled mineral dusts, especially in patients with asbestosis.

Treatment and Prognosis

There is no cure for pneumoconiosis, and treatment focuses on managing symptoms and preventing complications. The patient was advised to stop smoking to slow the progression of the disease. He was also given oxygen therapy and referred to a pulmonary rehabilitation clinic for respiratory muscle training. Despite these interventions, the patient's condition deteriorated and he experienced frequent exacerbations of his symptoms. 12 months after his initial presentation, he was diagnosed with a malignant mesothelioma of the lung and was referred to hospice care. He passed away 3 months later. A post-mortem was performed since the patient's family wished to seek compensation for his occupational exposure. Examination of the lungs at post-mortem revealed the presence of tumour with marked diffuse pulmonary interstitial fibrosis, and extensive deposition of black carbon

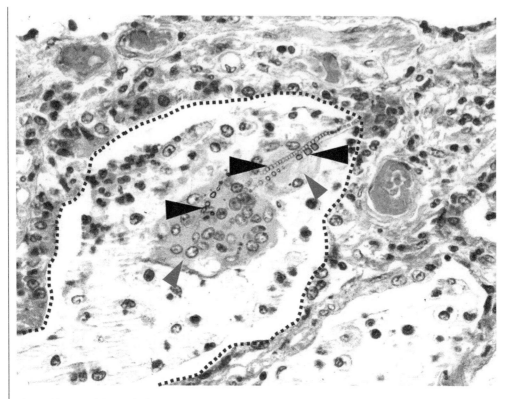

Figure CS69.1 Asbestosis. 61-year-old male who worked in the shipbuilding industry for many decades. This image shows at least three asbestos bodies (black arrowheads) engulfed by two giant cells (small red arrowheads) within an alveolus (lumen outlined in dotted line). Contributed by Dr. Mark Wick. *Source:* Yale Rosen / Wikimedia Commons / CC BY-SA 2.0.

with the presence of well-circumscribed plaques of dense collagen (pleural plaques), some of which were calcified. These plaques contained asbestos bodies, which were coated with iron that stained blue with the Perl stain. A photomicrograph of asbestos fibres is shown in Fig. CS69.1.

Questions

1 How is asbestos exposure linked to the development of lung cancer?
2 What is right ventricular heave?

Answers on p. 558.

Case Study 70   Fever and Leg Ulcer

Clinical Presentation

A 54-year-old man presented to the Accident and Emergency Department with a fever and a large ulcer on the sole of his right foot. This had been present and gradually enlarging over 2 weeks. Earlier action had not been taken because the ulcer was painless. He claimed to be otherwise fit and well, although on direct questioning, he had noticed increased thirst and urination over the last 18 months, a 10-kg weight loss over 2 years and non-specific lethargy of longer duration, which had been attributed to 'old age'. On examination, he was thin (BMI of 24), somewhat dehydrated and febrile (temperature 39°C). Examination of his retina revealed capillary haemorrhages. He was

Table CS70.1 Results of selected laboratory investigations.

Test	Result	Reference range
Blood		
Random glucose	**24.7 mmol/L**	3.5–8.9
Urea	**19.4 mmol/L**	2.5–6.7
Creatinine	**157 µmol/L**	59–104
Bicarbonate	23 mmol/L	22–30

Values outside the reference range are in bold.

unable to detect vibration sensation below his knees and had lost soft touch sensation below his ankles. There were no peripheral pulses below his femoral artery. There was a large deep malodorous ulcer under his great toe that was surrounded by spreading skin infection. Exploration of this ulcer revealed a cavity full of pus tracking deep along the sole of his foot. His dipstick urine analysis showed 3^+ glucose, 2^+ protein and 3^+ ketones.

Investigations

The results of laboratory investigations are shown in Table CS70.1. No islet cell or insulin autoantibodies were detected. A radiograph of the patient's right foot showed destruction of the bone in the great toe. Doppler ultrasound showed no signal in the arteries behind the knee. Microbiological analysis of the pus from the ulcer showed the presence of both aerobic (including *Staphylococcus aureus*, sensitive to flucloxacillin) and anaerobic bacteria.

Diagnosis

The patient had hyperglycaemia confirming diabetes. The raised blood urea and creatinine indicated poor renal function, almost certainly due to diabetic renal disease (*diabetic nephropathy*). This was further supported by the proteinuria (protein in the urine), which is the result of protein leaking through a damaged renal glomerular capillary basement membrane. The dehydration was the result of the osmotic loss of water induced by glycosuria resulting in increased urination and stimulation of thirst. His bicarbonate levels were normal, so the blood accumulation of ketone bodies (i.e. acetoacetate and 3-hydroxybutyrate, which are the breakdown products of non-esterified fatty acid metabolism) was not sufficient to cause acidosis (lowered blood pH), known as *ketoacidosis*, which is more common in type I diabetes (T1D).

The failure to detect islet and insulin autoantibodies is against a diagnosis of T1D, which usually has an autoimmune basis (Chapter 13). Clearly, this man has had diabetes for a long time. Microvascular complications involving the eyes (*diabetic retinopathy*, manifested by the presence of retinal capillary haemorrhages), kidney (*diabetic nephropathy*) and peripheral nerve damage (*diabetic neuropathy*) relate to a long exposure to high blood glucose levels. His onset of diabetes was insidious, possibly developing 10 years or so before presentation and so this is probably type 2 diabetes (T2D). The mean pre-diagnosis duration of unrecognised hyperglycaemia is estimated at 4 years. About 10% of patients with T2D have complications at diagnosis.

Diagnosis: T2D with diabetic nephropathy, retinopathy and neuropathy with an infected neuropathic foot ulcer.

Discussion

This patient has T2D, a metabolic disorder in which blood glucose concentrations are elevated. The patient's disease was complicated by retinopathy, nephropathy and neuropathy (Fig. CS70.1).

Conventionally, two major forms of diabetes mellitus are recognised (see also Chapter 13). T1D usually presents acutely in young patients, though it can present at any age, and is the result of an autoimmune attack on the insulin-producing pancreatic β-cells. This leads to destruction of the β-cells with a consequent lack of secretion of

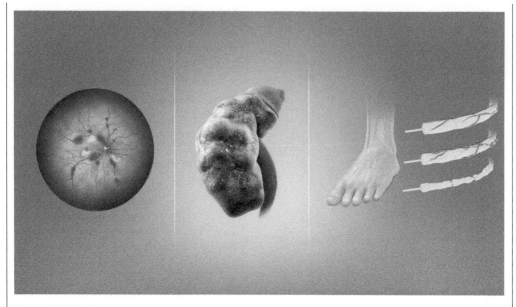

Figure CS70.1 Complications of Diabetes (from left to right): retinopathy, nephropathy and neuropathy. *Source:* Scientific Animations/ Wikimeida Commons / CC BY-SA 4.0.

insulin. There is a genetic predisposition to T1D, which is explained in part by the association with specific HLA types. However, it is likely that a triggering environmental factor such as a viral infection is involved in initiating the disease process. This hypothesis is supported by the finding that the rate of concordance for T1D between identical twins is less than 50% in most studies.

Insulin is a potent anabolic hormone that promotes the synthesis and storage of triacylglycerol, primarily in adipocytes and of glycogen in skeletal muscle and liver. Its major role, acting on these insulin-sensitive tissues, is to lower blood glucose levels by stimulating glucose uptake and storage. It also increases protein synthesis and inhibits lipolysis, including inhibition of the release of fatty acids from adipose cells. In the absence of insulin, these fatty acids are released from adipose cells and converted to ketones in the liver, resulting in the accumulation of ketones in the blood and their excretion in the urine (*ketonuria*). The excess ketoacids in the blood require buffering by the bicarbonate ion; this leads to a marked decrease in serum bicarbonate concentrations and thus acidosis. Patients with T1D therefore have a mandatory requirement for insulin injection therapy to prevent acidosis-induced coma and inevitable death.

In contrast, patients with T2D are usually older (although the prevalence of early-onset T2D in younger adults, teens and even children is rapidly increasing), often obese, and there is usually an insidious or gradual onset of symptoms. T2D is far more common than T1D, affecting some 90% of diabetic patients. Genetic factors are even more important than in T1D and among identical twins, there is a concordance rate of about 90%. T2D has a multifactorial aetiology with numerous genetic factors implicated, but there is no strong link with HLA type and no evidence that autoimmune mechanisms are involved. Post-mortem examination of sections of pancreas from patients who have died with T2D often reveals evidence of the accumulation of amyloid plaques around the β-cells. These plaques are composed of fibrils of β-pleated stacks of islet-amyloid polypeptide (IAPP or amylin), which is formed intracellularly in β-cells and deposits as aggregates in the islets when affected β-cells undergo programmed cell death (apoptosis).

The following two metabolic defects characterise T2D:

- A relative but not absolute, *insulin deficiency*. The cause of this insulin deficiency is related to a progressive loss of β-cell function and β-cell mass.
- An inability of peripheral tissues to respond to insulin, termed *insulin resistance*, due to a number of abnormalities of post-receptor stimulation of glucose transport. Insulin resistance alone is not sufficient to cause T2D; many obese individuals exhibit severe insulin resistance without any accompanying diabetes, since they are sufficiently hyperinsulinaemic to compensate for the decreased insulin sensitivity. Susceptible individuals, presumably with a genetic predisposition, are not able to sustain sufficient insulin secretion and once the loss of β-cell mass reaches a critical point (~50% reduction), diabetes becomes apparent.

In clinical practice, the distinction between T1D and T2D is not always clear cut and an appreciation of the underlying biochemical abnormalities often aids therapeutic decision making.

Treatment and Prognosis

Insulin resistance can be reduced by weight loss, exercise and medical therapy. Insulin secretion can be increased with medical therapy, but absolute or relative insulin deficiency can only be treated by insulin therapy.

Diet and lifestyle advice are combined with metformin (which reduces insulin resistance) as the first-line drug in T2D patients. If the HbA1c level does not show improvement or remains higher than the desired target (which varies based on the individual, with generally stricter targets such as an HbA1c level of 53 mmol/mol or

7% for younger, healthier individuals), then an additional medication is introduced, followed by another if the desired control is still not achieved. However, in cases of pronounced insulin deficiency and significant β-cell loss, these medications may not effectively manage T2D, necessitating the initiation of insulin therapy. Obviously, these drugs are not used in T1D in which insulin deficiency is absolute. Drugs that may be used in addition to metformin, include SGLT2 inhibitors (gliflozins) and GLP-1 receptor agonists/DPP4 inhibitors, sulphonylureas and insulin sensitisers (glitazones) (Chapter 13).

The patient was started on twice daily insulin injection therapy with rapid disappearance of urinary ketones and correction of blood glucose. Despite antibiotic therapy, he required surgical removal of the infected tissue, losing his first two toes and part of his forefoot. Vascular surgery to improve the blood flow to the foot was required. With healing, his mobility remained impaired and he never returned to employment. His retinopathy was stabilised using retinal laser photocoagulation to prevent blindness. Normal kidney function was not restored. Slow but progressive decline of renal function may be expected and he may eventually require dialysis. Six months after commencing therapy, he was given a trial of oral hypoglycaemic tablets (sulphonylureas) but had a brisk recurrence of symptomatic hyperglycaemia and ketonuria and was restarted on insulin therapy for life. This case underlines the importance of the early diagnosis and prompt treatment of patients with T2D.

Questions

1 Given that patients with T2D often respond well to dietary restriction and/or oral hypoglycaemic agents, why was the patient started on insulin therapy?
2 How can you explain the leg ulcer in this patient?

Answers on p. 558.

Case Study 71   Child with Unusual Presentation of Diabetes

Clinical Features

A 7-year-old girl of South Asian ethnicity was referred to paediatric endocrinology because of a presentation of significant obesity, accompanied by excessive tiredness, bedwetting and excessive thirst noted from age 6. She had a normal birthweight with no birth complications and her mother did not have gestational diabetes. On examination, there was severe acanthosis nigricans in the neck, axillae and groin regions. Random blood glucose was 10.1 mmol/L (reference range 3.5–8.9). Shortness of breath was noted on mild exertion. She had a BMI of 26.8 kg/m^2, placing her in the 99th percentile and in the obese category (see Chapter 12). Family history revealed two grandparents with a history of early-onset type 2 diabetes.

Investigations

The results of laboratory tests are shown in Table CS71.1. The reader is also referred to Chapter 13 for a detailed explanation of the specific tests used to make a diagnosis of diabetes.

In view of the raised alkaline phosphatase and alanine transaminase levels, an abdominal ultrasound scan was performed which revealed hepatomegaly and possible fatty liver disease. Immunological tests for islet cell antibodies (ICA) including GAD,

Table CS71.1   Results of laboratory investigations.

Test	Result	Reference range
HbA1c (glycated haemoglobin)	**51.9 mmol/mol**	<42 (to exclude IFG/pre-diabetes) <48 (to exclude diabetes)
Insulin	**3246 pmol/L**	11–220
C-peptide	**22.6 ng/mL**	0.5–5.5
Fasting plasma glucose	**7.9 mmol/L**	<6.0 (to exclude IFG/pre-diabetes) <7.0 (to exclude diabetes)
OGTT 2 hour plasma glucose	**14.4 mmol/L**	<7.8 (to exclude IGT/pre-diabetes) <11.1 (to exclude diabetes)
Alkaline phosphatase (ALP)	**159 U/L**	25–115
Alanine transaminase (ALT)	**63 U/L**	5–40
Gamma glutamyl transferase (GGT)	22 U/L	7–32
Total cholesterol	3.5 mmol/L	3.5–6.5
HDL	**0.5 mmol/L**	0.7–2.0
LDL	2.2 mmol/L	1.55–4.4
Triglycerides	**2.0 mmol/L**	<0.50–1.70
Bilirubin	6.0 μmol/L	<21

OGTT: Oral Glucose Tolerance Test. A post-OGTT level between 7.8 and 11.0 mmol/L is taken to indicate impaired glucose tolerance (IGT) or pre-diabetes. IFG (impaired fasting glucose) is a condition characterised by higher-than-normal blood glucose levels after an overnight fast, but in which levels are not high enough to be classified as diabetes. IFG also indicates prediabetes.
Values outside the reference range are in bold.

IAA (insulin autoantibodies), IA-2 and ZnT8, were negative. Genetic testing for known syndromic and non-syndromic monogenic diabetes and monogenic obesity variants was performed (including the following genes: ALMS, BBS, LEP, LEPR, MC4R, POMC, PPARG, ABCC8, GCK, HNF1A, HNF1B, HNF4A, KCNJ11, PDX1 and INS) and all were negative.

Diagnosis

The HbA1c, fasting and 2 hours post-OGTT glucose strongly suggest diabetes. Given the child's age, type 1 diabetes (T1D) or monogenic diabetes (MODY; see Chapter 13) may be considered as possibilities. However, the lack of relevant autoantibodies, the obesity (although obesity does not always rule out T1D) and the very high blood insulin and C-peptide levels, argue against a diagnosis of T1D. Genetic tests exclude known causes of monogenic diabetes (although an unknown variant might still be present). However, the extreme hyperinsulinaemia, hyperglycaemia, together with acanthosis nigricans, strongly indicate insulin-resistant type 2 diabetes (T2D). The liver changes are suggestive of non-alcoholic fatty-liver disease (NAFLD) and are consistent with a diagnosis of T2D.

Diagnosis: Type 2 diabetes (T2D).

Discussion

T2D is uncommon in young children, although young onset T2D is on the rise, very likely fuelled by the worldwide obesity pandemic. There is a higher incidence of T2D in particular ethnic groups, including many indigenous people, such as Native Americans (especially Pima Indian people), people of Afro-Caribbean origin and South Asians. Diabetic complications in young people diagnosed with T2D also develop more rapidly and are more severe; in this patient,

there was already evidence of NAFLD and changes in blood lipids (low HDL cholesterol, high triglycerides). This propensity to early and severe complications makes the correct differential diagnosis of young T2D even more important.

Whilst obesity makes T2D more likely, there has been an increase in T1D in children who are obese at the time of diagnosis (although the course of the disease suggests that diagnosis will be followed by weight loss). In this patient, genetic testing ruled out: monogenic obesity caused by mutations in genes encoding leptin, the leptin receptor, melanocortin 4 receptor or pro-opiomelanocortin (Chapter 12); non-syndromic monogenic diabetes (caused by mutations in GCK-encoding glucokinase, HNF1A, HNF1B, HNF4A or KCNJ11; Chapter 13); and syndromic monogenic diabetes (Alström, and Bardet-Biedl syndromes). Monogenic diabetes is rarely associated with obesity and hyperinsulinaemia although exclusion was important given the family history.

The diagnosis of T2D meant that the patient could be treated with metformin, although this had no effect on the underlying obesity. Other drugs for T2D could be used, including insulin secretagogues, such as sulphonylureas, or insulin itself, but these have the potential to worsen the obesity. Insulin sensitisers, such as pioglitazone, should also be considered. SGLT2 blockers (e.g. empagliflozin) or GLP-1 receptor agonists (e.g. semaglutide, which could also assist with weight loss) may be effective but are generally not licensed for use in children under the age of 12.

In conclusion, this is a rare case of T2D in a very young child. Given the dramatic rise in childhood obesity, such cases may well be more frequent in the future unless the public health crisis of obesity is tackled more effectively.

Treatment and Prognosis

Treatment with metformin was started and lifestyle modifications were advised, including dietary restrictions (although the parents asserted that the patient's calorie intake was not excessive) and exercise. Over the following year, good glycaemic control was achieved with regular HbA1c measurements well below 42 mmol/mol. However, there was no reduction in obesity. It is likely, without some reduction in obesity, the diabetes may recur despite the metformin and more aggressive therapy may be required.

Questions

1 What is the significance of the severe acanthosis nigricans observed in this case?
2 What is the mechanism of action of metformin in type 2 diabetes?

Answers on p. 559.

Case Study 72 Weight Loss, Excessive Thirst and Increased Urinary Frequency in a 17-year-old Girl

Clinical Features

A 17-year-old girl presented to the Emergency Department with a 2-day history of vomiting. Her mother was concerned because she also appeared drowsy and confused. The patient had been feeling unwell for the past 3 weeks with excessive thirst and increased urinary frequency associated with a weight loss of 4 kg. The patient's breathing was deep and sighing (known as '*Kussmaul*' breathing) and her breath smelled sweet with a pear-drop/acetone odour. She had no prior history of diabetes or other medical problems. Her oxygen saturation was 99% on room air, pulse was 110 bpm, blood pressure 90/40 mmHg and temperature 37.7°C. Her capillary refill time was increased at 3 seconds (indicating decreased peripheral perfusion) and her hands felt cool to the touch. Her mucous membranes were dry and there was increased skin turgor, both signs of dehydration.

Investigations

The results of laboratory tests are shown in Table CS72.1.

Urinalysis revealed 4+ glucose and 3+ ketones. A urine specimen was positive for leucocyte esterase and nitrite, suggesting urinary tract infection. Urine culture showed pure growth of *Escherichia coli*, (>10^6 colony forming units/mL) that was sensitive to the antibiotic, amoxycillin.

Diagnosis

The patient had a triad of hyperglycaemia, ketosis (indicated by the high β-hydroxybutyrate – a 'ketone body') and acidosis (low pH and bicarbonate levels) indicating diabetic ketoacidosis (DKA). DKA is an acute, life-threatening complication of diabetes characterised by;

- Increased blood ketones
- Decreased bicarbonate level
- Venous/arterial pH <7.3
- Hypokalaemia on admission
- Increased anion gap (anion gap is a calculated value used to assess the balance of electrolytes in the blood. It is determined by subtracting the sum of chloride [Cl^-] and bicarbonate [HCO_3^-] concentrations from the sum of sodium [Na^+] and potassium [K^+] concentrations. An anion gap above upper limit of the normal range suggests an increase in unmeasured anions in the blood. An elevated anion gap is often associated with metabolic acidosis).

Table CS72.1   Results of laboratory tests.

Test	Result	Reference range
Random blood glucose	**25.9 mmol/L**	3.5–8.9
pH	**7.01**	7.35–7.45
PaCO$_2$	**2.8 kPa**	4.8–6.1
PaO$_2$	**8.0 kPa**	10–13.3
Bicarbonate	**10.2 mmol/L**	22–30
Base excess	**−18.6 mEq/L**	−3 to +3
Capillary β-hydroxybutyrate	**4.5 mmol/L**	0–1
Plasma osmolality	**300 mosmol/kg**	280–296
Anion gap	**13 mEq/L**	<11
White cell count	**16.7 × 10^9/L**	4.0–11.0
Haematocrit	**0.48 L/L**	0.37–0.47
Haemoglobin	14.3 g/dL	11.5–16.5
Urea	**21 mmol/L**	2.5–6.7
Creatinine	**154 μmol/L**	45–84
Sodium	130 mmol/L	135–146
Potassium	3.2 mmol/L	3.5–5.0

Base excess is defined as the amount of strong acid that must be added to each litre of fully oxygenated blood to return the pH to 7.40 at a temperature of 37°C and a PCO$_2$ of 40 mmHg, while a base deficit (i.e. a negative base excess) is defined by the amount of strong base that must be added.

Plasma osmolality refers to the concentration of solutes in plasma. It represents the total amount of dissolved substances, including electrolytes (such as sodium, potassium and chloride), glucose, urea and other molecules. Elevated plasma osmolality can indicate DKA (see below).

Values outside the reference range are in bold.

Severe DKA is indicated by

- Venous/arterial pH <7.1
- Glasgow Coma Scale score <12 or abnormal AVPU scale (Appendix CS72.1)
- Oxygen saturation <92% on air (assuming normal baseline respiratory function)
- Systolic BP <90 mmHg, pulse >100 bpm or <60 bpm

Further diagnostic tests revealed that the patient had an extremely high level of glycated haemoglobin (HbA1c; 127.3 mmol/mol; normal <42). When blood glucose levels are high over an extended period, glucose molecules attach to haemoglobin in red blood cells. The higher the blood glucose concentration, the more glucose binds to haemoglobin, resulting in an increased percentage of HbA1c in the blood. An extremely high HbA1c level indicates that a person has experienced persistently elevated blood glucose levels over the past 3 months. It suggests poor glycaemic control and is often an indicator of uncontrolled diabetes. The patient also had a low C-peptide level (0.441 ng/mL; normal range 0.5–5.5). Immunological testing was positive for anti-GAD65, anti-IA2, anti-insulin, anti-ICA and anti-ZnT8 antibodies. These antibodies are commonly found in patients with autoimmune diabetes.

Diagnosis: Type 1 diabetes.

Discussion

DKA is a relatively frequent presentation of type 1 diabetes (T1D). Although conventionally considered pathognomonic for T1D, ketoacidosis is occasionally seen in type 2 diabetes (T2D), especially in younger patients. It is also worth noting that SGLT2 inhibitors (gliflozins; sodium-glucose co-transporter inhibitors) may themselves cause DKA (see Chapter 13).

The near absolute insulin deficiency seen in undiagnosed or untreated T1D results in substantial increases in lipolysis, with fat reserves being metabolised to release fatty acids. Lack of insulin, together with increases in counter-regulatory hormones, including glucagon and, potentially, adrenaline, increase gluconeogenesis and protein breakdown (loss of muscle mass being associated with untreated T1D). The fatty acids and amino acids released are normally metabolised to acetyl CoA (AcCoA), which then enter the tricarboxylic acid (TCA) cycle with oxaloacetate to generate reducing intermediates (nicotinamide adenine dinucleotide, or NADH; flavin adenine dinucleotide, or FADH2) which donate protons to the mitochondrial respiratory chain. With insufficient glucose entering insulin-sensitive cells in T1D (and sometimes in T2D as well), there is too little glycolysis and thus too much oxaloacetate is used for gluconeogenesis, preventing all the AcCoA from entering the TCA cycle; the excess is converted in the ketogenesis pathway to acetoacetate, which is then metabolised to β-hydroxybutyrate or spontaneously converted to acetone (giving the 'pear-drop' breath odour).

Once hyperglycaemia exceeds the renal threshold for reabsorption of glucose (around 10 mmol/L), glycosuria results in excessive fluid loss because of osmotic diuresis which then, especially with decreased fluid intake combined with the hyperglycaemia, results in hyperosmolarity and impaired renal function (indicated by the elevated creatinine in this patient). The increase in serum ketone bodies causes acidosis (low pH and $PaCO_2$) with depletion of alkali reserves (note low base reserve and low bicarbonate), leading to ketoacidosis. This, together with impaired kidney function, causes derangement of sodium and potassium ion balance, which then contributes to the loss of cognitive functions. The metabolic acidosis results in abnormal hyperventilation in an attempt to restore pH balance by losing CO_2, resulting in the deep sighing respiration known as Kussmaul breathing and which can be a precursor to imminent coma if the DKA is not treated rapidly. DKA may be precipitated by an infectious episode; in this case, the UTI might have been responsible.

In conclusion, this is a common and straightforward presentation of T1D, illustrating many of the features characteristic of this condition.

Treatment and Prognosis

The patient was started on antibiotics (IV cefuroxime). An IV infusion of normal saline was started together with an intravenous soluble insulin infusion and monitoring of her potassium levels with regular medical review. She made an uneventful recovery and was discharged with instructions to continue subcutaneous insulin injections for glycaemic control. Dose and insulin-formulation regimen would need to be appropriately adjusted over subsequent clinic visits to ensure good glycaemic control, which the patient would monitor (with advice from her physician) using home glucose kits or possibly CGM (continuous glucose monitoring) if needed.

Questions

1　What is the significance of the decreased levels of C-peptide?
2　If ketoacidosis is severe, this can lead to loss of consciousness/coma. How is this managed clinically?
3　What is capillary refill time?

Answers on p. 559

Appendix CS72.1: Glasgow Coma Scale and AVPU Scale

A Glasgow Coma Scale score of less than 12 or an abnormal AVPU scale indicates an alteration in the level of consciousness in an individual. Both the GCS and AVPU scales are commonly used clinical assessment tools to evaluate a person's neurological status and level of consciousness.

The Glasgow Coma Scale is a scoring system that assesses three components of neurological function: eye opening, verbal response and motor response. Each component is assigned a score and the total score ranges from 3 to 15, with 15 being the highest level of consciousness. A GCS score of less than 12 suggests a decreased level of consciousness and neurological impairment.

- Eye opening:
  Spontaneous: 4 points
  To verbal command: 3 points
  To pain: 2 points
  No response: 1 point

- Verbal response:
  Oriented conversation: 5 points
  Confused conversation: 4 points
  Inappropriate words: 3 points
  Incomprehensible sounds: 2 points
  No response: 1 point

- Motor response:
  Obeys commands: 6 points
  Localises pain: 5 points
  Withdraws from pain: 4 points
  Abnormal flexion (decorticate posture): 3 points
  Extensor response (decerebrate posture): 2 points
  No response: 1 point

The AVPU scale is a simpler test used to assess the level of consciousness based on four categories: Alert, Verbal, Pain and Unresponsive:

- *Alert (A):* The person is fully awake, aware and responsive.
- *Verbal (V):* The person responds to verbal stimuli but may not be fully alert.
- *Pain (P):* The person only responds to painful stimuli.
- *Unresponsive (U):* The person does not respond to any stimuli.

Answers

Case Study 1

1 Single-stranded positive-sense RNA viruses have a single strand of RNA that can be directly translated into viral proteins (similar to how cellular mRNA works). Single-stranded negative-sense RNA viruses have a single strand of RNA that must be converted into a positive-sense RNA intermediate before viral proteins can be made. Double-stranded RNA viruses have a genome consisting of two strands of RNA that are complementary to each other. The strands must be separated before viral proteins can be translated. Retroviruses have a genome consisting of RNA that is reverse-transcribed into DNA before being integrated into the host cell genome.

2 Kupffer cells are specialised macrophages found in the liver that constitute around 80–90% of the body's total population of tissue-resident macrophages. They play a crucial role in removing harmful substances and particulate matter from the portal blood. They are located in the sinusoidal lumen, adjacent to endothelial cells, and are able to perform phagocytosis and pinocytosis (a form of endocytosis, a process by which cells internalise substances from their external environment). Kupffer cells have an abundance of lysosomes that assist in the breakdown of substances taken up from the bloodstream. They can also secrete various toxic mediators that are vasoactive and are involved in host defence mechanisms. Their number and activity increase in response to liver injury.

Case Study 2

1 Yes, *Aspergillus* spp. express multiple virulence factors. For example, adhesins are fungal proteins that facilitate the attachment of *Aspergillus* to host cells or extracellular matrix. Adhesion is an important step in the establishment of infection, allowing the fungus to colonise and invade host tissues. *Aspergillus* also produces various hydrolytic enzymes, including proteases, phospholipases and ribonucleases that degrade host tissues, promoting tissue invasion. Mycotoxins are also important virulence factors. For example, the mycotoxin, *gliotoxin*, inhibits the migration of neutrophils, interferes with the production of reactive oxygen species (ROS) generated by neutrophils and other immune cells, and induces apoptosis in macrophages.

2 Airborne spread of *Aspergillus* spores is the predominant route of infection, and hospital air may contain significant numbers. For the severely immunocompromised, the

The Biology of Disease, Third Edition. Edited by Paul G. Murray, Simon J. Dunmore, and Shantha Perera.
© 2024 John Wiley & Sons Ltd. Published 2024 by John Wiley & Sons Ltd.
Companion website: www.wiley.com/go/murray/biologyofdisease3e

most effective protective measure is to provide their care in dedicated rooms with high-efficiency particulate air filters until such time as their immune function can recover. Prophylactic administration of anti-fungal drugs may also be considered.

Case Study 3

1 As of the time of writing (March 2023), the following drugs have been approved by the Food and Drug Administration (FDA) for the treatment of COVID-19:

- Tocilizumab, an IL-6 receptor inhibitor – for hospitalised adults receiving systemic corticosteroids and who require supplemental oxygen, non-invasive or invasive mechanical ventilation, or extracorporeal membrane oxygenation (ECMO).
- Remdesivir, a nucleoside analogue that inhibits the RNA-dependent RNA polymerase of coronaviruses – approved for use in adults and children who have mild-to-moderate disease and are at high risk for progression to severe COVID-19, including hospitalisation or death.
- Baricitinib, a JAK inhibitor – approved for hospitalised adults requiring supplemental oxygen, non-invasive or invasive mechanical ventilation, or ECMO.

Several other drugs, including Paxlovid, are approved by the FDA under Emergency Use authorisation (EUA).

2 The aim of a vaccine is to stimulate an adaptive immune response. Conventional vaccines usually involve administration of an attenuated or dead pathogen, or a recombinant antigen-encoding viral vector. In contrast, mRNA vaccines introduce mRNA into the vaccinated individual. Antigen-presenting cells take up the mRNA and produce the corresponding protein. The viral proteins are broken down

and presented to T and B cells (see also Chapter 6), resulting in T-cell-mediated and humoral (antibody) responses that target the foreign protein. The mRNA fragments are degraded within a few days and do not affect the body's genomic DNA.

Case Study 4

1 Drugs approved by the U.S. Food and Drug Administration (FDA) for the treatment of Hepatitis B include antivirals, for example, Adefovir, a nucleotide analogue. Adefovir is phosphorylated to its active metabolite, adefovir diphosphate, by cellular kinases. Adefovir diphosphate inhibits the HBV DNA polymerase by competing with the natural substrate, deoxyadenosine triphosphate, causing DNA chain termination after its incorporation into viral DNA. Interferon-alpha and Peg-IFN are also approved for the treatment of Hepatitis B infections. They stimulate type I interferon receptors, which activate the JAK/STAT pathway, increasing the expression of genes involved in the innate anti-viral response.

2 Hepatitis B-associated serum-sickness-like syndrome is a rare, immune-mediated, complication of acute HBV infection that occurs several weeks after the onset of symptoms. Serum-sickness-like syndrome (SSLS) is a type III hypersensitivity reaction (see also Chapter 6), in which there is deposition of immune complexes in blood vessels and tissues. During acute HBV infection, the immune system produces antibodies to fight the virus, and these antibodies can combine with viral antigens to form immune complexes. These immune complexes circulate in the blood and deposit in various tissues, such as the joints, skin and kidneys, triggering an inflammatory response. The symptoms of HBSSL include fever, rash, joint pain and swelling, which typically resolve within a

few weeks. Treatment is generally supportive and may include non-steroidal anti-inflammatory drugs (NSAIDs) or corticosteroids to relieve symptoms. In rare cases, complications such as vasculitis or glomerulonephritis can occur, which will require more aggressive treatment.

3 Individuals with chronic Hepatitis B infection have a significantly higher risk of developing HCC compared to those with acute or resolved infections. Frequently, HCC follows cirrhosis; the risk of HCC increasing substantially in individuals with cirrhosis, regardless of the underlying cause. Other factors, such as co-infection with Hepatitis C virus (HCV) or excessive alcohol consumption, can further increase the risk of HCC in individuals chronically infected with the Hepatitis B virus.

Case Study 5

1 The tissue cysts containing *Toxoplasma gondii* serve as a vehicle for parasite transmission by altering the behaviour of the intermediate host. In the case of rats and mice with latent toxoplasmosis, they become hyperactive and lose their instinctual fear of cats. This behavioural change increases the likelihood of the infected rodents being preyed upon by cats. In turn, this allows the parasite to complete its life cycle in the cat.

2 The worldwide prevalence of latent toxoplasmosis among humans is estimated to be around 30–70%. Seropositivity increases with age, typically peaking in the elderly. Seropositivity is highest in resource-poor countries. While most infected individuals do not display clinically relevant symptoms, there is growing evidence that latent toxoplasmosis may trigger cognitive changes in otherwise healthy 'asymptomatic' individuals.

Case Study 6

1 HIV infection is associated with a progressive increase in the risk of certain tumours, including aggressive lymphomas of B-lymphocytes, anal cancer (in individuals co-infected with human papillomavirus) and Kaposi sarcoma (in individuals co-infected with Human herpes virus 8/Kaposi sarcoma-associated herpesvirus). This suggests that the immune system is important in preventing the development of some tumours in immunocompetent individuals. Many of the tumours that arise during immunosuppression are associated with viruses, for example lymphoma (EBV), Kaposi sarcoma (KSHV) and CIN (HPV). The incidence of these tumours has declined in the era of ART.

2 Patients with AIDS-associated tumours present a therapeutic dilemma because of the risk of further bone marrow suppression and hence immunosuppression that results from treatment with cytotoxic drugs.

Case Study 7

1 Budd-Chiari syndrome, also known as hepatic venous outflow obstruction (HVOO), refers to the clinical picture that occurs when there is partial or complete obstruction of the hepatic veins.

2 Albendazole is a benzimidazole-type anthelmintic and antiprotozoal medication that has a broad-spectrum of activity. It is used for the treatment of several types of intestinal parasitic infections, including ascariasis, trichuriasis, strongyloidiasis, taeniasis, clonorchiasis, opisthorchiasis, giardiasis and gnathostomiasis, among others. Albendazole kills parasites by several mechanisms that include the inhibition of microtubule formation, impairment of glucose uptake and the disruption of mitochondrial function.

Case Study 8

1 The main reason for the high virulence of *P. falciparum* parasites is their ability to sequester in blood vessels of the deep viscera like the brain. The parasites adhere to the vascular endothelium in different organs, and also to surrounding uninfected red cells, causing mechanical obstruction and local induction of pathogenic levels of cytokines such as IFN-γ and TNFα, contributing to tissue damage that can lead to organ failure. Cerebral malaria is the most severe neurological complication; with approximately 575 000 cases annually, children in sub-Saharan Africa are most affected. Children who survive may have an increased risk of neurological and cognitive deficits, behavioural difficulties and epilepsy.

2 Persistence of parasites in blood with or without clinical symptoms despite anti-malarial treatment or, more commonly, initial clinical and parasitological response to treatment followed by re-appearance of symptoms and parasites within 28 days following initiation of treatment, is indicative of treatment failure. Symptoms that recur in these cases are commonly atypical and vague and repeated blood smears should be done for 28 days following treatment to ensure effective parasite elimination. RDTs, however, remain positive for variable periods following anti-malarial therapy, and therefore, are not used in treatment monitoring.

Case Study 9

1 Antibiotics destroy bacteria, resulting in the release of cell wall components. This leads to the generation of pro-inflammatory cytokines and a particularly marked, and potentially harmful, inflammatory response.

2 TI antigens can directly activate B cells without the need for T-cell help, whereas TD antigens require T-cell help for optimal B cell activation. TI antigens, including bacterial carbohydrate antigens, generally induce a lower degree of affinity maturation compared to TD antigens. Affinity maturation is the process by which B cells undergo mutations in their immunoglobulin genes, leading to the production of antibodies with increased affinity for the antigen (Chapter 6). TI responses often result in predominantly IgM antibodies. In contrast, TD responses allow for class switching to various antibody isotypes (such as IgG, IgA or IgE).

Case Study 10

1 The bacteria release toxins and trigger an inflammatory response that can lead to the release of cytokines, such as tumour necrosis factor-alpha (TNF-α) and interleukin-6 (IL-6). These cytokines activate the coagulation cascade, leading to the formation of blood clots in small blood vessels throughout the body. As the coagulation cascade is activated, the body may start to use up its clotting factors and platelets, leading to their depletion. This can cause bleeding symptoms, such as easy bruising, petechiae or even spontaneous bleeding in severe cases. At the same time, the formation of blood clots in small blood vessels can impair blood flow to vital organs, leading to organ dysfunction or failure.

2 Gas is produced through the action of bacteria, particularly anaerobic bacteria that invade the deeper layers of the skin and soft tissues. These bacteria can break down

tissues and release various gases, including hydrogen, nitrogen and carbon dioxide, as metabolic by-products. As the bacterial infection progresses, the gas produced by these bacteria can accumulate in the tissues leading to crepitus, a crackling sensation felt under the skin.

Case Study 11

1 D-dimer is a fibrin degradation product that is elevated in a variety of conditions associated with inflammation, coagulation, and fibrinolysis. In COVID-19, the elevation of D-dimer levels is thought to be due to a combination of factors, including activation of the coagulation system, endothelial injury and cytokine-mediated inflammation.

2 C-reactive protein is a protein produced by the liver in response to inflammation in the body. It is a non-specific marker of inflammation.

3 Antenatal corticosteroids are administered to pregnant women who are at risk of preterm birth to help accelerate foetal lung maturity between 24 and 35 weeks of gestation. However, this benefit does not last beyond 7 days, so babies born >7 days after administration will not benefit from the treatment. The lungs are one of the last organs to develop in the foetus. Corticosteroids mimic the action of natural hormones responsible for lung maturation. When administered before birth, they can stimulate the production of surfactant, which reduces surface tension in the lungs, allowing them to expand more easily. This is critical for effective oxygenation and carbon dioxide removal once the baby is born and begins to breathe air. Corticosteroids may also provide some benefits for brain development and help reduce other complications associated with preterm birth, such as intraventricular haemorrhage.

Case Study 12

1 TNF-α is a pro-inflammatory cytokine that plays a central role in the pathophysiology of sepsis. It is produced by macrophages and other immune cells. In sepsis, TNF-α can cause the release of other pro-inflammatory cytokines and chemokines, leading to a systemic inflammatory response that can damage tissues and organs and result in multiple organ dysfunction syndrome. TNF-α can also increase vascular permeability, leading to oedema and hypotension, which can contribute to the development of septic shock. Early clinical trials investigating the use of TNF-α inhibitors in sepsis showed promising results in animal models and in small human studies. However, larger clinical trials failed to show significant benefits in terms of mortality or other clinical outcomes, and some even reported harm. For example, a large randomised controlled trial of the anti-TNF-α antibody, etanercept, in sepsis patients, was terminated early due to increased mortality in the treatment group.

2 The innate immune system recognises PAMPs through pattern recognition receptors (PRRs), such as Toll-like receptors (TLRs) and nucleotide-binding oligomerisation domain-like receptors, which are expressed on the surface of immune cells. The recognition of PAMPs by PRRs leads to the activation of the innate immune response, including the production of pro-inflammatory cytokines and chemokines, the recruitment of immune cells to the site of infection, and the activation of complement. Examples of PAMPs include

lipopolysaccharide (LPS): LPS is a component of the cell wall of Gram-negative bacteria and is recognised by TLR4, which is expressed on the surface of immune cells. The binding of LPS to TLR4 leads to the activation of the NF-κB pathway and the production of pro-inflammatory cytokines, such as TNF-α and IL-1β.

bacillus in the faecal flora. Therefore, on a statistical basis, it is more likely to be involved in infections originating in the abdomen or after abdominal trauma or surgery. Certain serotypes of *E. coli* have been found to be more common in patients with UTI, for example those with certain acidic polysaccharide antigens (K1 and K5 strains).

Case Study 13

1   *Pseudomonas aeruginosa* is a significant cause of illness in immunocompromised individuals. It is an organism of low virulence, but in situations in which host defences are impaired, it can cause significant infection, i.e. it is an opportunistic pathogen. In individuals with healthy immune systems, *P. aeruginosa* is effectively eliminated or controlled by the immune response. *P. aeruginosa* does not commonly cause community-associated infection unless the patient has previously been on antibiotics. The source of *P. aeruginosa* infections can vary. In healthcare settings, it can be transmitted through contaminated equipment or surfaces, or via the hands of healthcare workers. *P. aeruginosa* is widespread in the environment and colonises moist areas, such as sinks and drains. Hence, tap water should not be used for cleaning compromised skin. Importantly, it can also colonise fluid solutions which may be used in healthcare settings. In the community, it can be found in various environments such as soil and water, and it can colonise various sites in the body without causing disease, until the individual becomes immunocompromised.

2   *Escherichia coli* is the commonest cause of UTI (both community and healthcare associated). It is an aerobic enteric Gram-negative bacillus. Numerically, *E. coli* is the predominant aerobic Gram-negative

Case Study 14

1   Erythrocyte mean cell volume (MCV) is often high in alcoholic liver disease. This is sometimes due to folate deficiency resulting from poor diet, but may also be due to liver damage. The precise mechanisms causing macrocytosis in alcoholic liver disease are not clear, but it has been suggested that acetaldehyde generated by oxidation of ethanol by bone marrow macrophages damages red cell precursors.

2   The liver synthesises most coagulation factors, so when there is hepatocellular damage, synthesis of blood coagulation factors is often reduced. In addition, if there is cholestasis (impaired flow of bile from the liver to the small intestine), there is also malabsorption of fat, thus reducing absorption of fat-soluble vitamins including vitamin K. Vitamin K is necessary for the synthesis of coagulation factors II, VII, IX and X, so in vitamin K deficiency their concentrations fall, thereby prolonging the prothrombin time.

3   Inflammation of the gastric mucosa and peptic ulceration are associated with alcohol abuse, and bleeding may occur in either condition. Furthermore, if the liver is damaged by cirrhosis, pressure in the hepatic portal vein rises and collateral circulation becomes established in other vessels to bypass the liver. One effect of this is that the veins of the lower oesophagus

become dilated and distended (a condition known as oesophageal varices) and may bleed. Any bleeding is often exacerbated by reduced levels of blood coagulation factors, so the administration of vitamin K or the replacement of missing coagulation factors by transfusion of fresh frozen plasma is indicated.

Case Study 15

1 The Mini-Mental State Examination (MMSE) is a commonly used screening tool to assess cognitive function in adults. The basis of the MMSE is to evaluate several cognitive domains, including orientation to time and place, attention and calculation, recall, language and visual construction. These domains are assessed through a series of questions and tasks, and a score is given based on the accuracy and completeness of the responses. The maximum score on the MMSE is 30, with a score of 24 or lower indicating possible cognitive impairment.

2 Yes, there are a number of new drugs approved for the treatment of people with AD:

- *Lecanemab:* This drug is a monoclonal antibody approved for medical use in the United States in January 2023. Lecanemab targets soluble Aβ aggregates (oligomers and protofibrils).
- *Aducanumab:* This drug was approved by the FDA in June 2021 for the treatment of early-stage AD. It is a monoclonal antibody that targets Aβ protein. Clinical trials have shown that aducanumab can reduce the accumulation of Aβ protein in the brain and may slow the progression of cognitive decline in some patients.
- *Pimavanserin:* This drug was approved by the FDA in 2021 for the treatment of dementia-related psychosis, which is a

common symptom of AD. Pimavanserin works by targeting serotonin receptors in the brain and has been shown to improve symptoms of psychosis without worsening cognitive function.
- *Donanemab:* This is a monoclonal antibody that targets Aβ protein. In clinical trials, donanemab was shown to reduce the accumulation of Aβ in the brain and improve cognitive function in patients with early-stage AD. The drug was granted breakthrough therapy designation by the FDA in 2020.

Case Study 16

1 Treatment of amyloidosis is dependent upon the type of disease. Thus, in AL amyloidosis, high-dose chemotherapy with melphalan followed by stem cell transplantation is recommended for eligible patients. For those not eligible for transplant, chemotherapy regimens including daratumumb or cyclophosphamide-bortezomib-dexamethasone are used. In AA amyloidosis, treating the underlying inflammatory condition can improve symptoms. Although TNF-α inhibitors, such as infliximab and etanercept, are not licensed specifically for amyloidosis, they can reduce the inducing inflammation and, in turn, limit amyloid deposition. For ATTR amyloidosis, tafamidis can be used, which stabilises transthyretin tetramers, reducing the number of monomers available for amyloidogenesis. Diflunisal and inotersen work by preventing the build-up of misfolded mutant TTR protein, while patisiran blocks the expression of TTR.

2 Answer: Hereditary ATTR amyloidosis is inherited in an autosomal dominant manner. Each child of an affected individual (who is heterozygous for one TTR pathogenic variant) has a 50% chance of inheriting the TTR variant.

Case Study 17

1 Neutrophils have three main functions: phagocytosis, degranulation with the release of soluble cytotoxic and pro-inflammatory mediators, and the release of nuclear material which forms so-called neutrophil extracellular traps (NETs; see also Chapter 5)

2 An abscess is a localised collection of pus composed of dead host tissue cells, including white cells (especially neutrophils), tissue fluid and microorganisms.

3 This refers to a clinical test used to evaluate irritation of the psoas muscle, which often indicates an underlying intra-abdominal condition. A positive psoas sign can suggest appendicitis, though it is not specific and can be associated with other intra-abdominal diseases. The test is performed by asking the patient to lie on their side with the knees flexed. The examiner extends the leg at the hip, stretching the psoas muscle. Alternatively, the patient may be asked to flex the hip against the examiner's resistance while lying on their back. If either of these manoeuvres causes abdominal pain, the psoas sign is considered positive.

Case Study 18

1 Mounier-Kuhn syndrome, also known as tracheobronchomegaly, is a rare condition that occurs due to congenital weakness/atrophy of the muscular and elastic tissue of the tracheal and bronchial walls.

2 Since cilia play an important role in clearing the nasal passages and sinuses, patients with PCD often suffer from chronic sinusitis. Cilia in the female reproductive system are also important for the movement of oocytes through the fallopian tubes, and in the male reproductive system, they are involved in the movement of spermatozoa. Therefore, PCD can cause infertility in both men and women. Cilia in the ear are responsible for moving earwax and other debris out of the ear canal. Patients with PCD are, therefore prone to recurrent ear infections. *Situs inversus* is a condition in which the organs in the chest and abdomen are mirror images of their normal positions. About 50% of patients with PCD have situs inversus, which is believed to be due to a defect in the way cilia direct the development of organs during embryonic development. In rare cases, PCD can lead to hydrocephalus, a condition in which there is an abnormal accumulation of cerebrospinal fluid in the brain.

Case Study 19

1 2.64 points
Further investigation needed.
Approximate fibrosis stage: Ishak 2–3 (Sterling et al., 2006)

2 The FibroScan is a non-invasive medical device used to assess the degree of liver fibrosis (and fatty change) in patients with chronic liver disease. The FibroScan uses a technique called transient elastography, which involves the use of a probe that sends low-frequency waves into the liver. The waves create a vibration that can be measured to determine the stiffness of the liver tissue. The stiffer the tissue, the greater the degree of fibrosis. The FibroScan is a quick and painless procedure that takes only a few minutes to complete.

Case Study 20

1 Malabsorption of a wide range of nutrients occurs in coeliac disease and deficiencies of nutrients with relatively low body stores

become apparent at an early stage. Iron and folate are absorbed mostly in the proximal small intestine, the area of the gut mainly affected by coeliac disease. Haemopoietic tissues have high demands for folate and iron, so anaemia soon follows the development of deficiency states. Anaemia may be predominantly due to iron deficiency (microcytic and hypochromic), or to folate deficiency (macrocytic, normochromic), or may show features of both types of anaemia (See also Chapter 7).

2 People with coeliac disease have an increased risk of developing non-Hodgkin lymphoma, particularly enteropathy-associated T-cell lymphoma, which arises in the small intestine. The risk is most pronounced among individuals with persistent or untreated coeliac disease. Of interest is the discovery that these tumours have mutations in JAK/STAT pathway genes, which might provide a mechanistic explanation for the association.

Case Study 21

1 Oxygen therapy is an important intervention in the management of respiratory failure as it can improve oxygenation and tissue perfusion. However, oxygen therapy should be administered with caution in patients with chronic obstructive pulmonary disease (COPD). This is because chronic retention of carbon dioxide can lead to a blunted respiratory drive, meaning the patient relies on low oxygen levels to stimulate their breathing. Administration of high levels of oxygen can suppress the respiratory drive and cause carbon dioxide retention, leading to hypercapnic respiratory failure.

2 Several mechanisms account for decreased renal function in right heart failure (RHF):

  • Reduced cardiac output: in RHF, the heart's ability to pump blood to the kidneys is compromised, resulting in decreased renal blood flow and a reduction in the glomerular filtration rate.

  • Increased venous pressure: RHF can cause an increase in venous pressure in the kidneys, leading to congestion.

  • Activation of the renin-angiotensin-aldosterone system (RAAS): RHF can stimulate the RAAS, leading to an increase in sodium and water retention, which exacerbates fluid overload.

  • Diuretic therapy: patients with RHF are often treated with diuretics; however, this can lead to volume depletion and electrolyte imbalance, which further impair renal function.

Case Study 22

1 The microcytic hypochromic anaemia, with low MCV and low mean cell haemoglobin, could indicate iron deficiency due either to chronic blood loss or malabsorption of iron. Iron deficiency is characterised by low serum iron, low serum ferritin and raised serum transferrin levels (Chapter 7). The causes of anaemia due to chronic inflammation are complex (see Case Study CS18). In anaemia associated with chronic inflammation, serum iron may also be low, but serum ferritin is normal or high, and serum transferrin is usually low. This patient's anaemia was probably due to a combination of chronic blood loss/iron malabsorption and inflammatory disease.

The ESR is a non-specific test, which is raised in a wide variety of conditions, notably inflammatory and malignant disorders. Occasionally, especially in elderly patients, the ESR is raised and despite extensive investigations, no cause is found. The measurement of plasma viscosity gives much the same information, and is subject to fewer variables, so it may be offered as an alternative to the ESR.

The raised leucocyte count, with the majority of the cells being neutrophils (neutrophilia), is consistent with an inflammatory disorder.

Patients with Crohn's disease may also present with a different form of anaemia. Thus, the terminal ileum, a key site for vitamin B12 absorption, is often affected by the inflammation or may be removed surgically. Vitamin B12 is important for DNA synthesis in red blood cell development. When B12 absorption is compromised, it disrupts the maturation of red cells, leading to megaloblastic anaemia. This condition is characterised by larger-than-normal, immature red blood cells that are less efficient in carrying oxygen. Because vitamin B12 is important in the nervous system, deficiency can lead to neurological symptoms, which can be helpful in distinguishing B12 deficiency from other causes of anaemia.

2   Clinically, Crohn's disease can be difficult to distinguish from ulcerative colitis (UC), since both diseases cause chronic inflammation and ulceration in the gastrointestinal tract. However, whereas Crohn's disease can affect any area of the gastrointestinal tract from the mouth to the anus, UC is confined to the colon. In addition, in Crohn's disease the inflammatory process typically extends through the entire thickness of the bowel wall (transmural), whereas, usually, in UC, only the mucosa and submucosa are affected. Non-caseating granulomas are also a characteristic histological finding in Crohn's disease and are not usually observed in UC.

Case Study 23

1   The Romberg test involves asking the patient to stand with their feet close together and their eyes closed while maintaining their balance. It evaluates a patient's ability to maintain an upright posture and steady themselves without visual input. The test is useful in the diagnosis of various conditions that affect proprioception and balance, including subacute combined degeneration of the spinal cord (associated with vitamin B12 deficiency), posterior cord syndrome (caused by a posterior spinal artery infarction), and hemisection of the spinal cord (Brown-Sequard syndrome).

2   The term anisopoikilocytosis comes from the Greek words 'aniso-', which means uneven, and 'poikilo-', which means of a great variety. Accordingly, anisopoikilocytosis encompasses both anisocytosis-red blood cells of uneven size, and poikilocytosis-red blood cells of variable shape.

Case Study 24

1   Patients with severe anaemia or cardiovascular disease may be ineligible for phlebotomy, but may benefit from either oral or intravenous iron chelation therapy. Iron chelation therapy uses drugs (e.g. deferasirox, deferiprone and deferoxamine) that bind to the excess iron in the body, which is then eliminated through the urine or stool.

2   Women typically present later because they have a lower risk of developing iron overload due to regular menstrual bleeding. Menstruation causes women to lose blood and, as a result, lose some of the excess iron.

Case Study 25

1   Uhthoff phenomenon, also known as Uhthoff symptom, is a transient worsening of neurological symptoms in patients with MS or other neurological conditions, triggered by an increase in body temperature.

The phenomenon is named after Wilhelm Uhthoff, a German ophthalmologist who first described it in the late 19th century. Uhthoff observed that some patients with optic neuritis experienced temporary visual blurring or dimming when their body temperature rose due to exercise, fever, or a hot bath. The exact mechanism of Uhthoff phenomenon is not fully understood, but it is thought to be related to the sensitivity of demyelinated nerve fibres to changes in temperature. As body temperature rises, nerve conduction is slowed or disrupted, leading to a temporary worsening of neurological symptoms. Uhthoff phenomenon is a well-recognised feature of MS but can also occur in other neurological conditions.

2 Rebound MS is a sudden worsening of MS disease activity, with the appearance of new, severe neurological symptoms, that occurs in some people following the abrupt withdrawal of disease-modifying therapies.

Case Study 26

1 Backwash ileitis is a condition that can occur in individuals with UC. UC is an inflammatory bowel disease that primarily affects the large intestine and rectum. In some cases, the inflammation can extend into the ileum. When this happens, it is called 'backwash' ileitis, referring to the fact that inflammatory cells and debris from the colon wash back into the ileum, causing inflammation in that area. The presence of backwash ileitis is often seen in people with advanced UC.

2 Post-colectomy pouchitis is an inflammatory condition that can occur after colectomy and the creation of an internal pouch (pouch-anal anastomosis) to restore intestinal continuity. The pouch is usually made from the last part of the ileum and is connected to the anus. Pouchitis is characterised by inflammation and ulceration of the internal pouch lining, leading to symptoms such as diarrhoea, abdominal pain and fever. It is believed to be caused by an overgrowth of harmful bacteria in the pouch due to changes in the gut microbiome after surgery.

Case Study 27

1 EBUS-TBNA is *endobronchial ultrasound transbronchial needle aspiration*, a minimally invasive procedure used to obtain tissue or fluid samples from the lungs and surrounding lymph nodes. EBUS-TBNA is performed using a flexible bronchoscope that has an ultrasound probe at its tip. The ultrasound allows the doctor to see the surrounding tissues and lymph nodes in real time, and then use a thin needle that passes through the bronchoscope to obtain a sample of the tissue or fluid. The samples are then sent to a pathologist for analysis. EBUS-TBNA is a relatively safe and effective procedure and can help in the diagnosis and staging of lung cancer, lymphoma, sarcoidosis and other lung diseases.

2 Asteroid bodies and Schaumann bodies are two types of intracellular inclusions that can be found in tissues affected by sarcoidosis. Asteroid bodies (Fig.CS27.2) are star-shaped, eosinophilic inclusions found within giant cells or in extracellular spaces. They are formed by the fusion of lysosomes with phagocytic vacuoles containing foreign substances. Schaumann bodies are calcified structures that can be found within epithelioid macrophages or in extracellular spaces. They are thought to be formed by the interaction of calcium and phospholipids. The presence of asteroid and Schaumann bodies is not specific to sarcoidosis, as they are found in other granulomatous diseases.

Case Study 28

1 An epithelioid granuloma is a mass or nodule comprised of chronically inflamed tissue and containing macrophages that look like epithelial cells, hence the term 'epithelioid'. Caseation refers to the presence of necrotic tissue that looks like soft cheese to the naked eye. Non-caseating granulomas lack this necrosis.

2 Healing by first intention is healing of a wound in which the edges are closely re-approximated and there is minimal requirement for granulation tissue. Healing by primary intention occurs in most surgical settings where a clean incision is possible. Healing by second intention is when the wound is allowed to heal by itself with granulation tissue occupying the gap between the edges. In some cases, assisted secondary intention can be performed by applying a vacuum-assisted closure dressing.

Case Study 29

1 The exact pathophysiology of Felty syndrome is poorly understood. The primary characteristic of Felty syndrome is the presence of neutropenia believed to be due to a combination of inadequate production of neutrophils in the bone marrow (potentially caused by infiltration of the bone marrow by cytotoxic lymphocytes), and increased sequestration of neutrophils in the spleen (splenectomy often results in an improvement in the neutropenia). Some studies have found the presence of antibodies against granulocyte-colony stimulating factor (G-CSF) in patients with neutropenia associated with Felty syndrome. Moreover, autoantibodies found in Felty syndrome have been shown to bind to deaminated histones and

neutrophil extracellular chromatin traps (NETs), which might also lead to the destruction and sequestration of neutrophils, contributing to neutropenia.

2 Some studies suggest a potential link between chronic large granular lymphocyte (LGL) leukaemia and Felty syndrome due to their similar clinical presentation and shared genetic susceptibility, including HLA-DR4. LGL leukaemia is associated with increased secretion of circulating Fas ligand, which induces neutrophil apoptosis. The presence of elevated Fas ligand levels in LGL leukaemia and the association of LGL leukaemia with other autoimmune diseases, like RA, further support a potential common pathogenic mechanism. The hypothesis that LGL leukaemia and Felty syndrome are part of the same disease spectrum has been strengthened by the recent discovery of somatic STAT3 mutations in T-cells in both disorders.

Case Study 30

1 Status asthmaticus is an acute exacerbation of asthma, which progresses in severity and is poorly responsive to standard therapeutic measures. All patients with bronchial asthma are at risk of developing an acute episode. Status asthmaticus is a medical emergency. Despite advances in pharmacotherapy and access to early diagnosis and newer treatments, asthma remains one of the most common causes of Emergency Room visits.

2 Corticosteroids are used to prevent the 'late phase' response in conditions like asthma or allergic reactions. They work by inhibiting the synthesis of arachidonic acid metabolites (see also Chapter 5), which are involved in the inflammatory

response. By doing so, corticosteroids effectively block the formation of new inflammatory mediators that contribute to the late phase of the allergic reaction, helping to reduce inflammation and its associated symptoms.

Case Study 31

1 Hypercholesterolaemia is associated with the development of atherosclerosis (Chapter 9). Moreover, the development of vascular damage associated with atherosclerosis is accelerated by hypertension.

2 Creatinine is released from skeletal muscle at a constant rate and excreted in the urine. In renal disease, creatinine may accumulate in the blood due to impaired excretion. However, measuring serum creatinine has limitations. For instance, creatinine levels can be influenced by factors such as muscle mass, age, sex, and diet. Additionally, serum creatinine may not increase until there is significant loss of kidney function. Estimated glomerular filtration rate (eGFR) is calculated using equations that consider serum creatinine levels, age, sex, and ethnicity. It provides an estimate of the kidney's glomerular filtration rate, which reflects overall kidney function more accurately than serum creatinine alone.

Case Study 32

1 In this case, there is acute blood loss due to haemoptysis, which has resulted in significant anaemia. In chronic kidney disease, anaemia may be the result of decreased erythropoietin production by the damaged kidney and the retention of waste products, which tend to shorten red cell lifespan and inhibit red cell production.

2 Two mechanisms are operative: (a) renal damage activates the renin-angiotensin-aldosterone system; (b) impaired renal excretion of sodium and water retention, an increase in blood volume and hence an increase in blood pressure.

Case Study 33

1 Immune reconstitution following allogeneic HSCT typically occurs in a predictable order over several months. Granulocytes, monocytes and NK-cells are the first cell types to recover, followed by T- and B-lymphocytes. Lymphocyte reconstitution results from the production of naïve lymphocytes derived from transplanted haemopoietic stem cells, or by the peripheral expansion of mature cells carried across in the stem cell graft. Donor T-lymphocytes are re-educated in the presence of recipient antigens, resulting in the development of immune tolerance. In patients affected by GvHD immune reconstitution is delayed, both as a consequence of the GvHD, and the immunosuppressive drugs used to treat it. Consequently, patients with GvHD experience a prolonged period of secondary immunodeficiency that makes them particularly vulnerable to infection.

2 A number of late-effects may complicate allogeneic HSCT, mainly resulting from the use of intensive chemotherapy and/or radiotherapy in the transplant process, or due to GvHD and its treatment. Infertility is common due to the gonadotoxic effects of chemotherapy or radiotherapy. There is an increased risk of secondary malignancies. Endocrine dysfunction is common, resulting in hypothyroidism and other hormone deficiencies. There is also an increased risk of cardiovascular disease and metabolic syndrome.

Case Study 34

1 The term 'sudden coronary death' may be applied when an individual previously in good health falls ill and dies within one hour of the onset of symptoms.

2 In most cases, sudden death is the result of the complications of coronary artery disease and is usually due to an arrhythmia caused by an acute myocardial infarction, myocardial ischaemia or scarring from an old myocardial infarct. Arrhythmia also accounts for a percentage of sudden deaths from acute myocarditis (inflammation of the myocardium) and hypertrophic cardiomyopathy (see also Chapter 9). Other non-cardiac causes of sudden death include cerebral haemorrhage.

Case Study 35

1 Improved early revascularisation, which refers to the timely restoration of blood flow to the heart after a myocardial infarction, reduces the extent of heart muscle damage and inflammation, which in turn reduces the risk of autoimmune reactions that can lead to Dressler syndrome.

2 Studies have shown there is a very small (around twofold) increased risk of myocarditis and/or pericarditis in people vaccinated with an mRNA vaccine against SARS-CoV-2 (e.g. Pfizer or Moderna) compared with unvaccinated people. Pericarditis and myocarditis after COVID-19 vaccines have mostly been reported in males under 40 years of age and mostly after the second dose. At the time of writing (October 2023), current advice, for example from the Centers for Disease Control and Prevention (CDC), is that everyone aged 6 months and older should be vaccinated for COVID-19. However, rates of post-vaccine pericarditis are still being reviewed.

Case Study 36

1 The Valsalva manoeuvre ('bearing down') is performed by exhaling forcefully with a closed glottis, resulting in increased abdominal pressure. The haemodynamic changes during the Valsalva manoeuvre are broken down into four stages, with the main effect occurring during phase 2 when there is decreased venous return to the heart and blood return to the left ventricle. The Valsalva manoeuvre is useful in cardiological examinations.

2 Epicardial adipose tissue (EAT) is regarded as a separate, anatomically distinct, organ. It is located between the pericardium and the myocardium and is in direct contact with the myocardium and coronary vessels. EAT has several important functions. For example, it provides a cushioning effect to the heart, protecting it from external mechanical stress and potentially limiting the effects of torsion during the cardiac cycle. EAT also serves as an energy reservoir, providing the myocardium with free fatty acids, which are the primary source of energy for cardiac muscle cells. EAT also helps maintain the optimal temperature within the heart by providing thermal insulation. EAT can also secrete various adipokines (see Chapter 12), cytokines and chemokines, which have local and systemic effects. Through the release of these factors, EAT also influences coronary artery function. However, excess EAT is associated with various cardiac conditions, including coronary artery disease, heart failure and atrial fibrillation. The precise mechanisms by which EAT contributes to these conditions are still not completely understood, but it is thought that alterations in the secretory profile and inflammatory state of EAT are involved. In current-day cardiology practice, EAT is easily recognised by cardiac CT (during CT for coronary arteries) and during cardiac MRI.

Case Study 37

1 During a TTE test, a small ultrasound probe (transducer) is placed on the chest wall and sends sound waves through the chest to the heart. The reflected sound waves bounce off the heart structures and are detected by the transducer, which then creates real-time images of the heart. TTE can provide information about the size and shape of the heart, the thickness and movement of the heart walls, the function of the heart valves, the blood flow through the heart and blood vessels, and other cardiac abnormalities.

2 Beta-haemolytic streptococci are capable of complete haemolysis of red blood cells, which is seen as a clear zone around the bacterial colony on blood agar. Beta-haemolytic streptococci produce various virulence factors (see also Chapter 4), including streptolysin O and streptolysin S, which are responsible for the tissue damage associated with infection.

Case Study 38

1 As with all emergencies, stabilising the patient is of paramount importance, with a focus on addressing the ABCDEs: **A**irway, **B**reathing, **C**irculation, **D**isability, and **E**xposure. See this link for a useful summary: https://oxfordmedicaleducation.com/emergency-medicine/abcde-assessment/ Specific interventions include applying 100% oxygen to maximise the oxygen saturation unless contraindicated (COPD patients), unfractionated heparin, IV fluids and analgesia. This is then followed by revascularisation procedures either endovascularly (catheter-directed thrombolysis, mechanical thrombectomy) or open surgical intervention (embolectomy/thrombectomy). In some cases, a hybrid endo/open approach is utilised. A hybrid endo/open approach refers to a procedure that combines elements of both minimally invasive endoscopic techniques (endo) and traditional open surgical techniques (open). This approach is often used when treating complex or challenging medical conditions that may benefit from the advantages of both approaches.

2 Delayed treatment of acute limb ischaemia can lead to severe complications, including necrosis due to prolonged hypoxia of the tissues, necessitating amputation. Additionally, revascularisation following delayed identification of acute limb ischaemia increases the risk of compartment syndrome, necessitating fasciotomy (a surgical procedure in which the fascia- the connective tissue surrounding muscles, blood vessels, and nerve- is cut to relieve tension or pressure), and/or reperfusion syndrome. This may result in hyperkalaemia and renal failure. *Compartment syndrome* is a potentially limb-threatening condition that arises when pressure builds up within the fascial compartments of a limb. It is more commonly observed in the legs than in the arms. Compartment syndrome can occur in response to bleeding, any soft tissue injury that causes oedema, tight dressing in a patient with a fracture, or following delayed revascularisation. Capillary blood flow is compromised when the pressure in the compartment rises beyond 20–30 mmHg; hence patients may develop muscular necrosis while having palpable distal pulses (as the pressure needs to exceed systolic blood pressure to cause pulselessness). Patients typically present with swelling, tightness of the limb and pain out of proportion to the injury that increases on passive stretching of the muscles. Paraesthesia is typically an early sign, whereas pulselessness (as explained above) is a late sign.

The treatment is an emergency fasciotomy. *Reperfusion syndrome* is a pathological response to revascularisation, which may occur particularly in the event of delayed revascularisation. The sudden return of blood produces localised and systemic reactions, including compartment syndrome, hyperkalaemia, renal failure and metabolic acidosis. The underlying mechanism involves the release of toxic metabolic by-products accumulated during the period of ischaemia, as well as ROS (see Chapter 3) that are generated when oxygen is suddenly reintroduced to the tissues.

Case Study 39

1 A pericardial rub is an abnormal sound that can be heard when listening to the heart with a stethoscope. It is caused by the rubbing together of the two layers (the outer parietal layer and the inner visceral layer) of the pericardium, the sac that surrounds and protects the heart. Pericardial rub typically has a scratching or grating quality and is often described as a 'leather-on-leather' sound. It is usually heard best when the patient is sitting up and leaning forward and may be more easily heard during expiration than inspiration. Pericardial rubs are most commonly associated with acute pericarditis-inflammation of the pericardium. Other causes include connective tissue disorders, such as systemic lupus erythematosus and rheumatoid arthritis.

2 Necrosis is common in GPA, especially in the lungs, and is due to in part to the ischaemia that is secondary to the vasculitis. Micro-abscesses form in blood vessels, which can partially or completely block them, causing ischaemia of distal tissues.

Case Study 40

1 The increased workload of the heart in pumping against prolonged resistance as a result of high blood pressure causes thickening of the walls of the left ventricle (left ventricular hypertrophy).

2 A reduction in the efficiency of the left side of the heart (left ventricular failure) increases the venous pressure in the pulmonary circulation. Hydrostatic pressure in the pulmonary capillary bed rises, forcing excess fluid into the interstitial spaces where it interferes with gaseous exchange. Eventually, some fluid may pass into the alveoli causing marked dyspnoea (shortness of breath).

Case Study 41

1 Yes, they include:

- Lynch syndrome – also known as hereditary non-polyposis colorectal cancer (HNPCC), an autosomal dominant condition that results from mutations in DNA mismatch repair genes (Chapter 11).
- MUTYH-associated polyposis (MAP), an autosomal recessive condition with mutations in the MUTYH gene. Individuals with MAP develop multiple polyps in the colon and rectum and have an increased risk of developing colon cancer.

2 Yes, FAP, Gardner syndrome, flat adenoma syndrome and Turcot syndrome are all distinct clinical conditions caused by mutations in the APC gene. They are characterised by different clinical features, including polyposis, osteomas and epitheliomas, as well as brain tumours (in the case of Turcot syndrome). The type and severity of these clinical features are associated with different types of mutations in the APC gene.

For example, FAP is typically caused by truncating mutations in APC, while Gardner syndrome is associated with specific missense mutations.

Case Study 42

1 If one or other parent is a known carrier of the deletion, the chance that any individual offspring will inherit that deletion is 50% irrespective of the number of deletion-negative offspring.

2 The hypocalcaemia seen in DiGeorge syndrome is due to an underdeveloped parathyroid gland which produces less than normal levels of parathyroid hormone (parathyroid hormone increases serum calcium levels). Clinically, hypocalcaemia manifests as neuromuscular irritability and convulsions.

Case Study 43

1 The CFTR gene encodes a cAMP-dependent membrane spanning chloride channel. Patients with cystic fibrosis have a defect in the CFTR gene, resulting in the absence or malfunction of the chloride ion channel and hence the deregulation of sodium and chloride ion transport.

2 In infants with CF, the pancreatic ducts can be blocked by thick mucus, preventing the passage of trypsinogen into the small intestine. As a result, levels of trypsinogen in the blood increase and are detected in the IRT test. A raised IRT alone is not diagnostic, but prompts further testing for common CF gene mutations. If one mutation is found, a sweat test is performed to confirm the diagnosis. If two mutations are identified, a diagnosis of CF is confirmed.

3 A second blood sample is taken after 21 days. If the IRT level is also elevated in this sample, the patient is referred for clinical assessment and sweat testing. This aspect of the screening programme is a safety net that helps to ensure that babies with CFTR variants that are not common in the United Kingdom or Europe are not missed.

Case Study 44

1 The more serious forms of gestational trophoblastic disease (GTD) include:

- Invasive mole: In this rare form of GTD, the abnormal placental tissue grows into the body of the uterus and can invade nearby organs.
- Choriocarcinoma: This is a malignant tumour that develops from the abnormal placental tissue after a molar pregnancy or other forms of GTD. Choriocarcinoma can spread to other organs, such as the lungs, liver or brain.
- Placental-site trophoblastic tumour (PSTT), a rarer form of GTD that arises from the placental implantation site after a normal pregnancy or a molar pregnancy. PSTT can invade the muscle of the uterus and can spread to other organs.
- Epithelioid trophoblastic tumour, the rarest form, it develops from the chorionic villi, the finger-like projections of the placenta and can invade the muscle of the uterus or spread to other organs.

These forms of GTD are more serious because they have a higher risk of complications. Whole-body imaging to exclude distant metastases and the possibility of more aggressive treatment, such as chemotherapy, surgery or radiotherapy, should be discussed at a multidisciplinary tumour board.

2 Yes, this type of choriocarcinoma, known as non-gestational choriocarcinoma, is very rare and can occur in both men and women. It can arise from primary germ cells of the ovary, testis, mediastinum or other sites in the body. Non-gestational choriocarcinomas have a poorer prognosis compared to gestational choriocarcinomas.

3 In a complete molar pregnancy, there is no embryo because fertilisation is abnormal and only paternal genetic material duplicates itself, resulting in a mass of abnormal cells. These cells grow within the uterus, resembling a pregnancy, but they do not form a viable foetus.

Case Study 45

1 Current evidence suggests that the incidence of balanced translocation in one partner of a couple who has experienced two or more miscarriages is 2–3%. This figure increases to 5–10% if the couple has previously had an abnormal or stillborn child, or there is a family history of recurrent miscarriage. These figures indicate that a careful clinical history may help in selecting appropriate couples for cytogenetic analysis.

2 The following options may be considered:

- *In vitro fertilisation (IVF):* This is often the first-line approach for couples with chromosomal abnormalities, especially when coupled with other techniques like preimplantation genetic testing.
- *Preimplantation genetic testing for structural rearrangements* (PGT-SR) allows for screening of embryos for the specific balanced translocation carried by one of the partners. Only embryos with a normal set of chromosomes or balanced translocations (like the parent) are implanted into the uterus.

- *Intracytoplasmic sperm injection (ICSI):* In cases in which male infertility is also an issue, ICSI may be used in conjunction with IVF to inject a single sperm directly into an egg. This is often followed by PGT-SR to screen for chromosomal issues.
- *Donor eggs or sperm:* If the balanced translocation leads to a very high likelihood of miscarriage or serious genetic disorders, and the couple is open to it, using eggs or sperm from a donor with normal chromosomes is an option.

Case Study 46

1 A detailed clinical review of that parent would be advisable. There are now a number of well-described cases in which unbalanced karyotypes are inherited from one generation to another with variable clinical effects.

2 *De novo* structural rearrangements are more likely to have been inherited from the father.

Case Study 47

1 The predominance of sickle cell disease in equatorial Africa is attributed to a phenomenon known as *balanced polymorphism* (heterozygote advantage). In this case, individuals who inherit one normal haemoglobin allele (HbA) and one sickle cell allele (HbS) have a survival advantage in regions where malaria is prevalent. This heterozygous state (HbAS) provides protection against severe forms of malaria, making it more advantageous than being homozygous for either HbA or HbS. Individuals with sickle cell trait exhibit resistance to severe malaria through several mechanisms. These

include increased phagocytosis of infected red blood cells by the spleen, premature haemolysis leading to parasite death, impaired haemoglobin digestion by the malaria parasite and reduced expression of *P. falciparum* erythrocyte membrane protein 1 (PfEMP1) on infected red cells, resulting in decreased parasite replication (see also Chapter 7).

2 As the kidneys lose their urine concentrating ability, patients with sickle cell disease pass large quantities of urine and hence drink increased amounts to prevent dehydration. The passage of large volumes of urine lowers the plasma urea.

3 HbS has a right-shifted oxygen dissociation curve which means that this haemoglobin variant more easily gives up its oxygen to the tissues than normal haemoglobin (HbA). The patient has also had chronic anaemia from childhood and the body can adapt by increasing cardiac output and reducing peripheral resistance, thereby increasing blood flow to the tissues.

4 CRISPR gene editing is a new approach to gene therapy. For example, CRISPR Therapeutics, in collaboration with Vertex Pharmaceuticals, is developing a drug called Exagamglogene Autotemcel (exa-cel) for the treatment of Beta thalassemia and sickle cell disease. Exa-cel has received Orphan Drug Designation from the U.S. FDA and the European Medicines Agency for its potential in treating these conditions. CRISPR gene editing is a revolutionary technology that allows scientists to precisely modify the DNA sequence.

prolongation is due to a deficiency of a coagulation factor in the patient's plasma. In other words, the patient's blood is lacking a specific clotting factor necessary for normal coagulation. This suggests a potential congenital or acquired coagulation factor deficiency. On the other hand, if the APTT remains prolonged even after mixing with normal plasma, it suggests the presence of an inhibitor of coagulation in the patient's plasma. In this case, the blood contains substances that interfere with the normal clotting process. In sum, the mixing APTT test helps differentiate between a deficiency of a coagulation factor and the presence of an inhibitor of coagulation.

2 During pregnancy, the levels of Factor VIII and VWF usually increase becoming normal in most patients with type 1 von Willebrand disease (VWD). Thus, for these patients, the symptoms of VWD may resolve or improve during pregnancy. In patients with type 3 disease, levels do not increase and replacement with coagulation factor concentrates will be necessary to cover delivery. It is important to remember that the baby will need to be investigated for VWD.

3 Acquired VWD may occur in the setting of autoimmune or lymphoproliferative disorders, when an autoantibody is produced that is directed against VWF. The diagnosis should be considered in elderly patients who present with the characteristic signs and symptoms of VWD, but without a prior history of bleeding. Patients with myeloproliferative neoplasms also have an increased risk of developing VWD.

Case Study 48

1 If the prolonged APTT corrects to normal after mixing with normal plasma (as in the case patient), it indicates that the

Case Study 49

1 The Apgar score is a quick assessment tool used to evaluate the physical condition of a new-born immediately after birth. It was

developed by Virginia Apgar in 1952. It is calculated based on five key indicators, each of which is assigned a score of 0, 1, or 2:

- *Appearance (skin colour):* A score of 0 is given if the baby is pale or blue, 1 if the body is pink but the extremities are blue, and 2 if the baby's entire body is pink.
- *Pulse:* A score of 0 may be assigned if there is no heartbeat, 1 if the heart rate is below 100 beats per minute, and 2 if the heart rate is above 100 beats per minute.
- *Grimace response:* Also known as *'reflex irritability'*, this evaluates the response to stimulation or a gentle pinch. A score of 0 is given if there is no response, 1 if there is a weak response, and 2 if there is a strong response, such as crying or pulling away.
- *Activity (muscle tone):* A score of 0 is given if the baby is limp, 1 if there is some flexion of the limbs, and 2 if the baby is active and moving.
- *Respiration:* A score of 0 is assigned if the baby is not breathing, 1 if the breathing is slow or irregular, and 2 if the baby is breathing well.

The assessment is typically performed twice: once at 1 minute after birth, and again at 5 minutes after birth, and at 5-minute intervals thereafter until 20 minutes for infants with a score less than 7.

2. It must be remembered that for all documented chromosomal syndromes, the range of features associated with a particular syndrome need not all be present, and that the severity can be variable. Cases published in the scientific literature are often the most badly affected. However, all cases of short arm 5 deletion involving the critical region for Cri du chat syndrome will be more severely affected than those deleted outside the region.

Case Study 50

1. Ehlers-Danlos syndrome (EDS) is a group of genetic disorders that affect connective tissues. Typical presenting features include joint hypermobility, skin hyperextensibility and tissue fragility. Up to one-quarter of EDS patients also show aortic aneurysmal disease. There are 13 different subtypes of EDS. Some subtypes are caused by mutations in genes that encode collagen subunits. For example, the classical subtype of EDS (cEDS) is caused by mutations in the COL5A1 or COL5A2 genes, which encode collagen type V subunits. The vascular subtype of EDS (vEDS) is caused by mutations in the COL3A1 gene, which encodes a subunit of collagen type III.

2. If either parent is affected by Marfan syndrome, there is a 50% chance that the child will have Marfan syndrome. However, because of variability in the expression of the mutation, it is difficult to predict how severely affected a child will be.

Case Study 51

1. Hypercalcaemia can cause significant symptoms of constipation, lethargy, nausea and abdominal pain. Her biochemistry should be kept under review since treatment may be required. Neurological signs of spinal cord compression may occur if tumour deposits or fractured vertebral bodies impinge on the cord. A history of either sudden or gradually increasing loss of function of limbs bilaterally in association with sensory loss, bladder and bowel disturbance should alert one to this possibility. Investigation and treatment is urgently required to avoid permanent neurological damage and disability. Further risk of bone pain and pathological fractures require palliation. Oral bisphosphonates inhibit osteoclast activity and may be given to patients

at high risk of these complications to control bone resorption.

2   Having a mother or sister with breast cancer increases a woman's risk of breast cancer by 1.5- to 3-fold compared to that of the general population. About 5–10% of breast cancers are hereditary – BRCA1 and BRCA2 germline mutations account for most cases with the other cases being associated with germline alterations in other breast cancer susceptibility genes including PALB2, CHEK2, ATM, CDH1, PTEN and TP53. Multigene panel testing for these mutations can inform risk-reducing strategies, including enhanced screening, prophylactic surgery as well as chemoprevention. BRCA germline testing is not only a strategy for surveillance and prevention but is also a predictive marker for PARP inhibitor treatment. Most women with a family history, however, do not have germline mutations in these genes and their risk of developing breast cancer is much lower. The daughter should be offered an upfront appointment at a geneticist before any genetic test is performed.

3   Yes, different factors affect outcome in breast cancer. These include the type of breast cancer, the grade, tumour size and lymph node status. The Nottingham Prognostic Index is one scoring system that uses these features to help a doctor work out the likelihood of survival for a given patient. Some tests analyse groups of genes found in breast cancer to provide information about the risk of recurrence. For example, the Oncotype DX test can predict how likely a cancer is to recur after surgery and the expected benefit of having chemotherapy, especially in post-menopausal women.

Case Study 52

1   Mutations in the TP53 gene, which encodes the p53 protein (a tumour suppressor gene, see Chapter 11), are frequently observed in blast crisis. These mutations disrupt the normal function of p53, contributing to uncontrolled cell division and resistance to apoptosis (Chapter 3). Ras pathway mutations are also involved.

2   Yes, two common examples are:
   - Chronic lymphocytic leukaemia (CLL), a leukaemia of mature B-lymphocytes, is typically an indolent disease. However, it can transform into a more aggressive form, known as *Richter transformation*. In Richter transformation, the CLL cells undergo a blast crisis-like change and transform into diffuse large B-cell lymphoma (Case Study CS55), a fast-growing type of non-Hodgkin lymphoma.
   - Myelodysplastic syndromes, which can transform into acute myeloid leukaemia (AML; see Case Study CS58), characterised by a rapid increase in the number of immature myeloid blasts.

3   The proliferation of large numbers of granulocytes and their precursors is a feature of CML. These cells produce transcobalamin-1, a vitamin B12-carrying protein, which is released into the plasma. High levels of transcobalamin-1 in the plasma are associated with raised vitamin B12 concentrations.

Case Study CS53

1   T2 N2a M0

2   Unlike most other solid tumours, there is no direct relationship between increasing size of the colorectal cancer and either spread to other tissues, lymphatic metastases or survival. The important feature about these tumours is how far they have spread through the bowel wall at the time of surgery. Some very small cancers can spread rapidly to other sites, while large tumours can sometimes be present for a long time without metastasising.

3 Although an attractive proposition, the answer is no. Screening programmes for colorectal cancer, which rely on testing the stools for 'occult' blood (blood in the stools that is 'hidden' or not clinically obvious), for example, may aid in reducing the incidence of the disease by detecting pre-malignant polyps, but this would by no means eliminate the disease. There are other pathways by which colorectal cancer can arise, for example as *de novo* lesions (with no prior evidence of polyps) or alternatively from so-called 'flat' adenomas. It has been estimated that at least 20 adenomas would need to be removed to prevent one cancer.

Case Study 54

1 H3K27me3 is an epigenetic modification in which there is tri-methylation of lysine 27 on histone H3 protein. H3K27me3 is a so-called *repressive mark,* which means that it generally suppresses the transcription of genes.

2 The blood–brain barrier is formed by endothelial cells, pericytes and astrocytes. It allows the passage of some small molecules into the brain but prevents the entry of larger molecules. Brainstem DMG shows reduced BBB permeability compared to similar tumours in the cerebral cortex, which is one of the reasons chemotherapies have failed to improve outcomes for patients with this disease.

Case Study 55

1 4 (high risk)

2 The Cell of Origin (COO) classification separates DLBCL into germinal centre B-cell-like (GCB) DLBCL and activated B-cell like (ABC) DLBCL subtypes on the basis of gene expression profiling. The GCB and ABC subtypes express genes that are characteristic of normal germinal centre B-cell differentiation or activated post-germinal centre B-cells, respectively. Each have distinct molecular features that drive different oncogenic signalling pathways reflected in different clinical behaviour and response to treatment; patients with the ABC subtype have a poorer prognosis. However, currently gene expression profiling to determine COO type is not done routinely since it does not affect the patient's management. 10–15% of DLBCL cannot be classified into any COO group. Almost all double-hit and triple-hit DLBCL are of GCB subtype. DEL tumours tend to be ABC-like and represent a more heterogeneous disease compared to DHL/THL. DEL are generally associated with poor patient outcomes, although with a slightly better clinical course compared with DHL/THL.

Case Study 56

1 Microvascular proliferation refers to the abnormal growth and multiplication of small blood vessels. This is a characteristic of GBM and contributes to its highly invasive nature and rapid growth. The newly formed blood vessels are often irregular and fragile, which can lead to areas of haemorrhage within the tumour.

2 Heterogeneously enhancing areas describe the appearance of some brain tumours on MRI or CT scans. The term indicates that the contrast agent used during the imaging study has highlighted areas within the tumour that appear different from the surrounding tissue. In the case of brain tumours, the contrast agent is typically injected into a vein, and will accumulate in areas where there is an increased

permeability of the blood–brain barrier. The heterogeneous enhancement may indicate necrosis, cysts or areas of varying cellularity within the tumour.

Case Study 57

1   cHL is indeed derived from B cells, but it is biologically and clinically distinct from most other B-cell lymphomas. One of the key differences lies in the expression of CD20, a B-cell surface marker that is the target of the monoclonal antibody rituximab. In most cases of cHL, the characteristic HRS cells do not express CD20 or other B-cell markers. This makes them poor targets for rituximab, which has been highly effective in treating other CD20-positive B-cell lymphomas, such as diffuse large B-cell lymphoma (DLBCL; see Case Study CS55) and follicular lymphoma.

2   The treatment of people with cHL has significantly improved over the years, leading to high survival rates. However, survivors are at risk for several late effects that may not manifest until years or even decades after treatment. These late effects can have a substantial impact on long-term health and quality of life. Some of the notable late effects include:

- *Second malignancies:* Patients treated for cHL are at a higher risk of developing secondary cancers, such as leukaemia, non-Hodgkin lymphoma, and solid tumours including breast, lung, and thyroid cancers. The risk is higher for those who have undergone radiotherapy or alkylating-agent chemotherapy.
- *Cardiovascular disease:* Both chemotherapy and radiotherapy can have long-term effects on the cardiovascular system. Patients may develop conditions such as coronary artery disease, congestive heart failure, and arrhythmias.

The risk is especially elevated for those who received mediastinal radiation.

- *Pulmonary disease:* Patients who have had thoracic radiation or certain types of chemotherapy may experience long-term lung disease, including pulmonary fibrosis, reduced lung capacity, and an increased risk of pneumonia.
- *Endocrine dysfunction:* Treatments can cause hypothyroidism, impaired fertility, and in some cases, early menopause in women.
- *Bone health:* Some chemotherapy agents and prolonged periods of physical inactivity during treatment can lead to decreased bone density and a heightened risk of fractures.
- *Neurological effects:* Peripheral neuropathy, a condition characterised by numbness, tingling and pain in the extremities, can occur after treatment with certain chemotherapy agents.
- *Psychosocial effects:* Survivors of cHL often experience long-term emotional and psychological effects, including anxiety and depression, often exacerbated by the physical late effects and the fear of recurrence or second malignancies.
- *Fatigue and cognitive impairment:* Many survivors report long-term fatigue and cognitive changes, commonly referred to as *chemo-brain* which can affect memory, concentration and problem-solving skills.

Given these potential late effects, long-term follow-up and regular monitoring are crucial for cHL survivors. Medical guidelines often recommend a multidisciplinary approach to monitor and manage the long-term risks.

Case Study 58

1   Auer rods are large, crystalline cytoplasmic inclusion bodies found in myeloid blast cells of some types of leukaemia and

myelodysplastic syndromes. They are composed of fused lysosomes and contain lysosomal enzymes. Auer rods often resemble needles. The presence of Auer rods can help distinguish AML, including APL, from other haematological malignancies.

2 Raised uric acid levels, known as hyperuricaemia, are commonly seen in patients with leukaemia. Hyperuricaemia occurs in a particular disorder, known as *tumour lysis syndrome* (TLS). This happens when large numbers of cancer cells are rapidly destroyed following chemotherapy. TLS is particularly common in patients with rapidly dividing leukaemias, such as acute lymphoblastic leukaemia and acute myeloid leukaemia.

Case Study 59

1 LEMS is a rare autoimmune disorder characterised by muscle weakness of the limbs. More than half of cases are associated with SCLC. LEMS is caused by antibodies against voltage-gated calcium channels (VGCC) on presynaptic nerve terminals, which result in a decrease in the neurotransmitter, acetylcholine.

2 Therapeutic monoclonal antibodies are designated '-mabs'. The name also indicates the type of monoclonal antibody. Thus, drugs ending in:

- -ximab (e.g. rituximab) are chimeric human-mouse antibodies.
- -zumab (e.g. bevacizumab) are humanised mouse antibodies.
- -mumab (e.g. ipilimumab) are fully human antibodies.

Small molecule inhibitors end with 'ib' and their function is further indicated by a sub-stem, for example

- -tinib (e.g. imatinib) are tyrosine kinase inhibitors

- -zomib (e.g. bortezomib) are proteasome inhibitors

Finally, an additional stem can be used to describe the target. For example, the 'tu' in rituximab indicates the target is a tumour.

Case Study 60

1 Yes, the hoarseness was caused by the tumour spreading outward and pressing on the recurrent laryngeal nerve which lies near the left main bronchus on its way to supply the larynx. This is an example of a local effect of a tumour.

2 There are two main causes of hypercalcaemia in malignant disease: (a) when malignant cells metastasize, they damage the tissues in which they settle. Thus, hypercalcaemia can be caused by metastatic destruction of bone leading to the release of calcium, and; (b) some tumours can influence distant sites in the body by producing hormones or their analogues. For example, some types of lung cancer can make antidiuretic hormone causing water retention. Similarly, hypercalcaemia can be caused by secretion of parathyroid hormone-related protein by the tumour cells. Other remote effects of tumours are less easy to understand, for example abnormalities in peripheral nerves (neuropathy) or muscle (myopathy). The clubbing of the patient's fingers was a remote effect of his lung disease, but the mechanism is not understood, and certain non-malignant conditions can also cause this.

Case Study 61

1 Seminomas may produce proteins that can be detected in the blood. Around 30% of people with testicular seminoma will show mild increases in β-hCG. Similarly, around

40–60% of people with a testicular germ cell tumour will also display increased levels of LDH. AFP is only increased in patients with a mixed tumour (meaning a tumour with seminoma and other types of testicular cancer). Pure seminomas do not produce AFP.

2   The staging of testicular cancer uses the TNM system, which assesses the extent of the primary tumour (T), regional lymph nodes involvement (N) and distant metastasis (M). However, an additional element, the 'S' value, is integrated into this staging system, taking into account the nadir values of tumour markers after the surgical removal of the affected testicle (orchiectomy). In this case, the patient's serum markers returned to normal after orchiectomy; hence the stage includes the additional designation 'S0'.

Case Study 62

1   There are many potential causes, including primary hyperparathyroidism (measure parathyroid hormone (PTH) levels), sarcoidosis (measure angiotensin-converting enzyme levels, see also Case Study CS27) and tuberculosis.

2   Sodium chloride to rehydrate the patient and increase urinary excretion of calcium. Monitor fluid balance to avoid overloading the patient. Bisphosphonates (e.g. pamidronate) to reduce bone turnover can be used for short-term treatment of hypercalcaemia after rehydration.

Case Study 63

1   Pembrolizumab is a monoclonal antibody that targets the programmed cell death protein 1 (PD-1) receptor on T cells. By binding to PD-1, pembrolizumab blocks the inhibitory signals that would normally prevent T-cells from killing the cancer cells.

2   This patient had four high-risk factors (age >60 years, stage IV disease, detectable plasma EBV DNA and distant lymph node metastasis). The estimated 3-year overall survival is 28%.

Case Study 64

1   Yes, for example, in November 2022, the FDA approved Elahere for patients with folate receptor alpha (FRa)-positive, platinum-resistant, epithelial ovarian, fallopian tube or primary peritoneal cancer who have received up to three prior systemic treatments. Elahere is a first-in-class *antibody-drug conjugate* (ADC) directed against FRα, a cell-surface protein that is highly expressed in ovarian cancer. This ADC is designed to deliver a potent chemotherapy agent, DM4, directly to ovarian cancer cells, minimising exposure of healthy tissues to the drug. The FDA simultaneously approved Roche's Ventana FOLR1 (FOLR1-2.1) RxDx Assay, a *companion diagnostic* to identify FRa-positive patients who are eligible for Elahere.

2   The most commonly used tumour marker for ovarian cancer is CA-125 (cancer antigen 125), a glycoprotein that is elevated in approximately 80% of women with ovarian cancer. However, CA-125 can also be elevated in other conditions such as endometriosis, uterine fibroids or pelvic inflammatory disease, as well as in some healthy women. Therefore, CA-125 is not specific for ovarian cancer and should not be used alone for diagnosis. Other tumour markers that may be elevated in ovarian cancer include HE4 (human epididymis protein 4), CA19-9 (carbohydrate antigen 19-9) and CA-72-4 (carbohydrate antigen 72-4).

These markers may be used in conjunction with CA-125 to improve the accuracy of diagnosis and monitoring of patients with ovarian cancer. These tumour markers can be useful in predicting response to treatment and overall prognosis. For example, it has been shown that women with high levels of CA-125 before treatment have a poorer prognosis than women with lower levels.

3 The term R0 resection refers to a surgical procedure in which the goal is the complete removal of all detectable cancerous tissue. In this context, 'R0' indicates that no microscopic or macroscopic tumour remains in the area left after surgery, based on pathological examination of the resected specimens. The 'R' stands for 'resection margin', and '0' signifies no detectable tumour cells at the edges of the tissue sample examined. An R0 resection is considered the optimal surgical outcome for many types of cancer, as it maximises the likelihood of a cure or long-term survival. An R1 resection is one in which microscopic amounts of tumour are detected at the margins of the resected tissue, while an R2 resection is one in which macroscopic tumour remains after the operation.

Case Study 65

1 Mature teratomas contain well-differentiated tissues that resemble normal adult tissues, for example hair, skin, teeth, and bone. These tumours are typically benign and rarely metastasise. Immature teratomas contain poorly differentiated tissues that resemble embryonic tissues. Immature teratomas are considered malignant and can metastasise.

2 An 'endobag' is a pouch made of a durable, impermeable material that is inserted into the body through a small incision. Once inside, the bag is opened and the tissue or organ to be removed is placed into it. The bag is then sealed and carefully pulled out through the incision, minimising the risk of spillage or rupture that could spread disease or infection within the body. Endobags are particularly useful in cancer surgery, where it is crucial to prevent any spillage of cancerous cells into the surrounding tissues. They are also useful in avoiding contamination from potentially infectious or hazardous material.

Case Study 66

1 Poly ADP ribose polymerase (PARP) inhibitors are available for the treatment of people with prostate cancer. They include Olaparib (Lynparza), which is approved for patients with metastatic castration-resistant prostate cancer in whom the disease has continued to grow and spread during treatment with abiraterone and/or enzalutamide. Rucaparib is another PARP inhibitor approved to treat metastatic castration-resistant prostate cancer in patients whose disease has not been stopped by treatment with abiraterone and/or enzalutamide and chemotherapy. Only patients whose cancers have mutations in DNA repair genes, such as BRCA2, are eligible for these drugs. PARP inhibitors kill by a mechanism known as *synthetic lethality* (see also Chapter 11, and Case Study CS64).

2 The BCL2 protein protects cells from apoptosis (see Chapter 3). Expression of BCL2, and the subsequent protection from apoptosis, is important for the progression of androgen-sensitive tumours to hormone independency.

Case Study 67

1 Several viruses are associated with lymphoma development:

- Human T-lymphotropic virus 1 (HTLV-1) is associated with the development of adult T-cell leukaemia/lymphoma.
- Human herpesvirus 8 (HHV-8, otherwise known as Kaposi sarcoma-associated herpesvirus, KSHV) contributes to the development of primary effusion lymphoma, a form of B-cell lymphoma, and multicentric Castleman disease, a rare type of lymphoproliferative disease.
- People with HIV have an increased risk of developing several types of lymphomas, including Burkitt lymphoma, Hodgkin lymphoma, and some forms of non-Hodgkin lymphoma.
- Hepatitis C virus (HCV) is associated with an increased risk of developing certain types of non-Hodgkin lymphoma, particularly marginal zone lymphoma and diffuse large B-cell lymphoma.

2 No, immunosuppression is associated with an increased risk of developing only certain tumours. For example, the risk of common solid tumours such as those of the lung, breast and colon, for example, is not typically increased in people who are immunosuppressed.

3 Thrombotic thrombocytopenic purpura (TTP) is a rare and life-threatening blood disorder characterised by the formation of small blood clots (thrombi) within the blood vessels, leading to a low platelet count (thrombocytopenia; see also Chapter 8) Classically, TTP is characterised by five features (pentad) consisting of thrombocytopenia, microangiopathic haemolytic anaemia, fever, variable neurological abnormalities (mostly confusion or severe headache) and renal failure. The condition is primarily caused by a deficiency or dysfunction of a protein called ADAMTS13, which plays a crucial role in regulating the size of von Willebrand factor (VWF) multimers in the blood. VWF is discussed in more detail in Chapter 8 and in Case Study CS48.

Case Study 68

1 Checkpoint inhibitors interfere with specific receptor-ligand interactions in the immune system, thereby disrupting the regulatory mechanisms that control immune responses. By blocking these interactions, checkpoint inhibitors help to unleash the body's immune system, enhancing its ability to recognise and attack cancer cells. Some well-known immune checkpoints include:

- *PD-1 (programmed cell death protein 1) and PD-L1 (programmed cell death ligand 1):* PD-1 is expressed on the surface of T-cells, while PD-L1 can be expressed on cancer cells and other immune cells. The interaction between PD-1 and PD-L1 can inhibit T-cell activation, leading to immune suppression.
- *CTLA-4 (cytotoxic T-lymphocyte antigen 4):* CTLA-4 is a receptor found on T-cells. It competes with CD28, another receptor, for binding to B7 molecules on antigen-presenting cells (APCs). When CTLA-4 binds to B7, it dampens T-cell activation, serving as a brake on the immune response.
- *LAG-3 (lymphocyte activation gene-3):* LAG-3 is another immune checkpoint protein expressed on T-cells. It can negatively regulate T-cell activation when it binds to its ligands, one of which is MHC class II molecules, on APCs.

- *TIM-3 (T-cell immunoglobulin and mucin domain-containing protein 3):* TIM-3 is expressed on T-cells and has multiple ligands, including galectin-9. TIM-3 activation contributes to T-cell exhaustion and immune suppression.
- *VISTA (V-domain Ig suppressor of T-cell activation):* VISTA is a relatively newly discovered immune checkpoint. It is expressed on various immune cells and can suppress T-cell responses when engaged by its ligands.
- *BTLA (B and T-lymphocyte attenuator):* BTLA is expressed on T-cells and can interact with its ligand, HVEM (herpesvirus entry mediator), to negatively regulate T-cell activation.
- *TIGIT (T-cell immunoreceptor with Ig and ITIM domains):* TIGIT is expressed on T-cells and can interact with its ligands, CD155 (PVR) and CD112 (PVRL2), which are expressed on various immune cells. TIGIT signalling can inhibit T-cell activation.

2 As well as inducing anti-cancer effects, immune checkpoint inhibitors prevent the normal negative feedback mechanisms that control immune cell activation. Side effects occur because of the immune system's heightened activity. Referred to as *immune-related adverse events* (irAEs), these side effects include inflammatory reactions of any body system, but most commonly cause colitis (diarrhoea) skin rash, pruritus, pneumonitis, hepatitis, endocrinopathies, myocarditis and arthralgias.

Case Study 69

1 The risk of developing lung cancer is greatly increased in individuals who have worked in industries in which asbestos exposure is common. Asbestos fibres are inhaled into the lungs, causing inflammation, fibrosis, and an increased risk of lung cancer (mainly adenocarcinoma and squamous carcinoma) and cancers of the pleural lining cells (mesothelioma). The risk of lung cancer associated with asbestos is greatly magnified by co-existing exposure to tobacco smoke. In addition to mesothelioma and lung cancer, asbestos exposure is associated with an increased risk of other cancers, including those of the larynx, ovary, and gastrointestinal tract. The relative risk for these cancers is generally lower than for mesothelioma and lung cancer but is still elevated compared to individuals without asbestos exposure.

2 *Right ventricular heave* refers to a visible or palpable pulsation in the right ventricular area of the chest during systole. It is typically caused by an increase in the size or workload of the right ventricle.

Case Study 70

1 The patient was symptomatic and losing weight, and had marked hyperglycaemia and ketonuria, suggesting that he had insulin deficiency requiring insulin replacement therapy. His diabetes was clearly well advanced at the time of diagnosis.

2 There were no pulses below the femoral artery indicating macrovascular disease of peripheral vascular disease. In addition, this man's foot had diminished sensation due to peripheral nerve damage (neuropathy) secondary to his diabetes. Often in T2D, foot ulceration is a combination of both peripheral vascular disease as well as neuropathic microvascular disease. A further consequence of neurological damage is an abnormal gait leading to very high pressures in localised areas of the foot during walking. The consequent trauma very

rapidly leads to ulcer formation, which is often not recognised by the patient. Once the skin is damaged, infection may ensue, causing further inflammation and necrosis. The patient's ability to respond to infection is impaired because of circulatory disturbances due to microvascular disease, together with other factors, such as impaired leucocyte function, which render patients with diabetes more susceptible to a variety of infections.

Case Study 71

1. Acanthosis nigricans (AN) is a thickening and increased pigmentation of the skin, particularly around skinfolds, which is a consequence of hyperinsulinaemia (and the associated insulin resistance and hyperglycaemia). When hyperinsulinaemia is very high, as in this case, insulin binds to, and activates, the IGF-1 receptor in keratinocytes resulting in the proliferation of skin cells and hyperpigmentation with the typical 'velvety' texture. Insulin can also compete with IGF-1 for binding to its binding proteins, leading to increased levels of free IGF-1 and IGF-2. Both IGF-1 and IGF-2 are growth factors that contribute to the cell proliferation and skin changes. T2D is a disease characterised by insulin resistance and relative insulin deficiency, although the insulin levels in the earlier stages may be very high as the body attempts to compensate for the insulin resistance. The presence of AN in this case helps to confirm the diagnosis of T2D.

2. Metformin is known for its ability to improve insulin sensitivity in peripheral tissues, such as muscle and adipose tissue. This means that cells become more responsive to the effects of insulin, leading to better glucose uptake and utilisation. Metformin is generally well-tolerated, with

gastrointestinal reactions (such as nausea, diarrhoea and stomach upset) being the most common side effects. Lactic acidosis, a rare but serious adverse effect associated with metformin use, can occur, particularly in individuals with kidney or liver impairment. Metformin acts in the liver to reduce gluconeogenesis, which is the process of glucose production from non-carbohydrate sources. By decreasing hepatic glucose release, metformin helps lower blood glucose levels. Additionally, metformin sensitises the liver to the effects of insulin by impacting lipid metabolism and mitochondrial respiratory chain activity. This leads to decreased expression of enzymes involved in gluconeogenesis. Metformin's effects on the liver and other tissues are in part mediated by the activation of AMP-activated protein kinase (AMPK), an enzyme that plays a crucial role in regulating cellular energy balance. Activating AMPK has several beneficial effects on metabolism, including reducing glucose production and increasing glucose uptake. Metformin also stimulates the secretion of glucagon-like peptide-1 (GLP-1), a hormone that promotes insulin secretion and helps regulate blood sugar levels. Benefits extend beyond glycaemic control. Metformin has anti-cancer properties, in part due to its ability to lower insulin levels, which can influence cancer cell growth. Metformin has also been shown to reduce the levels of various pro-inflammatory cytokines.

Case Study 72

1. C-peptide and insulin are both derived from pro-insulin, a precursor molecule produced in pancreatic β-cells. Pro-insulin is processed and converted into insulin and C-peptide (see also Chapter 13). The half-life of C-peptide is longer than that of insulin and therefore it is a good index of endogenous insulin secretion. Measurement

of insulin itself is unreliable if the patient is taking insulin. In this patient, the low levels of C-peptide indicated marked insulin deficiency caused by autoimmune destruction of β-cells.

2   Severe DKA requires urgent intervention including administration of intravenous insulin and fluids to correct the hyperglycaemia and ketosis and reduce acidosis.

3   Capillary refill time is a simple clinical test used to assess the circulatory status and perfusion of tissues. It measures the time it takes for blood to return to the capillaries after pressure is applied and then released from a small area, usually the fingertip or nail bed. In healthy individuals, capillary refill time is usually less than 2 seconds. Prolonged capillary refill time, exceeding 2 seconds, suggests poor peripheral perfusion.

Appendix 1

Reference Ranges

Table A1: Haematology

Typical reference ranges depend upon local factors, including techniques used and the population under investigation. Data presented in the case studies may reflect these variations. In each case, reference ranges are presented to indicate the analytical conditions at the time the investigations were performed. Note, only adult reference ranges relevant to cases present in the text are included. Where there are variations based on paediatric values, these are given within the relevant case study.

Red cell count		Lymphocytes	$1.5–4.0\times10^{9}$/L (20–45%)
Male	$4.5–6.5\times10^{12}$/L	Monocytes	$0.2–0.8\times10^{9}$/L (2–10%)
Female	$3.8–5.8\times10^{12}$/L	Eosinophils	$0.04–0.4\times10^{9}$/L (1–6%)
Haemoglobin			
Male	13.0–18.0 g/dL	Basophils	$0.02–0.1\times10^{9}$/L (1%)
Female	11.5–16.5 g/dL		
Packed cell volume (PCV; haematocrit value)		Platelet count	$150–400\times10^{9}$/L
Male	0.40–0.54 L/L	Prothrombin time	11–16 seconds
Female	0.37–0.47 L/L	Activated partial thromboplastin time	26–35 seconds
Mean corpuscular (cell) volume (MCV)	82–92 fL	VWF antigen	50–200 IU/dL
Mean cell haemoglobin (MCH)	27–32 pg	Erythrocyte sedimentation rate (ESR)	
Reticulocyte count	0.5–2.5%	Male<50 years	<15 mm/hour
Leucocyte count	$4.0–11.0\times10^{9}$/L	Male>50 years	<20 mm/hour
Differential leucocyte count		Female<50 years	<20 mm/hour
Neutrophils	$2.0–7.5\times10^{9}$/L (40–75%)	Female>50 years	<30 mm/hour

(*Continued on p. 568*)

The Biology of Disease, Third Edition. Edited by Paul G. Murray, Simon J. Dunmore, and Shantha Perera.
© 2024 John Wiley & Sons Ltd. Published 2024 by John Wiley & Sons Ltd.
Companion website: www.wiley.com/go/murray/biologyofdisease3e

Table A2: Clinical Biochemistry

Adrenocorticotrophic hormone (ACTH)	0–50 ng/L (9 a.m. sample)	Folate (serum)	3–20 µg/L
Alanine transaminase (ALT)	5–40 U/L	OGTT 2 hour glucose	<7.8 mmol/L (exclude IGT/ pre-diabetes)
Albumin	34–48 g/L		<11.1 mmol/L (exclude diabetes)
Alkaline phosphatase	25–115 U/L	Folate (red cell)	160–640 µg/L
Amylase	<220 U/L	Gamma glutamyltransferase (γ-GT)	
Angiotensin-converting enzyme	10–70 U/L	Male	11–50 U/L
Aspartate transaminase (AST)	10–40 U/L	Female	7–32 U/L
N-terminal pro-B-type natriuretic peptide (NT-proBNP)	<300 pg/mL	HbA1c (glycated haemoglobin)	20–42 mmol/mol (IFCC units; equates to 4–6% in old DCCT units)
Bicarbonate	22–30 mmol/L	β-human chorionic gonadotrophin	<5 mU/mL
Bilirubin			
Total	<21 µmol/L (<1.2 mg/dL)	Immunoglobulins (11 years and over)	
CA-125	<35 U/mL	IgA	0.8–4 g/L
Calcium	2.20–2.67 mmol/L	IgG	7.0–18.0 g/L
C-reactive protein	<10 mg/L	IgM	0.4–2.5 g/L
C-peptide	0.5–5.5 ng/mL	IgE	1.5–120 U/L
Creatinine		Immunoreactive trypsinogen	<57 ng/mL
Female	45–84 µmol/L	Insulin	11–220 pmol/L
Male	59–104 µmol/L	Iron	13–32 µmol/L (50–150 µg/dL)
Estimated glomerular filtration rate (eGFR)	>90 mL/min	Total iron binding capacity (TIBC)	42–80 µmol/L (250–400 µg/dL)
Creatine kinase		Lactate	0.50–2.0 mmol/L
Female	24–170 U/L	Lactate dehydrogenase	135–225 U/L
Male	24–195 U/L	Plasma osmolality	280–296 mosmol/kg
D-dimer	<500 ng/mL	Phosphate	0.8–1.5 mmol/L
Ferritin	5.8–144 nmol/L (15–300 µg/L)	Potassium	3.5–5.0 mmol/L
α-fetoprotein	<5.8 kU/L	Prostate-specific antigen	<4.0 ng/mL
Glucose (fasting)	4.5–5.6 mmol/L (70–110 mg/L)	Protein (total)	62–80 g/L
	<6.0 mmol/L (to exclude IFG/ pre-diabetes)	Sodium	135–146 mmol/L
		Transferrin	2.0–3.0 g/L
	<7.0 mmol/L (to exclude diabetes)	Transferrin saturation	15–50%
		Throid stimulating hormone	0.4–4.0 mU/L
Glucose (random)	3.5–8.9 mmol/L	Thyroxine (free)	9.0–24.0 pmol/L

(high-sensitivity cardiac)
Troponin-T (hs-TnT)

    Male $\leq$20 ng/mL

    Female $\leq$15 ng/mL

hs-TnI

    Male $\leq$20 pg/mL

    Female $\leq$15 pg/mL

Urate   0.18–0.42 mmol/L (3.0–7.0 mg/dL)

Urea   2.5–6.7 mmol/L (8–25 mg/dL)

Vitamin B_{12} (as cyanocobalamin)   160–760 ng/L

Vitamin D

    25-hydroxy   37–200 nmol/L (0.15–0.80 ng/L)

    1,25-dihydroxy   60–108 pmol/L (0.24–0.45 pg/L)

Lipids and lipoproteins

Cholesterol   3.5–6.5 mmol/L (ideal <5.2 mmol/L)

Lipids (total)   4.0–10.0 g/L

Lipoproteins

    VLDL   0.13–0.65 mmol/L

    LDL   1.55–4.4 mmol/L

    HDL

        Male   0.70–2.1 mmol/L (ideal >2.1)

        Female   0.50–1.70 mmol/L (ideal >1.7)

Lipoprotein A   <300 mmol/L (ideal)

Triglycerides

    Male   0.70–2.1 mmol/L (ideal <2.1)

    Female   0.50–1.70 mmol/L (ideal <1.7)

Blood gases

Partial pressure of CO_2 in arterial blood ($PaCO_2$)   4.8–6.1 kPa (36–46 mmHg)

Partial pressure of O_2 in arterial blood (PaO_2)   10–13.3 kPa (75–100 mmHg)

Arterial [H^+]   35–45 nmol/L

Arterial pH   7.35–7.45

Oxygen saturation (SaO_2)   >95%

Index

The Biology of Disease, Third Edition. Edited by Paul G. Murray, Simon J. Dunmore, and Shantha Perera.
© 2024 John Wiley & Sons Ltd. Published 2024 by John Wiley & Sons Ltd.
Companion website: www.wiley.com/go/murray/biologyofdisease3e